Vibration of solids and structures under moving loads

Monographs and textbooks on mechanics of solids and fluids

editor-in-chief: G. Æ. Oravas

Mechanics of structural systems

editor: J. S. Przemieniecki

Vibration of solids and structures under moving loads

Ladislav Frýba

Research Institute of Transport, Prague

Academia
Publishing House of
the Czechoslovak
Academy of Sciences
Prague

Noordhoff
International Publishing
Groningen

Noordhoff International Publishing
Groningen, The Netherlands

Translated by D. Hajšmanová

ISBN 978-94-011-9687-1 ISBN 978-94-011-9685-7 (eBook)
DOI 10.1007/978-94-011-9685-7

Library of Congress Catalog Card Number 70–151037

To Professor Vladimír Koloušek

on his sixtieth birthday

Contents

V Special problems

VI Appendix and bibliography

Preface

Transport engineering structures are subjected to loads that vary in both time and space. In general mechanics parlance such loads are called moving loads. It is the aim of the book to analyze the effects of this type of load on various elements, components, structures and media of engineering mechanics.

In recent years all branches of transport have experienced great advances characterized by increasingly higher speeds and weights of vehicles. As a result, structures and media over or in which the vehicles move have been subjected to vibrations and dynamic stresses far larger than ever before.

The author has studied vibrations of elastic and inelastic bodies and structures under the action of moving loads for many years. In the course of his career he has published a number of papers dealing with various aspects of the problem. On the strength of his studies he has arrived at the conclusion that the topic has so grown in scope and importance as to merit a comprehensive treatment. The book is the outcome of his attempt to do so in a single monograph.

The subject matter of the book is arranged in 27 chapters under six main Parts. The Introduction – a review of the history and the present state of the art – is followed by Part II, the most extensive of all, devoted to the discussion of dynamic loading of one-dimensional solids. The latter term refers to all kinds of beams, continuous beams, frames, arches, strings, etc. with a predominant length dimension – a typical feature of transport engineering structures. The exposition covers beams with several types of support and various alternatives of moving load, and presents the method of computing their deflections and stresses at different speeds of the moving objects.

Part III deals with two-dimensional solids such as rectangular plates and infinite plates on elastic foundation. Part IV is focused on stresses in three-dimensional space produced by moving forces. There, too, consideration is given to all types of speed, i.e. subsonic, transonic and supersonic.

Part V is a lengthy treatise devoted to special problems. It deals, for example, with the effect of variable speed of the load, the action of an axial force and the longitudinal vibration of beams. Study is also made of three-dimensional vibrations of thin-walled beams under moving loads, and of the effect of shear and rotatory inertia on beam stresses. Considerable attention is accorded to the inelastic properties of materials, i.e. to viscoelasticity and plasticity in connection with beam analyses. The concluding part examines the effects of random loads, a very topical problem at present.

The Appendix contains comprehensive tables of integral transformations frequently used throughout the book, and of practical importance in general.

The methodological approach adopted in the book is as follows: the exposition starts with a theoretical analysis of the problem at hand and solves it for all possible cases likely to be met with. The most important results established in this phase are expressed by formulas, represented by diagrams, etc. Many of the theoretical findings are verified experimentally and the test values compared with the computed ones. The conclusion of each chapter outlines the possible applications of the theory explained, and gives a list of recommended reading.

The broad range of problems discussed in the book makes the author hopeful that his work will be found equally useful in civil as in mechanical, transport, marine, aviation and astronautical engineering, for moving loads are present in all these fields. The publication may serve research scientists as an incentive to further development of an interesting and very modern branch of science, project engineers and designers as a guide to safer and more economic design of structures, and students as an advanced text in engineering mechanics and dynamics.

As the results presented in the book are deduced in detail, all that is necessary on the part of the reader is a knowledge of the fundamentals of mechanics, dynamics, vibration and elasticity theories, analysis, theory of differential and integral equations, functions of the complex variable and integral transformations.

In conclusion grateful acknowledgment is due to all those who in any way have contributed towards the successful termination of the book. In the first place the author wishes to thank his wife, Mrs. Dagmar Frýbová, for her rare understanding and support of his scientific work, as well as for her effort of typing most of the manuscript. He is indebted to Mr. Petr Chaloupka for his careful work of drawing all the illustrations and computing some of the diagrams, and is deeply grateful to his colleagues in the Research Institute of Transport in Prague where the book originated in the course of his theoretical and experimental research.

Very special thanks are extended to Professor Vladimír Koloušek, DrSc., corresponding member of the Czechoslovak Academy of Sciences, the scientific editor of the book, to Josef Henrych, DrSc., the referee, both from the Technical University in Prague, and to Professor A.D. de Pater, the reviewer, from the Technological University in Delft, for their thorough study of the manuscript and many valuable suggestions which the author has gratefully incorporated in the last revision.

Ladislav Frýba

Symbols

This is an alphabetical list of the basic symbols used throughout the book. Symbols specific to the matter at hand will be explained as they first come up in the text.

a	a constant
a	parameter expressing the depth of track uneveness
a	parameter expressing the variable stiffness of roadway
a	acceleration or deceleration of motion
a	distance
a	real coordinate in the complex plane
$a_i^2 =$	$\pm 1 \pm \alpha_i^2$
a_n	coefficients in a power series
a_1	parameter expressing a harmonic force
b	a constant
b	parameter expressing the length of track uneveness
b	parameter expressing the length of the variable stiffness of roadway
b	parameter of non-uniform motion
b	imaginary coordinate in the complex plane
b_0	track gauge
b_1	parameter expressing the frequency of a harmonic force
c	speed of motion
$c_i(t)$	speed of displacement
c_{cr}	critical speed
c_1, c_2	velocity of propagation of longitudinal or bending, and transverse or shear waves, respectively
d	a constant
d	wheel base parameter
f	frequency
f_i	frequency of sprung or unsprung parts of vehicle

$f_{(j)}$	natural frequency
$\bar{f}_{(j)}$	natural frequency of a loaded beam
$f(x)$	a function
$\mathring{f}(x, t)$	centred value of function $f(x, t)$
$f(t)$	equation of motion at non-uniform speed
$g =$	9·81 m/s² acceleration of gravity
$g_1(x)$	initial deflection of a beam
$g_2(x)$	initial speed of a beam
$g_{1,2}$	self-weight load of a beam, or of a cable, respectively
h	integration step
h	height difference
h	plate thickness
$h(x)$	beam depth
$h(x, t)$	influence function
$h_{(j)}(t)$	impulse function
$i =$	1, 2, 3, ...
$\mathrm{i}^2 =$	-1 imaginary unit
$j =$	1, 2, 3, ...
k	a constant
k	coefficient of Winkler elastic foundation
$k =$	1, 2, 3, ...
$\mathrm{kei}(x)$	Thomson function of the zero order
l	span
l	length
m	mass of load P
m_i	mass of sprung or unsprung parts of vehicle
m^2	expression (23.29) dependent on beam and foundation properties
$n =$	1, 2, 3
n	roots of the characteristic equation
n	half the number of vehicle axles
p	external load
p	complex variable in the Laplace-Carson integral transformation
$p_{(j)}(t)$	expansion of load in normal modes
q	complex variable in the Fourier integral transformation
q	continuous load
$q_{(j)}(t)$	generalized deflection
r	radius of a wheel, of an arch, of gyration, respectively

r	auxiliary variable
r	radius in polar coordinates
$r_{1,2} =$	$\Omega \pm \omega$
$r(x)$	ordinate of track uneveness
s	auxiliary variable
s	coordinate of point of load application
s	number of equations
t	time
u	displacement in the x-direction
$u(\varphi, t)$	tangential displacement
$u(t)$	auxiliary function
v	displacement in the y-direction
$v(\varphi, t)$	radial displacement
v_i	displacement of joints
$v_i(t)$	vertical displacement of a moving load, of sprung or unsprung parts of vehicle, respectively
$v(x, t)$	beam deflection
$v_{(j)}(x)$	natural mode of beam vibration
v_0	static deflection produced by load P
$v_n(x, t)$	n-th approximate solution of function $v(x, t)$
$v(s)$	dimensionless deflection of an infinite beam on elastic foundation
w	displacement in the z-direction
x	coordinate
x_i	point of contact between vehicle and beam
x_0	fixed point
x_p	coordinate of plastic hinge
y	coordinate
y, $\tilde{y}$	exact, approximate solution
y_h	solution at integration step h
$y(\xi, \tau)$	dimensionless beam deflection
$y_i(\tau)$	dimensionless vertical displacement of load, sprung or unsprung parts of vehicle, respectively
$y(x, t)$	cable sag
z	coordinate
z	auxiliary variable or complex variable
$z(x, t)$	cable deflection
$z(0, l, t)$	function dependent on boundary conditions

A	a constant
A	centre of flexure
$A^2 =$	$(1 - \alpha_1^2)(1 - \alpha_2^2)$
A_j	poles of the function of a complex variable
A, A_j	integration constant dependent on boundary conditions
A, A_i	distance between track irregularities
$A(t)$	reaction at beam left-hand end
B	a constant
B	parameter of acceleration or deceleration of motion
$B =$	$\alpha^2 - m^2(1 - \alpha_1^2)$
B, B_j	integration constant dependent on boundary conditions
$B(t)$	reaction at beam right-hand end
C	a constant
C	spring constant
C, C_j	integration constant dependent on boundary conditions
C_b	coefficient of viscous damping in vehicle springs
$\mathrm{C}(x)$	Fresnel integral
$C_f(x, t)$	coefficient of variation of function $f(x, t)$
D	a constant
D	vehicle base
D	bending stiffness of plate
D	operation of partial or ordinary differentiation
E	Young's modulus
$E^*(p)$	Laplace-Carson integral transformation of a time variable Young's modulus
$\mathrm{E}[f(x, t)]$	mean value of function $f(x, t)$
F	cross-sectional area
$F(q)$	Fourier integral transformation of function $f(x)$
$F_i(\lambda)$	functions tabulated in [130]
$\mathrm{F}(a, b, c, x)$	hypergeometric series
G	beam weight
G	modulus of elasticity in shear
$G(x, s)$	influence or Green's function
$G_{1,2}(j)$	Fourier transformation of initial functions $g_{1,2}(x)$
H	vehicle height
H	horizontal force of a string
$H(x)$	Heaviside function
$H_1(x)$	impulse function of the second order

$H(q, \omega)$	transfer function
I	impulse
I	mass moment of vehicle inertia
Im	imaginary part of the function of a complex variable
J	moment of inertia
$J_n(x)$	Bessel function of the first kind of index n
K	spring constant
K	material constant
K	beam curvature
$K_{ff}(x_1, x_2, t_1, t_2)$	correlation function (covariance)
L	differential operator
L	vehicle length
L	auxiliary datum
M	bending moment
M	torsion moment
M_0	static bending moment produced by load P
M_p	limit bending moment
N	horizontal force
N	normal force
N	printing after N steps
N	number of impulses
O	wheel circumference
O	vehicle centre of gravity
P	concentrated constant force
$P(t)$	concentrated force generally varying in time
$P^*(p)$	Laplace-Carson integral transformation of force $P(t)$
P_i	weight of sprung and unsprung parts of vehicle
$P_i(D)$	linear differential operator
Q	amplitude of harmonic force
Q_i	dimensionless static axle pressure
$Q(t)$	harmonic force
$Q_{(j)}(t)$	generalized force
$R_i(t)$	force acting between vehicle and beam
R	radius
Re	real part of the function of a complex variable
S	axial force
$S(x)$	Fresnel integral
$S_{ff}(q_1, q_2, \omega_1, \omega_2)$	spectral density of function $f(x, t)$

T	shear force
T	vertical force of a string
T	time of load traverse over beam
T_0	time interval
$T_{(j)}$	period of free vibration
U	Fourier integral transformation of function u
V	Fourier integral transformation of function v
V_j	expression (6.6)
$V(j, t)$	Fourier sine finite integral transformation of function $v(x, t)$
$V^*(j, p)$	Laplace-Carson integral transformation of function $V(j, t)$
W	Fourier integral transformation of function w
W	the Wronskian
X	coordinate axis of the centre of gravity
$X_{gh}(t)$	longitudinal force acting on bar gh at point g
X_i	force per unit volume along axis x_i
Y	coordinate axis of the centre of gravity
$Y_{gh}(t)$	transverse force acting on bar gh at point g
Z	coordinate axis of the centre of gravity
$Z(t)$	force acting along axis Z
$Z^*(p)$	Laplace-Carson integral transformation of function $z(0, l, t)$
$Z_i(t)$	force acting in spring C_i
$Z_{bi}(t)$	damping force acting in spring C_i
α	speed parameter
α'	parameter inversely proportional to speed
$\alpha_i =$	c/c_i
β	damping parameter
γ_i	frequency parameter of sprung and unsprung parts of vehicle
γ_{ij}	shear strain in plane $x_i x_j$
δ	dynamic coefficient (impact factor)
$\delta(x)$	Dirac delta function
δ_{ij}	Kronecker delta symbol
$\varepsilon_i =$	0 or 1
ε_i	relative elongation (strain)
ζ	cross-section rotation
η	error

η	imaginary coordinate in the complex plane
η	straight line along which a force moves
η	viscosity coefficient
η	auxiliary variable
ϑ	logarithmic decrement of damping
$\varkappa$	weight parameter
$\varkappa_i$	weight parameter of unsprung and sprung parts of vehicle
λ	coefficient expressing elastic foundation and stiffness of a beam or plate
λ	coefficient expressing rotation of sprung parts of vehicle
λ	Lamé's constant
λ, λ_j	value dependent on natural frequency
μ	mass per unit length of beam or unit area of plate
μ_p	mass appertaining to external load p
ν	Poisson's ratio $(\nu < 1)$
$\xi =$	x/l dimensionless length coordinate
ξ_i	dimensionless coordinate of the point of contact
$\xi =$	$x - ct$ length coordinate in the moving coordinate system
ξ	auxiliary variable
ξ	real coordinate in the complex plane
ϱ	mass per unit volume
ϱ	radius in polar coordinates
$\sigma_{i,j}$	stress component
$\bar{\sigma}$	Fourier integral transformation of stress σ
$\sigma_f(x, t)$	standard deviation of function $f(x, t)$
$\sigma_f^2(x, t)$	variance of function $f(x, t)$
τ	dimensionless time
τ	auxiliary time variable
τ_{ij}	tangential stress
$\bar{\tau}$	Fourier integral transformation of stress τ
φ	polar angle
$\varphi(t)$	rotation of sprung parts of vehicle
$\varphi(x)$	function
$\varphi_j(x)$	linearly independent functions satisfying boundary conditions
$\psi(x, t)$	beam section rotation
$\psi(x)$	Euler's psi-function
ω	circular frequency

ω	circular frequency expressing load motion
ω_b	circular frequency of damping
$\omega_{(j)}$	natural circular frequency
Δ	expression (2.13)
Δ	dynamic increment of deflection or stress
∇	Laplace's operator
Θ	relative volume change
$\Theta(x, t)$	rotation of transverse section about the centre of flexure
Θ	frequency
Σ	summation sign
Φ	central angle of an arc
Φ	angle between a straight line and axis x
Φ	polar angle
$\Psi(j, t)$	Fourier integral transformation of function $\psi(x, t)$
$\Psi^*(j, p)$	Laplace-Carson integral transformation of function $\Psi(j, t)$
$\Psi(q)$	Fourier integral transformation of function $\psi(s)$
Ω	circular frequency of a harmonic force
Ω	circular frequency of non-uniform motion

Subscripts

b	damping
cr	critical or limit value of a quantity
g	left-hand end of a beam
h	homogeneous
h	right-hand end of a beam
$i =$	1, 2, 3
$j =$	1, 2, 3, ...
$k =$	1, 2, 3, ...
p	particular
p	plastic
0	initial conditions
1	unsprung
1	left-hand end
1	first general linearly independent solution of the homogeneous differential equation
2	sprung or unsprung

2	right-hand end
2	second general linearly independent solution of the homogeneous differential equation
3	sprung

Superscripts

$' \; '' \; ''' \; {}^{IV}$	derivatives with respect to the length coordinate
$'$	damped vibration
$'$	in the oblique direction
$\cdot \; \cdot\cdot$	dots over the letter denote derivatives with respect to the time coordinate
(n)	n-th derivative
$-$	bar over the letter denotes a loaded quantity
$-$	bar over the letter denotes a quantity with a dimension

Units

	International System SI
length:	metre [m], centimetre [cm], kilometre [km]
force:	Newton [N], kilo Newton [kN], mega Newton [MN] 1 kilopond = 1 kilogram force = 9·806 65 N $\doteq$ $\doteq$ 10 N
mass:	kilogram [kg], metric ton [t]
time:	second [s], hour [h]
frequency:	cycle per second = Hertz [Hz]
circular frequency:	$[s^{-1}]$
speed:	metre per second [m/s], kilometre per hour [km/h]
stress:	Newton per square centimetre $[N/cm^2]$, kilo Newton per square centimetre $[kN/cm^2]$
coordinate system:	right-handed.

Part I

INTRODUCTION

Part I

Introduction

In this book we shall study in detail the effects of one type of load, i.e. of a moving load, on elastic and inelastic solids, elements and parts of structures, structures composed of those elements and parts, and on elastic media. Moving loads have a great effect on dynamic stresses in such bodies and structures, and cause them to vibrate intensively, especially at high velocities. Their peculiar feature is that they are variable in both time and space.

Vibration of a three-dimensional body may generally be described as an operator relation between the vector displacement $\mathbf{r}(x, y, z, t)$ at a point with coordinates x, y, z and time t, and the external load $\mathbf{p}(x, y, z, t)$ of the body, in the form

$$L[\mathbf{r}(x, y, z, t)] = \mathbf{p}(x, y, z, t) .$$

Symbol L denotes a linear or a nonlinear differential operator. Together with the boundary and the initial conditions, the above equation, or sets of such – usually partial differential – equations define the behaviour of the body.

If we think of a moving load as of a mass body moving in a generally curved path over the structure being examined, we see that according to d'Alembert's principle its effects are twofold: the weight, or gravitational, effect of the moving load, and the inertial effects of the load mass on the deformed structure.

If only the weight effect is considered, and the mass of the moving load neglected against the mass of the structure, the computation of strains in the solid is na easy enough matter. It becomes more complicated in the other extreme case, i.e. when the structure mass is assumed to be negligible against the load mass. But the most difficult of all is the problem involving both the gravitational and the inertial action of moving loads having masses commensurable with the mass of the structure. In this book, solutions of all three cases for several types of elastic as well as inelastic elements and structures are presented. Particularly, the first case is treated in Chapters 1–6 (with the exception of Chap. 3, where the effect of the mass of a continuous load is approximately taken into account), the second case in Chapter 7 and the general case in Chapters 8–10 for a simple beam.

A set of differential equations is not the only way of describing vibration of a solid. Sometimes the primary equation is in the form of the operator relation

$$\mathbf{r}(x, y, z, t) = L_1[\mathbf{p}(x, y, z, t), \mathbf{r}(x, y, z, t)]$$

which leads to linear or non-linear integro-differential equations. This method, though somewhat more laborious than the solution of differential equations, will be used in Chap. 5 for computing beam stresses.

From the historical viewpoint the problem of moving load is reviewed in detail in [219], Sects. 40 and 88. Its coming-into-being can be traced to the beginning of the nineteenth century, the time of erection of the early railway bridges. This makes it one of the original problems of structural dynamics in general. At that time, the engineering profession was split in two factions: one claiming that the effects of a moving load will resemble those of an impact, the other arguing that during rapid traverse of a locomotive over a bridge there will be not enough time for the structure to deform. It was the second class of thought that appealed to science fiction: in his well-known novel "Round the World in 80 Days" Jules Verne records the passage of a locomotive over a damaged bridge — a feat which has not been scientifically explained until recently, with the aid of the theory of plastic reserves in the material (see Chap. 25).

Theoretically, the problem of moving load was first tackled for the case in which the beam mass was considered small against the mass of a single, constant load. The original approximate solution is due to R. Willis [233], one of the early experimenters in the field. G. G. Stokes [207] and H. Zimmermann [236] approached the problem under similar assumptions.

The other extreme case, i.e. that of the load mass small against the beam mass, was originally examined for a simply supported beam and a constant concentrated force by A. N. Krÿlov [139] using the method of expansion of the eigenfunctions, and by S. P. Timoshenko [215].

A. N. Lowan [146] and N. G. Bondar' [23] solved it with the aid of Green's functions and integral equations, respectively.

S. P. Timoshenko [216] is also credited with the solution to the problem of the effects of a harmonic force moving over a beam at a constant speed – an idealization of the effects of counterweights on the locomotive driving wheels.

The problem involving both the load mass and the beam mass, considerably more complicated than the preceding special cases, was not solved until much later. It was first examined by H. Saller [196], then by H. H. Jeffcott [115] whose iterative method becomes divergent in some cases, and by H. Steuding [206] who studied several of its aspects. A satisfactory method (a Fourier series with unknown coefficients for the path of a single concentrated load of constant magnitude acting on a beam) was introduced by A. Schallenkamp [197]. V. M. Muchnikov [162] and M. Ya. Ryazanova [192] applied to the problem the method of integral equations, J. Naleszkiewicz [166] Galerkin's method, and V. V. Bolotin [21] the approximate method of asymptotic solutions in quadratures.

In a special class belongs the treatise by C. E. Inglis [111] who used harmonic analysis to solve all the practically important cases likely to come up in dynamic calculations of railway bridges traversed by steam locomotives (e.g. motion of a concentrated force, sprung and unsprung masses and harmonic forces acting on a beam, etc.). Its results – in excellent agreement with experimental findings – were later compared by A. H. Chilver [34] with those arrived at by K. Mise and S. Kunii [156] with the aid of elliptical functions.

General systems as well as statically complex systems have been studied by help of normal-mode analysis by S. T. Ödman [172] and worked out in full by V. Koloušek [130]. The results of the latter author, obtained by that method for continuous beams and arches, agree very well with experiments.

Problems specifically relating to the effects of moving loads on railway bridges have been treated by a number of authors. Next to the now classical treatises of Inglis [111] and Koloušek [130], [131], studies by B. Brückmann [29], I. I. Kazeĭ [121] and by the author [70, 73, 74, 76, 77, 84] may be quoted as examples of more recent work in the field.

In all the references mentioned so far, the vehicle was idealized by a single mass point. It goes without saying that for modern means of transport with distinctly differentiated unsprung and sprung masses such a simplification is no longer in order.*) The first solution of the motion of sprung masses on a beam is due to A. Hillerborg [107] who obtained it by means of Fourier's method and the method of numerical differences. Further advances in that direction were made possible by the arrival of digital computers. The topical problem was thus solved by J. M. Biggs, H. S. Suer and J. M. Louw [16] using Inglis' method, and by T. P. Tung, L. E. Goodman, T. Y. Chen and N. M. Newmark [224] using Hillerborg's method, and the solution was applied to vibration of highway bridges.

At the present time the problems of moving load are studied in technically advanced countries the world over. In Czechoslovakia, the theoretical and experimental foundations for their scientific treatment were laid by V. Koloušek in his authoritative series of books [130]. The work is now continued by the author [68 to 88], [247, 248].

In the USSR the effects of moving loads on solids and structures are followed in several institutions. One of those worthy of special mention is the Kharkov school of Professor A. P. Filippov [61, 62, 98, 127, 214] which has successfully dealt with many of the theoretical aspects of the problem. Theoretical studies are also conducted in Kiev [192 to 194]. A host of theoretical and experimental problems having to do with railway bridges has been investigated by the Dnepropetrovsk school of Professor N. G. Bondar' [23 to 25, 222]. The research institutes of Moscow and Leningrad have been engaged in extensive experimental studies of railway bridges traversed by diverse kinds of vehicles [121, 137, 141, 169]. Their work has a tradition of long standing [10, 58, 110, 176].

In the USA the work proceeded mostly along the experimental lines involving first railway bridges [191], [209], and [242], later also highway bridges dynamically tested up to failure [1], [268, 272]. Many of the theoretical studies that have been appearing of late in an ever greater number are aimed at application in naval, aircraft and astronautical engineering. Research into the effects of moving load is conducted at the universities in Urbana (University of Illinois) [168], [229], East Lansing (Michigan State University) [230 to 232], Evanston (Northwestern University) [3, 4], Stanford University [5 to 7], [205], [235], Massachusetts Institute of Technology [39], and elsewhere.

---

*) This fact was acknowledged by both Inglis [111] and Koloušek [130], Vol. II.

In Europe, Poland [92 to 95], [116 to 119], [138], the German Democratic Republic [181], [199], [261], Switzerland [190] and several other countries concentrate on the dynamics of highway bridges, while the German Federal Republic [28] and France [33], [46] carry out tests of railway bridges, especially at high speeds. Great Britain was the first to undertake large-scale experimental studies of railway bridges in the twenties [111], [157], [187]. Work on many of these problems under way in Japan and India is a source of much theoretical and experimental information on the effects of moving load on structures [173], [257], [27].

The dynamics of railway bridges has been greatly enhanced by international research assignments of several years' duration set up in part by the "Organization for Railways Cooperation" (OSShD) [24], [78], in part by the "Office for Research and Experiments" (ORE) of the International Union of Railways (UIC) [258]. Within the framework of the assignments, many European countries succeeded in collecting vast amounts of experimental data on stresses in a variety of railway bridges traversed by diverse types of locomotives, cars and trains at speeds up to 200 km/h. The experiments were supplemented by model research entrusted to Switzerland. The author had the privilege of contributing to the research of both organizations by several treatises, the gist of which is presented in Chaps. 8 to 10.

Since in each chapter the detailed explanation of the subject is preceded by a brief historical note, the list of references appended to the book quotes only the fundamental or more recent works from this as well as from allied fields to which the author has turned for information. It is clear from the extensive and far from complete bibliography on the effects of moving load that the pertinent questions are very actual, modern, and by no means answered satisfactorily. The aim of the book is to scientifically classify the broad theme, supply hitherto missing answers to the basic questions, and explain the methods best applicable to the problem at hand.

The load considered in the book is in the form of a moving force of constant magnitude, harmonic force, force generally variable in time, continuous load, moving impulses, etc., oftentimes with the inertial effects of the mass included. Attention is also accorded to special loads, such as those produced by multi-axle vehicles with unsprung and sprung masses, loads of random magnitude, etc. For the most, the load is assumed to move at constant speed; however, Chap. 19 also deals with uniformly accelerated and decelerated motions.

The solids discussed in the book are nearly all elements familiar from engineering mechanics and theory of elasticity and plasticity — namely beams of constant and variable cross section with various kinds of support, cantilever beams, infinite beams on elastic foundations, continuous beams, frames, arches, frame structures, strings, plates, elastic space, half-space, half-plane, thin-walled beams, rigid-plastic beams, etc. A particularly detailed examination is made of simply supported beams and of infinite beams on elastic foundations, under the action of all kinds of load, speed, damping, shear and rotatory inertia. The reason for this close attention are the immense possibilities of those two elements in actual structures.

The theoretical considerations are likely to find application in calculations relating to dynamic stresses in railway and highway bridges, suspension bridges, rails, sleepers, crane runways, cable and other types of cranes, cable railways, roadways and airport runways, underground railways, tunnels, foundations for all types of high- and railways, pipelines, etc.

The methods explained in the book can further be applied to calculations relating to motion of ground vehicles, such as automobiles, locomotives and railway cars, cranes, conveyors, etc., and they serve well in research of naval, aeronautical, and astronautical structures.

Modern means of transport are ever faster and heavier, while the structures over which they move are ever more slender and lighter. That is why the dynamic stresses they produce are larger by far than the static ones. On the strength of this the book devotes particular attention to the effects of speed, weight of the vehicle and structure, as well as of other parameters substantially bearing on the dynamic stresses in solids.

A theoretical examination of solids and elements subjected to a moving load may proceed along one of the following two lines: expansion in series, used for elements of finite length — a method whose advantage is a comparatively easy solution and whose shortcoming is slow convergence in some instances, especially at high speeds. The other method applicable to solids and elements of very large dimensions (infinite at the limit) considers only the steady-state vibration of the body as the load moves from infinity to infinity. This method is advantageous in that it supplies solutions in closed form; however, in many cases the solution is difficult to attain. An attempt to reconcile the two methods is made in Chap. 24.

Nearly all the problems contained in the book are solved by the methods of integral transformations which have proved unusually well suited and efficacious in studies of moving loads. The most frequently used are the Fourier transformation for the length coordinates and the Laplace-Carson transformation for the time coordinate. Comprehensive tables of integral transformations referred to in the text or applicable to solutions of other problems – most of them published for the first time – are at the end of the book.

With a very few exceptions, the vibrations of solids and elements of finite dimensions will be examined only during the period of the load traverse. Once the load departs from it, the structure begins to vibrate in free vibration, and this process no longer falls within the scope of our discussion. The attentuation of the whole phenomenon is greatly affected by the damping characteristics of both the structure and the material. However, in the course of the load traverse proper – an affair of relatively short duration – the qualitative effects of damping are not too intense and all they do – though sometimes substantially – is to reduce vibration amplitudes. That is the reason why in this book we have adopted the most elementary of all damping hypotheses – Voigt's – according to which damping is proportional to vibration velocity. Though we are well aware of the existence of more elaborate damping theories [140], we find this hypothesis wholly appropriate for our purposes, the more so that it is also in very good agreement with experiments conducted on actual structures subjected to moving loads.

Despite its fairly broad scope, the book can hardly be expected to do equal justice to all the aspects of the problem. Among the themes meriting but still awaiting lengthy treatment belong, for example, the effects of stress waves propagating from the point of action of the moving forces, the contact problem, the effects on layered homogeneous and non-homogeneous solids and structures, the relation between moving loads on the one hand and strength, fatigue and failure of materials on the other, etc.

What has been included as worthy of detailed exposition are the fundamentals of the most modern methods of structural design, i.e. those exploiting plastic reserves in the material when the structure is subjected to loads of a deterministic or in the case of elastic structures – a stochastic character (Chaps. 25 and 26). Those are highly advanced methods which – following further refinement – will be indispensable for design of structures with minimum weight or of structures which satisfy the theory of limit states. The final goal of such a design are structures optimal as regard safety, reliability, economy and long service life.

It is a sincere wish of the author that the book might become a dependable theoretical guide to economic design of new, and prolonged life of old, civil, mechanical and other engineering structures.

Part II

ONE-DIMENSIONAL SOLIDS

Part II

1

Simply supported beam subjected to a moving constant force

Of the wide range of problems involving vibration of structures and solids subjected to a moving load, the easiest one to tackle is that of dynamic stresses in a simply supported beam, traversed by a constant force moving at uniform speed. This classical case was first solved by A. N. Krȳlov [139], then by S. P. Timoshenko [215]. Other solutions worthy of mention are those by C. E. Inglis [111] and V. Koloušek [130]. In what follows we shall first outline the basic results arrived at through the application of the method of integral transformations, and then extend them to all possible cases of speed and viscous damping.

1.1 *Formulation of the problem*

In the solution we shall adopt the following assumptions (Fig. 1.1):

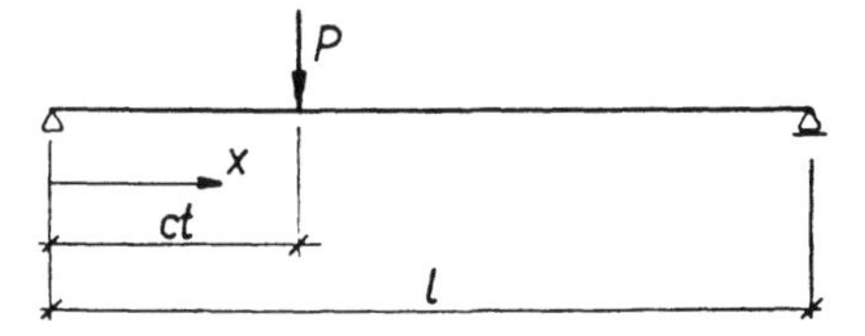

Fig. 1.1. Simple beam subjected to a moving force *P*.

1. The beam behaviour is described by Bernoulli-Euler's differential equation deduced on the assumption that the theory of small deformations, Hooke's law, Navier's hypothesis and Saint-Venant's principle can be applied. The beam is of constant cross-section and constant mass per unit length.

2. The mass of the moving load is small compared with the mass of the beam; this means that we shall consider only the gravitational effects of the load.

3. The load moves at constant speed, from left to right.
4. The beam damping is proportional to the velocity of vibration.
5. The computation will be carried through for a simply supported beam, i.e. a beam with zero deflection and zero bending moment at both ends. Further, at the instant of force arrival, the beam is at rest, i.e. possesses neither deflection nor velocity.

Under the above assumptions the problem is described by the equation

$$EJ\,\frac{\partial v^4(x, t)}{\partial x^4} + \mu\,\frac{\partial v^2(x, t)}{\partial t^2} + 2\mu\omega_b\,\frac{\partial v(x, t)}{\partial t} = \delta(x - ct)\,P\,; \quad (1.1)$$

the boundary conditions are

$$v(0, t) = 0\,; \qquad v(l, t) = 0\,,$$

$$\left.\frac{\partial^2 v(x, t)}{\partial x^2}\right|_{x=0} = 0\,; \quad \left.\frac{\partial^2 v(x, t)}{\partial x^2}\right|_{x=l} = 0\,, \quad (1.2)$$

and the initial conditions

$$v(x, 0) = 0\,; \quad \left.\frac{\partial v(x, t)}{\partial t}\right|_{t=0} = 0\,. \quad (1.3)$$

The symbols used in Eqs. (1.1) to (1.3) and throughout the subsequent chapters (see also Fig. 1.1) have the following meaning:

x – length coordinate with the origin at the left-hand end of the beam,
t – time coordinate with the origin at the instant of the force arriving upon the beam,
$v(x, t)$ – beam deflection at point x and time t, measured from the equilibrium position when the beam is loaded with own weight,
E – Young's modulus of the beam,
J – constant moment of inertia of the beam cross section,
μ – constant mass per unit length of the beam,
ω_b – circular frequency of damping of the beam,
P – concentrated force of constant magnitude,
l – span (length) of the beam,
c – constant speed of the load motion.

$$\delta(x) = \frac{\mathrm{d}H(x)}{\mathrm{d}x} \quad (1.4)$$

is the so-called Dirac (impulse, also delta) function that – as a generalized function – expresses the concentrated load as follows

$$p(x, t) = \delta(x)\, P\,. \tag{1.5}$$

The Dirac function is not a function in the conventional sense. It is a so-called generalized function and by Eq. (1.4) may be defined as the distributional derivative of the Heaviside function $H(x)$ (3.23) – for further details refer to [154]. In mechanics, the Dirac function $\delta(x)$ may be thought of as a unit concentrated force acting at point $x = 0$.

The following relations hold for the Dirac function (a, b, ξ denote constants and $f(x)$ is a continuous function in the interval $\langle a, b\rangle$)*)

$$\int_{-\infty}^{\infty} \delta(x)\, \mathrm{d}x = 1\,, \tag{1.6}$$

$$\int_{-\infty}^{\infty} \delta(x - a)\, f(x)\, \mathrm{d}x = f(a) \tag{1.7}$$

or

$$\int_a^b \delta(x - \xi)\, f(x)\, \mathrm{d}x = \begin{cases} 0 & \text{for} \quad \xi < a < b \\ f(\xi) & \text{for} \quad a < \xi < b \\ 0 & \text{for} \quad a < b < \xi \end{cases}$$

For the n-th generalized derivative of the Dirac function it holds more generally

$$\int_a^b \delta^{(n)}(x - \xi)\, f(x)\, \mathrm{d}x = \begin{cases} 0 & \text{for} \quad \xi < a < b \\ (-1)^n f^{(n)}(\xi) & \text{for} \quad a < \xi < b \\ 0 & \text{for} \quad a < b < \xi \end{cases}$$

Substitution of $\varphi(x)$ having zero value at some (single) point ξ $[\varphi(\xi) = 0]$ gives

$$\delta[\varphi(x)] = \frac{1}{|\varphi'(\xi)|}\, \delta(x - \xi)\,.$$

In the special case of $\varphi(x) = ax$

$$\delta(ax) = \frac{1}{a}\, \delta(x)\,. \tag{1.8}$$

---

*) The derivation of the following relations may be found in [154] and in: V. V. Novitskiĭ: Delta-function and its Application in Engineering Mechanics (in Russian). Raschet prostranstvennȳkh konstruktsiĭ. Volume 8. Gosstroĭizdat, Moscow, 1962, 207–244.

1.2 *Solution of the problem*

Eq. (1.1) together with conditions (1.2) and (1.3) will be solved by the methods of integral transformations, the theory of which is expounded at length in [48], [49], [204] and other well-known publications, e.g. [53].

Each term of Eq. (1.1) will first be multiplied by $\sin j\pi x/l$ and then integrated with respect to x between 0 and l. We shall make use of the following fundamental relations of the Fourier sine (finite) integral transformation [cf. (27.67)]:

$$V(j, t) = \int_0^l v(x, t) \sin \frac{j\pi x}{l} \, \mathrm{d}x \,, \quad j = 1, 2, 3, \ldots$$

$$v(x, t) = \frac{2}{l} \sum_{j=1}^{\infty} V(j, t) \sin \frac{j\pi x}{l} \tag{1.9}$$

where we speak of $V(j, t)$ as of the transform of the original $v(x, t)$. By the procedure just outlined, using the boundary conditions (1.2) and the properties of the Dirac function (1.7) we shall get by (27.69), (27.70) and (27.74)

$$\frac{j^4\pi^4}{l^4} EJ \, V(j, t) + \mu \, \ddot{V}(j, t) + 2\mu\omega_b \, \dot{V}(j, t) = P \sin \frac{j\pi ct}{l} \,. \tag{1.10}$$

We denote now the circular frequency at the j-th mode of vibration of a simply supported beam by

$$\omega_{(j)}^2 = \frac{j^4\pi^4}{l^4} \frac{EJ}{\mu} \,, \tag{1.11}$$

the corresponding natural frequency by

$$f_{(j)} = \frac{\omega_{(j)}}{2\pi} = \frac{j^2\pi}{2l^2} \left(\frac{EJ}{\mu}\right)^{1/2} \tag{1.12}$$

and the circular frequency by

$$\omega = \frac{\pi c}{l} \,. \tag{1.13}$$

Using this notation we rearrange Eq. (1.10) to give it the form

$$\ddot{V}(j, t) + 2\omega_b \dot{V}(j, t) + \omega_{(j)}^2 V(j, t) = \frac{P}{\mu} \sin j\omega t . \tag{1.14}$$

To solve Eq. (1.14) we apply the method of the Laplace-Carson integral transformation, i.e. multiply the equation by e^{-pt}, integrate each of its terms with respect to t between 0 and ∞, and then multiply it by p (p is a variable in the complex plane). The basic relations of this transformation are [cf. (27.1)]

$$V^*(j, p) = p \int_0^\infty V(j, t)\, e^{-pt}\, dt ,$$

$$V(j, t) = \frac{1}{2\pi i} \int_{a_0 - i\infty}^{a_0 + i\infty} e^{tp} \frac{V^*(j, p)}{p}\, dp \tag{1.15}$$

where a_0 in the second relation signifies that the integration is carried out along a straight line parallel to the imaginary axis lying to the right of all the singularities of the function of the complex variable $e^{tp} V(j, t)/p$ (the real argument of all the singularities is therefore less than a_0).

Transforming now Eq. (1.14) in accordance with (1.15) will give us – in view of the initial conditions (1.3) and the relations (27.2) to (27.4), (27.18) –

$$p^2 V^*(j, p) + 2\omega_b p\, V^*(j, p) + \omega_{(j)}^2 V^*(j, p) = \frac{Pj\omega}{\mu} \frac{p}{p^2 + j^2\omega^2} \tag{1.16}$$

from which it is not difficult to compute the transformed solution

$$V^*(j, p) = \frac{Pj\omega}{\mu} \frac{p}{p^2 + j^2\omega^2} \frac{1}{p^2 + 2\omega_b p + \omega_{(j)}^2} . \tag{1.17}$$

Depending on the position of the poles of the function of complex variable (1.17), distinction is made between several cases whose analysis is rendered easier by the introduction of the following two dimensionless parameters

$$\alpha = \frac{\omega}{\omega_{(1)}} = \frac{c}{2f_{(1)}l} = \frac{T_{(1)}}{2T} = \frac{cl}{\pi}\left(\frac{\mu}{EJ}\right)^{1/2} = \frac{c}{c_{cr}} , \tag{1.18}$$

$$\beta = \frac{\omega_b}{\omega_{(1)}} = \frac{\omega_b l^2}{\pi^2}\left(\frac{\mu}{EJ}\right)^{1/2} = \frac{\vartheta}{2\pi} \tag{1.19}$$

characteristic of the effect of speed — α and of the effect of damping — β. In the above the new symbols denote

$T_{(1)} = 1/f_{(1)}$ — period of the first free vibration,

$T \; = l/c$ — time of traverse of the force over the beam,

$$c_{cr} = 2f_{(1)}l = \frac{\pi}{l}\left(\frac{EJ}{\mu}\right)^{1/2} \text{ — critical speed,} \tag{1.20}$$

$\vartheta \; = \omega_b/f_{(1)}$ — logarithmic decrement of damping of the beam.

Symbol v_0 designates the deflection at mid-span of a beam loaded with static force P at point $x = l/2$(*)

$$v_0 = \frac{Pl^3}{48EJ} \approx \frac{2P}{\mu l \omega_{(1)}^2} = \frac{2Pl^3}{\pi^4 EJ}. \tag{1.21}$$

The circular frequency of a damped beam with light damping is

$$\omega_{(j)}'^2 = \omega_{(j)}^2 - \omega_b^2, \tag{1.22}$$

that of a damped beam with heavy damping

$$\omega_{(j)}'^2 = \omega_b^2 - \omega_{(j)}^2. \tag{1.23}$$

In the case of light damping the four poles of function (1.17) are $\pm ij\omega$, $-\omega_b \pm i\omega_{(j)}'$, where $\omega_{(j)}'$ is described by Eq. (1.22). Since $p^2 + 2\omega_b p + \omega_{(j)}^2 = (p + \omega_b)^2 + \omega_{(j)}'^2$, the original $V(j, t)$ may be computed with the aid of relation (27.41), so that following the inverse Fourier transformation (1.9) we get directly — after rearrangement — the solution of the basic case (for $t \leqq T$)

$$\begin{aligned} v(x, t) = v_0 \sum_{j=1}^{\infty} \frac{1}{j^2[j^2(j^2 - \alpha^2)^2 + 4\alpha^2\beta^2]} \Big[j^2(j^2 - \alpha^2) \sin j\omega t - \\ - \frac{j\alpha[j^2(j^2 - \alpha^2) - 2\beta^2]}{(j^4 - \beta^2)^{1/2}} e^{-\omega_b t} \sin \omega_{(j)}' t - \\ - 2j\alpha\beta(\cos j\omega t - e^{-\omega_b t} \cos \omega_{(j)}' t) \Big] \sin \frac{j\pi x}{l}. \end{aligned} \tag{1.24}$$

*) With a few exceptions in Chaps. 5, 9 and 10, the last two expressions of (1.21) are used for v_0 throughout the discussion.

The bending moment $M(x, t)$ and the shear force $T(x, t)$ are obtained from the relations

$$M(x, t) = -EJ \frac{\partial^2 v(x, t)}{\partial x^2}, \tag{1.25}$$

$$T(x, t) = -EJ \frac{\partial^3 v(x, t)}{\partial x^3}. \tag{1.26}$$

Denoting by

$$M_0 = \frac{Pl}{4}, \tag{1.27}$$

$$T_0 = P \tag{1.28}$$

the bending moment and the difference of shear forces, respectively, at mid-span of the beam, produced by static force P at point $x = l/2$ we get

$$M(x, t) = M_0 \sum_{j=1}^{\infty} \frac{8j^2}{\pi^2} \sin \frac{j\pi x}{l} \frac{1}{j^2[j^2(j^2 - \alpha^2)^2 + 4\alpha^2\beta^2]} \cdot \left[j^2(j^2 - \alpha^2) \sin j\omega t - \frac{j\alpha[j^2(j^2 - \alpha^2) - 2\beta^2]}{(j^4 - \beta^2)^{1/2}} e^{-\omega_b t} \sin \omega'_{(j)} t - - 2j\alpha\beta(\cos j\omega t - e^{-\omega_b t} \cos \omega'_{(j)} t) \right],$$

$$T(x, t) = T_0 \sum_{j=1}^{\infty} \frac{2j^3}{\pi} \cos \frac{j\pi x}{l} \frac{1}{j^2[j^2(j^2 - \alpha^2)^2 + 4\alpha^2\beta^2]} \cdot \left[j^2(j^2 - \alpha^2) \sin j\omega t - \frac{j\alpha[j^2(j^2 - \alpha^2) - 2\beta^2]}{(j^4 - \beta^2)^{1/2}} e^{-\omega_b t} \sin \omega'_{(j)} t - - 2j\alpha\beta(\cos j\omega t - e^{-\omega_b t} \cos \omega'_{(j)} t) \right]. \tag{1.29}$$

Series (1.24) for the beam deflection converges very fast, approximately like the series $\sum_{j=1}^{\infty} 1/j^4$; series (1.29) for the bending moment and shear force, on the other hand, converge far more slowly, approximately like the series $\sum_{j=1}^{\infty} 1/j^2$ and $\sum_{j=1}^{\infty} (-1)^{j+1}/j$, respectively – see Chap. 5 for further details.

1.3 *Special cases*

Let us now analyze Eqs. (1.17) and (1.24) for several special cases of the values of parameters α and β.

1.3.1 *Static case* $(\alpha = 0)$

If we set $\alpha = 0$ in Eq. (1.24), then

$$v(x, t) = v_0 \sum_{j=1}^{\infty} \frac{1}{j^4} \sin \frac{j\pi x}{l} \sin j\omega t\,. \tag{1.30}$$

This is the case of static deflection of a beam at point x if the beam is loaded with force P at point ct. Eq. (1.30) is therefore the equation of the influence line of beam deflection at point x expanded in the Fourier series (the deflections are magnified P-times). Fig. 1.2 shows the function for $x = l/2$.

1.3.2 *Case with no damping* $(\beta = 0)$

1.3.2.1 $\alpha \neq j,\ \beta = 0$

For this case we get from Eq. (1.24) for $\beta = 0$ or from Eq. (1.17) for $\omega_b =$ $= 0$ using (27.32)

$$v(x, t) = v_0 \sum_{j=1}^{\infty} \sin \frac{j\pi x}{l} \frac{1}{j^2(j^2 - \alpha^2)} \left(\sin j\omega t - \frac{\alpha}{j} \sin \omega_{(j)} t \right). \tag{1.31}$$

Fig. 1.2a shows Eq. (1.31) for $x = l/2$ and parameters $\alpha = 0{\cdot}5$, $\beta = 0$.

1.3.2.2 $\alpha = n,\ \beta = 0$

If α happens to be just equal to one of the numbers $j = 1, 2, 3, \ldots$, say to number n, then for $j = n$ the poles of function (1.17) will merge in two double poles $\pm in\omega$ since in this case $\omega_{(n)}^2 = n^4 \omega_{(1)}^2 = n^2 \omega^2$ and $n =$ $= \omega/\omega_{(1)} = \alpha$. For this term of the series, $j = n$, relation (27.35) must

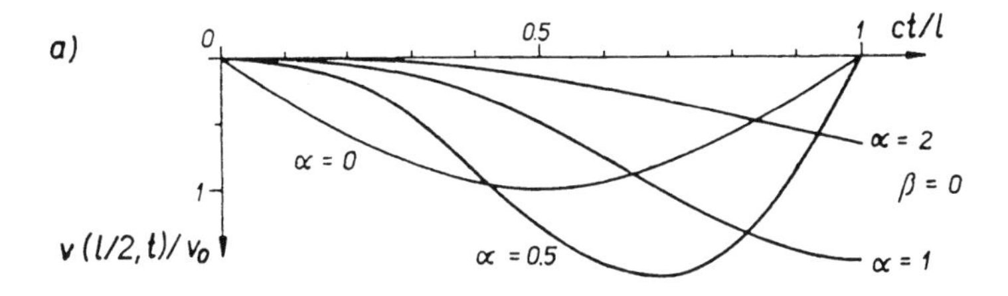

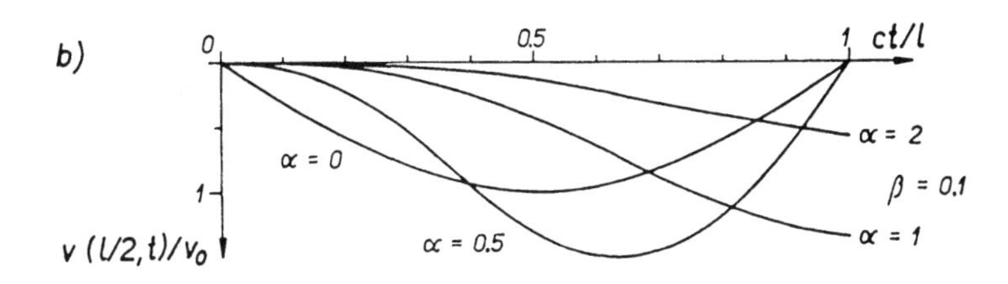

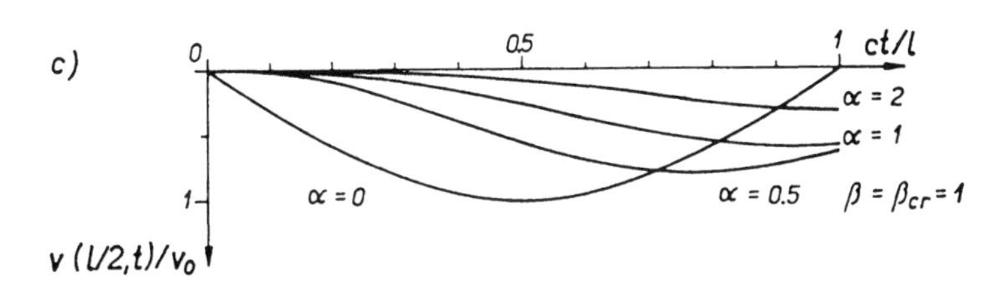

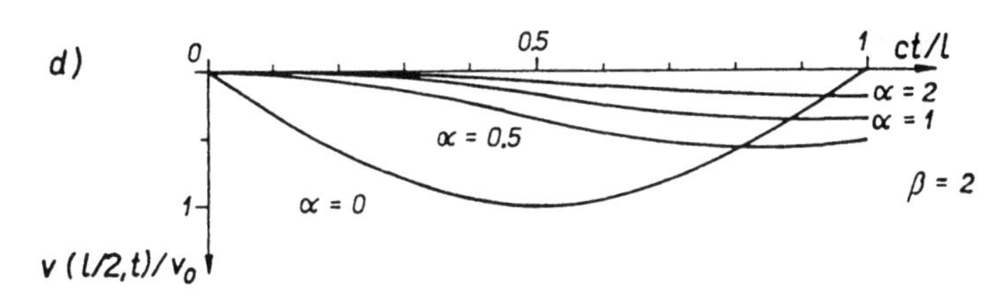

Fig. 1.2. Dynamic deflection at mid-span of a beam, $v(l/2, t)/v_0$, for various values of speed and damping. a) $\beta = 0$; $\alpha = 0, 0{\cdot}5, 1, 2$, b) $\beta = 0{\cdot}1$; $\alpha = 0, 0{\cdot}5, 1, 2$, c) $\beta = \beta_{cr} = 1$; $\alpha = 0, 0{\cdot}5, 1, 2$, d) $\beta = 2$; $\alpha = 0, 0{\cdot}5, 1, 2$.

be resorted to, whereas for the remaining ones, $j = 1, 2, \ldots, n - 1, n + 1, \ldots$ there holds the preceding solution (1.31)

$$v(x, t) = v_0 \frac{1}{2n^4} (\sin n\omega t - n\omega t \cos n\omega t) \sin \frac{n\pi x}{l} +$$

$$+ v_0 \sum_{j=1, j \neq n}^{\infty} \sin \frac{j\pi x}{l} \frac{1}{j^2(j^2 - \alpha^2)} \left(\sin j\omega t - \frac{\alpha}{j} \sin \omega_{(j)} t \right) \quad (1.32)$$

where the symbols underneath the summation sign indicate that the summation includes all j's except $j = n$.

In the case of $\alpha = n$, the displacements of one point of the beam grow with time; as the process is transient ($0 \leqq ct/l \leqq 1$), they do not, however, attain infinite values at $t = T = l/c$.

We may also think of this case as of one of a beam subjected — in addition to force P — to a centrally applied compressive force $S = \mu c^2$. Since the critical force of a simply supported beam $S_{cr} = (\pi^2/l^2) EJ$, it also holds [cf. (1.18)]

$$\alpha^2 = \frac{\mu c^2 l^2}{\pi^2 EJ} = \frac{S}{S_{cr}} . \quad (1.33)$$

For $\alpha = 1$, $\beta = 0$ the case discussed is illustrated in Fig. 1.2a for $x = l/2$. There $n = 1$, and the values for $j = 2, 3, \ldots$ are very small (or zero) against the first term of Eq. (1.32). Fig. 1.2a also shows the case of $\alpha = 2$, $\beta = 0$ ($n = 2$) in which for $x = l/2$ it is, of course, $\sin n\pi l/(2l) = 0$, and the expressions for $j = 3, 4, \ldots$ are very small (or zero) against the first term of the series at $j = 1$; therefore, the numerical computation was made using practically only the first term of the series in the second portion of Eq. (1.32).

1.3.3 *Light damping* ($\beta \ll 1$)

1.3.3.1 $\alpha \neq j$, $\beta \ll 1$

When the damping is very light, the terms with β and β^2 may be neglected in (1.24), and the expression then approximately gives

$$v(x, t) \approx v_0 \sum_{j=1}^{\infty} \sin \frac{j\pi x}{l} \frac{1}{j^2(j^2 - \alpha^2)} \left(\sin j\omega t - \frac{\alpha}{j} e^{-\omega_b t} \sin \omega_{(j)} t \right). \quad (1.34)$$

For $\alpha = 0{\cdot}5$, $\beta = 0{\cdot}1$, Eq. (1.34) with $x = l/2$ is plotted in Fig. 1.2b.

1.3.3.2 $\alpha = n,\ \beta \ll 1$

For this case of $n = \omega/\omega_{(1)} = \alpha$, Eq. (1.24) gives

$$v(x, t) \approx v_0 \frac{1}{2n^4}\left[e^{-\omega_b t} \sin n\omega t - \frac{n^2}{\beta} \cos n\omega t(1 - e^{-\omega_b t})\right] \sin \frac{n\pi x}{l} +$$

$$+ v_0 \sum_{\substack{j=1 \\ j \neq n}}^{\infty} \frac{1}{j^2(j^2 - \alpha^2)} \left(\sin j\omega t - \frac{\alpha e^{-\omega_b t}}{j} \sin \omega_{(j)} t\right) \sin \frac{j\pi x}{l}. \qquad (1.35)$$

Fig. 1.2b shows Eq. (1.35) with $x = l/2$ for $\beta = 0{\cdot}1$, $\alpha = 1$ and $\alpha = 2$. For reasons stated in connection with case 1.3.2.2, the numerical computation for $\alpha = 1$ was made using the first term of Eq. (1.35), for $\alpha = 2$, the first term of the series in the second of expressions (1.35).

1.3.4 *Critical damping* $(\beta = \beta_{cr} = n^2)$

The value of critical damping is attained whenever $\omega'_{(n)} = 0$ [cf. (1.22)] for some $j = n$. Then $\omega_b = \omega_{(n)} = \omega_{(1)} n^2$ and therefore

$$\beta_{cr} = \frac{\omega_b}{\omega_{(1)}} = n^2 . \qquad (1.36)$$

For $j = n$, the poles of expression (1.17) being $\pm in\omega$ and the double pole $-\omega_{(n)}$, we shall use for the inverse transformation relation (27.38) which after some handling gives

$$v(x, t) = v_0 \frac{1}{n^2(n^2 + \alpha^2)^2} \{(n^2 - \alpha^2) \sin n\omega t - 2n\alpha \cos n\omega t +$$

$$+ e^{-\omega_{(n)} t}[(n^2 + \alpha^2)\, n\omega t + 2n\alpha]\} \sin n\pi x/l . \qquad (1.37)$$

For $j < n$, we must add to expression (1.37) the solution for supercritical damping (1.39), and for $j > n$, the basic solution (1.24). Bearing this in mind we computed the deflection at mid-span $(x = l/2)$ of a beam with critical damping $\beta = \beta_{cr} = 1$ for speeds $\alpha = 0{\cdot}5$, 1, 2, and plotted it in Fig. 1.2c.

1.3.5 *Supercritical damping* $(\beta > \beta_{cr})$

Supercritical damping occurs whenever the damping is so heavy that for $j < n$, $\omega'_{(j)}$ according to Eq. (1.23) is positive. At the same time it must be

$\omega_b^2 > \omega_{(j)}^2 = j^4\omega_{(1)}^2$ and thus also $\omega_b^2 > n^4\omega_{(1)}^2$, and therefore

$$\beta = \frac{\omega_b}{\omega_{(1)}} > n^2 . \tag{1.38}$$

In such a case the poles of expression (1.17) are $\pm ij\omega$ and $-\omega_b \pm \omega'_{(j)}$.

The inverse transformation is computed using expression (27.39) which after some handling leads to

$$v(x, t) = v_0 \sum_{j=1}^{n} \frac{1}{j^2[j^2(j^2 - \alpha^2)^2 + 4\alpha^2\beta^2]} \left\{ j^2(j^2 - \alpha^2) \sin j\omega t - \right.$$

$$- 2j\alpha\beta \cos j\omega t + \frac{j\alpha e^{-\omega_b t}}{2(\beta^2 - j^4)^{1/2}} [(2\beta^2 - j^2(j^2 - \alpha^2) +$$

$$+ 2\beta(\beta^2 - j^4)^{1/2}) e^{\omega'_{(j)} t} - (2\beta^2 - j^2(j^2 - \alpha^2) - 2\beta(\beta^2 - j^4)^{1/2}) \times$$

$$\left. \times e^{-\omega'_{(j)} t}] \right\} \sin \frac{j\pi x}{l} . \tag{1.39}$$

For $j > n$, the basic solution (1.24) with the summation including $j = n + 1, n + 2, \ldots$ must be added to (1.39). If the critical damping exists at some higher j, expression (1.37), too, must appear in the solution. Note that (1.22) applies to $j > n$, (1.23) to $j \leqq n$(*). Expression (1.39) computed for $x = l/2$, $\beta = 2$ and $\alpha = 0{\cdot}5$, 1, 2 is shown in Fig. 1.2d.

1.4 *Application of the theory*

1.4.1 *The effect of speed*

It is clear to see in Figs. 1.2a to 1.2d that at subcritical speeds, $\alpha < 1$, the maximum deflection at mid-span of the beam is produced already during the load traverse, while at supercritical speeds, $\alpha \geqq 1$, it is not observed until instant $t = T$, i.e. when the moving force departs from the beam. The dynamic deflection is soon dampened out by damped free

*) This means that supercritical damping holds for lower, subcritical damping for higher natural modes. It is, however, of no practical significance to consider a larger number of terms in the expression of $v(x, t)$.

vibration that set in subsequently; that is the reason why the case of $t > T$ is of no further interest to us.

Fig. 1.3 shows the maximum deflection at mid-span of the beam, $\max v(l/2, t)/v_0$, as a function of parameter α (i.e. speed) for damping $\beta = 0$, 0·1, 0·3, $\beta_{cr} = 1$, 2. In the diagram the maximum dynamic de-

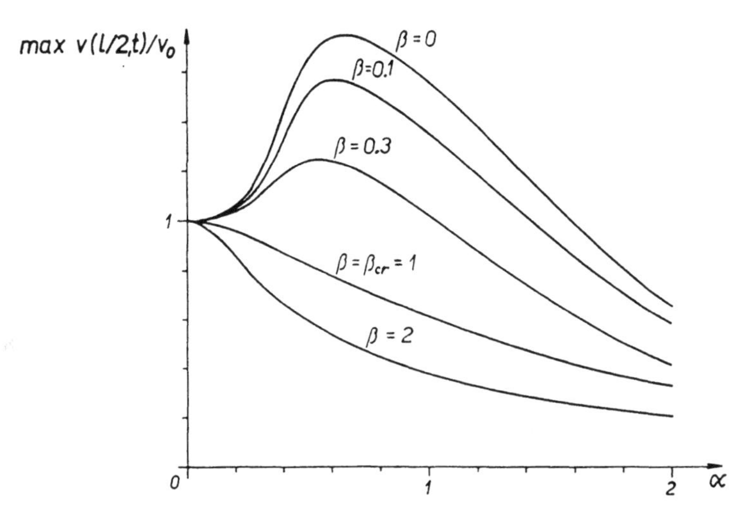

Fig. 1.3. Maximum dynamic deflection at mid-span of a beam, $\max v(l/2, t)/v_0$, in dependence on speed α, for various values of damping, $\beta = 0$, 0·1, 0·3, $\beta_{cr} = 1$, 2.

flection is associated with speeds $\alpha \approx 0{\cdot}5$ to 0·7. For large α's, the deflection rapidly tends to zero, for small α's it is practically equal to the static deflection.

The critical speed as defined by (1.20) is fairly high. By substituting in (1.20) the empirical formula (1.55) of the first natural frequency of steel railway bridges for $f_{(1)}$, we approximately get

$$c_{cr} = 2f_{(1)}l \approx 2l\,\frac{10^3}{4l} = 500\ \text{m/s} = 1800\ \text{km/h}\,. \tag{1.40}$$

1.4.2 *Application to bridges*

The theory expounded in the preceding paragraphs has found its widest field of application in calculations relating to large-span railway and highway bridges. Such bridges, simply supported in most instances, have a mass much larger than the mass of the vehicle, and very low first natural frequencies. That is why – if vibration of vehicles on own springs is neglected, too – the effects of moving vehicles may approximately be replaced by the effects of moving forces.

Since the damping of large-span bridges is light, the dynamic deflection may be computed from formula (1.34). For the very low speeds attainable in practice, i.e. for $\alpha \ll 1$, Eq. (1.34) further simplifies to (1.30) of which all we need to take is the first term because the terms with $j > 1$ are negligible compared to it. This results in

$$v(x, t) = v_0 \sin \omega t \sin \frac{\pi x}{l}, \tag{1.41}$$

which can be used as a dynamic deflection formula wholly satisfactory for practical purposes.

1.4.3 *Approximate solution of the effects of a moving mass*

In his approximate solution [111] of the effects of vehicles moving over large-span bridges, C. E. Inglis introduced an assumption according to which the gravitational effects of the load may be separated from the inertial ones. In the calculation the force is considered as moving along the beam (this is the case which we have solved in Sects. 1.1 to 1.3) while the mass of the vehicle acts at a definite, constant point x_0.

Let us now analyze whether or not the second part of the assumption is justified. Mass m of load P ($P = mg$, g – acceleration of gravity) acts on the beam at point x_0 by its intertial effects, a case that may be described by the differential equation

$$EJ \frac{\partial^4 v(x, t)}{\partial x^4} + \mu \frac{\partial^2 v(x, t)}{\partial t^2} + 2\mu\omega_b \frac{\partial v(x, t)}{\partial t} = -\delta(x - x_0)\, m \left.\frac{\partial^2 v(x, t)}{\partial t^2}\right|_{x = x_0}. \tag{1.42}$$

Assuming that the solution $v(x, t)$ will be in the form of the second of relations (1.9), and using the first of relations (1.9), conditions (1.2) and equations (27.69) and (27.74), we get the Fourier transform of Eq. (1.42)

$$\frac{j^4\pi^4}{l^4} EJV(j, t) + \mu \ddot{V}(j, t) + 2\mu\omega_b \dot{V}(j, t) = -\frac{2m}{l} \ddot{V}(j, t) \sin^2 \frac{j\pi x_0}{l}, \tag{1.43}$$

the transform on the right-hand side of the equation is but approximate, since it is assumed that

$$\left.\frac{\partial^2 v(x, t)}{\partial t^2}\right|_{x=x_0} = \frac{2}{l}\,\ddot{V}(j, t) \sin \frac{j\pi x_0}{l} .$$

Eq. (1.43) may then be given the form

$$\ddot{V}(j, t) + 2\bar{\omega}_b \dot{V}(j, t) + \bar{\omega}_{(j)}^2 V(j, t) = 0 \tag{1.44}$$

where

$$\bar{\omega}_{(j)}^2 = \omega_{(j)}^2 \frac{\mu}{\bar{\mu}} ; \quad \bar{\omega}_{(j)}'^2 = \bar{\omega}_{(j)}^2 - \bar{\omega}_b^2 , \tag{1.45}$$

$$\bar{\omega}_b = \omega_b \frac{\mu}{\bar{\mu}} , \tag{1.46}$$

$$\bar{\mu} = \mu \left(1 + \frac{2P}{G} \sin^2 \frac{j\pi x_0}{l}\right), \tag{1.47}$$

$$\bar{f}_{(j)} = f_{(j)} \left(1 + \frac{2P}{G} \sin^2 \frac{j\pi x_0}{l}\right)^{-1/2} \quad \text{and} \tag{1.48}$$

$$G = \mu l g \quad - \text{weight of the bridge.} \tag{1.49}$$

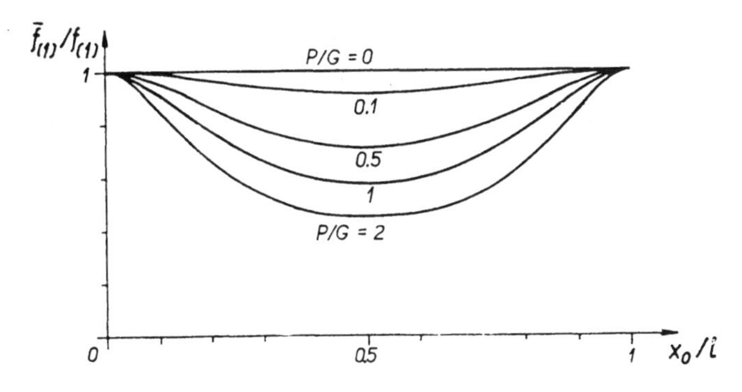

Fig. 1.4. First natural frequency of loaded beam, $\bar{f}_{(1)}/f_{(1)}$, dependence on mass position, x_0/l, for various values of ratio $P/G = 0, 0{\cdot}1, 0{\cdot}5, 1, 2$.

Eq. (1.44) is the same as the left-hand side of Eq. (1.14), except that $\omega_{(j)}^2$ and ω_b of the latter are replaced by quantities $\bar{\omega}_{(j)}^2$ and $\bar{\omega}_b$ expressing the fact that the beam is loaded at point x_0 with immobile mass m. Accordingly, if we are out to compute the effects of a moving force with an approximate expression of the mass effects, we shall use Eqs. (1.24) to (1.39) with $\omega_{(j)}$, ω_b, and $f_{(j)}$ replaced by the corresponding values with bar $\bar{\omega}_{(j)}$, $\bar{\omega}_b$, $\bar{f}_{(j)}$, $\bar{\omega}'^2_{(j)} = \bar{\omega}^2_{(j)} - \bar{\omega}^2_b$, etc.

All that remains to be done is to determine point x_0 where mass m is to be located. A means to this end is Fig. 1.4, plotting the dependence of the first natural frequency of a loaded beam $\bar{f}_{(1)}/f_{(1)}$ versus the point of action of mass m, x_0/l, described by Eq. (1.48), for several values of the P/G ratio. As the figure clearly shows, the natural frequency of a loaded bridge varies with the position of mass m. Near the centre of a simply supported beam this variation is rather small, particularly for small values of the P/G ratio. That is why, in cases of this sort, mass m is usually placed at mid-span of the beam, i.e. at $x_0 = l/2$; then Eqs. (1.47) and (1.48) approximately are

$$\bar{\mu} = \mu(1 + 2P/G)\,, \tag{1.50}$$

$$\bar{f}_{(1)} = f_{(1)}(1 + 2P/G)^{-1/2}\,. \tag{1.51}$$

1.4.4 *Experimental results*

Experimental data necessary for computations according to the theory explained above were collected in the course of extensive measurements made under author's leadership on various steel railway bridges (for further details refer to [70], [73], [74]). The measurements proved the approximate formula (1.41) to be valid for large-span bridges on the condition that both the track and the wheels are ideally smooth and the vehicle has no unbalanced rotating masses.

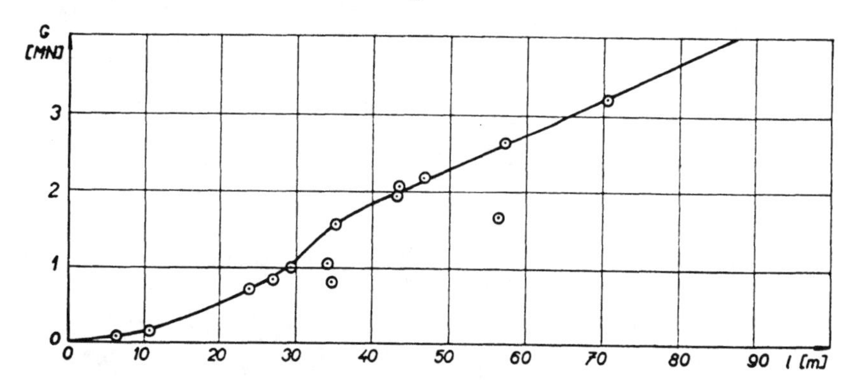

Fig. 1.5. Total weight of steel railway bridges G as a function of span l (the curve represents the variations smoothed according to empirical formulae given in [74]).

The total weight G, according to (1.49), of the most diverse types of steel railway bridges that have been subjected to the tests is shown in Fig. 1.5.

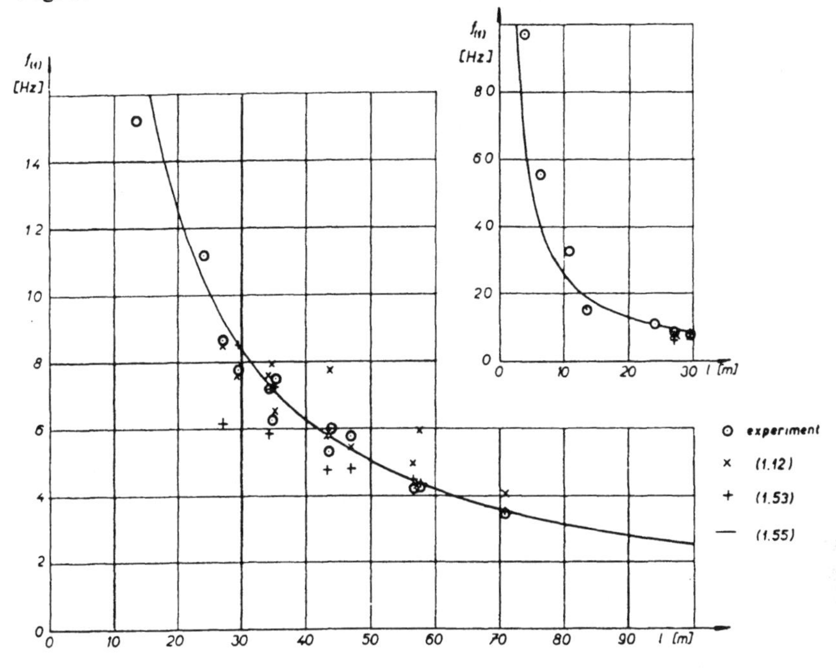

Fig. 1.6. First natural frequency $f_{(1)}$ of vertical vibration of steel railway bridges under no load, as function of span l.

The natural frequency of unloaded steel railway bridges as a function of span is plotted in Fig. 1.6. The graph shows the experimental data as well as the values computed using formula (1.12). The value of μ substituted in (1.12) was obtained from Eq. (1.49) and from data of Fig. 1.5. Even though the main girders of steel bridges are of no constant cross section, they may approximately be calculated as prismatic beams. That is why the moment of inertia J of trussed bridges was computed from the formula

$$J = \tfrac{1}{2}(F_h + F_s)\, h^2 \tag{1.52}$$

where F_h (F_s) is the cross-sectional area of one upper (lower) chord at mid-span of the bridge, and h is the theoretical height of the panel at mid-span of the bridge.

To describe the effect of the roadway, we computed the first natural frequency from the approximate formula

$$f_{(1)} \approx \frac{5{\cdot}62}{\bar{v}_{st}^{1/2}} \tag{1.53}$$

where $f_{(1)}$ is in [Hz], and $\bar{v}_{st}$ in [cm]. In the above formula $\bar{v}_{st}$ may be regarded as the measured deflection produced by bridge's own weight μg. Since this deflection cannot be measured directly we computed it from the relation $\bar{v}_{st} = v_{st}\bar{v}_L/v_L$ where v_{st} is the computed deflection produced by uniformly distributed own weight μg, $\bar{v}_L$ and v_L the measured and computed, respectively, deflection produced by the test load, for example by a locomotive. In so doing we took advantage of the approximate proportionality between the measured and the theoretical deflections produced by two kinds of load. The deflection at mid-span of the equivalent constant cross-section plate girder substituting for the bridge structure, subjected to uniform load μg, thus turns out to be

$$\bar{v}_{st} = \frac{5}{384}\frac{\mu g l^4}{EJ} \approx \frac{4}{\pi^5}\frac{\mu g l^4}{EJ}\,. \tag{1.54}$$

Computing $\mu l^4/(EJ)$ from this equation and substituting it in (1.12) give — after enumeration in the above units — the empirical formula (1.53).

The measured values of the first natural frequencies $f_{(1)}$ of the bridges were also smoothed out by the so-called group method(*) according to which the most satisfactory of all is the empirical dependence

$$f_{(1)} = \frac{1000}{4l} \tag{1.55}$$

where $f_{(1)}$ is in [Hz], l in [m]. Dependence (1.55) plotted in Fig. 1.6 works well for steel railway bridges over a broad range of spans, say from to 5 to 70 m, and agrees with the results reported abroad [24], [121].

Damping of bridge structures is a highly complicated problem (cf. [121]) in which both linear and nonlinear dependences come into play.

*) The group method (also the method of means) requires that the sum of errors should be zero in every group, of which there are as many as there are unknown parameters. See Z. Horák: Practical Physics (in Czech), Publishing House SNTL, Prague, 1958, p. 106.

Damping is usually considered as proportional to the velocity of vibration. This approximate damping theory is quite suitable for cases with a small range of possible frequencies. And that is just the case of simply supported railway bridges, where at present-day speeds the frequencies of the locomotive driving wheels practically do not reach but the first natural frequencies of the bridges.

Damping proportional to the velocity of vibration is characterized by the logarithmic decrement of damping ϑ defined as the natural logarithm of the ratio of two succeeding amplitudes (cf. [130]). This ratio was measured and ϑ computed from it. The experimental data thus obtained are plotted in Fig. 1.7 relative to span. For spans up to about 70 m, the

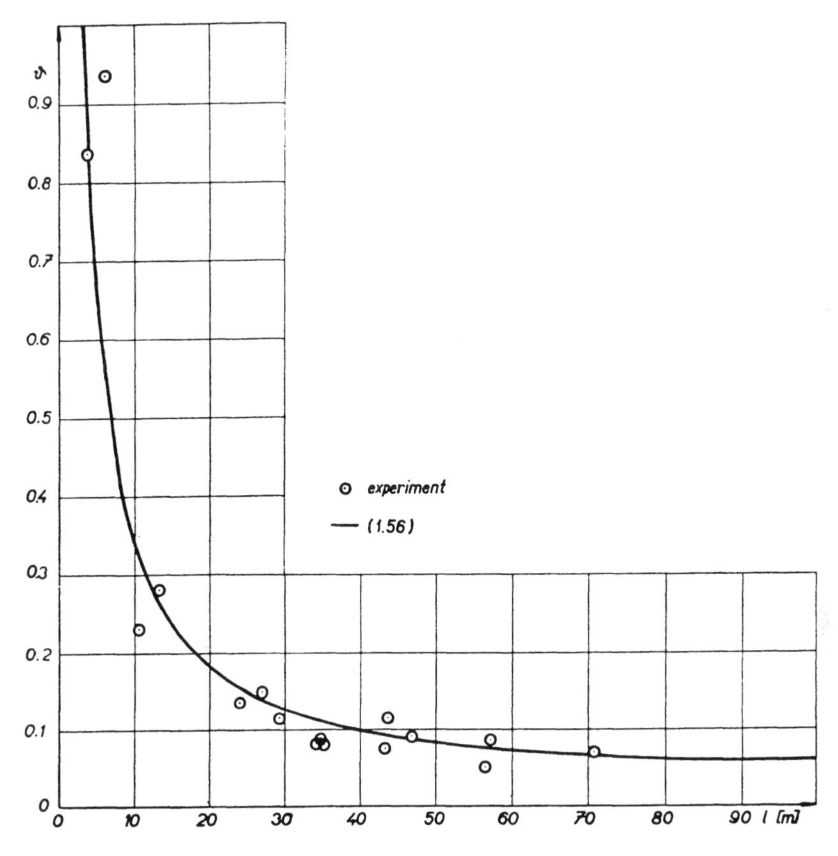

Fig. 1.7. Logarithmic decrement of damping ϑ of steel railway bridges as function of span l.

measured values are well represented by the empirical formula

$$\vartheta = \frac{1}{0{\cdot}3l - 1{\cdot}2 \times 10^{-3}l^2} \tag{1.56}$$

where ϑ is a dimensionless number, l in [m]. This dependence as arrived at by the group method is also shown in Fig. 1.7.

The circular frequency of damping is calculated from the logarithmic decrement with the aid of the formula (cf. [130])

$$\omega_b = \vartheta f'_{(1)} \tag{1.57}$$

where $f'_{(1)} = \omega'_{(1)}/(2\pi) \approx f_{(1)}$; see (1.22).

1.5 *Additional bibliography*

[1, 7, 15, 19, 24, 28, 46, 65, 67, 70, 73, 74, 94, 111, 114, 116, 121, 130, 131, 135, 139, 145, 150, 173, 182, 215, 220, 258, 261]

2

Moving harmonic force

Another of the problems of fundamental importance is that of a harmonic force moving along a simply supported beam. It was first solved by S. P. Timoshenko [216] and worked out in detail by C. E. Inglis [111] and V. Koloušek [130]. What we are going to do here is to deduce the basic results, set them forth in the form of useful formulae and compare them with experimental data.

2.1 *Formulation and solution of the problem*

The solution of the problem of a harmonic concentrated force moving at constant speed c over a simply supported beam with span l is carried out under the same assumptions as that discussed in Chap. 1. The time-variable concentrated force is of the form

$$P(t) = Q \sin \Omega t \tag{2.1}$$

where Q is the amplitude and Ω the circular frequency of the harmonic force. Vibration of the beam is then described by the equation

$$EJ \frac{\partial^4 v(x, t)}{\partial x^4} + \mu \frac{\partial^2 v(x, t)}{\partial t^2} + 2\mu\omega_b \frac{\partial v(x, t)}{\partial t} = \delta(x - ct)\, Q \sin \Omega t\,, \tag{2.2}$$

by the boundary conditions (1.2) and by the initial conditions (1.3). The symbols used in (2.2) have the same meaning as those of Chap. 1.

Eq. (2.2) together with conditions (1.2) and (1.3) will again be solved by the method of integral transformations. Following the Fourier sine transfomation according to (1.9), Eqs. (2.2) and (1.2) give

$$\frac{d^2 V(j, t)}{dt^2} + 2\omega_b \frac{dV(j, t)}{dt} + \omega_{(j)}^2 V(j, t) = \frac{Q}{\mu} \sin \Omega t \sin j\omega t\,. \tag{2.3}$$

Solving the above with (1.3) by the Laplace-Carson transformation (1.15) – making use of Eq. (27.24) in doing so and of the notation

$$r_1 = \Omega + j\omega\,; \quad r_2 = \Omega - j\omega \tag{2.4}$$

we get

$$V^*(j, p) = \frac{Q}{2\mu}\left(\frac{1}{p^2 + r_2^2} - \frac{1}{p^2 + r_1^2}\right)\frac{p^2}{(p + \omega_b)^2 + \omega_{(j)}'^2}\,. \tag{2.5}$$

After inverse transformations of Eq. (2.5) according to (27.24) and (1.9) the required result for $t \leqq T$ is

$$v(x, t) = \sum_{j=1}^{\infty} \frac{Q}{\mu l}\left\{\frac{1}{(\omega_{(j)}^2 - r_2^2)^2 + 4\omega_b^2 r_2^2}\left[(\omega_{(j)}^2 - r_2^2)\,.\right.\right.$$

$$.\,(\cos r_2 t - \mathrm{e}^{-\omega_b t}\cos \omega_{(j)}'\, t) + 2\omega_b r_2 \sin r_2 t - \frac{\omega_b}{\omega_{(j)}'}(\omega_{(j)}^2 + r_2^2)\,.$$

$$\left..\,\mathrm{e}^{-\omega_b t}\sin \omega_{(j)}' t\right] - \frac{1}{(\omega_{(j)}^2 - r_1^2)^2 + 4\omega_b^2 r_1^2}\left[(\omega_{(j)}^2 - r_1^2)\,.\right.$$

$$.\,(\cos r_1 t - \mathrm{e}^{-\omega_b t}\cos \omega_{(j)}' t) + \;2\omega_b r_1 \sin r_1 t -$$

$$\left.\left.- \frac{\omega_b}{\omega_{(j)}'}(\omega_{(j)}^2 + r_1^2)\,\mathrm{e}^{-\omega_b t}\sin \omega_{(j)}' t\right]\right\}\sin\frac{j\pi x}{l}\,. \tag{2.6}$$

We shall now simplify Eq. (2.6) to fit the cases most frequently met with in practical applications. Thus, for example, it is entirely satisfactory to use only the first of its terms ($j = 1$); further, as we know from Chap. 1, parameters α and β are usually much smaller than 1 ($\alpha = \omega/\omega_{(1)} \ll 1$, $\beta = \omega_b/\omega_{(1)} \ll 1$). And finally, since in practice a harmonic force is always accompanied by a constant force P, we shall introduce in (2.6) also the deflection v_0 according to (1.21). Following these simplifications Eq. (2.6) takes on the form

$$v(x, t) = v_0\,\frac{Q}{P}\,\frac{\omega_{(1)}^2}{\Omega^2}\,\frac{1}{\left(\dfrac{\omega_{(1)}^2}{\Omega^2} - 1\right)^2 + 4\left(\dfrac{\omega^2}{\Omega^2} + \dfrac{\omega_b^2}{\Omega^2}\right)}\,.$$

$$.\left\{\left[\left(\frac{\omega_{(1)}^2}{\Omega^2} - 1\right)^2 + 4\,\frac{\omega_b^2}{\Omega^2}\right]^{1/2}\sin(\Omega t + \varphi)\sin\omega t +\right.$$

$$\left.+\; 2\,\frac{\omega}{\Omega}(\cos \Omega t \cos \omega t - \mathrm{e}^{-\omega_b t}\cos \omega_{(1)} t)\right\}\sin\frac{\pi x}{l} \tag{2.7}$$

where

$$\operatorname{tg} \varphi = - \frac{2\omega_b/\Omega}{\omega_{(1)}^2/\Omega^2 - 1} \, .$$

The beam reaches the state of highest dynamic stressing in the region of resonance, i.e. whenever Ω is close or just equal to $\omega_{(1)}$, i.e.

$$\Omega = \omega_{(1)} \, . \tag{2.8}$$

In such a case Eq. (2.7) can further be simplified to

$$v(x, t) = v_0 \frac{Q\omega_{(1)}}{2P} \frac{\cos \omega_{(1)}t}{\omega^2 + \omega_b^2} \left[\omega(\cos \omega t - e^{-\omega_b t}) - \omega_b \sin \omega t\right] \sin \frac{\pi x}{l} \, . \tag{2.9}$$

2.2 *The dynamic coefficient*

In practice the dynamic coefficient is oftentimes defined as the ratio of the maximum dynamic deflection to the static deflection at mid-span of a beam (see [74])

$$\delta = \frac{\max v(l/2, t)}{v_0} \, . \tag{2.10}$$

The above is written on the consideration that the beam is traversed simultaneously by forces $P + Q \sin \Omega t$; its resultant vibration is therefore given by the sum of Eqs. (1.41) and (2.7) where v_0 is described by Eq. (1.21).

The maximum deflection of the first component of motion at centre $x = l/2$ of the beam is v_0 (according to Eq. (1.41)), i.e. a deflection at the instant of the force passing over the centre of the beam.The maximum dynamic deflection produced by the motion of the harmonic force occurs at the instant the force is past the centre of the beam. However, we assume approximately that the maximum deflection of the second component of motion also occurs at instant $t = T/2 = l/(2c)$. In this case $\sin \omega t = 1$, $\cos \omega t = 0$, cf. (1.13). The maximum deflection in Eq. (2.7) is at $\cos \omega_{(1)}t = -1$, and in saying so we assume, too, that $\sin (\Omega t + \varphi) = 1$. On substituting these values in Eq. (2.7), we get from (2.10) – with (1.41) and (2.7) included – the dynamic coefficient

$$\delta = 1 + \frac{Q}{P}\frac{\omega_{(1)}^2}{\Omega^2}\frac{1}{\left(\frac{\omega_{(1)}^2}{\Omega^2} - 1\right)^2 + 4\left(\frac{\omega^2}{\Omega^2} + \frac{\omega_b^2}{\Omega^2}\right)} \cdot$$

$$\cdot\left\{\left[\left(\frac{\omega_{(1)}^2}{\Omega^2} - 1\right)^2 + 4\frac{\omega_b^2}{\Omega^2}\right]^{1/2} + 2\frac{\omega}{\Omega}\,\mathrm{e}^{-\omega_b l/(2c)}\right\}. \tag{2.11}$$

Eq. (2.11), expressing the dependence of the dynamic coefficient on speed, is sometimes called the resonance curve.

The dynamic coefficient attains its maximum at resonance, i.e. at $\Omega = \omega_{(1)}$. Then Eq. (2.11) becomes

$$\delta = 1 + \frac{Q}{2P}\frac{\omega_{(1)}}{\omega^2 + \omega_b^2}\left(\omega\mathrm{e}^{-\omega_b l/(2c)} + \omega_b\right) = 1 + \frac{Q}{P}\Delta \tag{2.12}$$

where Δ, with the substitution of the speed and damping parameters α and β according to Eqs. (1.18) and (1.19), is

$$\Delta = \frac{1}{2}\frac{1}{\alpha^2 + \beta^2}\left(\alpha\mathrm{e}^{-\pi\beta/(2\alpha)} + \beta\right). \tag{2.13}$$

Fig. 2.1 shows Δ as a function of the speed parameter α for several values of the damping parameter β.

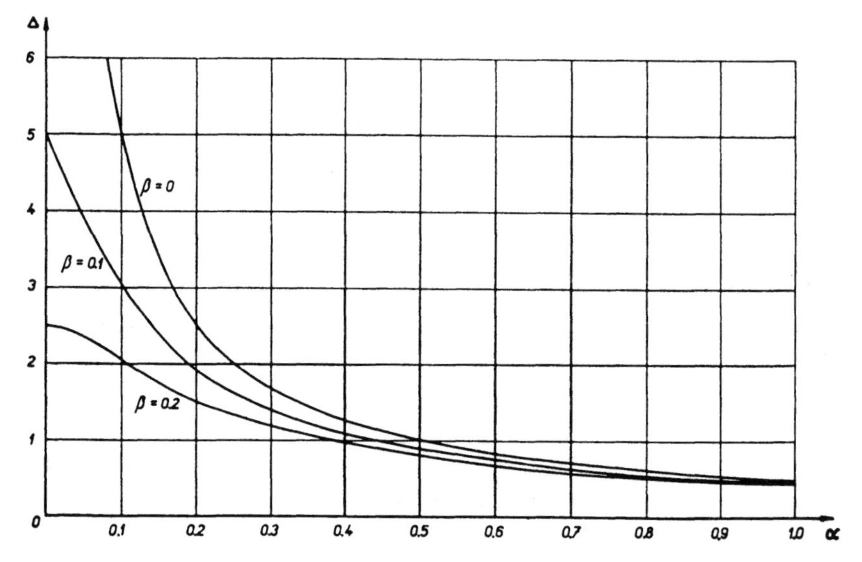

Fig. 2.1. Dependence of expression Δ [Eq. (2.13)] on speed α for various values of damping β.

2.3 *Application of the theory*

A classical example of the application of the theory expounded above are the dynamic effects of steam locomotives on large-span railway bridges [11], [130], [70], [74]. Steam two-cylinder locomotives have driving wheels provided with unbalanced counterweights, the right-hand ones being placed 90° against the left-hand ones. In motion the counterweights produce a harmonic force on either side of the locomotive; their resultant is likewise a harmonic force of the form of (2.1). The amplitude of this additional force depends on the velocity of wheel rotation. Thus, for example, according to tests, the resultant amplitude of the Czechoslovak steam locomotive ČSD 524.1 is

$$Q = 2{\cdot}4 \text{ to } 3 \times N^2 \tag{2.14}$$

where Q – amplitude of the force in kN,
$N = c/O$ – revolutions per second of the driving wheels,
$\Omega = 2\pi N = c/r$ – circular frequency of force (2.1),
r, O – radius, circumference of the driving wheels, respectively.

The whole passage of a steam locomotive over a large-span bridge may be idealized by a moving concentrated force P representing the force action of the locomotive, and by a harmonic force $Q \sin \Omega t$ expressing the dynamic effects of the counterweights. According to the deductions of paragraph 1.4.3, the locomotive mass is assumed to be placed stationary at mid-span of the bridge. In this way we may describe – approximately at least – the effects of the moving mass by using the equations evolved in Chaps. 1 and 2 with the quantities $\omega_{(j)}$, ω_b and μ replaced by those with bar, i.e. $\bar{\omega}_{(j)}$, $\bar{\omega}_b$, $\bar{\mu}$ according to Eqs. (1.45), (1.46) and (1.50).

In bridges with spans over 30 or 40 m, resonance according to (2.8) is likely to set in even at speeds attainable in practice. In that case Eq. (2.8) becomes

$$\Omega = \bar{\omega}_{(1)} \quad \text{or} \quad N = \bar{f}_{(1)} \,. \tag{2.15}$$

From there follows the equation for determining the critical speed at which the dynamic effects of a steam locomotive reach their maximum, namely

$$c_{cr} = \bar{f}_{(1)} O \tag{2.16}$$

where $\bar{f}_{(1)}$ is the first natural frequency of a bridge loaded at mid-span with a locomotive – cf. Eq. (1.51).

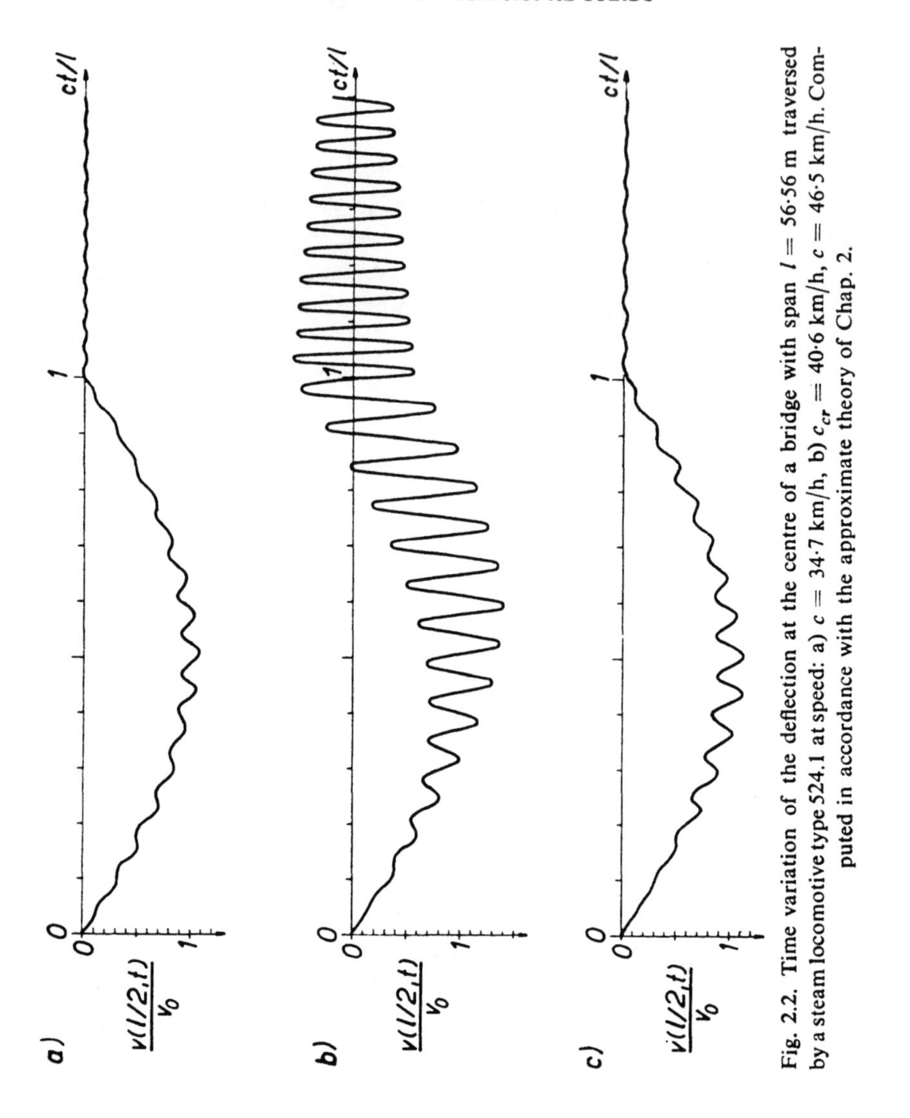

Fig. 2.2. Time variation of the deflection at the centre of a bridge with span $l = 56{\cdot}56$ m traversed by a steam locomotive type 524.1 at speed: a) $c = 34{\cdot}7$ km/h, b) $c_{cr} = 40{\cdot}6$ km/h, $c = 46{\cdot}5$ km/h. Computed in accordance with the approximate theory of Chap. 2.

2.3.1 *Comparison of theory with experiments*

In practical instances the resultant beam deflection is described by Eqs. (1.41) and (2.7).

To illustrate, we shall now compute the deflection at mid-span ($x = l/2$) of a steel railway bridge with span $l = 56{\cdot}56$ m, weight $G =$

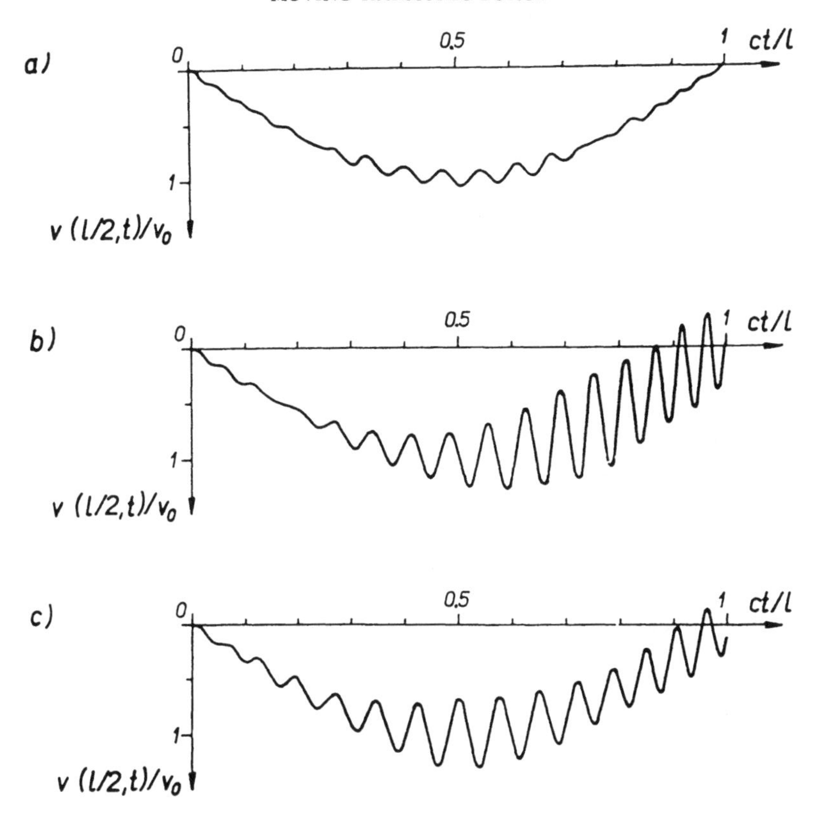

Fig. 2.3. Time variation of the deflection at the centre of a bridge with span $l =$ $= 56{\cdot}56$ m traversed by a steam locomotive type 524.1 at speed: a) $c = 34{\cdot}7$ km/h, b) $c_{cr} = 40{\cdot}6$ km/h, c) $c = 46{\cdot}5$ km/h. Computed according to the theory expounded in Chap. 8.

$= 1{\cdot}64$ MN, first natural frequency (measured) $f_{(1)} = 4{\cdot}2$ Hz, logarithmic decrement of damping $\vartheta = 0{\cdot}05$ (see [74]), traversed by a steam locomotive type 524.1, with weight $P = 0{\cdot}97$ MN, $r = 0{\cdot}63$ m, $O =$ $= 3{\cdot}96$ m, $Q = 3 \times N^2$.

The first natural frequency of the loaded bridge (1.51) is $\bar{f}_{(1)} = 2{\cdot}85$ Hz, the critical speed (2.16), $c_{cr} = 40{\cdot}6$ km/h. From Eqs. (1.41) and (2.7) we get the time variation of the deflection at mid-span of the bridge when traversed by a 524.1 locomotive at speeds: a) $c = 34{\cdot}7$ km/h, b) $c_{cr} =$ $= 40{\cdot}6$ km/h, c) $c = 46{\cdot}5$ km/h. The deflections computed with the aid of the approximate theory are shown in Fig. 2.2, while Fig. 2.3 depicts

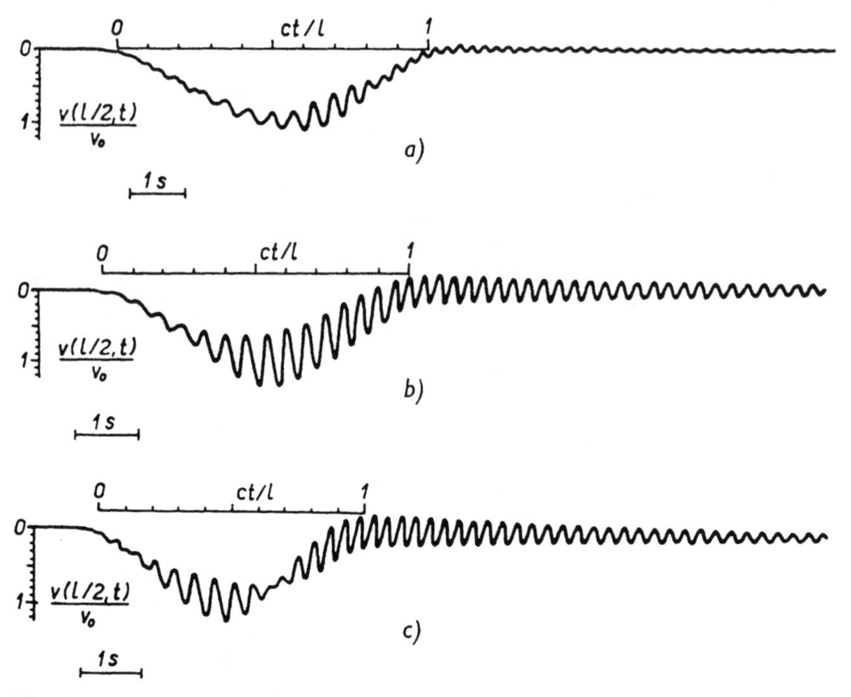

Fig. 2.4. Time variation of the deflection at the centre of a bridge with span $l =$ $= 56{\cdot}56$ m traversed by a steam locomotive type 524.1 at speed: a) $c = 34{\cdot}7$ km/h, b) $c = 38{\cdot}8$ km/h (speed close to $c_{cr} = 40{\cdot}6$ km/h), c) $c = 46{\cdot}5$ km/h. Experiments.

the same deflections obtained by the application of the theory of Chap. 8, in which consideration is also given to the motion of the locomotive mass. Fig. 2.4 reproduces the corresponding deflections arrived at experimentally – see [74].

Comparing Figs. 2.2, 2.3 and 2.4 we see that both the approximate (Chaps. 1 and 2) and the exact (Chap. 8) theoretical results agree very well with experimental records, especially at subcritical and critical speeds, with the locomotive near the centre of the bridge.

Fig. 2.5 shows the resonance curve computed for our case from Eq. (2.12). The figure also shows the experimental results and the theoretical dependence of the dynamic coefficient on speed computed using the exact theory of Chap. 8. It is clear that both the exact and the approximate theoretical results compare well with the experiments. The resonance curve according to the exact theory is somewhat flatter than that according to the approximate theory, and describes the values measured outside the resonance range more fittingly. In the resonance range both theories agree nearly perfectly with the experimental results.

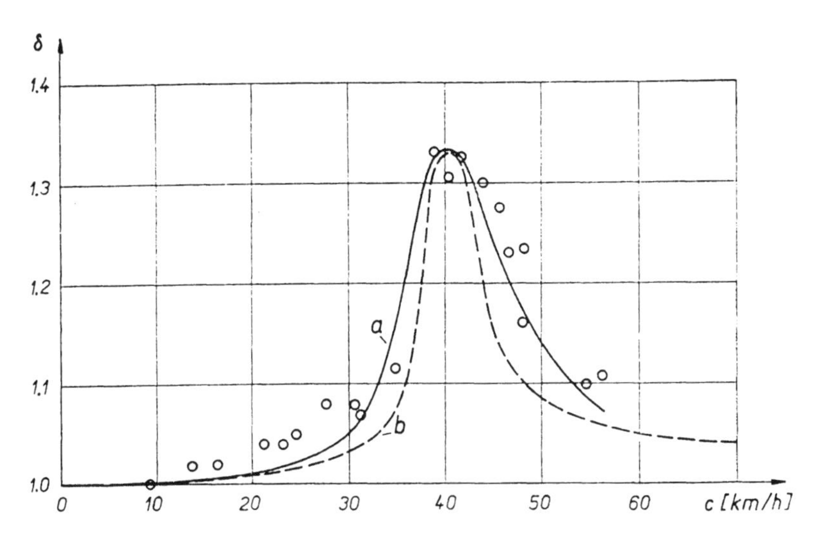

Fig. 2.5. Theoretical and experimental dependences of dynamic coefficient δ on speed c. Bridge with span $l = 56{\cdot}56$ m, steam locomotive type 524.1. a) exact theory according to Chap. 8, b) approximate theory according to Chap. 2, ○ — experiment.

2.3.2 *Critical speed and maximum dynamic coefficient for bridges of various spans*

We have applied the approximate theory to bridges with spans larger than 30 m traversed by a steam locomotive type 524.1, the worst balanced of all Czechoslovak Railways' engines. The pertinent calculations were made using the smoothed values of weight, first natural frequency and logarithmic decrement of damping of the respective bridges (Figs. 1.5, 1.6 and 1.7).

The critical speeds thus computed [Eqs. (2.16) and (1.51)] relative to the bridge span are plotted in Fig. 2.6 (curve $p = 0$). As the figure suggests, the critical speed falls off with growing span.

Fig. 2.7 shows the measured and the computed [Eq. (2.12)] values of maximum dynamic coefficient resulting in bridges of various spans from

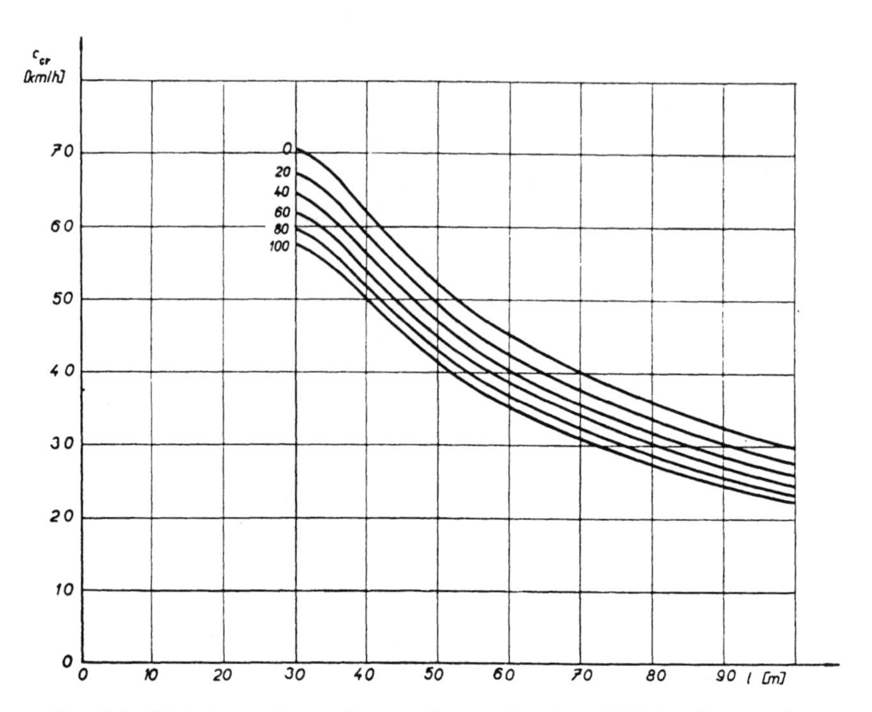

Fig. 2.6. Critical speed c_{cr} of steam locomotive type 524.1 hauling continuous load $p = 0, 20, 40, 60, 80, 100$ kN/m, in dependence on span l.

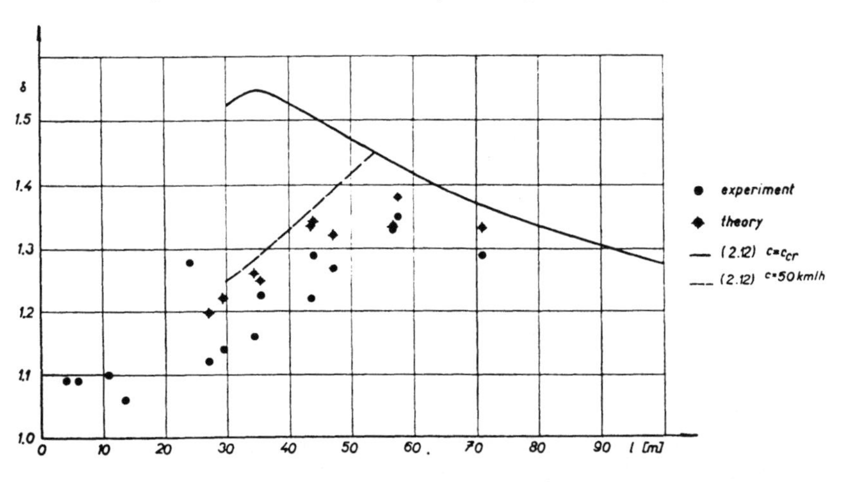

Fig. 2.7. Maximum theoretical and experimental dynamic coefficients δ produced by a steam locomotive type 524.1 in bridges with various spans l. -◆- theory, ● experiment, —— according to Eq. (2.12) at $c = c_{cr}$, - - - according to Eq. (2.12) at $c = 50$ km/h.

the action of the 524.1 locomotive. It also plots the dependence of the dynamic coefficient of span, computed from Eq. (2.12) and the smoothed values in Figs. 1.5 to 1.7, for the critical speeds indicated in Fig. 2.6, $p = 0$ (solid line). Since in the bridge tests the locomotive failed to attain speeds over 50 km/h, we also computed the dynamic coefficient-span dependence (in the same way as before) for $c = 50$ km/h (dashed line).

As we can see from Fig. 2.7, the maximum dynamic coefficient grows with growing span up to about $l = 54$ m. Up to this span the critical speed according to Fig. 2.6, $p = 0$, is namely higher than 50 km/h, a value the test locomotive was unable to reach. In the subsequent tests in which the critical speeds were attained, the dynamic effects fall off with span. The measured values, though somewhat lower than the computed ones, follow the two dependences with a fair degree of fit.

2.4 *Additional bibliography*

[24, 34, 70, 74, 111, 114, 121, 130, 131, 156, 198, 216, 220].

3

Moving continuous load

For large-span bridges the problem of the effects of a moving continuous load is of equal – if not greater – importance than the problem of a moving concentrated force. It has been treated at length by several authors, I. I. Gol'denblat [97] and V. V. Bolotin [20] among them. In the next sections we shall deduce a stationary solution for the case of a moving endless strip of load also with the approximate effect of its mass, and an approximate solution for the load arrival to and departure from the beam. In conclusion we shall review some of our experimental results.

3.1 *Steady-state vibration*

Consider a strip of load moving at constant speed c over a simply supported beam. Relative to the moving coordinate system

$$\xi = x - ct \tag{3.1}$$

the strip of load has magnitude $q(\xi, t)$ and mass $\mu_q(\xi) = q(\xi, t)/g$ per unit length of the beam, see Fig. 3.1.

Adopting the assumptions set forth in Sect. 1.1 we may describe this case by the differential equation

$$EJ \frac{\partial^4 v(x, t)}{\partial x^4} + \mu \frac{\partial^2 v(x, t)}{\partial t^2} + 2\mu\omega_b \frac{\partial v(x, t)}{\partial t} = p(x, t) \tag{3.2}$$

where the symbols are as those used in Sect. 1.1 and $p(x, t)$ denotes the load per unit length of the beam at point x and time t. If the mass of the moving strip is considered, too, the load becomes

$$p(x, t) = q(\xi, t) - \mu_q(\xi) \frac{d^2 v(x, t)}{dt^2}. \tag{3.3}$$

The second term on the right-hand side of (3.3) describes the inertial action of mass $\mu_q(\xi)$ on a deformed beam. According to d'Alembert's principle, it is the product of mass and acceleration. The acceleration is calculated from the total differential (see [186], p. 366)

$$\mathrm{d}^2 v(x, t) = \mathrm{d}t^2 \frac{\partial^2 v(x, t)}{\partial t^2} + 2\,\mathrm{d}x\,\mathrm{d}t \frac{\partial^2 v(x, t)}{\partial x\,\partial t} + \mathrm{d}x^2 \frac{\partial^2 v(x, t)}{\partial x^2}. \quad (3.4)$$

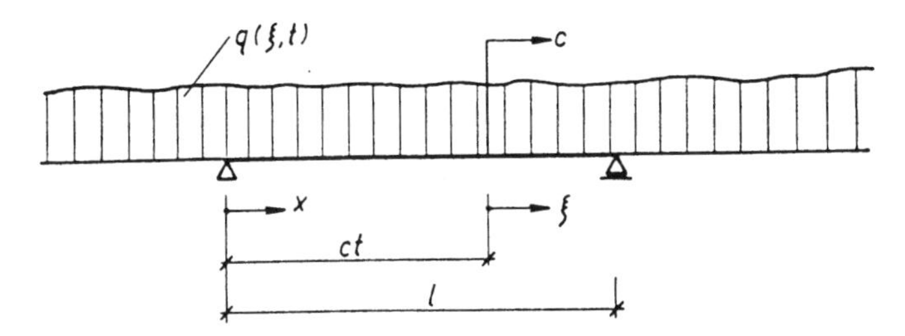

Fig. 3.1. Simple beam subjected to a moving strip of load.

Since the independent variables are related through (3.1) and in uniform motion $\mathrm{d}\xi/\mathrm{d}t = 0$, the differentiation of Eq. (3.1) $\mathrm{d}\xi/\mathrm{d}t = \mathrm{d}x/\mathrm{d}t - c$ gives

$$\mathrm{d}x = c\,\mathrm{d}t\,. \quad (3.5)$$

With this, the total differential (3.4) may be written as

$$\mathrm{d}^2 v(x, t) = \mathrm{d}t^2 \frac{\partial^2 v(x, t)}{\partial t^2} + 2c\,\mathrm{d}t^2 \frac{\partial^2 v(x, t)}{\partial x\,\partial t} + c^2\,\mathrm{d}t^2 \frac{\partial^2 v(x, t)}{\partial x^2} \quad (3.6)$$

and from there the acceleration

$$\frac{\mathrm{d}^2 v(x, t)}{\mathrm{d}t^2} = \frac{\partial^2 v(x, t)}{\partial t^2} + 2c \frac{\partial^2 v(x, t)}{\partial x\,\partial t} + c^2 \frac{\partial^2 v(x, t)}{\partial x^2}. \quad (3.7)$$

Starting from the equality sign, the terms on the right-hand side of (3.7) represent: the effect of acceleration in the direction of deflection $v(x, t)$, the effect of the so-called Coriolis force (complementary acceleration), and the effect of the path curvature (centripetal acceleration). In practical instances the effects of the third, and in particular of the second, term are smaller by far than those of the first term.

Eq. (3.2) with the right-hand side (3.3) including (3.7) is readily solved for the case of steady-state vibration of a beam subjected to continuous moving load $q(\xi, t) = q$, $\mu_q(\xi) = \mu_q = q/g$. The boundary conditions are the same as (1.2) while the initial conditions may be chosen at will — for example zero as in (1.3) because what is actually determined in the calculation of steady-state vibration is the limit $v(x, t)$ for $t \to \infty$, when the effect of the initial conditions is already damped out.

Let us first neglect the effect of the second term of Eq. (3.7)*). The equation to be solved is then

$$EJ \frac{\partial^4 v(x, t)}{\partial x^4} + \mu_q c^2 \frac{\partial^2 v(x, t)}{\partial x^2} + (\mu + \mu_q) \frac{\partial^2 v(x, t)}{\partial t^2} + \\ + 2\mu\omega_b \frac{\partial v(x, t)}{\partial t} = q \,. \tag{3.8}$$

Introduce the following notation

$$\varkappa = \frac{\mu_q}{\mu}, \tag{3.9}$$

$$\bar{\mu} = \mu + \mu_q = \mu(1 + \varkappa), \tag{3.10}$$

$$\bar{\omega}_{(j)}^2 = \omega_{(j)}^2 \mu/\bar{\mu}, \tag{3.11}$$

$$\bar{\omega}_b = \omega_b \mu/\bar{\mu}, \tag{3.12}$$

$$\bar{\bar{\omega}}_{(j)}^2 = \bar{\omega}_{(j)}^2 - \frac{j^2\pi^2c^2}{l^2} \frac{\mu_q}{\bar{\mu}} = \omega_{(j)}^2 \frac{\mu}{\bar{\mu}} \left(1 - \frac{1}{j^2} \alpha^2 \varkappa\right) \tag{3.13}$$

where the speed parameter α is the same as in (1.18).

Transforming (3.8) in accordance with (1.9) and using boundary conditions (1.2) and relations (27.68), (27.69), (27.71), give

$$\frac{j^4\pi^4}{l^4} EJ\, V(j, t) - \frac{j^2\pi^2}{l^2} \mu_q c^2\, V(j, t) + \bar{\mu}\, \ddot{V}(j, t) + 2\mu\omega_b\, \dot{V}(j, t) = \\ = \begin{cases} \dfrac{2ql}{j\pi} & \text{for} \quad j = 1, 3, 5, \ldots \\ 0 & \text{for} \quad j = 2, 4, 6, \ldots \end{cases} \tag{3.14}$$

---

*) This in fact is the case of a two-track torsionally rigid bridge traversed on one track by load $q/2$ with mass $\mu_q/2$ at speed c, and on the other track by the same load at speed $-c$. Under such conditions the second term of (3.7) drops out.

Transforming (3.14) in accordance with (1.15) and using (1.3) and (27.4) give

$$p^2 V^*(j, p) + 2\bar{\omega}_b p\, V^*(j, p) + \bar{\bar{\omega}}_{(j)}^2 V^*(j, p) = \begin{cases} \dfrac{2ql}{j\pi\bar{\mu}} & \text{for } j = 1, 3, 5, \ldots \\ 0 & \text{for } j = 2, 4, 6, \ldots \end{cases} \tag{3.15}$$

so that the transformed solution is

$$V^*(j, p) = \begin{cases} \dfrac{2ql}{j\pi\bar{\mu}} \dfrac{1}{p^2 + 2\bar{\omega}_b p + \bar{\bar{\omega}}_{(j)}^2} & \text{for } j = 1, 3, 5, \ldots \\ 0 & \text{for } j = 2, 4, 6, \ldots \end{cases} \tag{3.16}$$

For the inverse transformation according to (1.15) we use expression (27.20) or directly the limit of the expression $V^*(j, p)$ for $p \to 0$ according to (27.14)

$$\lim_{p \to 0+} V^*(j, p) = \begin{cases} \dfrac{2ql}{j\pi\bar{\mu}\bar{\bar{\omega}}_{(j)}^2} & \text{for } j = 1, 3, 5, \ldots \\ 0 & \text{for } j = 2, 4, 6, \ldots \end{cases} \tag{3.17}$$

Following the inverse transformation in accordance with (1.9) and some handling, the required solution turns out to be

$$\lim_{t \to \infty} v(x, t) = v_0 \sum_{j=1,3,5,\ldots}^{\infty} \frac{1}{j^5(1 - \alpha^2\varkappa/j^2)} \sin \frac{j\pi x}{l} \tag{3.18}$$

where

$$v_0 = \frac{5}{384} \frac{ql^4}{EJ} \approx \frac{4ql^4}{\pi^5 EJ} = \frac{4q}{\pi\mu\omega_{(1)}^2} \tag{3.19}$$

is the deflection at mid-span ($x = l/2$) of the beam produced by continuous load q.

Solution (3.18) satisfies the differential equation (3.2) with (3.3) substituted in, as well as the boundary conditions (1.2). It is, therefore, the solution of steady-state vibration of a beam traversed by an infinite strip of continuous load. It is of interest to note that expression (3.18) is independent of damping and for

$$\alpha^2\varkappa^2 = j^2 \tag{3.20}$$

tends to infinity. The dependence of $v(l/2, \infty)/v_0$ on speed α is shown in Fig. 3.2 for several values of load $\varkappa$.

For low speeds, $\alpha \ll 1$, Eq. (3.18) may be simplified to

$$v(x, \infty) = v_0 \sum_{j=1,3,5,\ldots}^{\infty} \frac{1}{j^5} \sin \frac{j\pi x}{l} \approx v_0 \sin \frac{\pi x}{l} \tag{3.21}$$

when computed with the first term of the series only.

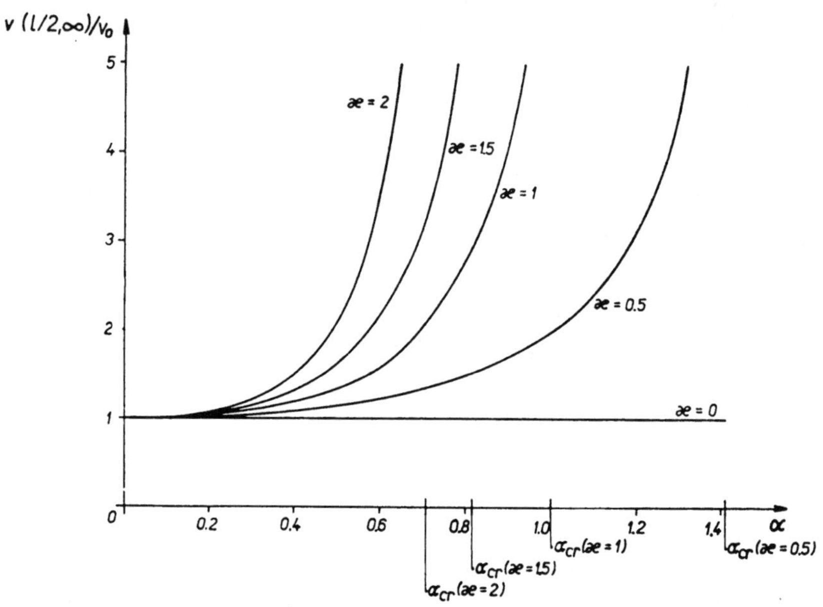

Fig. 3.2. Dependence of $v(l/2, \infty)/v_0$ on speed α for several values of load $\varkappa$ (for $j = 1$).

3.2 *Arrival of a continuous load on a beam*

To obtain an approximate solution of this case we neglect the effect of the second term on the right-hand side of (3.3) and write

$$p(x, t) = q[1 - H(x - ct)] \tag{3.22}$$

where function $H(x)$ is the so-called Heaviside unit function defined as follows:

$$H(x) = \begin{cases} 0 & \text{for} \quad x < 0 \\ 1 & \text{for} \quad x \geqq 0 . \end{cases} \tag{3.23}$$

In expression (3.22) function (3.23) describes the arrival of a continuous load on a beam (Fig. 3.3a). At the instant of arrival, the simply supported beam is at rest.

Eq. (3.2) with (3.22) for the right-hand side, is again solved by the transformation in accordance with (1.9). For boundary conditions (1.2) the use of (27.69) to (27.73) and some handling give

$$\ddot{V}(j,t) + 2\omega_b \dot{V}(j,t) + \omega_{(j)}^2 V(j,t) = \frac{ql}{j\pi\mu}(1 - \cos j\omega t) \qquad (3.24)$$

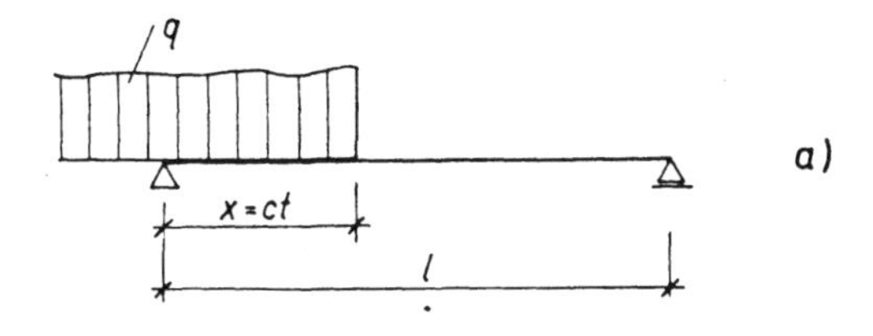

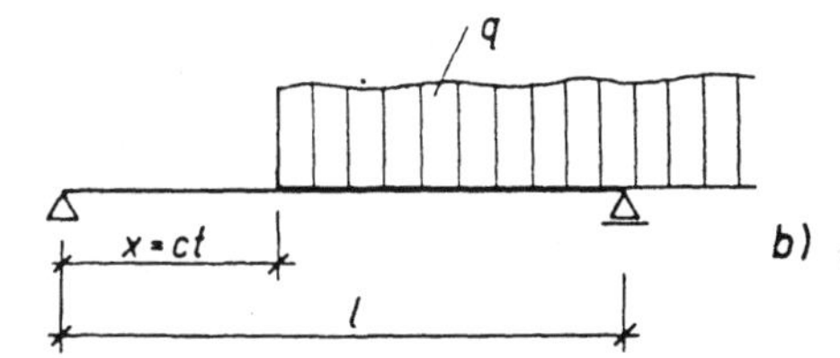

Fig. 3.3. a) Arrival of a continuous load on a beam, b) Departure of a continuous load from a beam.

and transformation (1.15), together with initial conditions (1.3) and the use of relations (27.4), (27.17) result in

$$V^*(j,p) = \frac{qlj\omega^2}{\pi\mu} \frac{1}{(p^2 + j^2\omega^2)\left[(p + \omega_b)^2 + \omega_{(j)}'^2\right]}. \qquad (3.25)$$

Following the inverse transformations in accordance with (1.9) and (27.40), Eq. (3.25) and some manipulation give

$$v(x,t) = \frac{v_0}{2}\sum_{j=1}^{\infty}\frac{1}{j^5}\sin\frac{j\pi x}{l} \cdot \frac{\alpha^2}{j^2(j^2-\alpha^2)^2 + 4\alpha^2\beta^2}\left\{\frac{j^4(j^2-\alpha^2)}{\alpha^2}\,.\right.$$

$$.(1 - \cos j\omega t) - \left[j^2(j^2-\alpha^2) - 4\beta^2\right](1 - e^{-\omega_b t}\cos\omega_{(j)}' t) -$$

$$\left. - \frac{2j^3\beta}{\alpha}\sin j\omega t + \frac{\beta}{(j^4-\beta^2)^{1/2}}\left[2j^4 + j^2(j^2-\alpha^2) - 4\beta^2\right] e^{-\omega_b t}\sin\omega_{(j)}' t\right\}. \qquad (3.26)$$

In the above expression, (1.18) and (1.19) were used for the speed parameter α and the damping parameter β, respectively.

For low speeds, $\alpha \ll 1$, light damping, $\beta \ll 1$, and $j = 1$, Eq. (3.26) becomes quite simple, namely

$$v(x, t) \approx \frac{v_0}{2} (1 - \cos \omega t) \sin \frac{\pi x}{l} . \tag{3.27}$$

This last relation is suitable for calculations relating to large-span bridges, the case that has been considered when deducing the theory in this chapter. Eqs. (3.26) and (3.27) apply only to the arrival of a continuous load, i.e. at $0 \leqq t \leqq l/c$.

3.3 *Departure of a continuous load from a beam*

This case is solved analogously to the one just discussed (see Fig. 3.3b). The initial conditions now are those given by (3.21)

$$v(x, 0) = v_0 \sum_{j=1,3,5\ldots}^{\infty} \frac{1}{j^5} \sin \frac{j\pi x}{l} ,$$
$$\dot{v}(x, 0) = 0 \tag{3.28}$$

because at the beginning of the process the beam is assumed to be performing steady-state vibration in the sense of Sect. 3.1.

The load now is

$$p(x, t) = q\, H(x - ct) . \tag{3.29}$$

Writing (3.2) with (3.29) on the right-hand side, and solving it in accordance with (1.9) with the use of boundary conditions (1.2) and relations (27.69) and (27.72) give after a bit of handling

$$\ddot{V}(j, t) + 2\omega_b \dot{V}(j, t) + \omega_{(j)}^2 V(j, t) = \frac{ql}{j\pi\mu} (\cos j\omega t - \cos j\pi) . \tag{3.30}$$

In view of initial conditions (3.28) and relations (27.4), (27.19) transformation (1.15) will give

$$V^*(j, p) = \frac{1}{(p + \omega_b)^2 + \omega_{(j)}'^2} \Big[V(j, 0)\, p(p + 2\omega_b) + + \frac{ql}{j\pi\mu} \left(\frac{p^2}{p^2 + j^2\omega^2} - \cos j\pi \right) \Big] \tag{3.31}$$

where $V(j, 0)$ is the Fourier transform of initial conditions (3.28)

$$V(j, 0) = \int_0^l v(x, 0) \sin \frac{j\pi x}{l} \, dx = \frac{v_0 l}{2} \sum_{j=1,3,5\ldots}^{\infty} \frac{1}{j^5} . \tag{3.32}$$

Applying the inverse transformations (1.9) and (27.22), (27.42), (27.20) we get from (3.31)

$$v(x, t) = \sum_{j=1}^{\infty} \frac{v_0}{j^5} \sin \frac{j\pi x}{l} \left[e^{-\omega_b t} \left(\cos \omega'_{(j)} t + \frac{\beta}{(j^4 - \beta^2)^{1/2}} \sin \omega'_{(j)} t \right) \right|_{j=1,3,5\ldots} +$$

$$+ \frac{1}{2} \left\{ \frac{j^4}{j^2(j^2 - \alpha^2)^2 + 4\alpha^2\beta^2} \left[(j^2 - \alpha^2) (\cos j\omega t - e^{-\omega_b t} \cos \omega'_{(j)} t) + \right. \right.$$

$$\left. + \frac{2\alpha\beta}{j} \sin j\omega t - \frac{\beta}{(j^4 - \beta^2)^{1/2}} (j^2 + \alpha^2) e^{-\omega_b t} \sin \omega'_{(j)} t \right] - \cos j\pi \, .$$

$$\left. \left. \cdot \left[1 - e^{-\omega_b t} \left(\cos \omega'_{(j)} t + \frac{\beta}{(j^4 - \beta^2)^{1/2}} \sin \omega'_{(j)} t \right) \right] \right\} \right] . \tag{3.33}$$

As in Chap. 1, expressions (1.18) and (1.19) were again used for the speed parameter α and the damping parameter β, respectively. For $\alpha \ll 1$, $\beta \ll 1$ and $j = 1$, Eq. (3.33) simplifies to

$$v(x, t) = \frac{v_0}{2} (1 + \cos \omega t) \sin \frac{\pi x}{l} , \tag{3.34}$$

a formula suitable for the calculation of large-span bridges.

Eqs. (3.33) and (3.34) again apply only during the time of departure of a continuous load from a beam, i.e. for $0 \leqq t \leqq l/c$.

3.4 *Application of the theory*

The foregoing theory is frequently used in the calculation of dynamic stresses in large-span railway bridges, resulting from the traverse of a train of cars hauled by a locomotive [74]. According to experience, the bridge vibration produced by the cars is very irregular, with a lot of damping coming into play, too. The reasons for this are track irregularities, the condition and lateral movements of the cars, damping of the car springs, damping in the connecting rods and other, mostly random causes.*) More often than not the dynamic effects of the locomotive are damped out by the train of cars.

3.4.1 *Approximate calculation of large-span railway bridges*

In the calculations that follow the load of the cars is assumed to be transferred to the main girders of the bridge as a continuous load, q, which moves behind a locomotive producing effects described in Sect. 2.3. The locomotive is assumed to be heavier than the cars, with unbalanced counterweights producing large dynamic forces.

Let us first examine free vibration of a bridge loaded with a locomotive stationed at mid-span, and a continuous load, q, extended from the left-hand end to the centre of the bridge. In the case of a heavy locomotive and not so heavy cars, the position is decisive for the calculation of free vibration of the bridge, and analogous to that of the case discussed in paragraph 1.4.3.

For free vibration of a beam thus loaded the right-hand side of Eq. (1.42) is

$$-\delta(x - l/2)\, m \left.\frac{\partial^2 v(x,t)}{\partial t^2}\right|_{x=l/2} - \mu_q \frac{\partial^2 v(x,t)}{\partial t^2}\left[1 - H(x - l/2)\right]. \quad (3.35)$$

Proceeding as in paragraph 1.4.3 we find that in the first approximation the differential equation of the beam vibration has the form of (3.2) in which the coefficients

$$\bar{\mu} = \mu\left(1 + \frac{2P}{G} + \frac{ql}{2G}\right), \quad (3.36)$$

$$\bar{\omega}_b = \omega_b \mu/\bar{\mu} \quad (3.37)$$

have been substituted for μ and ω_b.

Similarly, the frequencies are

$$\bar{\omega}_{(j)}^2 = \omega_{(j)}^2 \mu/\bar{\mu}\,; \quad \bar{\omega}_{(j)}'^2 = \bar{\omega}_{(j)}^2 - \bar{\omega}_b^2\,, \quad (3.38)$$

$$\bar{f}_{(j)} = f_{(j)}[1 + 2P/G + ql/(2G)]^{-1/2}\,. \quad (3.39)$$

Depending on the length of the train being considered, the calculation of forced vibration may be divided into the following stages:

*) The random concept of the whole phenomenon is examined in detail in Chap. 26.

Stage 1 – The motion of the locomotive followed by the cars (a concentrated force and a continuous load). It lasts from the instant the locomotive arrives on the bridge until it departs from it. In this stage the free vibration is computed from formulae (3.35) to (3.39).

Stage 2 – The motion of the continuous load. It lasts from the instant the locomotive departs from the bridge until the end car arrives on it. The bridge deflection is approximately described by Eq. (3.18).

Stage 3 – Departure of the continuous load from the bridge. It lasts from the instant the end car arrives on the bridge until it departs from it. The bridge deflection is approximately described by Eq. (3.34).

Stage 4 – The bridge is under no load, free vibration sets in (this case does not concern us).

In the course of each stage the bridge behaves in conformance with the character of the load imposed, and its vibration is also affected by the pertinent initial conditions. We shall presently return to stage 1 because – as we have assumed – vibration in its course is the most intensive of all, and the other two stages have in fact been solved in Sects. 3.1 and 3.3.

In the course of stage 1 the beam is loaded with a moving force, P (cf. Chap. 1), a harmonic force, $Q \sin \Omega t$ (cf. Chap. 2), and a continuous load, q (cf. Sect. 3.2). In view of the linear character of the problem, the total deflection of the beam is described by the equation

$$v(x, t) = v_P(x, t) + v_Q(x, t) + v_q(x, t) \tag{3.40}$$

where $v_P(x, t)$, $v_Q(x, t)$ and $v_q(x, t)$ are described by Eqs. (1.41), (2.7) and (3.27), respectively, with values $\bar{\mu}$, $\bar{\omega}_b$, $\bar{\omega}_{(j)}$ according to Eqs. (3.36) to (3.39) substituted for μ, ω_b, $\omega_{(j)}$.

3.4.2 *Comparison of theory with experiments*

The theory expounded in the foregoing will next be applied to the examination of a bridge having span $l = 56{\cdot}56$ m traversed by a 524.1 type locomotive (for data see paragraph 2.3.1) hauling three cars; the continuous load produced by them is $q = 39{\cdot}5$ kN/m. From (3.36) and (3.39)

$$\mu/\bar{\mu} = 0{\cdot}351 \; ; \quad \bar{f}_{(1)} = 4{\cdot}20 \times 0{\cdot}351^{1/2} = 2{\cdot}49 \text{ Hz} .$$

The measured first natural frequency of the bridge under load was $\bar{f}_{(1)} = 2{\cdot}44$ Hz, a value in very good agreement with the computed one.

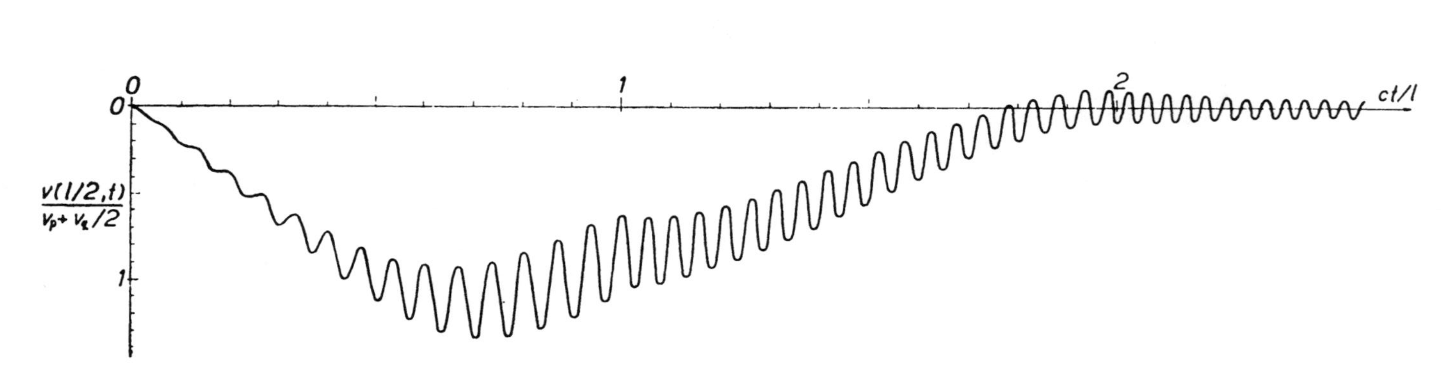

Fig. 3.4. Theoretical time variation of the deflection at the centre of a beam with span $l = 56{\cdot}56$ m traversed by a steam locomotive type 524.1 drawing three cars, $q = 39{\cdot}5$ kN/m at speed $c_{cr} = 35{\cdot}5$ km/h.

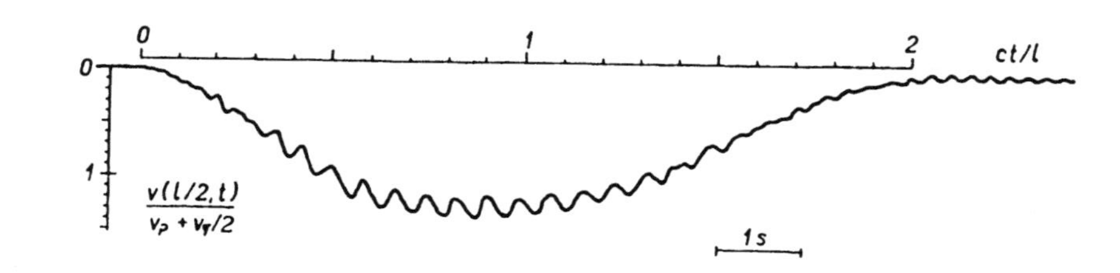

Fig. 3.5. Experimental time variation of the deflection at the centre of a beam with span $l = 56{\cdot}56$ m traversed by a steam locomotive type 524.1 drawing three cars, $q = 39{\cdot}5$ kN/m at speed $c = 34{\cdot}0$ km/h.

Resonant vibration sets in at speed [see (3.44)] $c_{cr} = 2{\cdot}49 \times 3{\cdot}96 = 9{\cdot}85$ m/s $= 35{\cdot}5$ km/h. The test speed nearest to the critical was $c = 34{\cdot}0$ km/h. The deflection at mid-span of the bridge, corresponding to c_{cr} was computed from Eqs. (3.40), (3.18) and (3.34). It is shown in Fig. 3.4, the measured deflection in Fig. 3.5. In the latter $v_P = 1{\cdot}7$ cm [$v_P = v_0$; for v_0 see Eq. (1.21)], $v_q = 2{\cdot}52$ cm [$v_q = v_0$; for v_0 see Eq. (3.19)], and $v[l/2,\ l/(2c)]_{\text{stat}} = v_P[l/2, l/(2c)] + v_q[l/2, l/(2c)] = v_P + v_q/2 = 2{\cdot}96$ cm. Comparing Figs. 3.4 and 3.5 we see that the theory is in qualitative agreement with experiments. Actually, however, the damping was far in excess of that used in the computation.

3.4.3 *The dynamic coefficient*

The most intensive vibration is found to arise at the instant the locomotive is in the neighbourhood of three fourths of the bridge span. To make up for the effect of damping, let us calculate the dynamic coefficient according to (2.10) and (3.40) for the instant the locomotive is at midspan, i.e. $t = T/2 = l/(2c)$:

$$\delta = \frac{v_P(l/2, T/2) + v_Q(l/2, T/2) + v_q(l/2, T/2)}{v_P(l/2, T/2) + v_q(l/2, T/2)} =$$

$$= 1 + \frac{v_Q(l/2, T/2)}{v_P(l/2, T/2)} \frac{1}{1 + \dfrac{v_q(l/2, T/2)}{v_P(l/2, T/2)}} = 1 + \varDelta_{PQ}\, \varDelta_{Pq}\,. \tag{3.41}$$

Coefficient $\varDelta_{PQ}$ expresses the effect of the harmonic force and equals the second term on the right-hand side of Eq. (2.11) in which $\bar{\omega}_{(1)}$ and $\bar{\omega}_b$ according to (3.38) and (3.37) are substituted for $\omega_{(1)}$ and ω_b. Coefficient $\varDelta_{Pq}$ describes the effect of the continuous load and is deduced via calculation from Eqs. (1.41), (1.21), (3.27) and (3.19):

$$\varDelta_{Pq} = \frac{1}{1 + ql/(\pi P)}\,. \tag{3.42}$$

In the case of resonance, $\Omega = \bar{\omega}_{(1)}$, Eq. (3.41) simplifies to

$$\delta = 1 + \frac{Q}{2P} \frac{\bar{\omega}_{(1)}}{\omega^2 + \bar{\omega}_b^2} \left(\omega e^{-\bar{\omega}_b l/(2c)} + \bar{\omega}_b\right) \frac{1}{1 + ql/(\pi P)}\,. \tag{3.43}$$

Eq. (3.41) represents the resonance curve computed for the 56·56 m-span bridge, type 524.1 locomotive with three cars producing load

$q = 39{\cdot}5$ kN/m. The result of the computation is shown in Fig. 3.6 together with the measured values. It is clear to see from the figure that the theory agrees well with the experiments. The bridge vibration is also affected by damping in the car springs, a factor hard to express correctly.

Maximum effects again arise at resonance, i.e. at speed

$$c_{cr} = \bar{f}_{(1)} O \tag{3.44}$$

analogous to (2.16) except that in the above $\bar{f}_{(1)}$ is given by (3.39).

Fig. 2.6 plots the critical speed (3.44) versus the span for loads $p = q = 0$, 20, 40, 60, 80 and 100 kN/m. As the figure indicates, continuous load exerts a strong effect on the critical speed.

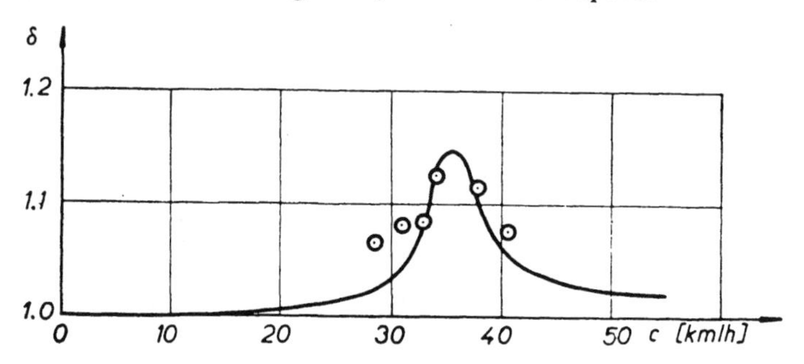

Fig. 3.6. Theoretical (—) and experimental (○) dependences of dynamic coefficient δ on speed c. Bridge with span $l = 56{\cdot}56$ m, steam locomotive type 524.1 with three cars, $q = 39{\cdot}5$ kN/m.

3.4.4 *Pipelines carrying moving liquid*

The method of calculation outlined above may also be employed in analyses relating to vibration of pipelines conveying liquid flowing with velocity c. The pipeline is taken for a beam, the liquid is assumed to be ideal and incompressible, and its velocity distribution across the pipeline cross section approximately uniform. On these assumptions we can use Eqs. (3.2) to (3.7) and the approximate solution deduced in Sect. 3.1, with q and μ_q denoting respectively the weight and mass of the liquid per unit length of pipeline.

3.5 *Additional bibliography*

[19, 20, 25, 74, 97, 132, 159, 213].

4

Moving force arbitrarily varying in time

We shall now turn to the case of a simply supported beam with span l traversed at constant speed c by concentrated force $P(t)$ arbitrarily varying in time. The calculation will also be carried out for several special cases of $P(t)$.

With the notation of Chap. 1 the problem is described by the differential equation

$$EJ\frac{\partial^4 v(x, t)}{\partial x^4} + \mu\frac{\partial^2 v(x, t)}{\partial t^2} + 2\mu\omega_b\frac{\partial v(x, t)}{\partial t} = \delta(x - ct)\, P(t) \qquad (4.1)$$

and by the boundary and initial conditions (1.2) and (1.3). Eq. (4.1) will again be solved by the Fourier integral transformation according to (1.9) and (27.74)

$$\ddot{V}(j, t) + 2\omega_b\,\dot{V}(j, t) + \omega_{(j)}^2\, V(j, t) = \frac{1}{\mu}\, P(t)\sin j\omega t \qquad (4.2)$$

and by the Laplace-Carson transformation (1.15). In so doing we shall also use relation (27.11) so that in view of (1.3) we get the transformed solution in the form

$$V^*(j, p) = \frac{1}{\mu}\frac{1}{(p + \omega_b)^2 + \omega_{(j)}'^2}\frac{p}{2\mathrm{i}}\left[\frac{P^*(p - \mathrm{i}j\omega)}{p - \mathrm{i}j\omega} - \frac{P^*(p + \mathrm{i}j\omega)}{p + \mathrm{i}j\omega}\right] \qquad (4.3)$$

where

$$P^*(p) = p\int_0^\infty P(t)\,\mathrm{e}^{-pt}\,\mathrm{d}t \qquad (4.4)$$

is the Laplace-Carson integral transform of concentrated force $P(t)$, and the other symbols are as in Chap. 1. On computing the inverse transform with the aid of relations (27.5) and (27.21) we get the result

$$v(x, t) = \sum_{j=1}^{\infty} \frac{2}{\mu l \omega'_{(j)}} \sin \frac{j\pi x}{l} \int_0^t P(\tau) \sin j\omega\tau \, e^{-\omega_b(t-\tau)} \sin \omega'_{(j)}(t-\tau) \, d\tau \, . \tag{4.5}$$

In this way we can calculate the deflection of, and stresses in, a beam subjected to arbitrary force $P(t)$ either specified in advance or obtained, say, experimentally. The integral of Eq. (4.5) can even be evaluated numerically.

In what follows we shall discuss several special cases of force $P(t)$.

4.1 *Force linearly increasing in time*

Consider force $P(t)$ to be of the form

$$P(t) = P \left(\frac{t}{T} \right)^n \tag{4.6}$$

where P is the constant force of Chap. 1,

T — a definite reference time,

$n = 0, 1, 2, 3, \ldots$

According to (27.15) the transform of Eq. (4.6) is

$$P^*(p) = \frac{P}{T^n} \frac{n!}{p^n} \, . \tag{4.7}$$

Substitution of (4.7) in (4.3) gives

$$V^*(j, p) = \frac{Pn!}{\mu T^n} \frac{1}{(p + \omega_b)^2 + \omega'^2_{(j)}} \frac{p}{2i} \frac{(p + ij\omega)^{n+1} - (p - ij\omega)^{n+1}}{(p^2 + j^2\omega^2)^{n+1}} \, . \tag{4.8}$$

We shall solve Eq. (4.8) for the special case of force $P(t)$ linearly increasing in time, i.e. for $n = 1$ in Eq. (4.6) — see Fig. 4.1a. Then (4.8) gives

$$V^*(j, p) = \frac{2Pj\omega}{\mu T} \frac{p^2}{[(p + \omega_b)^2 + \omega'^2_{(j)}] (p^2 + j^2\omega^2)^2} \, . \tag{4.9}$$

The original of (4.9) is obtained by help of relation (27.66). For $\omega_b = 0$ the solution is

$$v(x, t) = \sum_{j=1}^{\infty} v_0 \frac{2}{T\omega_{(1)}} \frac{\alpha}{j^3(j^2 - \alpha^2)^2} \left(\frac{j^2 - \alpha^2}{2\alpha^2} j\omega t \sin j\omega t - \cos j\omega t + \right.$$
$$\left. + \cos \omega_{(j)} t \right) \sin \frac{j\pi x}{l} . \tag{4.10}$$

Eq. (4.10) is written in terms of (1.18) and (1.21).

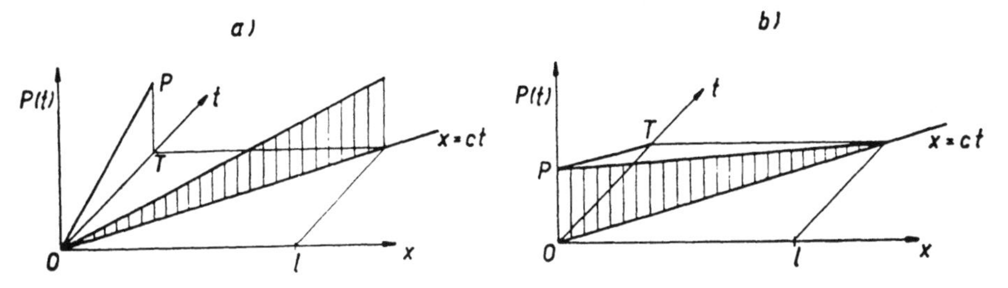

Fig. 4.1. a) Moving force linearly increasing in time.
b) Moving force linearly decreasing in time.

For a linearly decreasing force (Fig. 4.1b)

$$P(t) = P\frac{T - t}{T} = P\left(1 - \frac{t}{T}\right) \tag{4.11}$$

the procedure is analogous and the solution consists of that of Chap. 1 minus Eq. (4.10).

4.2 *Force exponentially varying in time*

According to (27.16), the transform of a force exponentially varying in time (Fig. 4.2)

$$P(t) = Pe^{-dt} \tag{4.12}$$

is

$$P^*(p) = P\frac{p}{p + d} . \tag{4.13}$$

Substitution of (4.13) in (4.3) leads to the transformed solution

$$V^*(j, p) = \frac{Pj\omega}{\mu} \frac{p}{[(p + \omega_b)^2 + \omega_{(j)}'^2] [(p + d)^2 + j^2\omega^2]} \tag{4.14}$$

and the use of (27.45) and (1.9) to the original solution. For $\omega_b = 0$, for example, the beam deflection

$$v(x, t) = \sum_{j=1}^{\infty} \frac{2Pj\omega}{\mu l} \sin\frac{j\pi x}{l} \frac{1}{(d^2 + j^2\omega^2 - \omega_{(j)}^2)^2 + 4d^2\omega_{(j)}^2} \cdot$$

$$\cdot \left[\frac{d^2 + j^2\omega^2 - \omega_{(j)}^2}{\omega_{(j)}} \sin \omega_{(j)}t + \frac{d^2 - j^2\omega^2 + \omega_{(j)}^2}{j\omega} e^{-dt} \sin j\omega t - \right.$$

$$\left. - 2d(\cos \omega_{(j)}t - e^{-dt} \cos j\omega t)\right] . \tag{4.15}$$

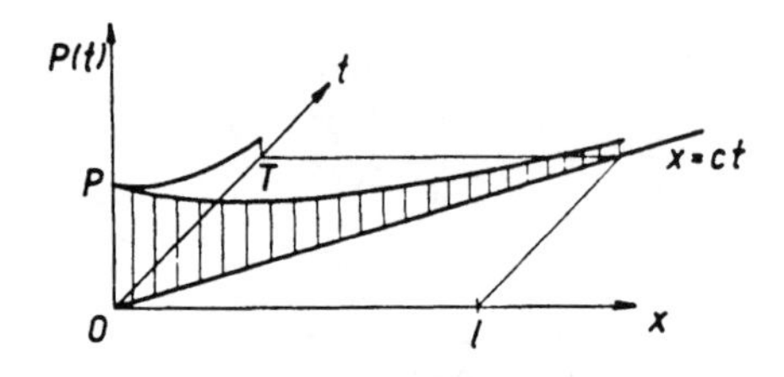

Fig. 4.2. Moving force exponentially decreasing in time.

4.3 *Moving impulses*

Equally interesting is the case of a beam subjected to moving instantaneous impulses acting at regular time intervals (see Fig. 4.3). In such a case force $P(t)$ is

$$P(t) = \sum_{n=1}^{N} I\delta(t - nT_0) \tag{4.16}$$

where I — force impulse [force × time],
T_0 — time interval,
N — finite number of impulses acting on the beam during the traverse.

By (27.7) the transform of force (4.16) is

$$P^*(p) = \sum_{n=1}^{N} Ipe^{-nT_0p} . \tag{4.17}$$

Substitution of (4.17) in (4.3) and a bit of handling give the transformed solution

$$V^*(j, p) = \sum_{n=1}^{N} \frac{I}{\mu} \frac{p}{(p + \omega_b)^2 + \omega_{(j)}'^2} e^{-nT_0p} \sin jnT_0\omega \tag{4.18}$$

whence – according to (27.10), (27.21) and (1.9) –

$$v(x, t) = \sum_{j=1}^{\infty} \sum_{n=1}^{N} \frac{2I}{\mu l \omega'_{(j)}} \sin \frac{j\pi x}{l} \sin jnT_0\omega \, e^{-\omega_b(t - nT_0)} \, .$$

$$. \sin \omega'_{(j)}(t - nT_0) \, H(t - nT_0) \, . \tag{4.19}$$

In view of the definition of unit function (3.23), the form of writing of (4.19) indicates that after each impulse there is added to the solution another component characterized in (4.19) by the next number n.

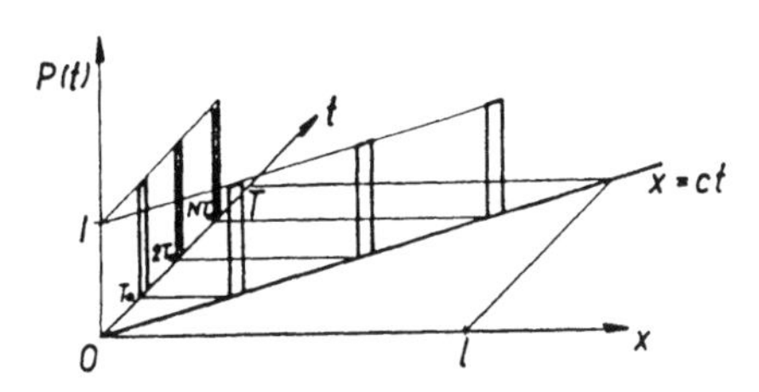

Fig. 4.3. Moving impulses.

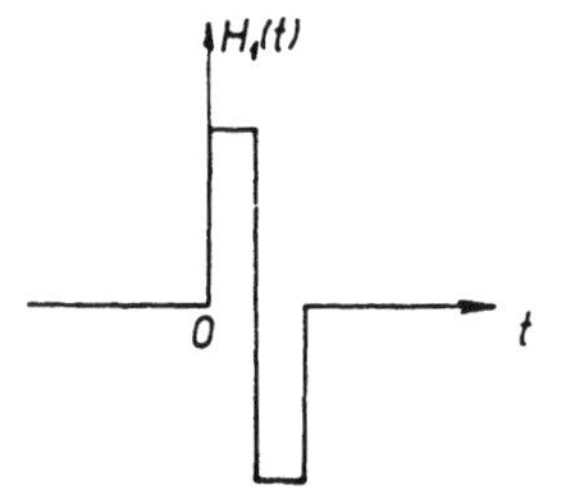

Fig. 4.4. Impulse function of the second order.

The case is solved analogously also for the impulse function of the second order.*) To that end, the function $H_1(t)$ is defined as the generalized derivative of the Dirac function $\delta(t)$ in the distributive sense

$$H_1(t) = \frac{d\delta(t)}{dt} \, . \tag{4.20}$$

The moving force is written in the form

$$P(t) = \sum_{n=1}^{N} I_1 H_1(t - nT_0) \tag{4.21}$$

*) The impulse function of the second order, $H_1(t)$, imparts unit displacement to unit mass (Fig. 4.4). In terms of variable x, $H_1(x)$ denotes a concentrated unit bending moment. If at point $x = a$ from interval $\langle 0, l \rangle$ function $f(x)$ has the first derivative, it also holds that

$$\int_0^l H_1(x - a) f(x) \, dx = - \left. \frac{df(x)}{dx} \right|_{x=a} \, . \tag{4.20a}$$

where I_1 is the impulse of the second order [force × square of time]. By (27.8) the transform of force $P(t)$ is

$$P^*(p) = \sum_{n=1}^{N} I_1 p^2 e^{-nT_0 p} . \tag{4.22}$$

Substitution of (4.22) in (4.3) gives

$$V^*(j, p) = \sum_{n=1}^{N} \frac{I_1}{\mu} \frac{p}{(p + \omega_b)^2 + \omega_{(j)}'^2} (p \sin jnT_0\omega - j\omega \cos jnT_0\omega) e^{-nT_0 p} \tag{4.23}$$

and the subsequent application of (27.10), (27.21), (27.22) and (1.9)

$$v(x, t) = \sum_{j=1}^{\infty} \sum_{n=1}^{N} \frac{2I_1}{\mu l} \sin \frac{j\pi x}{l} e^{-\omega_b(t - nT_0)} \left\{ \left[\cos \omega_{(j)}'(t - nT_0) - \right.\right.$$
$$\left. - \frac{\omega_b}{\omega_{(j)}'} \sin \omega_{(j)}'(t - nT_0) \right] \sin jnT_0\omega - \frac{j\omega}{\omega_{(j)}'} \sin \omega_{(j)}'(t - nT_0) \, .$$
$$\left. . \cos jnT_0\omega \right\} H(t - nT_0) . \tag{4.24}$$

4.4 *Application of the theory*

Forces in the form of Eqs. (4.6), (4.11) or (4.12) are used in calculations of structures exposed to the effects of explosion, pressure waves, etc. Moving impulses such as those described in Sect. 4.3 can be made to represent, for example, impacts of flat wheels on large-span railway bridges (see also Chap. 9).

But the problem in which the theory expounded above has found its widest application is that relating to impacts of rolling stock wheels on rail joints on bridge structures [121], [24].*) It differs from the one just discussed by that the rail joint (Fig. 4.5) is at a fixed point x_0 on the bridge, and the wheels produce in it — in more or less regular time intervals T_0 — impact loads that may be represented by simple impulses. The load acting on the beam owing to the impacts is then

$$p(x, t) = P(t)\, \delta(x - x_0) \tag{4..25}$$

*) The problem of rail joints is very complicated and the solution given in the present Section 4.4 is only an approximate approach.

where – similarly as in (4.16) the force

$$P(t) = \sum_{n=0}^{\infty} I\,\delta(t - nT_0)\,. \tag{4.26}$$

The above equations are written on the assumption that the number of impacts is large, i.e. $N \to \infty$; in practice this means a train with a large number of wheels, which is much longer than the bridge span.

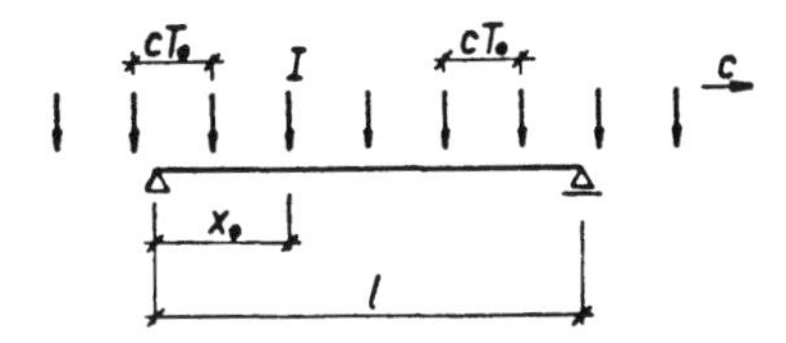

Fig. 4.5. Wheel impacts on rail joint at point x_0.

Substitution of (4.25) in the right-hand side of Eq. (4.1) and a procedure analogous to that used in handling the latter lead to the transformed solution [cf. Eq. (27.74)]

$$V^*(j, p) = \frac{1}{\mu} \frac{P^*(p)}{(p + \omega_b)^2 + \omega'^2_{(j)}} \sin \frac{j\pi x_0}{l}\,. \tag{4.27}$$

The transform of force (4.26) can now be summed like a geometric series [cf. Eq. (27.7)]

$$P^*(p) = \sum_{n=0}^{\infty} I p\,e^{-nT_0 p} = \frac{Ip}{1 - e^{-pT_0}}\,; \tag{4.28}$$

hence – by (27.28) and (1.9) – the inverse transformation of (4.27) becomes

$$v(x, t) = \sum_{j=1}^{\infty} \frac{2I}{\mu l \omega'_{(j)}} \sin \frac{j\pi x}{l} \sin \frac{j\pi x_0}{l} \left\{ \frac{e^{-\omega_b t}}{1 - 2e^{\omega_b T_0} \cos \omega'_{(j)} T_0 + e^{2\omega_b T_0}} \,. \right.$$

$$. \left[\left(1 - e^{\omega_b T_0} \cos \omega'_{(j)} T_0\right) \sin \omega'_{(j)} t - e^{\omega_b T_0} \sin \omega'_{(j)} T_0 \cos \omega'_{(j)} t \right] +$$

$$+ \frac{2\omega'_{(j)}}{T_0} \sum_{n=0}^{\infty} \frac{1}{\left(\omega^2_{(j)} - 4n^2\pi^2/T_0^2\right)^2 + 16n^2\pi^2\omega_b^2/T_0^2} \left[\left(\omega^2_{(j)} - 4n^2\pi^2/T_0^2\right) . \right.$$

$$\left. \left. . \cos \frac{2n\pi t}{T_0} + \frac{4n\pi\omega_b}{T_0} \sin \frac{2n\pi t}{T_0} \right] \right\} . \tag{4.29}$$

Eq. (4.29) is the general solution to the problem of a beam with zero initial conditions (1.3) on which there starts to act an infinite number of impulses I in regular time intervals T_0. After a time the beam will evidently be set to vibrate in forced steady-state vibration whose maximum amplitude

$$A = \sum_{j=1}^{\infty} \frac{2I}{\mu l \omega'_{(j)}} \sin \frac{j\pi x}{l} \sin \frac{j\pi x_0}{l} \frac{1}{(1 - 2e^{\omega_b T_0} \cos \omega'_{(j)} T_0 + e^{2\omega_b T_0})^{1/2}} \tag{4.30}$$

derives from the largest possible amplitude of the term in the first brackets on the right-hand side of Eq. (4.29).

Amplitude (4.30) attains its maximum whenever interval T_0 is equal to some period of beam free vibration $T_{(j)} = 2\pi/\omega_{(j)}$ or to an integer multiple thereof.

It might happen in practice that the train passing over a rail joint consists of cars with equal wheel bases or of groups of identical cars. The rail joint is then subjected to so-called group impacts that are apt to set the bridge structure even to resonant vibration.

Such cases have been observed and given detailed theoretical treatment in the USSR [121], [24]. The value of impulses I was also determined empirically as ranging from 1 to 2 kN s, and expressed analytically in dependence on speed by the formula

$$I = ac(1 - 0{\cdot}012c)\, P_1^{1/2} \tag{4.31}$$

or for short-span bridges by the formula

$$I = 0{\cdot}01 l P_1^{1/2} \,. \tag{4.32}$$

In the above the old unit of force (megapond = 1 Mp ≐ 10 kN) has been used:

I is in [Mp s],
a — an empirical constant,
c — speed in [m/s],
P_1 — weight of the unsprung mass of the axle in [Mp],
l — span in [m].

In a like manner one can also obtain the amplitude of forced steady-state vibration in the case of second-order impulses. Then

$$P^*(p) = I_1 p^2/(1 - e^{-pT_0})$$

and the maximum amplitude resulting from (4.27) and (27.29) via the procedure applied to (4.29) and (4.30) is

$$A = \sum_{j=1}^{\infty} \frac{2I_1\omega_{(j)}}{\mu l \omega'_{(j)}} \sin\frac{j\pi x}{l} \sin\frac{j\pi x_0}{l} \frac{1}{(1 - 2e^{\omega_b T_0}\cos\omega'_{(j)} T_0 + e^{2\omega_b T_0})^{1/2}}. \tag{4.33}$$

4.5 *Additional bibliography*

[24, 88, 121, 146].

5

Beam stresses

Up to now we have been satisfied with finding the dynamic deflection from its expression in the form of a series. The comparatively fast convergence of the series enabled us to obtain adequately accurate results even with the use of only the first term. Sometimes, however, we might wish to establish also the other quantities of interest, i.e. the rotations of cross sections

$$\frac{\partial v(x, t)}{\partial x}, \tag{5.1}$$

the bending moments

$$M(x, t) = -EJ \frac{\partial^2 v(x, t)}{\partial x^2}, \tag{5.2}$$

and the shear forces

$$T(x, t) = -EJ \frac{\partial^3 v(x, t)}{\partial x^3}. \tag{5.3}$$

From the point of beam stress calculations, the most important of the above quantities are the bending moments. This chapter will show several methods of computing them for a simply supported, assumably undamped beam traversed by a constant force. Though illustrated by way of this simple example, some of the methods are equally well applicable to computations involving beams with other loads and support conditions, damped vibration, as well as the remaining quantities – i.e. the deflection, rotation and shear forces.

5.1 *Calculation by expansion in series*

As we have pointed out at the end of Sect. 1.2, differentiation of the deflection expression with respect to x impairs the convergence of the series.

We shall now demonstrate the truth of this statement with a numerical example*).

According to the theory of the Fourier series, Eq. (1.31), the solution of the motion of constant force P over a simply supported beam, is a uniformly convergent series on intervals $0 \leqq x \leqq l$, $0 \leqq t \leqq l/c$. Like a Fourier series the equation, rewritten to a somewhat different form, may be differentiated term by term according to (5.2) and (5.3). The operation will give us respectively the deflection, the bending moment, and the shear force of an undamped beam in the form

$$v(x, t) = v_0 \sum_{j=1}^{\infty} \frac{96}{\pi^4} \frac{1}{j^4(1 - \alpha^2/j^2)} \left(\sin j\omega t - \frac{\alpha}{j} \sin \omega_{(j)} t \right) \sin \frac{j\pi x}{l} ,$$

$$M(x, t) = M_0 \sum_{j=1}^{\infty} \frac{8}{\pi^2} \frac{1}{j^2(1 - \alpha^2/j^2)} \left(\sin j\omega t - \frac{\alpha}{j} \sin \omega_{(j)} t \right) \sin \frac{j\pi x}{l} ,$$

$$T(x, t) = P \sum_{j=1}^{\infty} \frac{2}{\pi} \frac{1}{j(1 - \alpha^2/j^2)} \left(\sin j\omega t - \frac{\alpha}{j} \sin \omega_{(j)} t \right) \cos \frac{j\pi x}{l} \tag{5.4}$$

where according to (1.21) and (1.27) $v_0 = Pl^3/(48EJ)$ and $M_0 = Pl/4$ are the deflection and the static moment at the centre of the beam span under the action of load P at point $x = l/2$. (These expressions are used throughout in Chap. 5.) The speed parameter is defined by Eq. (1.18).

Expressions (5.4) were computed for the centre of the beam span, $x = l/2$, and the instant when the force passes over it, $t = l/(2c)$ with speed $\alpha = 1/2$. Only the shear force was determined for instant $t = 3l/(4c)$ – i.e. for the load at 3/4 of the span – because all the terms of the series turn out zero for $t = l/(2c)$.

---

*) Generally, this is what is being merely proved: If series $\sum_{j=1}^{\infty} f'_{(j)}(x)$ is uniformly convergent on the interval $\langle 0, l \rangle$ and if series $\sum_{j=1}^{\infty} f_{(j)}(x)$ is convergent for at least one $x_0 \in \langle 0, l \rangle$, then series $\sum_{j=1}^{\infty} f_{(j)}(x)$ is uniformly convergent on the whole interval $\langle 0, l \rangle$, differentiable term by term and satisfying the condition $\mathrm{d}[\sum_{j=1}^{\infty} f_{(j)}(x)]/\mathrm{d}x = \sum_{j=1}^{\infty} [\mathrm{d}f_{(j)}(x)]/\mathrm{d}x$. So far as the Fourier series are concerned, the sufficient condition for function $f(x)$ to be differentiable term by term is that $f(x)$ should be continuous and $f'(x)$ in parts continuous over the interval $\langle 0, l \rangle$ (cf. [186], pp. 557 and 592).

The partial sums for several terms j of the series are set out in Tables 5.1, 5.2 and 5.3. As the tables indicate, the convergence of series (5.4) is comparatively very fast for the deflection but much less so for the bending moment, and in particular for the shear force.

Table 5.1 Deflection $v(l/2, l/(2c))/v_0$

Method	Number of terms			
series Eq. (5.4)	$j = 1$ 1·314 046	$j = 1, 3$ 1·326 561	$j = 1, 3, 5$ 1·328 154	$j = 1, 3, 5, 7$ 1·328 573
combined Eq. (5.72)	1	$j = 1$ 1·328 511	$j = 1, 3$ 1·328 859	$j = 1, 3, 5$ 1·328 875

Table 5.2 Bending moment $M(l/2, l/(2c))/M_0$

Method	Number of terms			
series Eq. (5.4)	$j = 1$ 1·080 759	$j = 1, 3$ 1·173 396	$j = 1, 3, 5$ 1·206 146	$j = 1, 3, 5, 7$ 1·222 773
integro- differential Eq. (5.60) error η	1	$n = 0$ 1·205 617 0·040 954	1·256 351	$n = 1$ 1·269 018 0·002 560
combined Eq. (5.71)	1	$j = 1$ 1·270 190	$j = 1, 3$ 1·272 763	$j = 1, 3, 5$ 1·273 091

Table 5.3 Shear force $T(l/2, 3l/(4c))/(P/4)$

Method	Number of terms			
series Eq. (5.4)	$j = 1, 2$ 1·358 122	$j = 1$ to 6 0·930 741	$j = 1$ to 10 1·186 027	$j = 1$ to 16 1·003 903
combined Eq. (5.73)	1	$j = 1, 2$ 1·084 883	$j = 1$ to 6 1·078 947	$j = 1$ to 10 1·080 223

5.2 *The integro-differential equation*

Using the influence function $G(x, s)$ (called Green's function in mathematics – see [8], [186]) we may write the integro-differential equation

$$v(x, t) = \int_0^l G(x, s)\, p(s, t)\, \mathrm{d}s - \int_0^l G(x, s)\, \mu(s) \frac{\partial^2 v(s, t)}{\partial t^2}\, \mathrm{d}s \tag{5.5}$$

describing – as does the partial differential equation (3.2) – vibration of a beam with variable mass $\mu(x)$ subjected to load $p(x, t)$ per unit length. Function $G(x, s)$ represents the beam deflection at point x produced by static load $P = 1$ placed at point s.

Analogously to the above we may write the equations of the bending moment

$$M(x, t) = \int_0^l G_M(x, s)\, p(s, t)\, \mathrm{d}s - \int_0^l G_M(x, s)\, \mu(s) \frac{\partial^2 v(s, t)}{\partial t^2}\, \mathrm{d}s \tag{5.6}$$

and of the shear force

$$T(x, t) = \int_0^l G_T(x, s)\, p(s, t)\, \mathrm{d}s - \int_0^l G_T(x, s)\, \mu(s) \frac{\partial^2 v(s, t)}{\partial t^2}\, \mathrm{d}s\,. \tag{5.7}$$

In the above, $G_M(x, s)$ and $G_T(x, s)$ are the influence functions of the bending moment and the shear force, respectively (that is to say their values at point x produced by static load $P = 1$ placed at point s).

Eq. (5.5), i.e. the deflection equation, was solved directly by N. G. Bondar' [23] for the case of a constant force moving over a simply supported uniform beam with the influence functions

$$G(x, s) = \begin{cases} \dfrac{1}{6EJl}\left[-(l - s)\, x^2 + (2l^2 - 3ls + s^2)\, s\right] x & \text{for} \quad x \leqq s \\ \dfrac{1}{6EJl}\left[-(l - x)\, s^2 + (2l^2 - 3lx + x^2)\, x\right] s & \text{for} \quad x \geqq s \end{cases} \tag{5.8}$$

$$G_M(x, s) = -EJ\, \frac{\partial^2 G(x, s)}{\partial x^2} = \begin{cases} \left(1 - \dfrac{s}{l}\right) x & \text{for} \quad x \leqq s \\ \left(1 - \dfrac{x}{l}\right) s & \text{for} \quad x \geqq s \end{cases} \tag{5.9}$$

$$G_T(x, s) = -EJ \frac{\partial^3 G(x, s)}{\partial x^3} = \begin{cases} 1 - \dfrac{s}{l} & \text{for} \quad x < s \\ -\dfrac{s}{l} & \text{for} \quad x > s \end{cases} \tag{5.10}$$

We shall now show how to compute directly the bending moment according to Eq. (5.6) for the same case. Let us first write the expression of the quasi-static bending moment produced by load $p(x, t) = \delta(x - ct) P$

$$M_{st}(x, t) = \int_0^l G_M(x, s)\, p(s, t)\, ds = \int_0^l G_M(x, s)\, \delta(s - ct)\, P\, ds = = P\, G_M(x, ct)\,. \tag{5.11}$$

In Eq. (5.6) we shall also substitute the expression

$$v(x, t) = \frac{1}{EJ} \int_0^l G_M(x, s)\, M(s, t)\, ds \tag{5.12}$$

which we have found — by differentiating twice with respect to x — to be analogous to the expression

$$\frac{\partial^2 v(x, t)}{\partial x^2} = -\frac{1}{EJ} M(x, t) \tag{5.13}$$

because

$$\partial^2 G_M(x, s)/\partial x^2 \doteq -\delta(x - s)\,. \tag{5.14}$$

Next we write Eq. (5.6) in the form

$$M(x, t) = M_{st}(x, t) - \frac{\mu}{EJ} \int_0^l \int_0^l G_M(x, r)\, G_M(r, s) \frac{\partial^2 M(s, t)}{\partial t^2}\, dr\, ds\,. \tag{5.15}$$

On substituting (5.9) we get the integral

$$\int_0^l G_M(x, r)\, G_M(r, s)\, dr = EJ\, G(x, s) \tag{5.16}$$

and using (5.16) are ready to write Eq. (5.15) in the form

$$M(x, t) = M_{st}(x, t) - \mu \int_0^l G(x, s) \frac{\partial^2 M(s, t)}{\partial t^2}\, ds\,. \tag{5.17}$$

Eq. (5.17) is an integro-differential equation for computing the bending moment under the boundary conditions

$$M(0, t) = 0\,; \quad M(l, t) = 0 \tag{5.18}$$

and the initial conditions

$$M(x, 0) = 0\,; \quad \dot{M}(x, 0) = 0\,, \tag{5.19}$$

provided the influence function $G(x, s)$ also satisfies the boundary conditions of a simply supported beam.

The solution of Eq. (5.17) consists of the particular solution $M_p(x, t)$ of the non-homogeneous equation (5.17) and of the solution of the appertaining homogeneous equation $M_h(x, t)$, i.e.

$$M(x,t) = M_p(x, t) + M_h(x, t)\,. \tag{5.20}$$

5.2.1 *Particular solution of the non-homogeneous integro-differential equation*

The particular solution of the non-homogeneous integro-differential equation (5.17) will be carried out by using the method of successive approximations, each subsequent approximation being computed from the preceding one via the recurrent formula

$$M_{n+1}(x, t) = M_{st}(x, t) - \mu \int_0^l G(x, s)\, \frac{\partial^2 M_n(s, t)}{\partial t^2}\, \mathrm{d}s\,. \tag{5.21}$$

Taking the quasi-static bending moment (5.11) for the zero approximation

$$M_0(x, t) = M_{st}(x, t) = P\, G_M(x, ct) \tag{5.22}$$

we get from (5.21) the first approximation

$$\begin{aligned} M_1(x, t) &= M_{st}(x, t) + \mu c^2 P\, G(x, ct) = \\ &= P\, G_M(x, ct) + \mu c^2 P\, G(x, ct)\,. \end{aligned} \tag{5.23}$$

In computing the integral in Eq. (5.21) we make use of the symmetry of the influence function $G_M(x, r) = G_M(r, x)$ so that by (5.14) and at $r = ct$

$$\frac{\partial^2 G_M(x, r)}{\partial t^2} = \frac{\partial^2 G_M(x, r)}{\partial r^2}\left(\frac{\mathrm{d}r}{\mathrm{d}t}\right)^2 = c^2\, \frac{\partial^2 G_M(r, x)}{\partial r^2} = -c^2 \delta(r - x)\,. \tag{5.24}$$

Substitution of (5.23) in (5.21) gives the second approximation

$$M_2(x, t) = P\, G_M(x, ct) + \mu c^2 P\, G(x, ct) - \\ - \mu^2 c^2 P \int_0^l G(x, s) \frac{\partial^2 G(s, ct)}{\partial t^2}\, ds \tag{5.25}$$

and of (5.25) in (5.21) the third approximation

$$M_3(x, t) = P\, G_M(x, ct) + \mu c^2 P\, G(x, ct) - \\ - \mu^2 c^2 P \int_0^l G(x, s) \frac{\partial^2 G(s, ct)}{\partial t^2}\, ds + \frac{\mu^3 c^6}{EJ} P \int_0^l G(x, s)\, G(s, ct)\, ds\,. \tag{5.26}$$

Proceeding in this manner up to the limit we obtain the particular solution of the non-homogeneous equation in the form of an infinite series, while assuming that the successive approximations converge

$$M_p(x, t) = P\mu c^2 \sum_{n=0}^{\infty} \left[\left(\frac{\mu^2 c^4}{EJ}\right)^n \int_0^l G(x, s)\, G_n(s, ct)\, ds - \\ - \mu \left(\frac{\mu^2 c^4}{EJ}\right)^{n-1} \int_0^l G_n(x, s) \frac{\partial^2 G(s, ct)}{\partial t^2}\, ds\right]. \tag{5.27}$$

In the above

$$G_0(x, s) = \delta(x - s)$$

$$G_1(x, s) = G(x, s)$$

$$\cdots\cdots\cdots\cdots\cdots\cdots$$

$$G_n(x, s) = \int_0^l G_{n-1}(x, r)\, G(r, s)\, dr \tag{5.28}$$

and according to (5.9)

$$\frac{\partial^2 G(x, ct)}{\partial t^2} = - \frac{c^2}{EJ} G_M(x, ct)\,. \tag{5.29}$$

Of course, we now have to see whether the successive approximations (5.27) converge. Let us denote by C_n^2 the least upper bound of the integral

$$C_n^2 \geqq \int_0^l |G_n(x, s)|^2\, ds \tag{5.30}$$

that exists except for the value C_0^2 for $n = 0$. For integral (5.28) we shall use the Cauchy-Bunyakovski-Schwarz inequality

$$|G_n(x, s)|^2 \leqq \int_0^l |G_{n-1}(x, r)|^2 \, dr \, . \int_0^l |G(r, s)|^2 \, dr \tag{5.31}$$

and integrate it with respect to s:

$$\int_0^l |G_n(x, s)|^2 \, ds \leqq \int_0^l |G_{n-1}(x, r)|^2 \, dr \frac{B^2}{\mu^2} \leqq C_{n-1}^2 \frac{B^2}{\mu^2} \, . \tag{5.32}$$

In the above we have used the notation

$$B^2 = \mu^2 \int_0^l \int_0^l |G(x, s)|^2 \, dx \, ds \, . \tag{5.33}$$

In view of (5.30), inequality (5.32) results in the recurrent inequality

$$C_n^2 \leqq C_{n-1}^2 \frac{B^2}{\mu^2} \tag{5.34}$$

which – if we start from C_1^2 – may be rearranged as follows:

$$C_n^2 \leqq C_1^2 \left(\frac{B}{\mu}\right)^{2n-2} \, . \tag{5.35}$$

Now we are in a position to apply the Cauchy-Bunyakovski-Schwarz inequality to the first of the integrals in Eq. (5.27)

$$\left|\int_0^l G(x, s) \, G_n(s, ct) \, ds\right|^2 \leqq \int_0^l |G(x, s)|^2 \, ds \, . \int_0^l |G_n(s, ct)|^2 \, ds \leqq C_1^2 C_n^2 \leqq$$
$$\leqq C_1^4 \left(\frac{B}{\mu}\right)^{2n-2} \, . \tag{5.36}$$

The general term of the first of the addends in the brackets of (5.27) will therefore be less than

$$\left(\frac{\mu^2 c^4}{EJ}\right)^n C_1^2 \left(\frac{B}{\mu}\right)^{n-1} = \frac{\mu C_1^2}{B} \left(\frac{\mu c^4}{EJ} B\right)^n \, . \tag{5.37}$$

This means that the first part of series (5.27) will converge faster than a geometric series, provided its quotient will meet the condition

$$\left|\frac{\mu c^4}{EJ} B\right| < 1 . \tag{5.38}$$

Because of (1.18) and the estimate of B (see [11]),

$$B > \frac{1}{\omega_{(1)}^2} , \tag{5.39}$$

inequality (5.38) also implies the condition

$$\alpha < 1 . \tag{5.40}$$

In a similar way we get the value of the second of the integrals in Eq. (5.27)

$$\left|\int_0^l G_n(x, s) \frac{\partial^2 G(s, ct)}{\partial t^2} \, ds\right|^2 \leqq \int_0^l |G_n(x, s)|^2 \, ds \, . \int_0^l \left|\frac{\partial^2 G(s, ct)}{\partial t^2}\right|^2 ds \leqq$$

$$\leqq C_n^2 D^2 \leqq C_1^2 \left(\frac{B}{\mu}\right)^{2n-2} D^2 \tag{5.41}$$

where the upper estimate D^2 of the integral is

$$D^2 \geqq \int_0^l \left|\frac{\partial^2 G(x, ct)}{\partial t^2}\right|^2 dx . \tag{5.42}$$

The general term of the second of the addends (5.27) is therefore less than

$$\mu \left(\frac{\mu^2 c^4}{EJ}\right)^{n-1} C_1 \left(\frac{B}{\mu}\right)^{n-1} D = \mu C_1 D \left(\frac{\mu c^4}{EJ} B\right)^{n-1} . \tag{5.43}$$

This means that the second term of the series, too, will converge faster than a geometric series, provided the quotient of the former meets the the conditions (5.38) and (5.40).

Since its upper bound is a geometric series, series (5.27) is a uniformly convergent one for $0 \leqq x \leqq l$, $0 \leqq s \leqq l$, $0 \leqq t \leqq l/c$ by the Weierstrass criterion. As proved for the general case of integral equations in [153], solution (5.27) converges solely for subcritical speeds (5.40) and is unique.

Knowing the properties of the majorant geometric series we may also estimate the error resulting from our taking from series (5.27) only $n+1$ terms, i.e. summing terms $0, 1, 2, \ldots, n$. The error of the first series of (5.27) is

$$\eta_1 = P\frac{\mu^2 c^2 C_1^2}{B}(S - S_n) = P\frac{\mu^2 c^2 C_1^2}{B}\frac{q^{n+1}}{1-q} \tag{5.44}$$

where

$$S = \sum_{j=0}^{\infty} q^j = \frac{1}{1-q}, \quad S_n = \sum_{j=0}^{n} q^j = \frac{1-q^n}{1-q} + q^n = \frac{1-q^{n+1}}{1-q},$$

and of the second series

$$\eta_2 = P\mu^2 c^2 C_1 D(S - S_n) = P\mu^2 c^2 C_1 D\frac{q^n}{1-q} \tag{5.45}$$

where

$$S = \sum_{j=0}^{\infty} q^{j-1} = q^{-1} + \frac{1}{1-q}, \quad S_n = \sum_{j=0}^{n} q^{j-1} = q^{-1} + \frac{1-q^n}{1-q}$$

and where by (5.38) and (5.40) the quotient is approximately $q = \alpha^4$.

According to (5.27), (5.44) and (5.45) the absolute error of the bending moment will thus be less than

$$\eta = |\eta_1 - \eta_2| = P\mu^2 c^2 C_1 \frac{\alpha^{4n}}{1-\alpha^4}\left|\frac{C_1\alpha^4}{B} - D\right|. \tag{5.46}$$

5.2.2 *General solution of the homogeneous integro-differential equation*

As indicated in [8] the general solution of the homogeneous integro-differential equation [see (5.17)]

$$M(x, t) = -\mu \int_0^l G(x, s)\frac{\partial^2 M(s, t)}{\partial t^2}\, ds \tag{5.47}$$

is a uniformly convergent series for $0 \leqq x \leqq l$, $0 \leqq t \leqq l/c$

$$M_h(x, t) = \sum_{j=1}^{\infty} (A_j \sin \omega_{(j)}t + B_j \cos \omega_{(j)}t)\, v_{(j)}(x) \tag{5.48}$$

where $v_{(j)}(x)$ are the normal modes of vibration, and coefficients A_j and B_j are determined from the initial conditions (5.19). Since $M_p(x, 0) = 0$, the first of the initial conditions (5.19) results in $B_j = 0$. The second initial condition (5.19) and (5.48) gives

$$\dot{M}(x, 0) = \sum_{j=1}^{\infty} A_j \omega_{(j)} v_{(j)}(x) + \dot{M}_p(x, 0) = 0 . \qquad (5.49)$$

On multiplying Eq. (5.49) by $v_{(k)}(x)$, integrating with respect to x from 0 to l and taking advantage of the orthogonal properties of the normal modes of vibration $\int_0^l v_{(j)}(x)\, v_{(k)}(x)\, dx = 0$ for $j \neq k$, we get after re-arrangement

$$A_j = - \frac{1}{\omega_{(j)} \int_0^l v_{(j)}^2(x)\, dx} \int_0^l \dot{M}_p(x, 0)\, v_{(j)}(x)\, dx . \qquad (5.50)$$

For a simply supported beam

$$v_{(j)}(x) = \sin \frac{j\pi x}{l}, \quad \int_0^l v_{(j)}^2(x)\, dx = \frac{l}{2} . \qquad (5.51)$$

If we take but one term of Eq. (5.27) – see (5.22) – then according to (5.9) we get for $s \leqq x$

$$\dot{M}_p(x, 0) = \frac{Pc}{l}(l - x) , \qquad (5.52)$$

and on substituting (5.50) to (5.52) in (5.48), the solution of the homogeneous equation

$$M_h(x, t) = \sum_{j=1}^{\infty} -M_0 \frac{8\alpha}{j^3 \pi^2} \sin \omega_{(j)} t \sin \frac{j\pi x}{l} . \qquad (5.53)$$

It is plain to see from (5.53) that the solution of the homogeneous equation converges very rapidly, even with just one term taken for the calculation of (5.50) and (5.52). Moreover, (5.53) attenuates very quickly on account of damping which has not been included in the calculation at all.

In accord with (5.20), the solution of Eq. (5.17) is given by the sum of Eqs. (5.27) and (5.53).

5.2.3 *Numerical example*

For our case of a simply supported uniform beam subjected to a constant moving force, Eq. (5.8) gives the following relations

$$\int_0^l G(x, s)\, G(s, r)\, \mathrm{d}s = \frac{1}{(6EJ)^2\, l} \left\{ \frac{1}{140} \left[rx(x^6 + r^6) - lx^7 \right] + \right.$$

$$+ \frac{8}{105} l^6 rx + \frac{1}{10} l^2 rx(x^4 + r^4) + \frac{1}{20} \left[r^3 x^3 (x^2 + r^2) - lr^2 x \,. \right.$$

$$\left. . (r^4 + 3x^4) \right] + lr^3 x^3 \left(\frac{l}{3} - \frac{r}{4} \right) - \left. \frac{2}{15} l^4 rx(x^2 + r^2) \right\} \quad \text{for} \quad x \leqq r \,, \quad (5.54)$$

$$\int_0^l G^2(x, s)\, \mathrm{d}s = \frac{x^2}{945 l (EJ)^2} (3x^6 - 12lx^5 + 14l^2 x^4 - 7l^4 x^2 + 2l^6) \,, \quad (5.55)$$

$$\int_0^l G(x, s) \frac{\partial^2 G(s, r)}{\partial t^2}\, \mathrm{d}s = \frac{c^2 x}{360 l (EJ)^2} \left[10x^2 r(r^2 - 3lr + 2l^2) + \right.$$

$$\left. + 3x^4(r - l) + r^3(3r^2 - 15lr + 20l^2) - 8l^4 r \right] \quad \text{for} \quad x \leqq r, r = ct \,. \quad (5.56)$$

On account of the symmetry $G(s, x) = G(x, s)$, the necessary integrals for $x \geqq r$ are obtained from (5.54) and (5.56) by interchanging x for r and r for x.

For $x = l/2$, $t = l/(2c)$, i.e. $x = r = ct = l/2$, (5.54) or (5.55) gives

$$\left. \int_0^l G(x, s)\, G(s, r)\, \mathrm{d}s \right|_{x = l/2,\, r = l/2} = \frac{17 l^7}{80\,640 (EJ)^2} \,; \quad (5.57)$$

from (5.56)

$$\left. \int_0^l G(x, s) \frac{\partial^2 G(s, r)}{\partial t^2}\, \mathrm{d}s \right|_{x = l/2,\, r = l/2} = -\frac{c^2 l^5}{480 (EJ)^2} \quad (5.58)$$

and also

$$\left. G_M(x, ct) \right|_{x = l/2,\, t = l/(2c)} = \frac{1}{4} l \,,$$

$$\left. G(x, ct) \right|_{x = l/2,\, t = l/(2c)} = \frac{l^3}{48 EJ} \,. \quad (5.59)$$

In our case of $\alpha = 1/2$, $t = l/(2c)$ and $x = l/2$ the solution of the homogeneous equation (5.53) turns out zero. From the particular solution (5.27) we take only two terms of the series, $n = 0, 1$, and consider merely Eq. (5.26). In view of (5.20) and (1.18), the latter – with (5.57) to (5.59) substituted in – gives

$$M(l/2,\ l/(2c)) = M_0 \left(1 + \frac{\mu c^2 l^2}{12EJ} + \frac{\mu^2 c^4 l^4}{120(EJ)^2} + \frac{17}{20\,160} \cdot \right.$$

$$\left. \cdot \frac{\mu^3 c^6 l^6}{(EJ)^3} + \dots \right) = M_0 \left(1 + \frac{\pi^2}{12}\alpha^2 + \frac{\pi^4}{120}\alpha^4 + \frac{17\pi^6}{20\,160}\alpha^6 + \dots\right). \quad (5.60)$$

For $\alpha = 1/2$ the successive sums of (5.60) are set out in Table 5.2. As the table indicates, the convergence is now much faster than it was with the first method. So far as practical computations are concerned it is sufficient to take but two terms of series (5.60) for which no time-consuming quadratures are needed either – see Eqs. (5.23).

To compute the error, we first find from (5.30), (5.33), (5.42) and (5.55) the quantities

$$C_1^2 \geqq \int_0^l |G(x, s)|^2 \, ds\,; \quad C_1^2 = \int_0^l G^2(x, s) \, ds \bigg|_{x = l/2} = \frac{17 l^7}{80\,640(EJ)^2}, \quad (5.61)$$

$$B^2 = \frac{\mu^2}{945 l (EJ)^2} \int_0^l (3x^8 - 12 l x^7 + 14 l^2 x^6 - 7 l^4 x^4 + 2 l^6 x^2) \, dx =$$

$$= \frac{\mu^2 l^8}{9450(EJ)^2}, \quad (5.62)$$

$$D^2 = \int_0^l \left[\frac{\partial^2 G(x, r)}{\partial t^2}\right]^2 dx \bigg|_{r = ct = l/2} = \frac{c^4 l^3}{48(EJ)^2}, \quad (5.63)$$

and then substitute them in (5.46). For $n = 0$ and $n = 1$ the errors are set out in Table 5.2. As we see from the table, already for $n = 0$ the error is so small (less than 4%) as to be negligible in practice..

5.3 *Combined method*

Either of the two methods described in Sects. 5.1 and 5.2 has advantages as well as shortcomings. The first method is simple but converges very slowly. The fast convergence of the second method, on the other hand, is more than outweighed by the laboriousness of the computation. We shall now show how to combine the two methods to get a solution at once simple and quickly convergent and applicable to any load or support conditions.

We start out from Eq. (5.6) in the second integral, on the right-hand side of which we substitute the solution $v(x, t)$ expanded in normal modes $v_{(j)}(x)$ and generalized coordinates $q_{(j)}(t)$:

$$v(x, t) = \sum_{j=1}^{\infty} v_{(j)}(x)\, q_{(j)}(t) \tag{5.64}$$

which is a series uniformly convergent in $0 \leqq x \leqq l$, $0 \leqq t \leqq l/c$. The influence function $G(x, s)$, too, may be expanded in normal modes to form a uniformly convergent series in $x \in \langle 0, l \rangle$, $s \in \langle 0, l \rangle$, (see [8])

$$G(x, s) = \sum_{j=1}^{\infty} \frac{v_{(j)}(x)\, v_{(j)}(s)}{\omega_{(j)}^2 V_j} \tag{5.65}$$

where

$$V_j = \int_0^l \mu\, v_{(j)}^2(x)\, \mathrm{d}x\,.$$

Series (5.64) and (5.65) may therefore be differentiated term by term; the order of integration and summation may be interchanged, too. Further, by making use of the orthogonal properties of normal modes $\int_0^l v_{(j)}(x)\, v_{(k)}(x)\, \mathrm{d}x = 0$ for $j \neq k$, and of (5.9), (5.64) and (5.65) we get for the second of the integrals in (5.6)

$$\int_0^l G_M(x, s)\, \mu\, \frac{\partial^2 v(s, t)}{\partial t^2}\, \mathrm{d}s = \int_0^l \sum_{k=1}^{\infty} - \frac{EJ v''_{(k)}(x)\, v_{(k)}(s)}{\omega_{(k)}^2 V_k}\, \mu \sum_{j=1}^{\infty} v_{(j)}(s)\, \ddot{q}_{(j)}(t)\, \mathrm{d}s =$$

$$= - \sum_{j=1}^{\infty} \frac{EJ}{\omega_{(j)}^2}\, v''_{(j)}(x)\, \ddot{q}_{(j)}(t)\,. \tag{5.66}$$

In view of (5.11) the solution of Eq. (5.6) will thus become

$$M(x, t) = M_{st}(x, t) + \sum_{j=1}^{\infty} \frac{EJ}{\omega_{(j)}^2}\, v''_{(j)}(x)\, \ddot{q}_{(j)}(t)\,, \tag{5.67}$$

the boundary conditions being already automatically satisfied in expressions $M_{st}(x, t)$ and $v_{(j)}(x)$, and the initial conditions in expression $q_{(j)}(t)$.

These expressions must of course be known, for example, from a previous solution of the deflection (5.64) and from the static solution.

The above method can also be used for other quantities, such as deflection, shear forces, support reactions, cross section rotation, etc. For the deflection and the shear forces it gives

$$v(x, t) = v_{st}(x, t) - \sum_{j=1}^{\infty} \frac{1}{\omega_{(j)}^2} v_{(j)}(x)\, \ddot{q}_{(j)}(t)\,,$$

$$T(x, t) = T_{st}(x, t) + \sum_{j=1}^{\infty} \frac{EJ}{\omega_{(j)}^2} v'''_{(j)}(x)\, \ddot{q}_{(j)}(t) \tag{5.68}$$

where $v_{st}(x, t)$ and $T_{st}(x, t)$ are the respective static quantities

$$v_{st}(x, t) = \int_0^l G(x, s)\, p(s, t)\, ds\,,$$

$$T_{st}(x, t) = \int_0^l G_T(x, s)\, p(s, t)\, ds\,. \tag{5.69}$$

Next to uniform beams it can handle beams of variable cross section as well as other structures.

As an *example* of its application consider our case of a constant force moving along a simply supported uniform beam. By (5.11), (5.9), (1.31)

$$M_{st}(x, t) = P\, G_M(x, ct)\,,$$

$$v''_{(j)}(x) = -\frac{j^2\pi^2}{l^2} \sin\frac{j\pi x}{l}\,,$$

$$\ddot{q}_{(j)}(t) = \frac{2P}{\mu l} \frac{\alpha}{j(1 - \alpha^2/j^2)} \left(\sin \omega_{(j)}t - \frac{\alpha}{j} \sin j\omega t\right) \tag{5.70}$$

and the bending moment according to (5.67)

$$M(x, t) = M_0 \left[\frac{4}{l} G_M(x, ct) + \sum_{j=1}^{\infty} \frac{8}{\pi^2} \frac{\alpha}{j^3(1 - \alpha^2/j^2)} \cdot \right.$$

$$\left. \cdot \left(\frac{\alpha}{j} \sin j\omega t - \sin \omega_{(j)}t\right) \sin\frac{j\pi x}{l}\right]. \tag{5.71}$$

Similarly, the deflection

$$v(x, t) = v_{st}(x, t) + v_0 \sum_{j=1}^{\infty} \frac{96}{\pi^4} \frac{\alpha}{j^5(1 - \alpha^2/j^2)} \cdot$$

$$\cdot \left(\frac{\alpha}{j} \sin j\omega t - \sin \omega_{(j)} t\right) \sin \frac{j\pi x}{l}, \tag{5.72}$$

$$T(x, t) = T_{st}(x, t) + P \sum_{j=1}^{\infty} \frac{2}{\pi} \frac{\alpha}{j^2(1 - \alpha^2/j^2)} \cdot$$

$$\cdot \left(\frac{\alpha}{j} \sin j\omega t - \sin \omega_{(j)} t\right) \cos \frac{j\pi x}{l}. \tag{5.73}$$

Expressions (5.71) to (5.73) were enumerated for $x = l/2$, $t = l/(2c)$ [or $t = 3l/(4c)$ for the shear force] and $\alpha = 1/2$, and set out in Tables 5.1, 5.2, 5.3. Consulting them we see that with the combined method the solution converges much faster than with the series method. Thus, for example, the bending moment computed with only the first term of series (5.71) differs from that computed with five terms by as little as $3^0/_{00}$. The faster convergence of expressions (5.71) to (5.73) is due to the fact that the power of j in the denominator is higher by one than that of the analogous expressions in (5.4).

The combined method outlined above takes the quasi-statical effects of load for the basic ones and expands in normal modes merely the difference between the actual solution and the quasi-statical effects*). That is why it is particularly well suited for cases in which the quasi-statical effects predominate over the inertial forces, i.e. when the first terms on the right-hand sides of (5.5) to (5.7) are larger by far than the second ones.

5.4 *Application of the theory*

Of particular importance for practice is the answer to the question of which is the larger of the two: the dynamic coefficient computed from the deflection or that obtained from the bending moment. Going by the analysis of the motion of a constant force along a simply supported beam we

*) In contrast to the first method (Sect. 5.1) that immediately expands the whole solution in normal modes.

find that the dynamic coefficient of bending moment, δ_M, is somewhat smaller than the dynamic coefficient of deflection, δ_v, (cf. Tables 5.1 and 5.2). For low speeds ($\alpha \ll 1$) of the force motion, and referring to (5.60) and [23] we may write the approximate formulae of the dynamic coefficients for the centre of the beam span and the instant the force passes through that point, as follows*):

$$\delta_v = \frac{v(l/2, l/(2c))}{v_{st}(l/2, l/(2c))} = 1 + \frac{\pi^2}{10}\alpha^2 ,$$

$$\delta_M = \frac{M(l/2, l/(2c))}{M_{st}(l/2, l/(2c))} = 1 + \frac{\pi^2}{12}\alpha^2 . \qquad (5.74)$$

For other beam cross sections, other loads, other definitions of the dynamic coefficients (of which there exists a whole number – see [107]), and particularly if the track irregularities are taken into consideration, we may expect results even totally different than (5.74).

Another important conclusion that can be drawn from this chapter, is the finding concerning the high effectiveness of the combined method which has proved equally suitable for the deflection as for the bending moment. Frequently both quantities are obtained accurately enough with just the first terms of its series taken into account.

5.5 *Additional bibliography*

[8, 11, 23, 62, 111, 117, 162, 174, 192, 193, 235, 258].

*) Similar situation has been observed to also arise in more complex systems moving along a simply supported beam without track irregularities (see Chaps. 9 and 10, especially Figs. 9.7, 9.8, 9.10 to 9.15, 10.6). There, too, the relative dynamic increment of stress is somewhat less than the analogous increment of deflection, even though the dynamic coefficients are defined in a slightly different way and Eqs. (5.74) naturally holds no longer.

6

Beams with various boundary conditions subjected to a moving load

The only problems considered so far have been those involving beams on simple supports. However, we may just as readily solve the problem of the effects of a moving load for uniform bars with general support conditions. The method we plan to use for that purpose is the generalized method of finite integral transformations which leads to results formally identical with the method of expansion in normal modes.

6.1 *Generalized method of finite integral transformations*

This method is defined by the following relations between the original $v(x, t)$ and its transform $V(j, t)$:

$$V(j, t) = \int_0^l v(x, t)\, v_{(j)}(x)\, \mathrm{d}x\,; \quad j = 1, 2, 3, \ldots \tag{6.1}$$

$$v(x, t) = \sum_{j=1}^{\infty} \frac{\mu}{V_j}\, V(j, t)\, v_{(j)}(x)\,. \tag{6.2}$$

The first relation defines the transformation of function $v(x, t)$, the second the inverse of this transformation. In Eqs. (6.1) and (6.2),

$$v_{(j)}(x) = \sin\frac{\lambda_j x}{l} + A_j \cos\frac{\lambda_j x}{l} + B_j \sinh\frac{\lambda_j x}{l} + C_j \cosh\frac{\lambda_j x}{l} \tag{6.3}$$

is the j-th normal mode of vibration of a uniform beam (in this particular arrangement it is also called "the beam function" in literature),

l – is the bar length,

λ_j, A_j, B_j, C_j are constants obtained by substituting (6.3) in the boundary conditions*).

*) Tables of these constants for various modes of support are to be found, for example, in [8].

The natural modes (6.3) satisfy the homogeneous differential equation

$$EJ \frac{d^4 v_{(j)}(x)}{dx^4} - \mu\omega_{(j)}^2 v_{(j)}(x) = 0 \tag{6.4}$$

where the natural circular frequency $\omega_{(j)}$ is in the form

$$\omega_{(j)}^2 = \frac{\lambda_j^4}{l^4} \frac{EJ}{\mu} . \tag{6.5}$$

Further, in (6.1) and (6.2):

μ denotes the mass per unit length of beam,

$$V_j = \int_0^l \mu \, v_{(j)}^2(x) \, dx . \tag{6.6}$$

Following substitution in (6.3) and integration, the last expression may be rearranged to

$$V_j = \frac{\mu l}{2} \Big\{ 1 + A_j^2 - B_j^2 + C_j^2 + \frac{1}{\lambda_j} [2C_j - 2A_jB_j - B_jC_j - \tfrac{1}{2} \, .$$
$$. (1 - A_j^2) \sin 2\lambda_j + 2A_j \sin^2 \lambda_j + (B_j^2 + C_j^2) \sinh \lambda_j \cosh \lambda_j +$$
$$+ 2(B_j + A_jC_j) \cosh \lambda_j \sin \lambda_j + 2(-B_j + A_jC_j) \sinh \lambda_j \, .$$
$$. \cos \lambda_j + 2(C_j + A_jB_j) \sinh \lambda_j \sin \lambda_j + 2(-C_j + A_jB_j) \, .$$
$$. \cosh \lambda_j \cos \lambda_j + B_jC_j \cosh 2\lambda_j] \Big\} . \tag{6.7}$$

The proof to establish the validity of the mutual transformation relations, (6.1) and (6.2), will be made on the assumption that series (6.2) is uniformly convergent in the interval $\langle 0, l \rangle$; therefore, the integral of such series will also be uniformly convergent in that interval and the order of integration and summation may be changed. Further, it is assumed that the normal modes possess orthogonal properties, i.e. that

$$\int_0^l \mu \, v_{(j)}(x) \, v_{(k)}(x) \, dx = \begin{cases} 0 & \text{for } j \neq k \\ V_j & \text{for } j = k . \end{cases} \tag{6.8}$$

Assuming the above, the first expression, (6.1), is proved by substituting in its right-hand side the right-hand side of (6.2) in which the summation is now made with respect to k

$$\int_0^l \sum_{k=0}^{\infty} \frac{\mu}{V_k} V(k, t)\, v_{(k)}(x)\, v_{(j)}(x)\, \mathrm{d}x = \sum_{k=0}^{\infty} \frac{1}{V_k} V(k, t) \,.$$

$$\cdot \int_0^l \mu\, v_{(j)}(x)\, v_{(k)}(x)\, \mathrm{d}x = V(j, t) \,. \tag{6.9}$$

The validity of the second equation, (6.2), is proved by first multiplying both its left-hand and its right-hand side by expression $v_{(k)}(x)$ and then integrating with respect to x from 0 to l:

$$\int_0^l v(x, t)\, v_{(k)}(x)\, \mathrm{d}x = \int_0^l \sum_{j=1}^{\infty} \frac{\mu}{V_j} V(j, t)\, v_{(j)}(x)\, v_{(k)}(x)\, \mathrm{d}x =$$

$$= \sum_{j=1}^{\infty} \frac{1}{V_j} V(j, t) \int_0^l \mu\, v_{(j)}(x)\, v_{(k)}(x)\, \mathrm{d}x = V(k, t) \,. \tag{6.10}$$

Accordingly, the transformation relations (6.1) and (6.2) hold good for functions $v_{(j)}(x)$ with the orthogonal properties (6.8), uniformly convergent in $\langle 0, l \rangle$, for all $j = 1, 2, 3, \ldots$, with notation (6.6) applying. Function $v_{(j)}(x)$ need not even be of the "beam" form (6.3). Formally, of course, the required solutions (6.2) are obtained expanded in normal modes, a result that can also be arrived at by another method known from the vibration theory.

Integral transformations (6.1) and (6.2) are actually a generalization of the Fourier finite transforms hitherto used (see, for example [204]) in either the sine or the cosine form. The Fourier finite sine transform is a special case of our transformations (6.1) and (6.2) in which the normal mode is used in the form (6.3) where $A_j = B_j = C_j = 0$, $\lambda_j = j\pi$ [cf. Eq. (1.9)]. The cosine transform is another special case if all that remains of (6.3) is $\cos j\pi x/l$.

When this method is applied to beam vibration, the point of importance is the calculation of the transform of the fourth derivative of the deflection with respect to x. It is effected by integration by parts used four times:

$$\int_0^l \frac{\partial^4 v(x, t)}{\partial x^4} v_{(j)}(x)\, \mathrm{d}x = \left[\frac{\partial^3 v(x, t)}{\partial x^3} v_{(j)}(x) - \frac{\partial^2 v(x, t)}{\partial x^2} \frac{\mathrm{d}v_{(j)}(x)}{\mathrm{d}x} + \right.$$

$$\left. + \frac{\partial v(x, t)}{\partial x} \frac{\mathrm{d}^2 v_{(j)}(x)}{\mathrm{d}x^2} - v(x, t) \frac{\mathrm{d}^3 v_{(j)}(x)}{\mathrm{d}x^3}\right]_0^l + \int_0^l v(x, t) \frac{\mathrm{d}^4 v_{(j)}(x)}{\mathrm{d}x^4}\, \mathrm{d}x \,. \tag{6.11}$$

The function of time in the brackets is dependent in variable x only on the boundary conditions, and as such may be written as follows:

$$z(0, l, t) = \left[\frac{\partial^3 v(x, t)}{\partial x^3} v_{(j)}(x) - \frac{\partial^2 v(x, t)}{\partial x^2} \frac{\mathrm{d}v_{(j)}(x)}{\mathrm{d}x} + \right.$$
$$\left. + \frac{\partial v(x, t)}{\partial x} \frac{\mathrm{d}^2 v_{(j)}(x)}{\mathrm{d}x^2} - v(x, t) \frac{\mathrm{d}^3 v_{(j)}(x)}{\mathrm{d}x^3}\right]_0^l . \tag{6.12}$$

The right-hand side of (6.12) is evaluated by establishing the difference between the values of the expression for $x = l$ and $x = 0$. It is to be noted that in conventional cases (self-adjoint differential operators with homogeneous boundary conditions), the function $z(0, l, t)$ is usually equal to zero because at least one of the geometric*) or dynamic boundary conditions of the deflection or of the normal mode is zero.

For beams, the fourth derivative of the normal modes can be computed from Eq. (6.4); the integral on the right-hand side of (6.11) then becomes

$$\int_0^l v(x, t) \frac{\mathrm{d}^4 v_{(j)}(x)}{\mathrm{d}x^4} \mathrm{d}x = \frac{\mu}{EJ} \omega_{(j)}^2 \int_0^l v(x, t) v_{(j)}(x) \mathrm{d}x . \tag{6.13}$$

With reference to (6.1), (6.12) and (6.13), the transform (6.11) of the fourth derivative of function $v(x, t)$ with respect to x may be written in the form

$$\int_0^l \frac{\partial^4 v(x, t)}{\partial x^4} v_{(j)}(x) \mathrm{d}x = z(0, l, t) + \frac{\mu}{EJ} \omega_{(j)}^2 V(j, t) . \tag{6.14}$$

6.2 *Motion of a force generally varying in time, along a beam*

With damping neglected, the differential equation of vibration of a straight uniform beam subjected to a moving force $P(t)$ is in the well-known form

$$EJ \frac{\partial^4 v(x, t)}{\partial x^4} + \mu \frac{\partial^2 v(x, t)}{\partial t^2} = \delta(x - ct) P(t) . \tag{6.15}$$

---

*) The geometric boundary conditions are $v(x, t)|_{x=0, x=l}$, $\left.\frac{\partial v(x, t)}{\partial x}\right|_{x=0, x=l}$, the dynamic boundary conditions $\left.\frac{\partial^2 v(x, t)}{\partial x^2}\right|_{x=0, x=l}$, $\left.\frac{\partial^3 v(x, t)}{\partial x^3}\right|_{x=0, x=l}$ and similarly so for the normal modes $v_{(j)}(x)$.

The boundary conditions will be considered wholly general but uniquely specifying expression $z(0, l, t)$. The initial conditions, too, will be considered generally different from zero

$$v(x, t)\Big|_{t=0} = g_1(x)\,; \quad \frac{\partial v(x, t)}{\partial t}\Big|_{t=0} = g_2(x) \tag{6.16}$$

and it will be assumed that there exist generalized transforms of these functions in the sense of (6.1) and (6.2)

$$G_1(j) = \int_0^l g_1(x)\, v_{(j)}(x)\, \mathrm{d}x\,, \quad G_2(j) = \int_0^l g_2(x)\, v_{(j)}(x)\, \mathrm{d}x\,,$$

$$g_1(x) = \sum_{j=1}^{\infty} \frac{\mu}{V_j}\, G_1(j)\, v_{(j)}(x)\,, \quad g_2(x) = \sum_{j=1}^{\infty} \frac{\mu}{V_j}\, G_2(j)\, v_{(j)}(x)\,. \tag{6.17}$$

We are in a position to solve Eq. (6.15) by the generalized method of finite integral transformations. On multiplying both sides of (6.15) by $v_{(j)}(x)$ and integrating with respect to x from 0 to l we get – in view of (6.14) and (1.7) –

$$EJ\left[z(0, l, t) + \frac{\mu}{EJ}\,\omega_{(j)}^2 V(j, t)\right] + \mu\, \ddot{V}(j, t) = P(t)\, v_{(j)}(ct)\,, \tag{6.18}$$

and after rearrangement

$$\ddot{V}(j, t) + \omega_{(j)}^2 V(j, t) = \frac{1}{\mu}\, P(t)\, v_{(j)}(ct) - \frac{EJ}{\mu}\, z(0, l, t)\,. \tag{6.19}$$

By (6.16) and (6.17) the boundary conditions of this ordinary differential equation are

$$V(j, t)\Big|_{t=0} = G_1(j)\,; \quad \dot{V}(j, t)\Big|_{t=0} = G_2(j) \tag{6.20}$$

and the equation is solved by the Laplace-Carson integral transformation (1.15) and the use of (27.4) to give

$$p^2\, V^*(j, p) - p^2\, G_1(j) - p\, G_2(j) + \omega_{(j)}^2\, V^*(j, p) =$$
$$= \frac{1}{\mu}\, P^*(p) - \frac{EJ}{\mu}\, Z^*(p) \tag{6.21}$$

in which

$$P^*(p) = p \int_0^\infty P(t)\, v_{(j)}(ct)\, e^{-pt}\, dt\,,$$

$$Z^*(p) = p \int_0^l z(0, l, t)\, e^{-pt}\, dt \qquad (6.22)$$

denote the Laplace-Carson transforms of expressions $P(t)\, v_{(j)}(ct)$ and $z(0, l, t)$.

With respect to unknown $V^*(j, p)$, Eq. (6.21) may be rearranged to

$$V^*(j, p) = \frac{1}{p^2 + \omega_{(j)}^2} \left[\frac{1}{\mu} P^*(p) - \frac{EJ}{\mu} Z^*(p) + p^2\, G_1(j) + p\, G_2(j)\right]. \qquad (6.23)$$

With regard to (27.5), (27.18), (27.19) and (6.22), the inverse Laplace-Carson transformation of Eq. (6.23) gives

$$V(j, t) = \frac{1}{\mu\omega_{(j)}} \int_0^t [P(\tau)\, v_{(j)}(c\tau) - EJ\, z(0, l, \tau)] \sin \omega_{(j)}(t - \tau)\, d\tau + \\ + G_1(j) \cos \omega_{(j)}t + \frac{1}{\omega_{(j)}}\, G_2(j) \sin \omega_{(j)}t\,. \qquad (6.24)$$

Following the use of relations (6.17) the inverse transformation of Eq. (6.24) according to (6.2) of our generalized method becomes

$$v(x, t) = \sum_{j=1}^{\infty} \left\{\frac{1}{V_j\omega_{(j)}}\, v_{(j)}(x) \int_0^t [P(\tau)\, v_{(j)}(c\tau) - EJ\, z(0, l, \tau)]\,. \right. \\ . \sin \omega_{(j)}(t - \tau)\, d\tau + \frac{\mu}{V_j}\, G_1(j)\, v_{(j)}(x) \cos \omega_{(j)}t + \\ \left. + \frac{\mu}{V_j\omega_{(j)}}\, G_2(j)\, v_{(j)}(x) \sin \omega_{(j)}t\right\}. \qquad (6.25)$$

Eq. (6.25) describes the deflection of a beam with general boundary and initial conditions, traversed by a concentrated force variable in time. To get the stresses in the beam we would apply the procedure outlined in Chap. 5.

In the special case of constant force $P(t) = P$ and the beam function (6.3), the integral in Eq. (6.25) is easy to evaluate. To simplify still further, assume that $z(0, l, t) = 0$ and the initial conditions are also zero. Then using the notation

$$\omega = \frac{\lambda_j c}{l} . \tag{6.26}$$

Eq. (6.25) turns out as follows:

$$v(x, t) = \sum_{j=1,\dots,k-1,k+1,\dots}^{\infty} \frac{P}{V_j} v_{(j)}(x) \left\{ \frac{1}{\omega_{(j)}^2 - \omega^2} \left[\sin \omega t - \frac{\omega}{\omega_{(j)}} . \right.\right.$$
$$\left. \sin \omega_{(j)} t + (\cos \omega t - \cos \omega_{(j)} t) A_j \right] + \frac{1}{\omega_{(j)}^2 + \omega^2} \left[\left(\sinh \omega t - \right.\right.$$
$$\left.\left.\left. - \frac{\omega}{\omega_{(j)}} \sin \omega_{(j)} t \right) B_j + (\cosh \omega t - \cos \omega_{(j)} t) C_j \right] \right\} +$$
$$+ \frac{P}{2 V_k \omega_{(k)}^2} v_{(k)}(x) \left[\sin \omega_{(k)} t - \omega_{(k)} t \cos \omega_{(k)} t + \right.$$
$$+ A_k \omega_{(k)} t \sin \omega_{(k)} t + (\sinh \omega_{(k)} t - \sin \omega_{(k)} t) B_k +$$
$$\left. + (\cosh \omega_{(k)} t - \cos \omega_{(k)} t) C_k \right] . \tag{6.27}$$

The term within the summation sign is the solution for $\omega_j \neq \omega$ while the next one is the solution for $j = k$ at which $\omega_{(k)} = \omega$ (the resonant case).

Next we shall obtain the deflection of a beam over which force P moves from right to left. This means that load $p(x, t)$ is in the form

$$p(x, t) = \delta(x - l + ct) P . \tag{6.28}$$

Applying the procedure outlined in connection with the deduction of Eqs. (6.25) and (6.27), Eq. (6.15) with (6.28) for its right-hand side gives

$$v(x, t) = \sum_{j=1}^{\infty} \frac{P}{V_j} v_{(j)}(x) \left\{ \frac{1}{\omega_{(j)}^2 - \omega^2} \left[\left(\sin \omega t - \frac{\omega}{\omega_{(j)}} \sin \omega_{(j)} t \right) . \right.\right.$$
$$. (-\cos \lambda_j + A_j \sin \lambda_j) + (\cos \omega t - \cos \omega_{(j)} t) (\sin \lambda_j +$$
$$\left. + A_j \cos \lambda_j) \right] + \frac{1}{\omega_{(j)}^2 + \omega^2} \left[\left(\sinh \omega t - \frac{\omega}{\omega_{(j)}} \sin \omega_{(j)} t \right) . \right.$$
$$. (-B_j \cosh \lambda_j - C_j \sinh \lambda_j) + (\cosh \omega t - \cos \omega_{(j)} t) .$$
$$\left.\left. . (B_j \sinh \lambda_j + C_j \cosh \lambda_j) \right] \right\} . \tag{6.29}$$

Expression (6.29) will be made use of in Chap. 11. Similarly as in (6.27) the resonant case would occur at $\omega_{(k)} = \omega$.

If the force in (6.25) is a harmonic one

$$P(t) = Q \sin \Omega t ,$$

the expression obtained after evaluation of the integrals is

$$\begin{aligned} v(x, t) = \sum_{j=1}^{\infty} \frac{Q}{2V_j\omega_{(j)}} v_{(j)}(x) \Bigg\{ & \frac{1}{r_1^2 - \omega^2} [\omega \cos \omega_{(j)}t - r_1 \sin \omega t \sin \Omega t - \\ & - \omega \cos \omega t \cos \Omega t + A_j(r_1 \sin \omega_{(j)}t - r_1 \cos \omega t \sin \Omega t + \\ & + \omega \sin \omega t \cos \Omega t)] + \frac{1}{r_1^2 + \omega^2} [B_j(\omega \cos \omega_{(j)}t - \\ & - r_1 \sinh \omega t \sin \Omega t - \omega \cosh \omega t \cos \Omega t) + \\ & + C_j(r_1 \sin \omega_{(j)}t - r_1 \cosh \omega t \sin \Omega t - \omega \sinh \omega t \cos \Omega t)] + \\ & + \frac{1}{r_2^2 - \omega^2} [-\omega \cos \omega_{(j)}t + r_2 \sin \omega t \sin \Omega t + \omega \cos \omega t \cos \Omega t + \\ & + A_j(r_2 \sin \omega_{(j)}t + r_2 \cos \omega t \sin \Omega t - \omega \sin \omega t \cos \Omega t)] + \\ & + \frac{1}{r_2^2 + \omega^2} [B_j(-\omega \cos \omega_{(j)}t + r_2 \sinh \omega t \sin \Omega t + \\ & + \omega \cosh \omega t \cos \Omega t) + C_j(r_2 \sin \omega_{(j)}t + r_2 \cosh \omega t \sin \Omega t + \\ & + \omega \sinh \omega t \cos \Omega t)] \Bigg\} \end{aligned} \tag{6.30}$$

where

$$r_1 = \Omega - \omega_{(j)}, \quad r_2 = \Omega + \omega_{(j)} . \tag{6.31}$$

The resonant cases $r_1 = \omega$ or $r_2 = \omega$ would be computed in an analogous manner. Another case of interest is that in which the frequency of the harmonic force equals some of the frequencies of free vibration, $\Omega = \omega_{(k)}$. Then naturally, $r_1 = 0$ and the respective k-th addend in Eq. (6.30) is obtained by simple substitution of $r_1 = 0$.

Expressions (6.25), (6.27), (6.29) and (6.30) represent the generalization of the results of Chaps. 1, 2 and 4 to bars of constant cross section on any kind of supports. In the main, series (6.25), (6.27), (6.29) and (6.30) converge more slowly than their counterparts for simply supported bars.

6.3 *Motion of a force along a cantilever beam*

We shall illustrate the theory expounded in the foregoing by computing the vibration of a cantilever bar along which moves a constant force P.

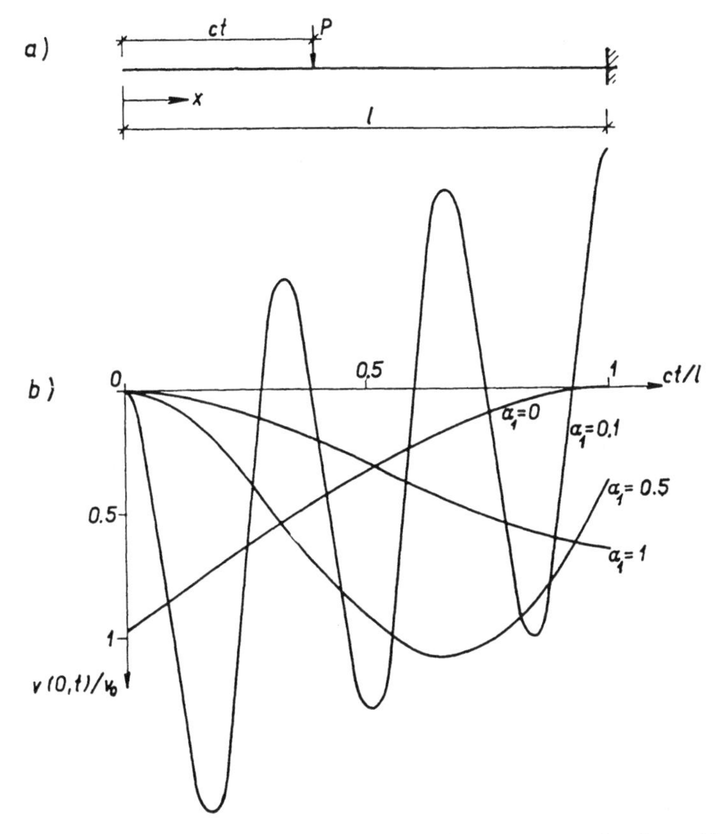

Fig. 6.1. a) Cantilever with free left-hand end subjected to a moving force. b) Deflection of the cantilever free end, $v(0, t)/v_0$, for various values of speed, $\alpha_1 = 0, 0{\cdot}1, 0{\cdot}5, 1$.

The cantilever is free on the left-hand and clamped on the right-hand end (Fig. 6.1a). Accordingly, the boundary conditions are

$$v''(0, t) = 0\,; \quad v'''(0, t) = 0\,; \quad v(l, t) = 0\,; \quad v'(l, t) = 0$$

and hence also

$$v''_{(j)}(0) = 0\,; \quad v'''_{(j)}(0) = 0\,; \quad v_{(j)}(l) = 0\,; \quad v'_{(j)}(l) = 0\,. \tag{6.32}$$

On substituting (6.32) in (6.12) we see that $z(0, l, t) = 0$. The initial conditions are assumed to be zero, i.e. $g_1(x) = g_2(x) = 0$. Substitution of the last four boundary conditions from (6.32) in Eq. (6.3) leads to four equations for the unknowns λ_j, A_j, B_j, C_j and after rearrangement to

$$1 + \cos\lambda_j \cosh\lambda_j = 0\,; \quad \lambda_1 = 1{\cdot}875; \ldots$$

$$A_j = C_j = -\frac{\sin\lambda_j + \sinh\lambda_j}{\cos\lambda_j + \cosh\lambda_j}\,; \quad A_1 = C_1 = -1{\cdot}362; \ldots$$

$$B_j = 1\,. \tag{6.33}$$

V_j is computed from Eq. (6.27) and for $j = 1$, $V_1 = 1{\cdot}856\mu l$.

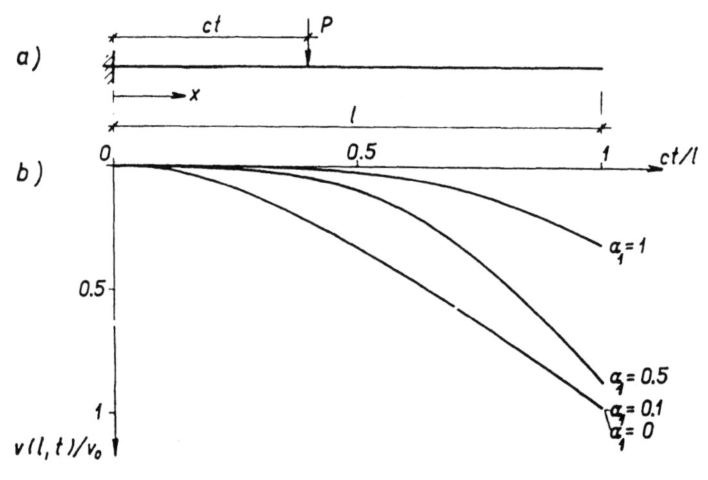

Fig. 6.2. a) Cantilever with free right-hand end subjected to a moving force. b) Deflection of the cantilever free end, $v(l, t)/v_0$, for various values of speed, $\alpha_1 = 0, 0{\cdot}1, 0{\cdot}5, 1$.

A cantilever clamped on the left-hand and free on the right-hand end (Fig. 6.2a) is computed in a like manner. The boundary conditions now are

$$v(0, t) = 0\,; \quad v'(0, t) = 0\,; \quad v''(l, t) = 0\,; \quad v'''(l, t) = 0 \tag{6.34}$$

and the constants turn out to be

$$\lambda_1 = 1{\cdot}875\,; \; A_1 = -C_1 = -1{\cdot}362\,; \; B_1 = -1\,; \; V_1 = 1{\cdot}856\mu l\,. \tag{6.35}$$

The free-end deflections produced by constant force P traversing the cantilever from left to right (Figs. 6.1b and 6.2b) were computed for both cases considered. They were referred to the static free-end deflection

$$v_0 = \frac{Pl^3}{3EJ} . \tag{6.36}$$

The computation was made with only the first term of the series in Eq. (6.27), that is to say for $j = 1$ at $\omega_1 \neq \omega$, and $k = 1$ at $\omega_1 = \omega$. Figs. 6.1b and 6.2b show the free end deflections of the respective cantilevers at several speeds

$$\alpha_1 = \frac{\lambda_1}{\pi} \frac{c}{2lf_{(1)}} \tag{6.37}$$

where

$$\alpha_j = \frac{\omega}{\omega_{(j)}} . \tag{6.38}$$

6.4 *Application of the theory*

It is evident from Figs. 6.1 and 6.2 that stresses in a cantilever beam are far more unfavourable if the force applied starts to move from the free end than if it does from the clamped one. At low speeds of motion, the former case results in very intensive, the latter in very light beam vibrations.

This fact ought to be borne in mind when designing or building bridges with overhung ends, hinged beams, etc.

6.5 *Additional bibliography*

[8, 99, 142, 172, 232, 235].

7

Massless beam subjected to a moving load

In Chap. 1 we have solved the case of a beam traversed by a load with mass negligible against the beam mass. Now we shall tackle the other extreme case, i.e. that of the beam mass very small compared to the mass of the moving load. As this primary assumption implies, our computation will have to take account not only of the force effects of the moving load but according to d'Alembert's principle, of its inertia effects as well.

It is of interest to recall that this case – though actually more difficult than that discussed in Chap. 1 – was studied much earlier, in fact in the first half of the nineteenth century, in connection with the erection of the first railway bridges in England. The problem was originally formulated and approximately solved by R. Willis [233], the first experimenter in this field. Its exact solution was offerd by G. G. Stokes [207] and later on, by H. Zimmermann [236].

7.1 *Formulation of the problem*

Consider a simply supported beam with span l and negligible mass traversed by load P with mass $m = P/g$ moving at constant speed c, see Fig. 7.1. The load acts on the beam by force P and according to d'Alembert's principle, also by inertia force $-m\mathrm{d}^2v(ct, t)/\mathrm{d}t^2$ dependent on vertical acceleration at point $x = ct$ at which the load is situated at the time considered.

Accordingly, by Eq. (5.5) in which we set $\mu(x) = 0$ and $p(x, t) = \delta(x - ct)\left[P - m\mathrm{d}^2v(ct, t)/\mathrm{d}t^2\right]$, the beam deflection at point x and time t is described by the equation

$$v(x, t) = \left[P - m\,\frac{\mathrm{d}^2v(ct, t)}{\mathrm{d}t^2}\right] G(x, ct) \tag{7.1}$$

and by the boundary and initial conditions (1.2) and (1.3). In the above $G(x, ct)$ is the influence function according to (5.8). The vertical displacement $v_1(t)$ of mass m is equal to the beam deflection at the point of load action

$$v_1(t) = v(ct, t) \tag{7.2}$$

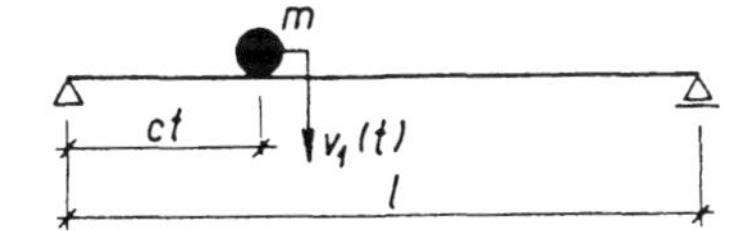

Fig. 7.1. Massless beam with a moving mass.

so that after substituting (7.2) in (7.1)

$$v_1(t) = \left[P - m\,\frac{\mathrm{d}^2 v_1(t)}{\mathrm{d}t^2}\right] G(ct, ct) \tag{7.3}$$

while the initial conditions are

$$v_1(t)\big|_{t=0} = 0\,; \quad \left.\frac{\mathrm{d}v_1(t)}{\mathrm{d}t}\right|_{t=0} = 0\,. \tag{7.4}$$

We can solve Eq. (7.1) provided we know the solution of Eq. (7.3).

7.2 *Exact solution*

If expression (5.8) is taken in place of $G(ct, ct)$, Eq. (7.3) will assume the following form

$$v_1(t) = \left[P - m\,\frac{\mathrm{d}^2 v_1(t)}{\mathrm{d}t^2}\right] \frac{c^2 t^2 (l - ct)^2}{3EJl}\,. \tag{7.5}$$

This is an ordinary linear differential equation of the second order with variable coefficients. It can be written somewhat more clearly mathematically with a new independent variable

$$\tau = ct/l\,, \tag{7.6}$$

a dependent variable using the static deflection v_0 at the centre of the beam according to (1.21)

$$y(\tau) = v_1(t)/v_0\,, \tag{7.7}$$

and the constant

$$\alpha' = \frac{12EJg}{Plc^2}\,. \tag{7.8}$$

With expressions (7.6) to (7.8) in, Eq. (7.5) becomes

$$\tau^2(1-\tau)^2\frac{\mathrm{d}^2y(\tau)}{\mathrm{d}\tau^2} + \frac{\alpha'}{4}\,y(\tau) = 4\alpha'\tau^2(1-\tau)^2\,. \tag{7.9}$$

Eq. (7.9) has the initial conditions

$$y(\tau)\big|_{\tau=0} = 0\,; \quad \frac{\mathrm{d}y(\tau)}{\mathrm{d}\tau}\bigg|_{\tau=0} = 0\,, \tag{7.10}$$

and we shall solve it in the interval $0 \leqq \tau \leqq 1$, the solution naturally having regular singular points at $\tau = 0$ and $\tau = 1$.

Eq. (7.9) was solved by G. G. Stokes [207], G. Boole [26] and H. Zimmermann [236]. According to [120] (Eqs. Nos. 2.380 and 2.381), the general solutions of the homogeneous equation associate to (7.9) are the expressions

$$y_1(\tau) = \tau^k(1-\tau)^{1-k}\,, \tag{7.11}$$

$$y_2(\tau) = \tau^{1-k}(1-\tau)^k \tag{7.12}$$

where

$$k(k-1) + \frac{\alpha'}{4} = 0\,, \quad \text{i.e.}$$

$$k_{1,2} = \tfrac{1}{2}[1 \pm (1-\alpha')^{1/2}]\,. \tag{7.13}$$

Solutions (7.11) and (7.12) are linearly independent one of another because the Wronskian

$$W(\tau) = y_1(\tau)\,\dot{y}_2(\tau) - \dot{y}_1(\tau)\,y_2(\tau) = 1 - 2k \tag{7.14}$$

is different from zero over the whole interval of τ. The only exception is the case of $k = 1/2$ for which it is necessary to set up a new pair of linearly independent solutions of the homogeneous equation associate to (7.9):

$$y_1(\tau) = \tau^{1/2}(1-\tau)^{1/2}\,, \tag{7.15}$$

$$y_2(\tau) = \tau^{1/2}(1-\tau)^{1/2}\ln\frac{t}{1-t}\,. \tag{7.16}$$

The Wronskian of expressions (7.15) and (7.16) is really different from zero in the whole interval because

$$W(\tau) = y_1(\tau)\,\dot{y}_2(\tau) - \dot{y}_1(\tau)\,y_2(\tau) = 1\,. \tag{7.17}$$

The particular solution of the non-homogeneous Eq. (7.9) is obtained, for example, by the method of variation of parameters – see [186], p. 648 – on the basis of our knowledge of the fundamental systems (7.11), (7.12), or (7.15), (7.16).

The general solution of (7.9) then is

$$y(\tau) = \left[A_1 - \frac{4\alpha'}{W(\tau)}\int y_2(\tau)\,\mathrm{d}\tau\right]y_1(\tau) + \left[A_2 + \frac{4\alpha'}{W(\tau)}\int y_1(\tau)\,\mathrm{d}\tau\right]y_2(\tau) \tag{7.18}$$

which satisfies both Eq. (7.9) and the initial conditions (7.11) with any choice of integration constants A_1, A_2. Depending on whether $\alpha' \neq 1$ or $\alpha' = 1$, the values of $y_1(\tau)$, $y_2(\tau)$, $W(\tau)$, k to be substituted in Eq. (7.18) are those defined by (7.11) to (7.14) or by (7.15) to (7.17).

Eq. (7.9) implies directly the solutions of the following two extreme cases:

a) $c \to 0$, i.e. $\alpha' \to \infty$ (static case)

$$y(\tau) = 16\tau^2(1 - \tau)^2\,, \tag{7.19}$$

b) $c \to \infty$, i.e. $\alpha' \to 0$ [in view of (7.10)]

$$y(\tau) = 0\,. \tag{7.20}$$

Expression (7.18) is the exact solution of Eq. (7.9); of course, it contains quadratures not amenable to further closed-form solution*). As suggested by H. Zimmermann in [236], the solutions of homogeneous Eqs. (7.11) (7.12) as well as of (7.15), (7.16) may also be expressed in terms of geometric and hyperbolic functions. However, expression (7.18) is very inconvenient for numerical evaluation in any case.

---

*) The exception to this statement is the case of $\int y_1(\tau)\,\mathrm{d}\tau$ with (7.15) substituted for $y_1(\tau)$. Then

$$\int y_1(\tau)\,\mathrm{d}\tau = \frac{1}{4}\left\{[1 - 2(1-\tau)]\,\tau^{1/2}(1-\tau)^{1/2} + \operatorname{arctg}\left(\frac{\tau}{1-\tau}\right)^{1/2}\right\}.$$

7.3 *Approximate solutions*

For the reasons just explained Eq. (7.9) is better solved numerically or approximately. The numerical method of Runge-Kutta was applied by A. P. Filippov and S. S. Kokhmanyuk [61], [62] to find the deflection underneath the moving load. Fig. 7.2 representing their results shows the dependence of maximum deflection underneath the load on parameter $1/\alpha'$, i.e. also on speed — see (7.8). Comparing Figs. 1.3 and 7.2 we see that if the beam mass is considered and the load mass neglected, dynamic stresses in the beam first grow but very slowly with growing speed (Fig. 1.3). If, on the other hand, the beam mass is neglected and the load mass considered (Fig. 7.2), this growth is first very fast, the effects are at their maximum at about $1/\alpha' = 0{\cdot}2$, then slowly fall off and become zero at infinite speed.

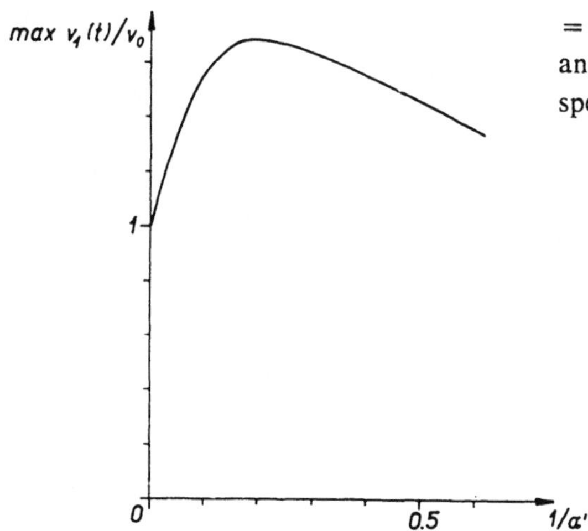

Fig. 7.2. Dependence of the maximum deflection of beam underneath the load, $v_1(t)/v_0$, on parameter $1/\alpha'$; α' defined by (7.8).

7.3.1 *The perturbation method*

This method is applied to advantage at low speeds, i.e. at $1/\alpha' \ll 1$, for then we can put to good use the fact that we know the solution of (7.9) for $1/\alpha' = 0$, which is Eq. (7.19). We shall therefore assume that the solution of (7.9) is in the form

$$y(\tau) = 16\tau^2(1-\tau)^2\left[1 + \frac{1}{\alpha'}\,y_1(\tau) + \frac{1}{\alpha'^2}\,y_2(\tau) + \ldots\right]. \qquad (7.21)$$

On substituting (7.21) in (7.9) and comparing the coefficients of terms containing the same powers of $1/\alpha'$ we get very simple differential equations in functions $y_1(\tau)$, $y_2(\tau)$, ... which we solve successively. This gives

$$\begin{aligned} y_1(\tau) &= -8(1 - 6\tau + 6\tau^2), \\ y_2(\tau) &= 64(1 - 24\tau + 102\tau^2 - 156\tau^3 + 78\tau^4), \\ &\dots\dots\dots \end{aligned} \tag{7.22}$$

If $1/\alpha'$ is small enough and the number of equations thus solved n, Eq. (7.21) describes quite accurately the required function (except for the terms of the $n + 1$ and higher orders).

Thus, for example, for the deflection of the centre $x = l/2$ of the beam, $v(l/2, l/(2c))$ at the instant of the load passing over it, $t = l/(2c)$ we get by (7.1)

$$\frac{v(l/2, l/(2c))}{v_0} = 1 + \frac{4}{\alpha'} - \frac{8}{\alpha'^2} + \dots . \tag{7.23}$$

We have evaluated Eq. (7.23) by computing from (7.1) – after substitution of (7.6) to (7.8) and (5.8) –

$$\frac{v(l/2, t)}{v_0} = 1 - \frac{1}{4\alpha'} \frac{d^2 y(\tau)}{d\tau^2} . \tag{7.24}$$

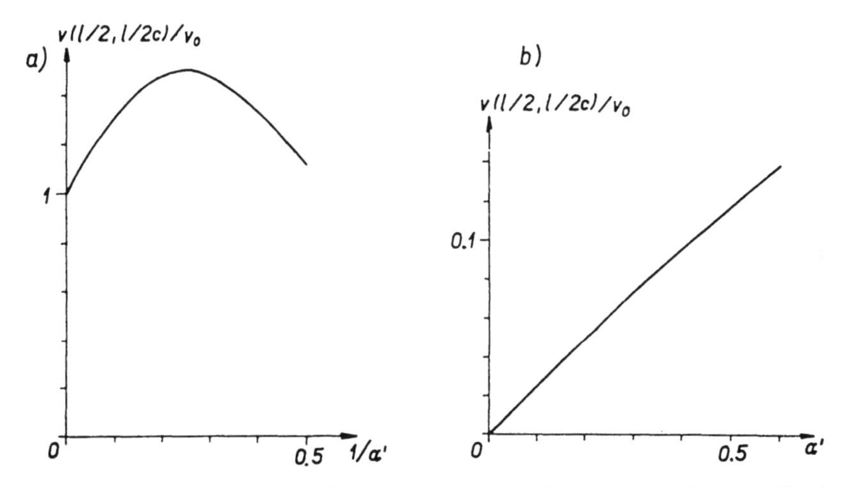

Fig. 7.3. Dependence of the deflection at the centre of a beam at the instant of load passing over it. Approximate solution for a) low speeds, b) high speeds.

The derivative in (7.24) was found from (7.9)

$$\frac{d^2y(\tau)}{d\tau^2} = 4\alpha'\left[1 - \frac{y(\tau)}{16\tau^2(1-\tau)^2}\right] \tag{7.25}$$

and the approximate solution (7.21) together with (7.22) substituted in (7.25). Eq. (7.23) in dependence on parameter $1/\alpha'$, i.e. on speed, is graphically represented in Fig. 7.3a.

7.3.2 *Method of successive approximations*

Results similar to the foregoing are also obtained by the method of successive approximations. With the latter method we start out from Eq. (7.9) and compute the approximations according to the following procedure

$$y_n(\tau) = 16\tau^2(1-\tau)^2\left[1 - \frac{1}{4\alpha'}\frac{d^2y_{n-1}(\tau)}{d\tau^2}\right]. \tag{7.26}$$

Assuming that the zero approximation has the second derivative equal to zero, $\ddot{y}_0(\tau) = 0$, we get successively from (7.26)

$$y_1(\tau) = 16\tau^2(1-\tau)^2\,,$$

$$y_2(\tau) = 16\tau^2(1-\tau)^2\left[1 - \frac{8}{\alpha'}(1 - 6\tau + 6\tau^2)\right],$$

$$y_3(\tau) = 16\tau^2(1-\tau)^2\left[1 - \frac{8}{\alpha'}(1 - 6\tau + 6\tau^2) + \frac{64}{\alpha'^2}(1 - 24\tau + 114\tau^2 - 180\tau^3 + 90\tau^4)\right], \tag{7.27}$$

....

Sequence $y_n(\tau)$ is assumed to converge uniformly in the interval $0 \leqq \tau \leqq 1$ toward the solution of Eq. (7.9). In practice this method is applicable only at low speeds, i.e. for $1/\alpha' \ll 1$. In that case the first and second approximations give wholly identical, and the third approximation only slightly different results than does the preceding method (paragraph 7.3.1) – cf. Eqs. (7.22) and (7.27).

7.3.3 *Expansion in power series*

In solutions involving high speeds, i.e. for $\alpha' \ll 1$, it is useful to expand the particular integral of the non-homogeneous equation (7.9) in a power series. With regard to the regular singular points $\tau = 0$ and $\tau = 1$, we shall consider the particular solution of Eq. (7.9) to be in the form

$$y_p(\tau) = \tau^2(1 - \tau)^2 \sum_{n=0}^{\infty} a_n \tau^n . \qquad (7.28)$$

Coefficients a_n are determined by substituting (7.28) in (7.9) and comparing the coefficients of terms containing the like powers of τ. This operation leads to equations from which one can successively compute a_n. The computation results in

$$\begin{aligned} a_0 &= \frac{16\alpha'}{8 + \alpha'} , \\ a_1 &= \frac{32\alpha'}{24 + \alpha'} \frac{24}{8 + \alpha'} , \\ &\dots\dots\dots\dots\dots\dots \end{aligned} \qquad (7.29)$$

The general solution of Eq. (7.9) is the sum of the fundamental system of solutions of the homogeneous equation and the particular solution [cf. Eq. (7.18); the equation that follows is in fact an expression of the terms with integrals in (7.18) written with the aid of a power series]

$$y(\tau) = A_1\, y_1(\tau) + A_2\, y_2(\tau) + \tau^2(1 - \tau)^2 \sum_{n=0}^{\infty} a_n \tau^n . \qquad (7.30)$$

Substituting the above solution in (7.25) and (7.24) and choosing $A_1 = -A_2$ we get

$$\frac{v(l/2,\, l/(2c))}{v_0} = \frac{\alpha'}{8 + \alpha'} \left(1 + \frac{24}{24 + \alpha'} + \dots \right) . \qquad (7.31)$$

The graph in Fig. 7.3b represents Eq. (7.31) in dependence on parameter α', i.e. according to (7.8) in indirect dependence on speed.

7.4 *Application of the theory*

The theory outlined above has found application in dynamic computations of short-span bridges [131] and crane runways [71]. In both instances the beam mass is really very small against the moving load mass, and as such may be neglected.

It should be noted, however, that in the two cases quoted the effect of the moving mass is fairly small compared to other factors producing high dynamic stresses in those structures. Thus, for example, in short-span railway bridges, it is the effect of impacts of flat wheels, rail joints, etc. — in crane runways the effect of sudden lifting and braking of the load, of track irregularities, etc. that predominate over that of the moving load. Moreover, in short-span bridges, the vehicle can no longer be idealized by a single mass point.

7.5 *Additional bibliography*

[7, 26, 46, 61, 62, 71, 83, 150, 207, 233, 236].

8

Beam subjected to a moving system with two degrees of freedom

The problem in which the moving load mass and the beam mass are both taken into account is far more complicated than the special cases analyzed in Chaps. 1 and 7. It is described by the differential equation

$$EJ\frac{\partial^4 v(x,t)}{\partial x^4} + \mu\frac{\partial^2 v(x,t)}{\partial t^2} + 2\mu\omega_b\frac{\partial v(x,t)}{\partial t} = $$

$$= \delta(x - ct)\left[P - m\frac{\mathrm{d}^2 v(ct,t)}{\mathrm{d}t^2}\right] \tag{8.1}$$

the right-hand side of which expresses the motion of force P with mass m including the inertia effects. Because of the second derivative on the rigth-hand side, the solution to Eq. (8.1) is fairly difficult compared to that of the special cases in which either mass m (Chap. 1) or the beam mass μ (Chap. 7) is neglected.

The first authors to tackle the solution of Eq. (8.1) were H. Saller [196], H. H. Jeffcott [115] whose iterative method fails to converge in some cases, and H. Steuding [206] who treated several specific cases. A satisfactory method (Fourier series with unknown coefficients for the trajectory of the moving mass) was devised by A. Schallenkamp [197]. V. V. Muchnikov [162] and M. Ya. Ryazanova [192] solved the problem by the method of integral equations, J. Naleszkiewicz [166] by Galerkin's method and V. V. Bolotin [21] by a method that leads to approximate asymptotic solutions in quadratures. C. E. Inglis [111] and V. Koloušek [130] studied the problem in relation to vibrations of railway bridges.

Eq. (8.1) describes the case in which the vehicle is idealized by a single mass point. However, such a simplification is no longer satisfactory for modern vehicles with clearly differentiated unsprung and sprung masses. That is why A. Hillerborg [107] made a study of the motion of a sprung mass along a beam, some earlier attempts in that direction having been made by C. E. Inglis [111] and V. Koloušek [130]. Further advances in the solution of the problem were contingent on the arrival of automatic computers. With their help J. M. Biggs, H. S. Suer, J. M. Louw [16] and T. P. Tung, L. E. Goodman, T. Y. Chen, N. M. Newmark [224] solved the problem originally treated by Hillerborg, and applied the solution to vibrations of highway bridges.

In modern bridges where the theory finds its widest field of application the actual conditions are more complicated still. It is now a well-known fact that track irregularities, elastic properties of roadways on railway bridges and tires on highway bridges, unbalanced components of unsprung masses and other factors are apt to bear considerable effect on dynamic stresses in the respective structures. In order that all those effects might be accounted for, the problem was theoretically generalized and is now divided in two classes differing by the ratio between vehicle length and bridge span. To adhere to this classification we will discuss one class — large-span bridges — in this chapter, and the other — short-span bridges in Chaps. 9 and 10.

8.1 *Formulation of the problem*

In the analysis that follows we will consider the mechanical model shown schematically in Figs. 8.1 and 8.2, and make the following assumptions (next to assumptions 1, 3, 4 and 5 of Sect. 1.1):

1. The moving vehicle is idealized by a system with two degrees of freedom. Unsprung mass m_1 is in indirect contact with the beam*); m_2 denotes the sprung mass. The total weight of the vehicle is

$$P = P_1 + P_2 \tag{8.2}$$

where $P_1 = m_1 g$ is the weight of unsprung parts, $P_2 = m_2 g$ the weight of sprung parts of the vehicle.

The two masses are connected by means of a linear spring with spring constant C and coefficient of viscous damping C_b. Their vertical displace-

*) According to Figs. 8.1 and 8.2, mass m_1 is also elastically supported. The term "unsprung mass" — m_1 — is used here and in the next chapters in the sense of literature dealing with vehicle vibrations.

ments are denoted by $v_1(t)$ and $v_2(t)$ – Figs. 8.1 and 8.2. Displacement $v_2(t)$ is measured from the position in which spring C is deformed by force P_2, displacement $v_1(t)$ from that marked in dashed lines in Figs. 8.1, 8.2, left, where springs $K(x)$ are undeformed.

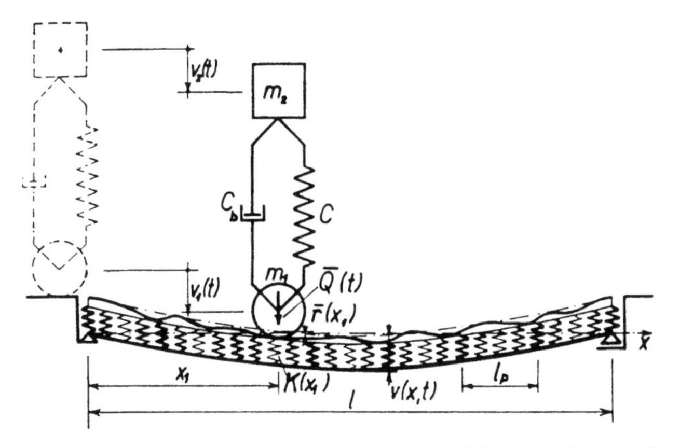

Fig. 8.1. Model of a beam with an elastic layer and irregularities, subjected to a moving system with two degrees of freedom and force $\bar{Q}(t)$.

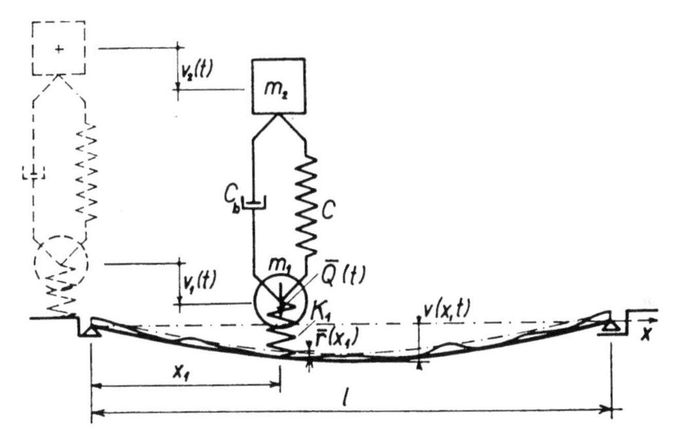

Fig. 8.2. Model of a beam with irregularities, subjected to a moving system with two degrees of freedom and force $\bar{Q}(t)$.

Since the system is assumed to be moving from left to right at constant speed c along the beam, the coordinate of the contact point is

$$x_1 = ct\,. \tag{8.3}$$

2. Consider the unsprung mass to be acted on by a harmonic force, e.g. that expressing the effect of counterweights on the driving wheels of a steam locomotive,

$$\bar{Q}(t) = Q \sin \Omega t \tag{8.4}$$

where

Q — amplitude of the force ,
$\Omega = 2\pi c/O$ — circular frequency of the force ,
O — circumference of the driving wheels.

3. Assume that the top surface of the beam is covered with an elastic layer of variable stiffness

$$K(x) = K_1 + K_2 \cos 2\pi x/l_p \; ; \tag{8.5}$$

the meaning of constants K_1 and K_2 is evident from Fig. 8.3a, l_p is the length equal to the span of longitudinal beams, spacing of sleepers, etc.

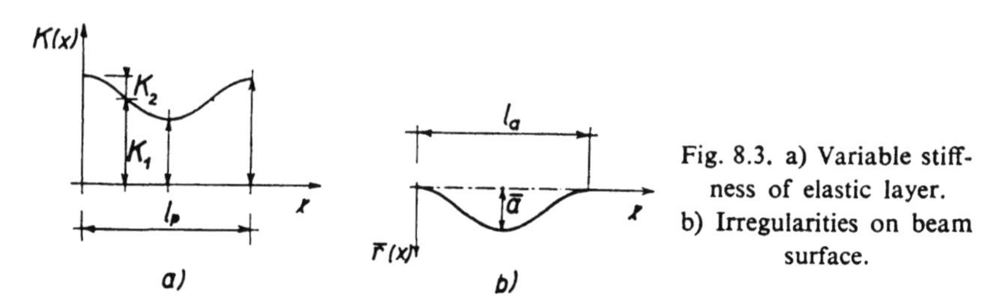

Fig. 8.3. a) Variable stiffness of elastic layer. b) Irregularities on beam surface.

Eq. (8.5) expresses in fact the idealized elastic properties of a roadway which may be expected to vary harmonically along the length of the bridge. This is so, for example, in the "cross-beam effect" known from steel railway bridges, or in the "sleeper spacing effect", etc. If $K_2 = 0$, Eq. (8.5) can also be used for describing other cases met with in practice: thus, e.g. in calculations relating to highway bridges, K_1 is the spring constant of tires, Fig. 8.2.

4. Track irregularities (Figs. 8.1 and 8.2) are assumed to vary harmonically along the bridge span:

$$\bar{r}(x) = \tfrac{1}{2}\bar{a}(1 - \cos 2\pi x/l_a) \tag{8.6}$$

where $\bar{a}$ — maximum depth of track uneveness,
l_a — length of track irregularity (Fig. 8.3b).

Assuming the above we may formulate the problem as a set of three simultaneous differential equations with variable coefficients (because of (8.5), (8.6)) describing respectively the vertical displacements of sprung and unsprung masses, and the beam vibration. Within the interval $0 \leqq t \leqq l/c$, $0 \leqq x \leqq l$ the equations are of the form

$$-m_2 \frac{d^2 v_2(t)}{dt^2} - C[v_2(t) - v_1(t)] - C_b \left[\frac{dv_2(t)}{dt} - \frac{dv_1(t)}{dt}\right] = 0\,, \tag{8.7}$$

$$P + \bar{Q}(t) - m_1 \frac{d^2 v_1(t)}{dt^2} + C[v_2(t) - v_1(t)] + $$

$$+ C_b \left[\frac{dv_2(t)}{dt} - \frac{dv_1(t)}{dt}\right] - \bar{R}(t) = 0\,, \tag{8.8}$$

$$EJ \frac{\partial^4 v(x, t)}{\partial x^4} + \mu \frac{\partial^2 v(x, t)}{\partial t^2} + 2\mu\omega_b \frac{\partial v(x, t)}{\partial t} = \delta(x - x_1)\, \bar{R}(t) \tag{8.9}$$

where

$$\bar{R}(t) = K(x_1)\,[v_1(t) - v(x_1, t) - \bar{r}(x_1)] \geqq 0 \tag{8.10}$$

is the force by which a moving system acts on a beam at the point of contact x_1.

The set of equations (8.7) to (8.9) should satisfy the boundary conditions of a simply supported beam (1.2) and the initial conditions

$$v(x, 0)\,; \qquad v_1(0) = v_{10}\,; \qquad v_2(0) = v_{20}\,;$$

$$\left.\frac{\partial v(x, t)}{\partial t}\right|_{t=0}; \quad \left.\frac{dv_1(t)}{dt}\right|_{t=0} = \dot{v}_{10}\,; \quad \left.\frac{dv_2(t)}{dt}\right|_{t=0} = \dot{v}_{20}\,. \tag{8.11}$$

Eqs. (8.7) to (8.9) are a very general statement of the problem of vibrations excited by a system of masses moving along a beam; hence all the solutions referred to in the introduction to this chapter are but special cases of our formulation.

8.2 *Solution of the problem*

8.2.1 *Dimensionless parameters*

Dimensionless parameters are used to advantage as input data in computer calculations. In our discussion we shall make use of the following:

1. Speed parameter α introduced earlier by Eq. (1.18)

$$\alpha = \frac{c}{2f_{(1)}l} \tag{8.12}$$

where $f_{(1)}$ is the first natural frequency of the unloaded beam [cf. Eq. (1.12)].

2. The ratio between the weights of vehicle and beam

$$\varkappa = \frac{P}{G} \tag{8.13}$$

where $G = \mu g l$ is the total weight of beam.

3. The ratio between the weights of unsprung and sprung vehicle parts

$$\varkappa_0 = \frac{P_1}{P_2}\,. \tag{8.14}$$

4. The frequency parameter of unsprung mass

$$\gamma_1 = \frac{f_1}{f_{(1)}} \tag{8.15}$$

or possibly also

$$\gamma_1' = \frac{K_1}{C_0} \tag{8.16}$$

where f_1 is the natural frequency of unsprung mass

$$f_1 = \frac{1}{2\pi}\left(\frac{K_1}{m_1}\right)^{1/2}, \tag{8.17}$$

C_0 — the total bridge stiffness

$$C_0 = \frac{P}{v_0} = \frac{G\omega_{(1)}^2}{2g}, \tag{8.18}$$

and the deflection at mid-span of the beam loaded with force P at point $x = l/2$ [Eq. (1.21)]

$$v_0 = \frac{2Pg}{G\omega_{(1)}^2} . \tag{8.19}$$

The relation between γ_1 and γ_1'

$$\gamma_1^2 = \frac{1 + \varkappa_0}{2\varkappa\varkappa_0} \gamma_1' . \tag{8.20}$$

5. The frequency parameter of sprung mass

$$\gamma_2 = \frac{f_2}{f_{(1)}} \tag{8.21}$$

or possibly also

$$\gamma_2' = \frac{C}{C_0} \tag{8.22}$$

where the natural frequency of sprung mass

$$f_2 = \frac{1}{2\pi} \left(\frac{C}{m_2} \right)^{1/2} . \tag{8.23}$$

The relation between γ_2 and γ_2'

$$\gamma_2^2 = \frac{1 + \varkappa_0}{2\varkappa} \gamma_2' . \tag{8.24}$$

6. Parameter a expressing the variable stiffness of the elastic layer of the roadway (Fig. 8.3a and Eq. (8.5))

$$a = \frac{K_2}{K_1} . \tag{8.25}$$

7. Parameter b, a function of beam span and length l_p

$$b = \frac{2l}{l_p} . \tag{8.26}$$

8. Parameter a_1, a function of the amplitude of harmonic force [Eq. (8.4)]

$$a_1 = \frac{Q}{P} . \tag{8.27}$$

9. Parameter b_1, a function of the frequency of harmonic force [Eq. (8.4)]

$$b_1 = \frac{2l}{O} . \tag{8.28}$$

10. Parameter a_2 expressing the depth of track uneveness [Eq. (8.6)]

$$a_2 = \frac{\bar{a}}{v_0} . \tag{8.29}$$

11. Parameter b_2, a function of the length of track irregularity [Eq. (8.6)]

$$b_2 = \frac{2l}{l_a} . \tag{8.30}$$

12. The logarithmic decrement of beam damping

$$\vartheta \approx \frac{\omega_b}{f_{(1)}} . \tag{8.31}$$

13. The logarithmic decrement of vehicle spring damping

$$\vartheta_2 \approx \frac{C_b}{2m_2 f_2} . \tag{8.32}$$

14. to 20. The initial parameters

$$\tau_0 \, ; \quad y_{20} \, ; \quad \dot{y}_{20} \, ; \quad y_{10} \, ; \quad \dot{y}_{10} \, ; \quad q_0 \, ; \quad \dot{q}_0 \tag{8.33}$$

to be defined by Eqs. (8.47).

21. The speed at which the system moves along the beam may be such as to bring about resonant vibration owing to the cross-beam effect, harmonic force or track irregularities. In resonant vibration, the frequency of the exciting force is approximately equal to the natural frequency of a beam loaded at point $x = l/2$ by a system with two degrees of freedom. This natural frequency was derived in [76] and [78] on the assumption that only the first normal mode of beam vibration is taken into account and the unsprung mass is in direct contact with the beam at point $x = l/2$ (i.e. that K_1 tends to infinity because in practical cases

K_1 is always much larger than C). Accordingly, the natural frequency of a beam subjected to an immobile system with two degrees of freedom – $\bar{f}_{(1)}$ – has two values

$$\bar{f}_{(1)1,2} = \alpha_{1,2} b^* f_{(1)} \tag{8.34}$$

where

$$\alpha_{1,2} = \frac{1}{b^*}\left[A \mp (A^2 - B)^{1/2}\right]^{1/2}, \tag{8.35}$$

$$A = \frac{1}{2}\left[\frac{1}{(2\varkappa\varkappa_0)/(1+\varkappa_0)+1}(1+\gamma_2') + \gamma_2^2\right],$$

$$B = \frac{1}{(2\varkappa\varkappa_0)/(1+\varkappa_0)+1}\,\gamma_2^2,$$

$$b^* = b \quad \text{or} \quad b_1 \quad \text{or} \quad b_2 .$$

The condition that the exciting force frequency should equal frequency $\bar{f}_{(1)1,2}$ then gives the critical speeds

$$c_{1,2} = \bar{f}_{(1)1,2} l^* = 2\alpha_{1,2} f_{(1)} l \tag{8.36}$$

where $l^* = l_p$ or O or l_a.

The values of speed parameter α_1 and α_2 are computed from Eq. (8.35). In order to find out which of parameters b^* we have to substitute in (8.35), we introduce among the input parameters a handy auxiliary quantity, L, to be used as follows:

$$\text{when} \quad L\begin{cases} =0 \\ >0 \\ <0 \end{cases} \text{one substitutes in (8.35) for} \quad b^* = \begin{cases} b \\ b_1 \\ b_2 \end{cases}. \tag{8.37}$$

If $L = 0$ and $b = 0$, α_1 and α_2 are not computed at all.

22. to 23. If we intend to obtain a solution of our problem for α equal to α_1 or α_2 or the multiples thereof, we introduce yet two other parameters, k_1 and k_2, to be used as follows in the computation of α:

$$\alpha = \begin{cases} k_1\alpha_1 & \text{for} \quad k_1 \neq 0 \\ k_2\alpha_2 & \text{for} \quad k_2 \neq 0, \quad k_1 = 0 . \end{cases} \tag{8.38}$$

If $k_1 = k_2 = 0$, then parameter α must of course be given among the input parameters.

24. The last of our input parameters is the length of the integration step, h, which we shall discuss in detail in paragraph 8.2.3.

Constants 1. to 24. form a set of input data for computer calculations and uniquely specify the problem to be solved. An example set of input parameters is in Table 8.1.

8.2.2 *Reduction of the equations to the dimensionless form*

For the purposes of numerical computations we shall introduce the dimensionless independent variables

$$\xi = x/l\,; \quad \tau = \pi c t/l \tag{8.39}$$

and dependent variables

$$\begin{aligned} y(\xi, \tau) &= v(x, t)/v_0\,, \\ y_i(\tau) &= v_i(t)/v_0\,, \quad i = 1, 2 \end{aligned} \tag{8.40}$$

where v_0 is defined by Eq. (8.19).

Because of the second of expressions (1.9), the boundary conditions of a simply supported beam – (1.2) – are satisfied by the function

$$y(\xi, \tau) = \sum_{j=1}^{\infty} q_{(j)}(\tau) \sin j\pi\xi\,. \tag{8.41}$$

As demonstrated in the case of a constant moving force (Sect. 1.2) and in the case of a harmonic force (Sect. 2.1), series (8.41) converges at a rapid rate.

That is why – when dealing with large-span bridges where the moving force $\bar{R}(t)$ differs but slightly from the above two cases – it will be sufficient to consider only the first term of series (8,41), $q_{(1)}(\tau) = q(\tau)$, and write

$$y(\xi, \tau) \approx q(\tau) \sin \pi\xi\,. \tag{8.42}$$

Substitution of (8.12) to (8.42) in (8.7) to (8.10) and some manipulation will lead to a set of ordinary differential equations of the second order expressed in terms of the dimensionless parameters. Eq. (8.9) will again be solved by the method of finite Fourier (sine) integral transformations in accordance with (1.9), and because of (8.40) to (8.42) only the first term, $j = 1$, will be retained. In the interval $0 \leqq \tau \leqq \pi$, Eqs. (8.7) to (8.9) will then take on the form

$$\ddot{y}_2(\tau) = \frac{1}{\alpha^2} \gamma_2^2 [y_1(\tau) - y_2(\tau)] + \frac{\vartheta_2}{\pi\alpha} \gamma_2 [\dot{y}_1(\tau) - \dot{y}_2(\tau)] ,$$

$$\ddot{y}_1(\tau) = \frac{1}{\alpha^2} \frac{1 + \varkappa_0}{2\varkappa\varkappa_0} [1 + Q(\tau) - R(\tau)] - \frac{1}{\alpha^2} \frac{\gamma_2^2}{\varkappa_0} [y_1(\tau) - y_2(\tau)] - $$
$$- \frac{\vartheta_2}{\pi\alpha} \frac{\gamma_2}{\varkappa_0^{1/2}} [\dot{y}_1(\tau) - \dot{y}_2(\tau)] ,$$

$$\ddot{q}(\tau) = \frac{1}{\alpha^2} R(\tau) \sin \tau - \frac{1}{\alpha^2} q(\tau) - \frac{\vartheta}{\pi\alpha} \dot{q}(\tau) \tag{8.43}$$

where

$$R(\tau) = \frac{\bar{R}(t)}{P} = \gamma_1'(1 + a \cos b\tau) [y_1(\tau) - q(\tau) \sin \tau - r(\tau)] \geqq 0 \tag{8.44}$$

is the dimensionless force according to (8.10), by which a moving system acts on a beam at the point of contact,

$$Q(\tau) = \frac{\bar{Q}(t)}{P} = a_1 \sin b_1 \tau \tag{8.45}$$

the dimensionless force (8.4), and

$$r(\tau) = \frac{\bar{r}(x_1)}{v_0} = \tfrac{1}{2} a_2 (1 - \cos b_2 \tau) \tag{8.46}$$

the dimensionless ordinate of track irregularities (8.6).

The initial conditions are

$$q(0) = q_0 ; \quad \left.\frac{\mathrm{d}q(\tau)}{\mathrm{d}\tau}\right|_{\tau=0} = \dot{q}_0 ,$$
$$y_i(0) = y_{i0} ; \quad \left.\frac{\mathrm{d}y_i(\tau)}{\mathrm{d}\tau}\right|_{\tau=0} = \dot{y}_{i0} ; \quad i = 1, 2 . \tag{8.47}$$

The initial conditions $v(x, 0)$ and $\partial v(x, t)/\partial t|_{t=0}$ [cf. (8.11)] are assumed to be expandable in Fourier series of which only the first term is taken into account. In the dimensionless form this first term is q_0 or $\dot{q}_0$.

8.2.3 *Numerical solution*

In the numerical solution we made use of Runge-Kutta-Nyström's method [37] for which Eqs. (8.43) are already suitably arranged. The procedure was programmed for an Ural 2 computer and later on also for a Leo 360 computer.

The solution starts with the printing of input data (see the examplary set in Table 8.1) and the calculation and printing of α_1 and α_2. Following this, Runge-Kutta-Nyström's method is applied to the computation of the various functional values in the specified time steps. The computed values of

$$\tau\,, \quad y_2(\tau)\,, \quad y_1(\tau)\,, \quad q(\tau)\,, \quad R(\tau) \tag{8.48}$$

are printed at every N-th step, the number N having been chosen in advance. If $N = 0$, the intermediate results are not printed. At the end of the computation the machine also prints the largest values of q and R, i.e. max q and max R, and the values of τ at which they occur.

To verify the correctness of the computer programme we made check calculations of an example (described in [78]) using an electric desk calculator and the methods of Runge-Kutta-Nyström, Taylor's series and linear acceleration with iterations. The first step computed via these three methods was found to be in excellent agreement with the computer results. This is a proof that Runge-Kutta-Nyström's method may be considered well suited for the solution of problems of the sort discussed.

The choice of the integration step length, h, was made by Collatz's method [36] adapted to our case (cf. [78]). The results obtained at steps $h = \pi/1000$ or $h = \pi/2000$ chosen in that way were accurate enough.

The errors of the numerical solution may be found by running through the computation twice: first, at integration step h giving the approximate values $\tilde{y}_h$, then at step $h^* = \nu h$ giving $\tilde{y}_{h^*}$. In accordance with [78] and [37] the corrected solution then is

$$y = \tilde{y}_h + \frac{1}{\nu^n - 1}(\tilde{y}_h - \tilde{y}_{h^*}) = \tilde{y}_{h^*} + \frac{\nu^n}{1 - \nu^n}(\tilde{y}_{h^*} - \tilde{y}_h) \tag{8.49}$$

where n is the number of terms of the Taylor's series used by the numerical method. In the case of Runge-Kutta-Nyström's method $n = 4$. In conventional cases the error $|y - \tilde{y}|$ established by means of (8.49) was less than 0·005, which means that the results were correct to two decimal places at least.

Detailed information on the calculation of the check example, choice of the integration steps and estimate of errors in numerical evaluations may be found in [78].

8.3 *The effect of various parameters*

Using the set of input data shown in Table 8.1 we next made a study of the effect of some of the dimensionless parameters on the maximum value of $q(\tau)$, i.e. on the maximum beam deflection. The input data of Table 8.1 approximately correspond to the parameters of a 50 m-span steel railway bridge traversed by an electric or a Diesel-electric locomotive.

Table 8.1 Set of input parameters

$\tau_0 = 0$	$y_{20} = \frac{1}{13}$	$\dot{y}_{20} = 0$	$y_{10} = \frac{1}{13}$	$\dot{y}_{10} = 0$	$q_0 = 0$
$\dot{q}_0 = 0$	$a \;\; = 0{\cdot}3$	$\gamma_1' \;\; = 10$	$b \;\; = 20$	$\gamma_2' \;\; = 0{\cdot}2$	$a_1 = 0$
$a_2 = 0$	$b_1 \;\; = 0$	$b_2 \;\; = 0$	$\varkappa \;\; = 0{\cdot}5$	$\varkappa_0 \;\; = 0{\cdot}25$	$\vartheta \;\; = 0{\cdot}08$
$\vartheta_2 = 0{\cdot}5$	$k_1 \;\; = 0$	$k_2 \;\; = 0$	$\alpha \;\; = 0{\cdot}12$	$h \;\; = \pi/1000$	$L \;\; = 0$

In our study we varied only those parameters whose effect we wished to establish, and kept the other ones constant. Since the values of the dimensionless force $R(\tau)$ at the initial point were for the most equal to unity ($R_0 = R(0) = 1$), Eq. (8.44) resulted in $y_{10} = y_{20} = 1/[\gamma_1'(1 + a)] =$ $= 1/13$.

In the Figures referred to in our discussion we denote the maximum value of $q(\tau)$ by δ, i.e.

$$\delta = \max q(\tau) \tag{8.50}$$

and call it the dynamic coefficient.

8.3.1 *The effect of speed*

The effect of speed is described by parameter α [cf. (8.12)]; the $\delta - \alpha$ relation for $0 \leqq \alpha \leqq 0{\cdot}12$ growing in steps of $0{\cdot}005$ is drawn in Fig. 8.4. The diagram displays a generally ascending tendency with a number of

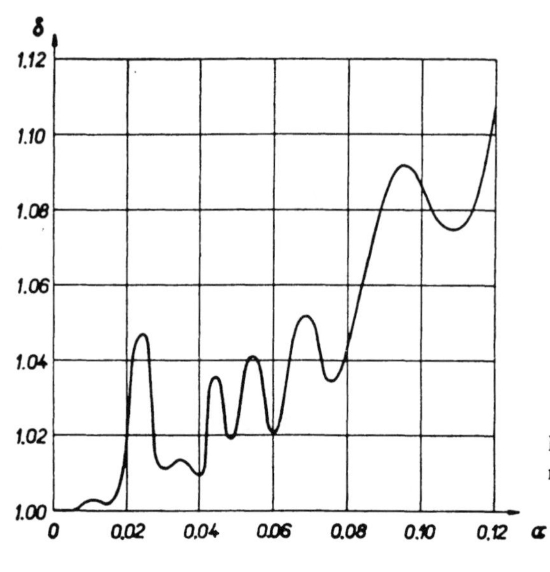

Fig. 8.4. Effect of speed parameter α (for input data set out in Table 8.1).

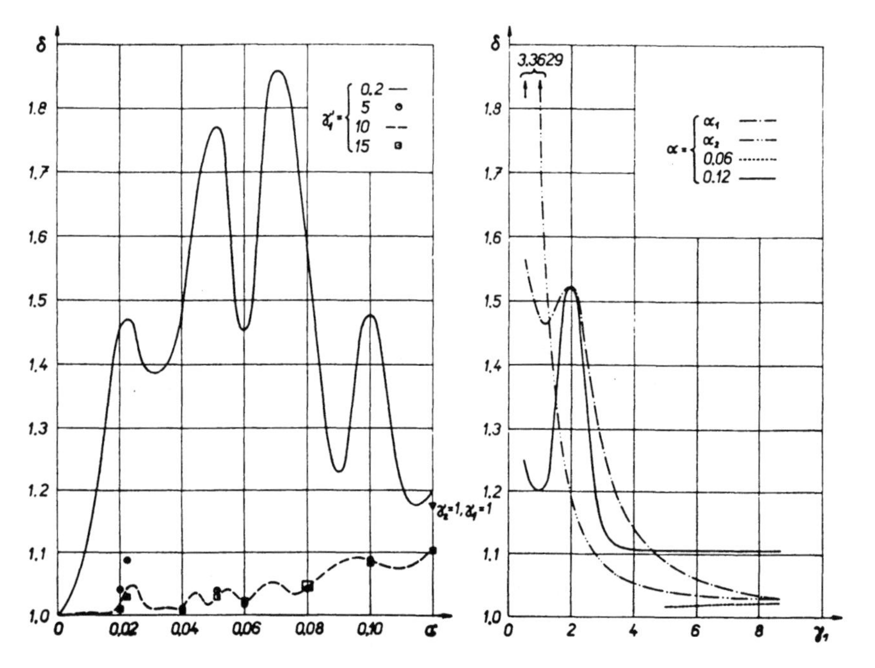

Fig. 8.5. Effect of the frequency parameter of unsprung mass, γ_1, (for input data set out in Table 8.1).

local peaks. One of the first, well defined peaks approximately corresponds to $\alpha = \alpha_1 = 0{\cdot}02225$, while at $\alpha = \alpha_2 = 0{\cdot}05128$ there is hardly any peak at all. The reason for this may be found in that the system under study is very complex compared to the case of a moving force (Fig. 1.3); for further details the reader is referred to paragraph 8.3.4.

8.3.2 *The effect of the frequency parameter of unsprung mass*

The dependence of δ on γ_1, the frequency parameter of unsprung mass [Eq. (8.15)], is illustrated in Fig. 8.5 for different values of α and $R_0 = 1$. As γ_1 falls off, δ grows larger very quickly. The greatest effects take place in the neighbourhood of $0 \leqq \gamma_1 \leqq 2$; at $\gamma_1 > 5$ and the same α, however, the dynamic effects are nearly constant. This means that a very hard elastic layer on the beam surface has virtually the same effect as an infinitely rigid layer (i.e. the case of unsprung mass in direct contact with the beam).

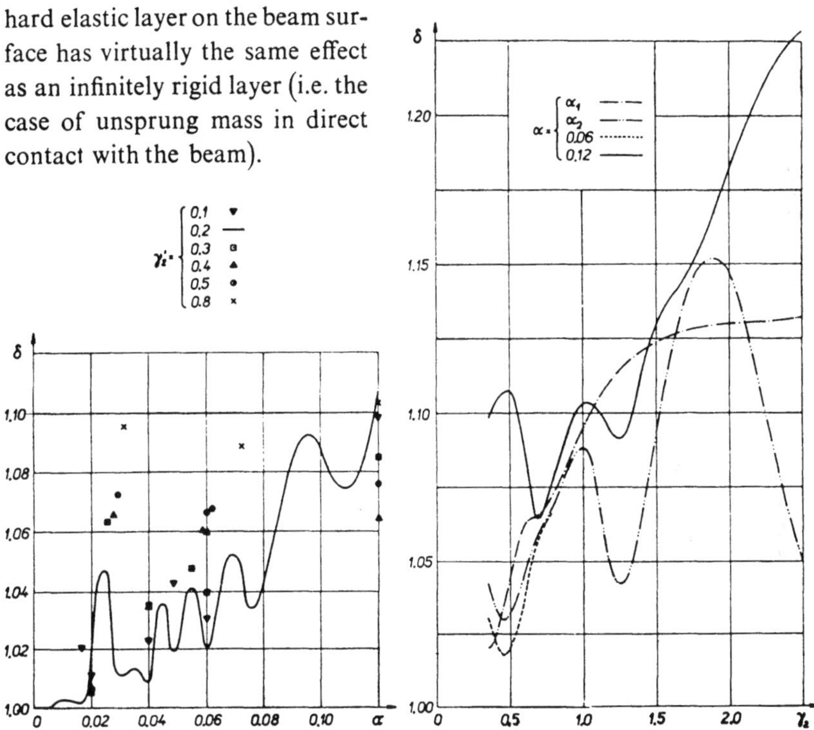

Fig. 8.6. Effect of the frequency parameter of sprung mass, γ_2, (for input data set out in Table 8.1).

8.3.3 *The effect of the frequency parameter of sprung mass*

According to (8.21), the effect of the frequency parameter of sprung mass is characterized by parameter γ_2. The dependence of δ on γ_2 for different α's is shown in Fig. 8.6 which indicates that the dynamic effects have a tendency to grow (at $a \neq 0$, $R_0 = 1$) with growing γ_2. The diagram again displays a number of local peaks.

8.3.4 *The effect of variable stiffness of the elastic layer*

The relation between δ and a, the variable stiffness of the elastic layer [Eq. (8.25)] is plotted in Fig. 8.7 for various values of α and $R_0 = 1$. Comparing the curve for $a = 0$ (constant stiffness of the elastic layer) with that for $a = 0{\cdot}3$ we see that both show a tendency to rise with growing α,

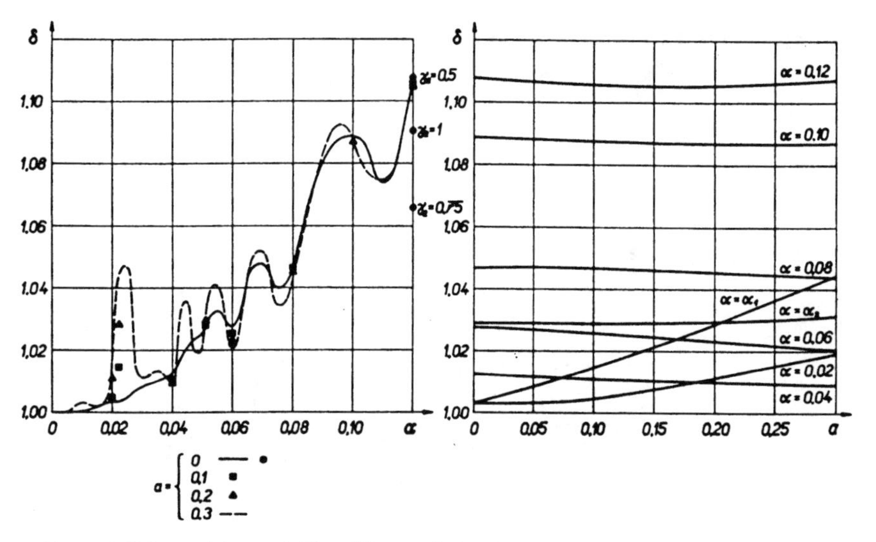

Fig. 8.7. Effect of the variable stiffness of elastic layer, a, (for input data set out in Table 8.1)

and the latter ($a = 0{\cdot}3$) displays moreover a number of marked local peaks, particularly in the neighbourhood of $\alpha = \alpha_1 = 0{\cdot}02225$. At constant α (with the exception of low values), δ is virtually independent of parameter a. At $\alpha = \alpha_1$, δ is nearly directly proportional to a.

We are now in a position to explain the occurrence of local peaks in the $\delta - \alpha$ relation to which we have referred in paragraph 8.3.1. Its primary cause is the motion of a system with two degrees of freedom along a beam at $a = 0$. If parameter $a \neq 0$, the peaks are more markedly defined, particularly when $\alpha = \alpha_1$.

8.3.5 *The effect of the ratio between the weights of vehicle and beam*

The dependence of δ on parameter $\varkappa$ [Eq. (8.13)] at various values of α is shown in Fig. 8.8. At constant α, the $\delta - \varkappa$ relation is fairly complicated, but generally speaking, δ grows with growing $\varkappa$. This is particularly noticeable at high speeds.

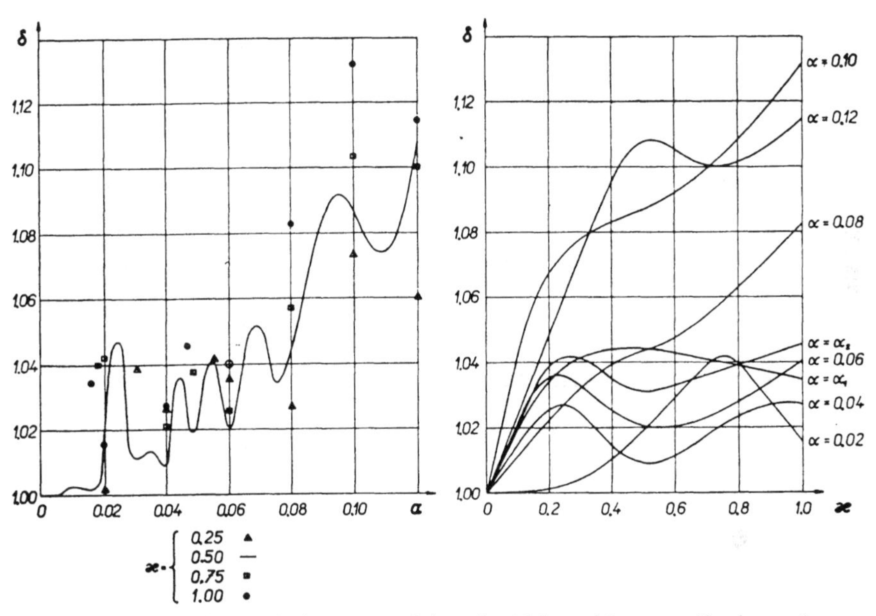

Fig. 8.8. Effect of the ratio between weights of vehicle and beam, $\varkappa$ (for input data set out in Table 8.1).

8.3.6 *The effect of the ratio between the weights of unsprung and sprung parts of vehicle*

The dependence of δ on parameter $\varkappa_0$ [Eq. (8.14)] at various values of α is in Fig. 8.9. Judging by the general pattern of the diagram, at constant speeds parameter $\varkappa_0$ bears but a small effect on the dynamic coefficient.

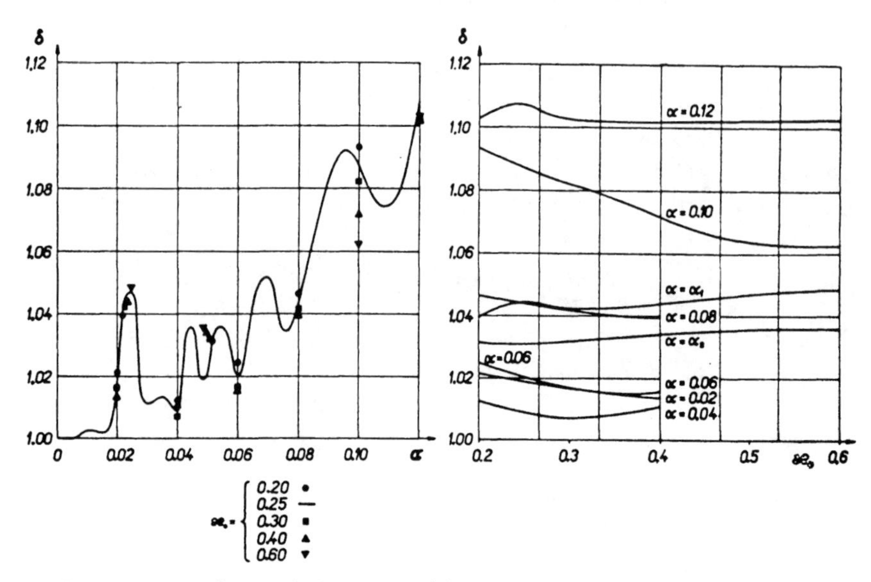

Fig. 8.9. Effect of the ratio between weights of unsprung and sprung vehicle parts, $\varkappa_0$ (for input data set out in Table 8.1).

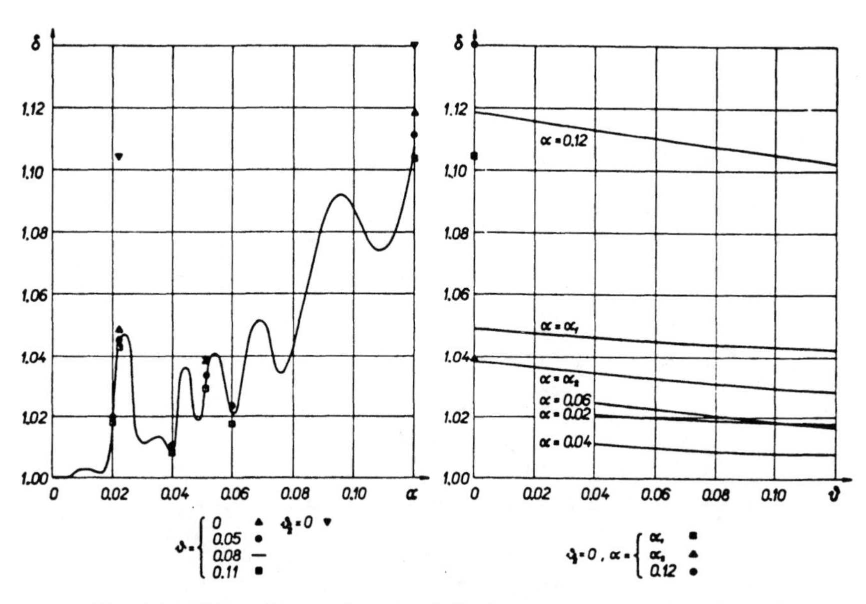

Fig. 8.10. Effect of beam damping ϑ (for input data set out in Table 8.1).

8.3.7 *The effect of beam damping*

The effect of the logarithmic decrement of beam damping, ϑ, [Eq. (8.31)] is shown in Fig. 8.10. It is plain to see that the dynamic coefficients, δ, gradually fall off with growing ϑ. The figure also shows the case of $\vartheta_2 = 0$.

8.3.8 *The effect of vehicle spring damping*

The effect of the logarithmic decrement of vehicle spring damping, ϑ_2, is indicated in Fig. 8.11. Even though the dynamic coefficients δ fall off with growing ϑ_2 in most cases, they have an interesting tendency to rise at some speeds. The figure also shows the case of $\vartheta = 0$.

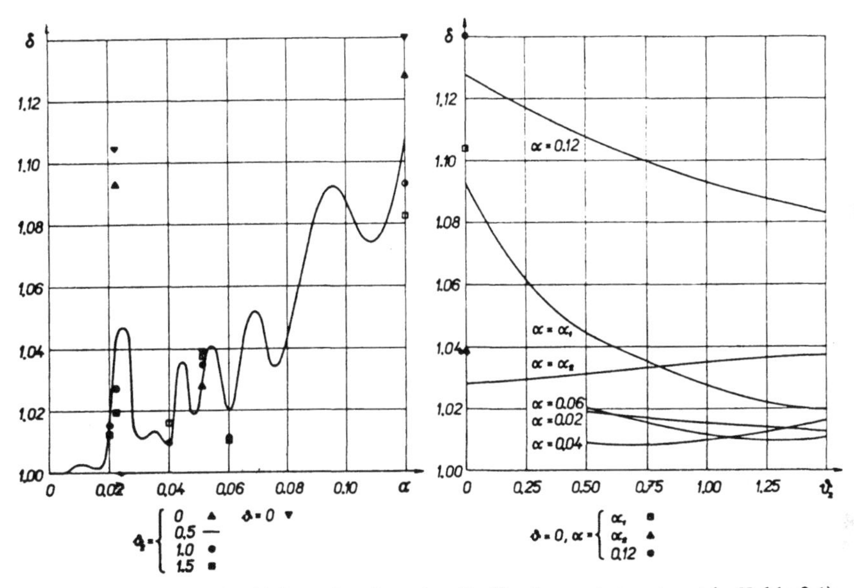

Fig. 8.11. Effect of vehicle spring damping ϑ_2 (for input data set out in Table 8.1).

8.3.9 *The effect of initial conditions*

A detailed analysis of the effect of initial conditions on dynamic stresses in a beam is presented in [78]. In the calculations contained there the beam was always assumed to be at rest, i.e. $q_0 = 0$, $\dot{q}_0 = 0$, at the instant the moving system started to traverse it. The question of the effect of initial conditions of the vehicle became one of primary importance because in traversing the beam, a vibrating system with two degrees of freedom imparts to it a portion of its kinetic energy.

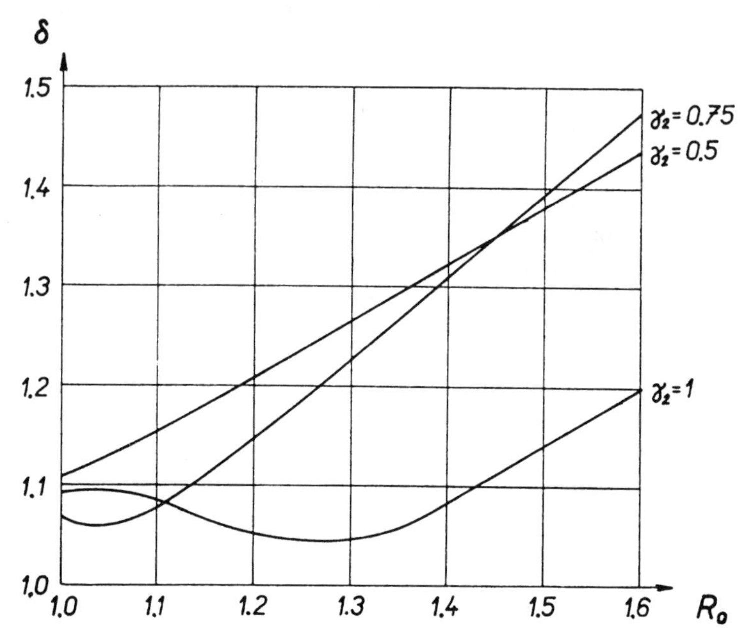

Fig. 8.12. Effect of initial force R_0 for various values of γ_2, $a = 0$, $\alpha = 0{\cdot}12$ (the remaining input data are as set out in Table 8.1).

In paragraphs 8.3.1 to 8.3.8 we have assumed that R_0 was always unity. Let us now examine the effect of initial force $R_0 = R(0)$ on beam stresses. The results of such an examination made for $\alpha = 0{\cdot}12$, $a = 0$, $\gamma_2 = 0{\cdot}5$, 0·75 and 1 are presented in Fig. 8.12. As the figure indicates, at fairly large R_0's the dynamic coefficients grow proportional to the initial force and attain high values. This is a very important finding inasmuch as it tells us that dynamic stresses in a beam are affected to a substantial degree by the amplitude of vibration of the sprung parts of the vehicle. This finding was confirmed by a study in which we investigated the effect of a single track irregularity placed ahead of the entrance to the beam, the effect of harmonic motion and its initial phases, of the sprung mass. Details of the investigation may be found in [78].

8.3.10 *The effect of other parameters*

We have also made a study of the effect of some of the remaining parameters. Thus, for example, we established the dependence of δ on b, and found that at constant speeds (with the exception of low values of α) the ratio of beam span to longitudinal beam length has but a slight effect on δ, see Fig. 8.13. However, according to (8.35), α_1 and α_2 fall off with increasing b.

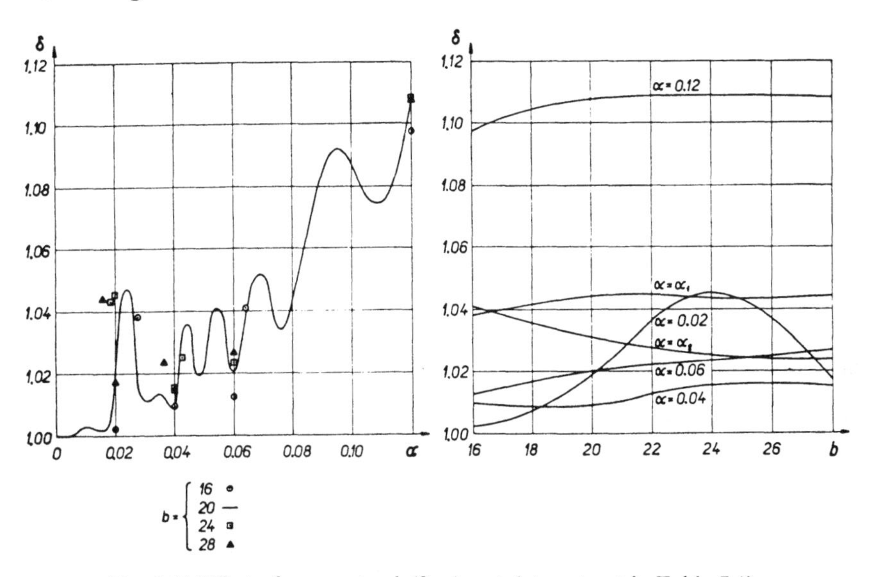

Fig. 8.13 Effect of parameter b (for input data set out in Table 7.1).

Next we examined the question of which of the two – variable stiffness of an elastic layer without irregularities ($a \neq 0$, $a_2 = 0$) or constant stiffness of an elastic layer with irregularities ($a = 0$, $a_2 \neq 0$) – has a larger effect on beam stresses. In the second case the irregularity depth a_2 was calculated from the condition that it should be equal to the difference between the maximum and the minimum static deflections of the elastic layer in the first case, divided by deflection v_0. Comparing the results of our examination (Fig. 8.14) we see that the two cases are nearly identical.

Fig. 8.15 is a graphical summary of our study involving the effect of all possible combinations of the following parameters:

$$R_0 = 1{\cdot}2,\ 1{\cdot}4\,, \qquad \varkappa = 0{\cdot}5,\ 1\,,$$
$$\gamma_2 = 0{\cdot}5,\ 0{\cdot}75,\ 1\,, \qquad \varkappa_0 = 0{\cdot}25,\ 0{\cdot}4\,.$$

As the figure suggests, δ falls off with growing γ_2 $(a = 0,\ R_0 \neq 1)$ – a result wholly contrary to that obtained in paragraph 8.3.3 and Fig. 8.6 for $a \neq 0$, $R_0 = 1$. Further, the dynamic effects are more pronounced at large $\varkappa_0$ and small $\varkappa$ than at small $\varkappa_0$ and large $\varkappa$. As in paragraph 8.3.9, here, too, an increase in R_0 brings about an increase in δ.

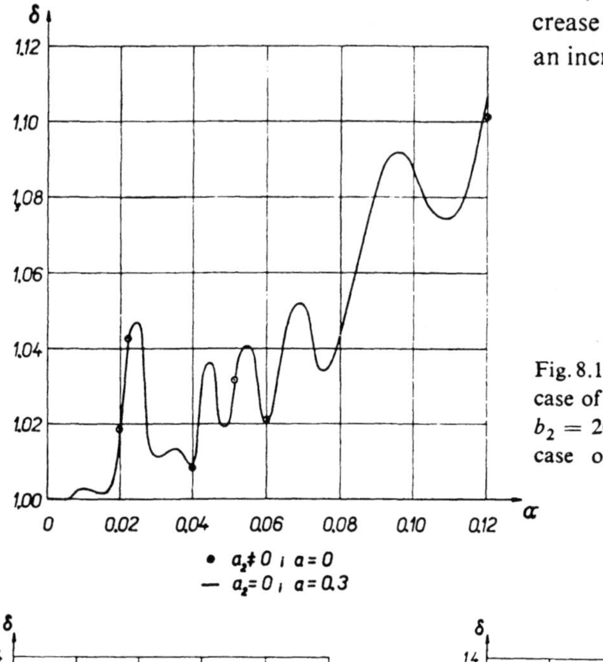

Fig. 8.14. Comparison of the case of $a_2 = 0{\cdot}065\,934\,066$, $b_2 = 20$, $a = 0$, with the case of $a = 0{\cdot}3$, $b = 20$, $a_2 = 0$.

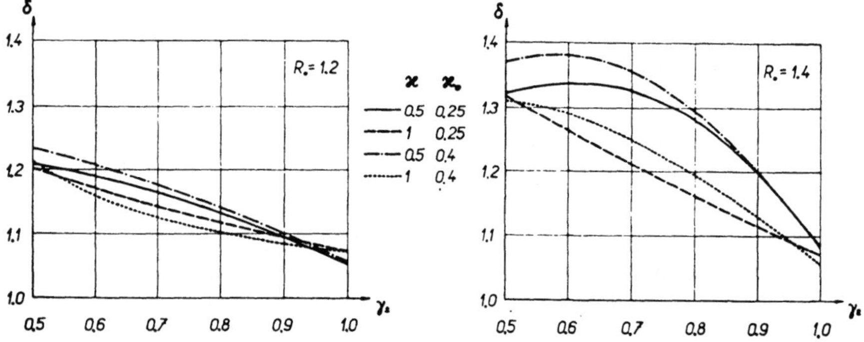

Fig. 8.15. Effect of the combination of parameters R_0, γ_2, $\varkappa$, $\varkappa_0$. $a = 0$, $\alpha = 0{\cdot}12$ and the remaining input data are as set out in Table 8.1.

8.4 *Application of the theory*

8.4.1 *Comparison of theory with experiments*

The theory evolved in the foregoing section was applied to dynamic calculations of large-span steel railway bridges (for details refer to [78]).

The effects of a steam two-cylinder locomotive type 524.1 on stresses in a 56·56 m-span bridge were already examined in paragraph 2.3.1. Fig. 2.3 shows the deflection at mid-span of the bridge for three different speeds of the locomotive, obtained through the application of our theory and a computer. The corresponding experimental records (Fig. 2.4) show a very good agreement with the theory. The largest dynamic coefficients appertaining to the different speeds are plotted in Fig. 2.5 together with the theoretical resonance curve, the result of computer calculations and the application of the approximate theory discussed in Chap. 2.

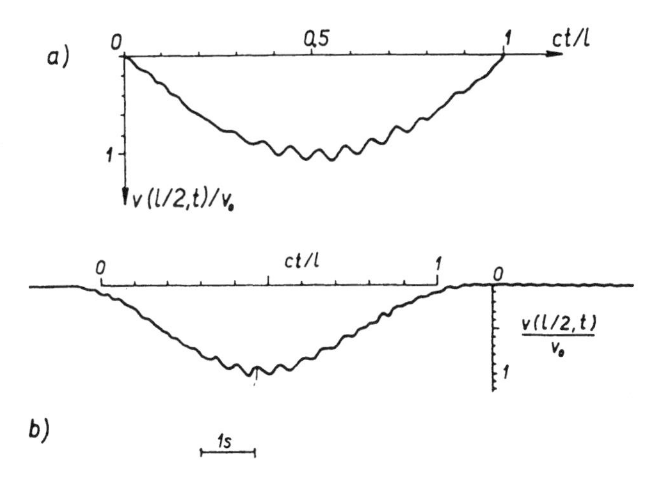

Fig. 8.16. Deflection at the centre of a beam with span $l = 56{\cdot}56$ m, traversed by a Diesel-electric locomotive type T 658.0 at speed $c = 33{\cdot}3$ km/h. a) theory, b) experiment.

The bridge was also tested with a Diesel-electric locomotive type T 658.0 traversing it. A comparison of the theoretical and experimental deflections at mid-span of the bridge at 33·3 km/h is in Fig. 8.16, a diagram plotting the dynamic coefficients, δ, and the frequency of forced vibra-

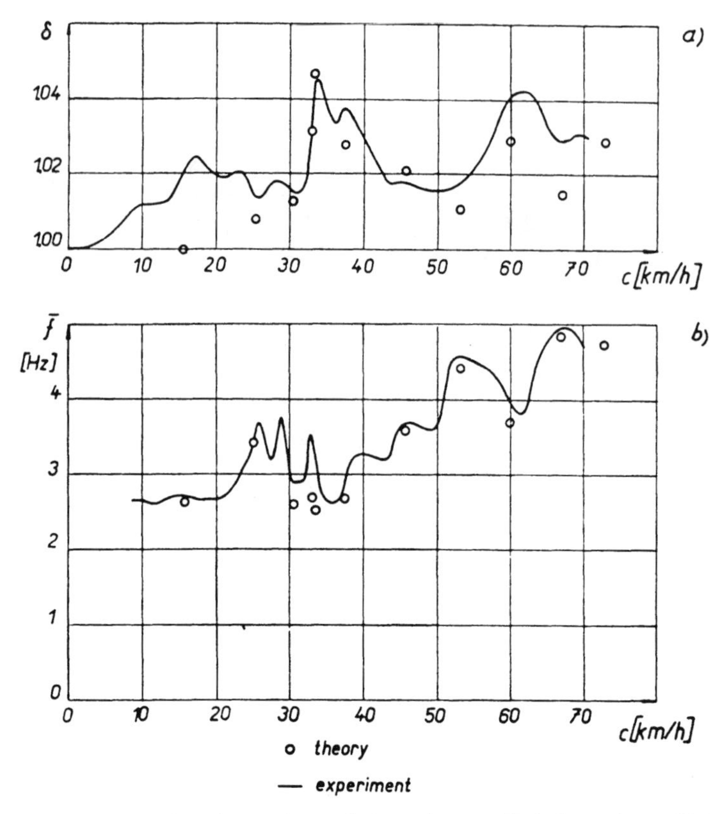

Fig. 8.17. Theoretical and experimental dependences of a) dynamic coefficient δ, b) frequency of forced vibration $\bar{f}$, on speed c. Bridge with span $l = 56{\cdot}56$ m, Diesel-electric locomotive T 658.0.

tions, $\bar{f}$, versus speed, in Fig. 8.17*). As both figures suggest, there is a reasonable agreement between the theory and experiments.

Next came tests of an electric locomotive type E 469.1 traversing a 34·8 m-span bridge. Fig. 8.18 shows a comparison of the computed and measured deflections at mid-span of the bridge at 40·7 km/h, Fig. 8.19*) the dynamic coefficients and frequencies of forced vibrations, $\bar{f}$, at different locomotive speeds. Again, the agreement is quite satisfactory.

*) The experimental dependences shown in Figs. 8.17 and 8.19 were obtained from measured data by the method of sliding means, see: A. M. Dlin: Mathematical Statistics in Engineering (Matematicheskaya statistika v tekhnike). 3rd edition, Sovetskaya nauka, Moscow, 1958.

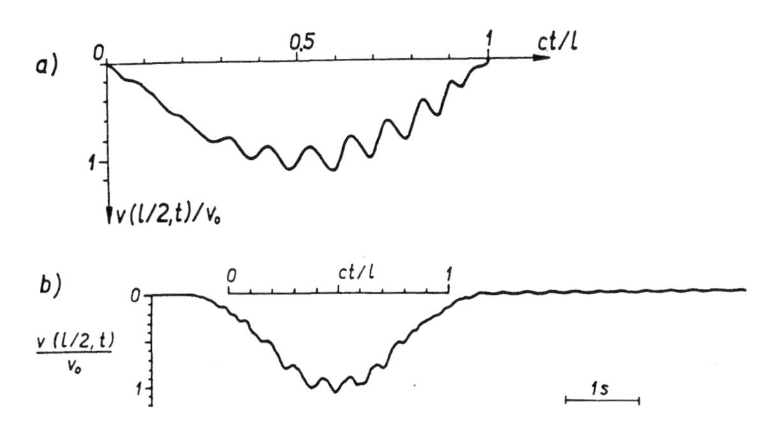

Fig. 8.18. Deflection at the centre of a beam with span $l = 34{\cdot}8$ m, traversed by an electric locomotive type E 469.1 at speed $c = 40{\cdot}7$ km/h. a) theory, b) experiment.

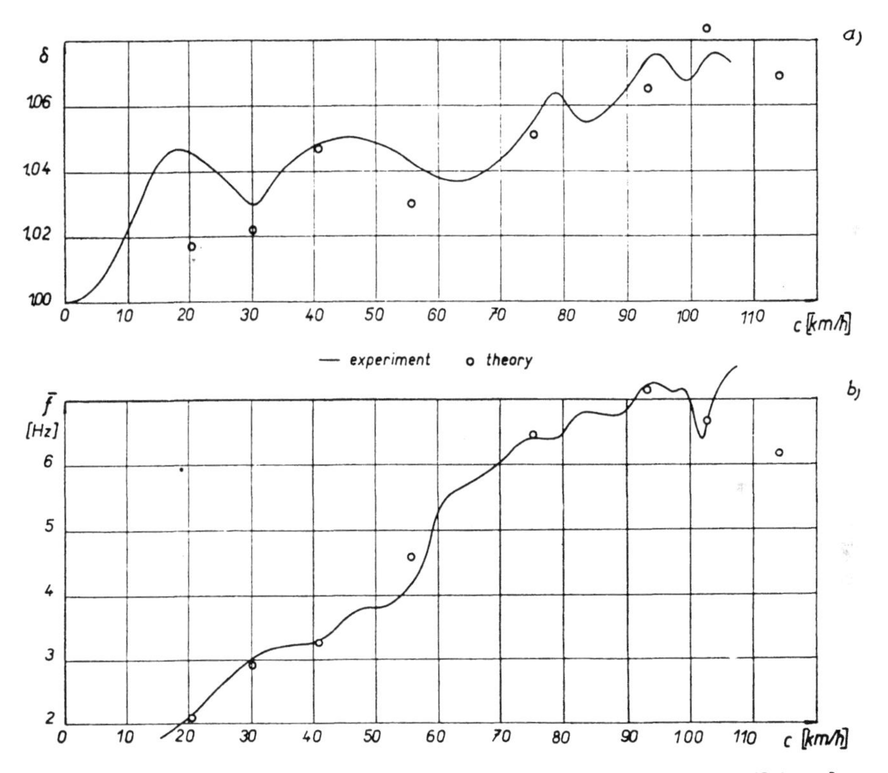

Fig. 8.19. Theoretical and experimental dependences of a) dynamic coefficient δ, b) frequency of forced vibration, $\bar{f}$, on speed c. Bridge with span $l = 34{\cdot}8$ m, electric locomotive type E 469.1.

A detailed report on a total of 11 steel railway bridges with spans ranging from 27 to 70 m, tested in verification of our theory can be found in [78].

8.4.2 *Dynamic stresses in large-span railway bridges*

It was found during the experimental verification of the theory that the calculation procedure as outlined is well suited for bridges with spans over 30 or 40 m, simply supported as has been conventional of late. In such cases the vehicle length can usually be neglected against the bridge span.

The variable quantity that most influences dynamic stresses in bridges of that sort is the vehicle speed. Speaking generally, high vehicle speeds always produce an increase in maximum bridge deflection and stresses. Second in importance are the cross beam effect, uniform sleeper spacing and other regular unevenesses that enlarge the local peaks in the dynamic coefficient – speed diagram. This holds true about the new types of locomotives with bogies. As to steam locomotives: the predominant cause of dynamic stresses in bridges produced by these machines are the unbalanced counterweights on their driving wheels. More often than not they are responsible for the single well-defined peak that appears in the resonance curve at critical speed. This holds true especially for two-cylinder units while multi-cylinder engines have but slight periodic forces.

The dynamic effects of railway vehicles grow larger approximately proportional to the frequency of sprung masses and vehicle weight. In large-span bridges a role is also played by the initial conditions of motion as the vehicle starts to traverse the bridge. That is why dynamic stresses in bridges are also affected by the state of track ahead of the entrance to the bridge.

The theory works equally well for large-span railway bridges as for highway bridges built of steel, reinforced or prestressed concrete.

8.5 ***Additional bibliography***

[1, 7, 15, 16, 19, 21, 25, 29, 34, 45, 65, 76, 77, 78, 80, 81, 82, 85, 91, 92, 106, 107, 111, 115, 116, 117, 130, 131, 142, 145, 156, 162, 166, 192, 196, 197, 206, 208, 224, 242, 245, 246, 258, 261, 271].

9

Beam subjected to a moving two-axle system

If the vehicle axle base is comparable with the beam span, it cannot be neglected as done in Chap. 8 but must be included in the analysis. The problem of motion of a two-axle vehicle along a beam was first treated by R. K. Wen [230] whose solution was based on the assumption that the vehicle wheels were continually in contact with the roadway. However, owing to track irregularities the contact between wheels and roadway is easily lost and an impact follows shortly thereafter.*)

Since it is precisely impacts that produce the highest dynamic effects in short-span bridges which we are going to analyze presently, we must also consider the impacts that arise between the moving system and the beam. In formulating the problem we shall take advantage of some of the findings established in Chap. 8, e.g. of those specifying that the effect of variable stiffness of the roadway may be replaced by the effect of track irregularities, and the harmonic force need not be considered, because in practical cases its frequency can never equal the frequency of a loaded short-span bridge. The first natural frequencies of short-span bridges are in fact fairly high.

9.1 *Formulation of the problem*

We shall consider the mechanical model schematically shown in Figs. 9.1 and 9.2, and make the following assumptions (next to assumptions 1, 3, 4 and 5 of Sect. 1.1):

1. The moving vehicle is a system with four degrees of freedom; its two unsprung masses m_i, $i = 1, 2$, perform vertical motion $v_i(t)$; its sprung mass m_3 is able to move vertically – $v_3(t)$ – as well as rotate –

*) Impact is here defined as collision of two moving bodies.

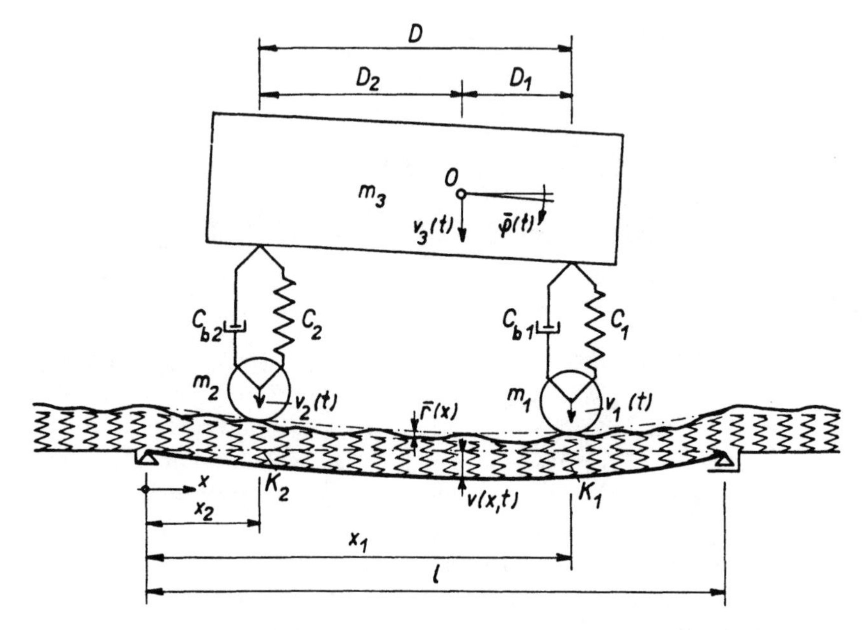

Fig. 9.1. Model of a beam with an elastic layer and irregularities. The beam is loaded with a moving system with four degrees of freedom.

$\bar{\varphi}(t)$. Quantity $\bar{\varphi}(t)$ is measured in the clockwise direction from the horizontal, $v_3(t)$ from the equilibrium position in which springs C_i are compressed by weight m_3, and $v_i(t)$ from the position in which springs K_i are undeformed. In order that our discussion might take care of both railway (Fig. 9.1) and highway (Fig. 9.2) vehicles, the linear vehicle springs have generally different spring constants C_i and coefficients of viscous damping C_{bi} for the two axles. The total weight of the vehicle with mass m is

$$P = P_1 + P_2 + P_3 = mg \tag{9.1}$$

where $P_i \;= m_i g$ — weight of unsprung parts of vehicle, $i = 1, 2$,

$P_3 = m_3 g = P_{31} + P_{32}$ — weight of sprung part of vehicle,

$P_{3i} = P_3 D_{\bar{i}}/D$ — weight of sprung part of vehicle per the i-th axle,

$$\bar{i} = \begin{cases} 1 \\ 2 \end{cases} \text{for } \; i = \begin{cases} 2 \\ 1 \end{cases}$$

D_i — horizontal distance between the centroid of sprung mass and unsprung mass m_i,

$D = D_1 + D_2$ — vehicle axle base.

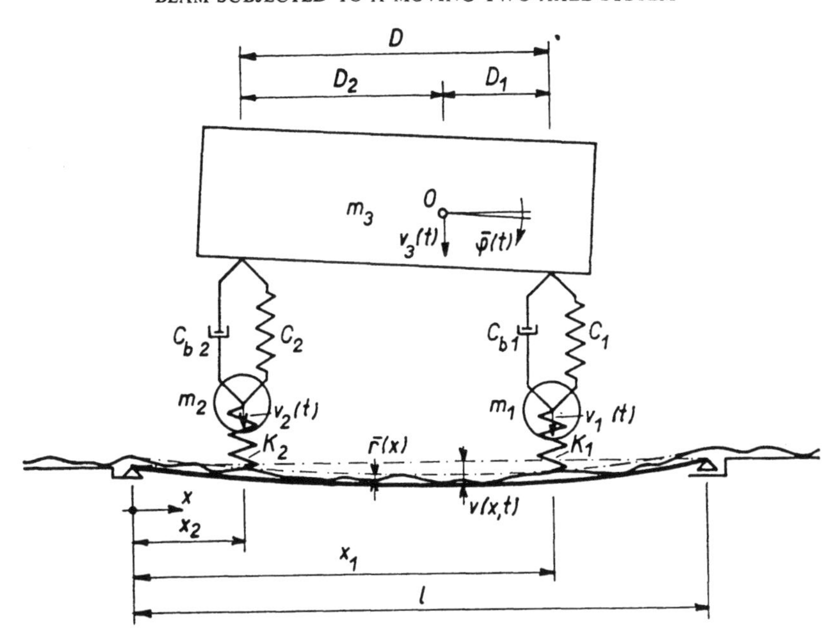

Fig. 9.2. Model of a beam with irregularities loaded with a moving system with four degrees of freedom.

The mass moment of inertia of the sprung part of the vehicle relative to the principal gravity axis O (horizontal transverse axis of vehicle – Figs. 9.1 and 9.2) is I(*).

2. There is on the beam surface an elastic layer with spring constant K_i underneath the i-th axle (Fig. 9.1). In the case of highway vehicles constants K_i represent the spring constants of tires, generally different underneath the two axles. However, unlike in Chap. 8, K_i are here considered constant along the beam length.

3. The track surface contains irregularities $\bar{r}_i(x)$, again generally different underneath the two axles. This makes it possible to describe the various types of track or wheel irregularities, such as flat wheels of railway vehicles, isolated unevenesses in the roadway, wavy surface of the whole roadway, etc. (Fig. 9.3).

*) If the sprung part of the vehicle is a homogeneous right parallelepiped with length L, height H and mass m_3, its mass moment of inertia is

$$I = m_3(L^2 + H^2)/12\,.$$

4. Since the vehicle moves along the beam at uniform speed c, the coordinates of the contact points are

$$x_1 = ct\,, \quad x_2 = ct - D \tag{9.2}$$

where $i = 1$ always denotes the front axle, and $i = 2$ the rear axle; the vehicle moves from left to right.

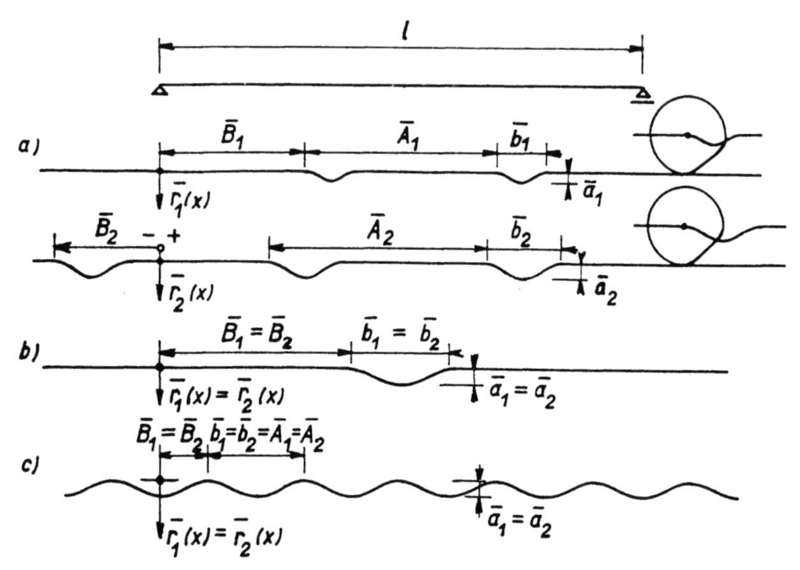

Fig. 9.3. Track irregularities: a) effect of flat wheels, b) isolated irregularity of roadway, c) wavy surface of roadway.

5. At the vehicle approach the beam is at rest; the vehicle, on the other hand may or may not have been set to vibration owing to track irregularities ahead of the entrance to the beam.

Assuming the above we can express the problem by a set of five differential equations that successively describe the rotation and vertical motion of sprung mass, vertical motions of unsprung masses and bending vibration of beam. In the intervals $0 \leqq x \leqq l$, $0 \leqq t \leqq l/c + D/c$ the equations are in the form

$$-I\,\frac{\mathrm{d}^2\bar{\varphi}(t)}{\mathrm{d}t^2} + \sum_{i=1}^{2}(-1)^i\,D_i[Z_i(t) + Z_{bi}(t)] = 0\,, \tag{9.3}$$

$$-m_3\,\frac{\mathrm{d}^2 v_3(t)}{\mathrm{d}t^2} - \sum_{i=1}^{2}[Z_i(t) + Z_{bi}(t)] = 0\,, \tag{9.4}$$

$$P_i + P_{3i} - m_i \frac{\mathrm{d}^2 v_i(t)}{\mathrm{d}t^2} + Z_i(t) + Z_{bi}(t) - \bar{R}_i(t) = 0\,; \quad i = 1, 2\,, \tag{9.5}$$

(9.6)

$$EJ \frac{\partial^4 v(x, t)}{\partial x^4} + \mu \frac{\partial^2 v(x, t)}{\partial t^2} + 2\mu\omega_b \frac{\partial v(x, t)}{\partial t} = \sum_{i=1}^{2} \bar{\varepsilon}_i \delta(x - x_i)\, \bar{R}_i(t)\,. \tag{9.7}$$

The new symbols in Eqs. (9.3) to (9.7) have the following meaning:

$$Z_i(t) = C_i[v_{3i}(t) - v_i(t)]\,, \quad i = 1, 2 \tag{9.8}$$

– force in spring C_i;

$$Z_{bi}(t) = C_{bi}[\dot{v}_{3i}(t) - \dot{v}_i(t)]\,, \quad i = 1, 2 \tag{9.9}$$

– damping force in spring C_i;

$$v_{3i}(t) = v_3(t) - (-1)^i\, D_i\, \bar{\varphi}(t)\,, \quad i = 1, 2 \tag{9.10}$$

– vertical displacement of sprung part of vehicle at the place of the i-th axle;

$$\bar{R}_i(t) = K_i\, u_i(t) \geqq 0\,, \quad i = 1, 2 \tag{9.11}$$

– moving force by which the i-th axle acts on beam at point x_i

where

$$u_i(t) = v_i(t) - \bar{\varepsilon}_i\, v(x_i, t) - \bar{r}_i(x_i) \tag{9.12}$$

– mutual approach of the i-th axle and the beam;

$$\bar{\varepsilon}_i = \begin{cases} 1 & \text{for} \quad 0 \leqq x_i \leqq l \\ 0 & \text{for} \quad x_i < 0;\ x_i > l\,, \end{cases} \tag{9.13}$$

and $\delta(x)$ is the Dirac function.

To the system of equations (9.3) to (9.7) we shall specify the boundary conditions of a simply supported beam, (1.2), and the initial conditions

$$v(x, 0) = 0\,, \qquad \left.\frac{\partial v(x, t)}{\partial t}\right|_{t=0} = 0\,,$$

$$v_i(0) = v_{i0}\,, \qquad \left.\frac{\mathrm{d}v_i(t)}{\mathrm{d}t}\right|_{t=0} = \dot{v}_{i0}\,, \quad i = 1, 2,$$

$$v_3(0) = v_{30}\,, \qquad \left.\frac{\mathrm{d}v_3(t)}{\mathrm{d}t}\right|_{t=0} = \dot{v}_{30}\,,$$

$$\bar{\varphi}(0) = \bar{\varphi}_0\,, \qquad \left.\frac{\mathrm{d}\bar{\varphi}(t)}{\mathrm{d}t}\right|_{t=0} = \dot{\bar{\varphi}}_0\,. \tag{9.14}$$

According to Hertz's law the contact force during impact is

$$\bar{R}_i(t) = k_2\, u_i^{3/2}(t) \tag{9.15}$$

where k_2 is a constant (cf. [96]) and $u_i(t)$ the mutual approach of the two bodies (9.12). In practical instances expression (9.15) is adequately linearized by Eq. (9.11). A physical interpretation of this consideration leads to the introduction of linear spring constants K_i – see Fig. 9.2. The linear contact deformation (9.11) is an approximate expression of the contact between two bodies, and corresponds to the case of two bodies bounded by planes touching one another in those planes. The linear approximation has been used with success in [69] in the solution of impact of flat wheels on rails. The reason for our introducing it here is that in practical cases of impact of wheels on bridges the assumptions of Hertz's theory are no longer satisfied (particularly because of large deformations occurring in parts of vehicle and roadway lying outside the narrow neighbourhood of the contact point).

According to (9.11), contact force $\bar{R}_i(t)$ must be either positive or zero. If $u_i(t)$ [Eq. (9.12)] is less than zero, we must substitute $\bar{R}_i(t) = 0$ in Eqs. (9.5) to (9.7) At that instant the vehicle loses contact with the beam, and both vehicle and beam start to vibrate in natural vibrations. After a short time, i.e. when again $u_i(t) = 0$, $i = 1, 2$, there will occur an impact accompanied by force $\bar{R}_i(t) > 0$. The case of $\bar{R}_i(t) < 0$ is excluded from our considerations because negative forces of contact between vehicle and beam are not possible.

9.2 *Solution of the problem*

9.2.1 *Dimensionless parameters*

As in Chap. 8 here, too, we shall introduce several dimensionless parameters, some identical, some slightly different from those discussed there. For the sake of completeness we will write them all together:

$$\alpha = \frac{c}{2f_{(1)}l}, \tag{9.16}$$

$$\gamma_i = \frac{f_i}{f_{(1)}}; \qquad i = 1, 2, \tag{9.17}$$

$$\gamma_{3i} = \frac{f_{3i}}{f_{(1)}}; \qquad i = 1, 2, \tag{9.18}$$

$$\varkappa = P/G, \tag{9.19}$$

$$\varkappa_i = P_i/P; \qquad i = 1, 2, \tag{9.20}$$

$$\lambda = I/(mD^2), \tag{9.21}$$

$$a_i = \bar{a}_i/v_0; \qquad i = 1, 2, \tag{9.22}$$

$$b_i = \bar{b}_i/l; \qquad i = 1, 2, \tag{9.23}$$

$$A_i = \bar{A}_i/l; \qquad i = 1, 2, \tag{9.24}$$

$$B_i = \bar{B}_i/l; \qquad i = 1, 2, \tag{9.25}$$

$$d = D/l, \tag{9.26}$$

$$d_1 = D_1/D, \tag{9.27}$$

$$\vartheta = \omega_b/f_{(1)}, \tag{9.28}$$

$$\vartheta_i = C_{bi}/(2mf_{3i}); \quad i = 1, 2. \tag{9.29}$$

In Eqs. (9.16) to (9.29)

$$v_0 = \frac{Pl^3}{96EJ} \tag{9.30}$$

is the deflection at mid-span of a beam loaded at point $x = l/2$ with force $P/2$ [the introduction of v_0 according to (9.30) is advantageous in view of the short-span bridges we are going to examine],

$$f_i = \frac{1}{2\pi}\left(\frac{K_i}{m}\right)^{1/2} \tag{9.31}$$

the frequency of unsprung mass, and

$$f_{3i} = \frac{1}{2\pi}\left(\frac{C_i}{m}\right)^{1/2} \tag{9.32}$$

the frequency of sprung mass above the i-th axle.

Lengths $\bar{a}_i$, $\bar{b}_i$, $\bar{A}_i$, $\bar{B}_i$ are shown in Fig. 9.3 and will be discussed in paragraph 9.2.2. The dimensionless quantities corresponding to them are a_i, b_i, A_i, B_i; $f_{(1)}$ is again the first natural frequency of the unloaded beam in accordance with (1.12).

Table 9.1 Example of set of input parameters

1	Case No.	1	30
2	y_{10}	1·1	10
3	y_{20}	0·9	10
4	y_{30}	1	10
5	φ_0	0·1	0
6	$\dot{y}_{10}$	0·01	0
7	$\dot{y}_{20}$	−0·01	0
8	$\dot{y}_{30}$	0·01	0
9	$\dot{\varphi}_0$	0·01	0
10	$\varkappa$	0·5	1
11	$\varkappa_1$	0·05	0·05
12	$\varkappa_2$	0·05	0·05
13	λ	0·1	0·2
14	α	0·1	0·02
15	γ_{31}	0·05	0·04
16	γ_{32}	0·05	0·04
17	γ_1	1	0·2
18	γ_2	1	0·2
19	a_1	4	0
20	a_2	4	20
21	b_1	0·04	0·02
22	b_2	0·04	0·02
23	A_1	1	0·5
24	A_2	1	0·5
25	B_1	0·5	0
26	B_2	−0·5	−1
27	d	1	1
28	d_1	0·5	0·5
29	ϑ	1	0·5
30	ϑ_1	1	0·5
31	ϑ_2	1	0·5
32	h_1	0·001	0·001
33	h_2	0·0005	0·0001
34	N	10	10
35	s	1	3

Next to quantities (9.16) to (9.29) the following data are also considered among input parameters:

the initial conditions according to (9.14)

$$y_{i0} = v_{i0}/v_0 \,, \qquad \dot{y}_{i0} = \frac{l}{cv_0}\,\dot{v}_{i0} \,; \quad i = 1, 2 \,,$$

$$y_{30} = v_{30}/v_0 \,, \qquad \dot{y}_{30} = \frac{l}{cv_0}\,\dot{v}_{30} \,,$$

$$\varphi_0 = \frac{D}{v_0}\,\bar{\varphi}_0 \,, \qquad \dot{\varphi}_0 = \frac{Dl}{cv_0}\,\dot{\bar{\varphi}}_0 \,; \tag{9.33}$$

the integration steps h_1 and h_2 [for further details see paragraph 9.2.4 and Eqs. (9.59)];

the auxiliary parameter N (the intermediate results are printed after every N-th step; they are not printed at all if $N = 0$);

number s, i.e. the number of equations specifying $q_{(j)}$ [for further details see paragraph 9.2.3 and Eqs. (9.54)].

An exemplary set of input parameters is shown in Table 9.1.

9.2.2 *Reduction of the equations to the dimensionless form*

For the purposes of numerical computations we will introduce the independent variables

$$\xi = x/l \,, \quad \tau = ct/l \tag{9.34}$$

and the dependent variables

$$y(\xi, \tau) = v(x, t)/v_0 \,, \quad y_i(\tau) = v_i(t)/v_0 \,; \quad i = 1, 2 \,,$$

$$y_3(\tau) = v_3(t)/v_0 \,, \quad \varphi(\tau) = \bar{\varphi}(t)\, D/v_0 \,. \tag{9.35}$$

Substitution of expressions (9.16) to (9.35) in Eqs. (9.3) to (9.7) and rearrangement lead to the following set of equations:

$$\ddot{\varphi}(\tau) = \frac{1}{\lambda\alpha^2} \sum_{i=1}^{2} (-1)^i d_i \gamma_{3i} [\pi^2 \gamma_{3i}\, z_i(\tau) + \vartheta_i \alpha\, \dot{z}_i(\tau)] \,, \tag{9.36}$$

$$\ddot{y}_3(\tau) = -\frac{1}{(1 - \varkappa_1 - \varkappa_2)\,\alpha^2} \sum_{i=1}^{2} \gamma_{3i} [\pi^2 \gamma_{3i}\, z_i(\tau) + \vartheta_i \alpha\, \dot{z}_i(\tau)] \,, \tag{9.37}$$

$$\ddot{y}_i(\tau) = \frac{48}{\pi^2 \varkappa_i \varkappa \alpha^2} \left[2Q_i + \frac{\pi^4}{48} \varkappa \gamma_{3i}^2 \, z_i(\tau) + \frac{\pi^2}{48} \vartheta_i \gamma_{3i} \varkappa \alpha \, \dot{z}_i(\tau) - R_i(\tau) \right]; \quad i = 1, 2, \tag{9.38}$$

$$\tag{9.39}$$

$$y^{IV}(\xi, \tau) + \pi^2 \alpha^2 \ddot{y}(\xi, \tau) + \pi^2 \vartheta \alpha \, \dot{y}(\xi, \tau) = 48 \sum_{i=1}^{2} \varepsilon_i \, \delta(\xi - \xi_i) \, R_i(\tau). \tag{9.40}$$

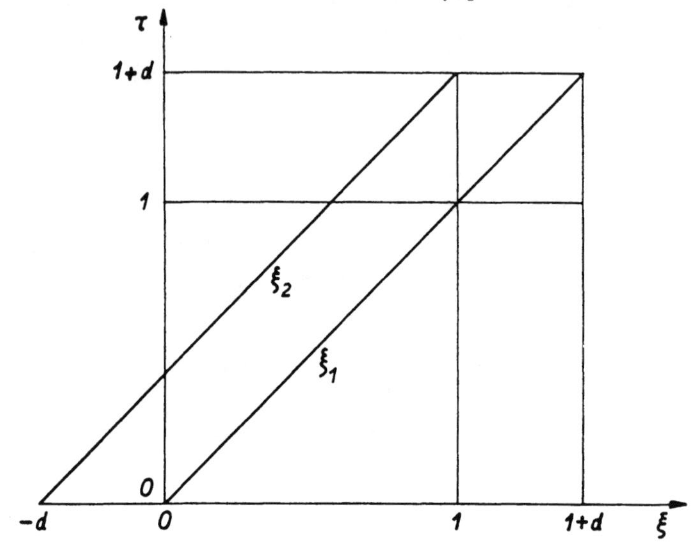

Fig. 9.4. Integration range.

The boundary conditions now are

$$\begin{aligned} y(0, \tau) &= 0, \quad y(1, \tau) = 0, \\ y''(0, \tau) &= 0, \quad y''(1, \tau) = 0 \end{aligned} \tag{9.41}$$

and the initial conditions

$$\begin{aligned} y(\xi, 0) &= 0, & \dot{y}(\xi, 0) &= 0, \\ y_i(0) &= y_{io}, & \dot{y}_i(0) &= \dot{y}_{io}, \\ y_3(0) &= y_{30}, & \dot{y}_3(0) &= \dot{y}_{30}, \\ \varphi(0) &= \varphi_0, & \dot{\varphi}(0) &= \dot{\varphi}_0. \end{aligned} \tag{9.42}$$

The integration ranges in which we shall solve our problem now are as shown in Fig. 9.4

$$0 \leqq \xi \leqq 1, \quad 0 \leqq \tau \leqq 1 + d. \tag{9.43}$$

According to (9.2) the coordinates of the points of contact between vehicle and beam are

$$\xi_1 = ct/l = \tau\,, \quad \xi_2 = ct/l - D/l = \tau - d \tag{9.44}$$

and their range (Fig. 9.4) is

$$0 \leqq \xi_1 \leqq 1 + d\,, \quad -d \leqq \xi_2 \leqq 1\,. \tag{9.45}$$

In Eqs. (9.36) to (9.40)

$$z_i(\tau) = y_3(\tau) - (-1)^i\, d_i\, \varphi(\tau) - y_i(\tau)\,, \quad i = 1, 2 \tag{9.46}$$

is the dimensionless approach of masses m_i and m_3,

$$Q_i = (P_i + P_{3i})/P = \varkappa_i + d_i(1 - \varkappa_1 - \varkappa_2)\,, \quad i = 1, 2 \tag{9.47}$$

the dimensionless static axle load, and

$$R_i(\tau) = 2\bar{R}_i(t)/P = \frac{1}{48}\,\pi^4 \varkappa \gamma_i^2 [y_i(\tau) - \varepsilon_i\, y(\xi_i, \tau) - r_i(\xi_i)] \geqq 0\,, \; i = 1, 2 \tag{9.48}$$

the dimensionless dynamic contact force, and ε_i and $\delta(\xi)$ are the respective dimensionless functions $\bar{\varepsilon}_i$ and $\bar{\delta}(x)$.

The dimensionless coordinates of track irregularities (Fig. 9.3) are

$$r_i(\xi) = \frac{1}{v_0}\,\bar{r}_i(x) = \begin{cases} \dfrac{1}{2}\, a_i \left[1 - \cos \dfrac{2\pi}{b_i}\,(\xi - kA_i - B_i)\right] \\ 0, \quad i = 1, 2 \end{cases} \tag{9.49}$$

$$\text{for} \quad \begin{cases} B_i + kA_i \leqq \xi \leqq B_i + kA_i + b_i \\ B_i + kA_i + b_i < \xi < B_i + (k + 1)\, A_i\,, \quad k = 0, 1, 2, \ldots \end{cases}$$

Eq. (9.49) is capable of expressing diverse types of irregularities (Fig. 9.3). Thus, for example:

a) The effects of flat wheels of railway cars can be thought of as changes in distance between wheel centre of gravity and beam neutral axis. According to measurements reported in [69], the coordinates of this irregularity are approximately equal to function (9.49) with the following dimensionless notation: a_i – depth of the flat spot, b_i – length of the flat spot, A_i – wheel circumference, B_i – place of first impact of flat wheel, measured from the origin $x = 0$, see Fig. 9.3a. If flat spots are on one axle only, say, on the first (second) axle, we substitute $a_2 = 0$ ($a_1 = 0$) in (9.49).

b) An isolated uneveness on the roadway surface (Fig. 9.3b) is expressed by Eq. (9.49) with the parameters

$$a_1 = a_2\,,\quad b_1 = b_2\,,\quad A_i > 1 + d\,,\quad B_1 = B_2\,.$$

c) Wavy surface of the roadway – Fig. 9.3c – can be expressed by Eq. (9.49) with the following parameters:

$$a_1 = a_2\,,\quad b_1 = b_2\,,\quad A_i = b_i\,,\quad B_1 = B_2\,.$$

Other irregularities likely to be met with in practice, may be expressed by suitable combinations of the values of parameters a_i, b_i, A_i, B_i. The positive sense of irregularities is downward, the negative upward.

9.2.3 *Reduction of the equations to a form suitable for numerical calculations*

We will again solve the partial differential equation (9.40) by the method of finite Fourier (sine) integral transformation defined in present symbols by the relations

$$q_{(j)}(\tau) = 2\int_0^1 y(\xi, \tau)\sin j\pi\xi \, d\xi\,, \tag{9.50}$$

$$y(\xi, \tau) = \sum_{j=1}^{\infty} q_{(j)}(\tau)\sin j\pi\xi \tag{9.51}$$

where $q_{(j)}(\tau)$ is the generalized dimensionless time coordinate of the beam. In view of the boundary conditions (9.41), transformation of Eq. (9.40) in accordance with (9.50) will after rearrangement give the following set of mutually independent equations

$$\ddot{q}_{(j)}(\tau) = \frac{96}{\pi^2\alpha^2}\sum_{i=1}^{2}\varepsilon_i\, R_i(\tau)\sin j\pi\xi_i - j^4\frac{\pi^2}{\alpha^2}\, q_{(j)}(\tau) - \frac{\vartheta}{\alpha}\,\dot{q}_{(j)}(\tau)\,,$$

$$j = 1, 2, 3, \ldots \tag{9.52}$$

Because of the first two of conditions (9.42), the initial conditions now turn out to be

$$q_{(j)}(0) = 0\,,\quad \dot{q}_{(j)}(0) = 0\,. \tag{9.53}$$

In numerical calculations we take a finite number s (s – the specified parameter) of Eqs. (9.52) so that in view of (9.51) the approximate solution is

$$y(\xi, \tau) \approx \sum_{j=1}^{s} q_{(j)}(\tau) \sin j\pi\xi\,, \quad j = 1, 2, 3, \ldots, s\,. \tag{9.54}$$

The set of equations (9.36) to (9.39) and (9.52) with the initial conditions (9.42) and (9.53) has already been written in a form suitable for numerical evaluation. The unknown functions are $\varphi(\tau)$, $y_3(\tau)$, $y_i(\tau)$ and $q_{(j)}(\tau)$. The last named function is used in the computation of dimensionless deflection $y(\xi, \tau)$ according to (9.54).

In calculations relating to stresses in beams subjected also to impact effects, consideration must be given to the bending moment proportional to the stress. The calculation is made by the combined method described in Sect. 5.3, particularly Eqs. (5.67), the first two of Eqs. (5.70) and Eq. (5.9).

Dividing the bending moment by bending moment M_0 of the centre of a beam loaded at point $x = l/2$ with force $P/2$

$$M_0 = \frac{Pl}{8}\,, \tag{9.55}$$

gives the dimensionless bending moment of Eq. (5.67) in the following form:

$$M(\xi, \tau) = \sum_{i=1}^{2} M_{Ri}(\xi, \tau) + M_{\mu}(\xi, \tau)\,. \tag{9.56}$$

In the above equation $M_{Ri}(\xi, \tau)$ is the dimensionless bending moment at point ξ produced by load $R_i(\tau)$ placed at point ξ_i. It corresponds to the quasistatic component $M_{st}(x, t)$ in Eq. (5.67), and is obtained from the first of Eqs. (5.70) [with $R_i(\tau)$ replacing P] and Eq. (5.9) on dividing them by M_0, in the form

$$M_{Ri}(\xi, \tau) = \begin{cases} \dfrac{8}{Pl}\dfrac{1}{l}\,\varepsilon_i\,\bar{R}_i(t)\,(l - x_i)\,x = 4\varepsilon_i\,R_i(\tau)\,(1 - \xi_i)\,\xi \quad \text{for } \xi_i \geqq \xi\,, \\[2ex] \dfrac{8}{Pl}\dfrac{1}{l}\,\varepsilon_i\,\bar{R}_i(t)\,(l - x)\,x_i = 4\varepsilon_i\,R_i(\tau)\,(1 - \xi)\,\xi_i \quad \text{for } \xi_i \leqq \xi\,. \end{cases} \tag{9.57}$$

The second term on the right-hand side of Eq. (9.56), $M_{\mu}(\xi, \tau)$, is the dimensionless bending moment produced by inertia forces $-\mu\,\ddot{v}(x, t)$.

It corresponds to the second term on the right-hand side of Eq. (5.67)*). On substituting in it the second of Eqs. (5.70) and our dimensionless variables and dimensionless parameters, and dividing by M_0 we get

$$M_\mu(\xi, \tau) = -\frac{1}{12}\alpha^2 \sum_{j=1}^{\infty} \frac{1}{j^2} \ddot{q}_{(j)}(\tau) \sin j\pi\xi \,. \tag{9.58}$$

In view of the chosen method of numerical evaluation, it is advantageous to compute the bending moment from Eqs. (9.56) to (9.58) because the necessary functions $R_i(\tau)$ and $\ddot{q}_{(j)}(\tau)$ must be calculated in any case [see Eqs. (9.38), (9.39), (9.48) and (9.52)]. The computation is accurate enough even with a small number of terms of series (9.58), particularly at low speeds of motion, i.e. in case of $M_\mu(\xi, \tau) \ll M_{Ri}(\xi, \tau)$ – cf. the concluding paragraphs of Sect. 5.3.

9.2.4 *Numerical solution*

The problem specified by the set of ordinary differential equations of the second order (9.36) to (9.39) and (9.52), with the initial conditions (9.42) and (9.53) was again solved by the method of Runge-Kutta-Nyström [37]. The solution was programmed for the Ural 2 automatic computer. The computation starts with the printing of input data, with the values of $M(1/2, \tau)$, $y(1/2, \tau)$, $R_1(\tau)$ and $R_2(\tau)$ printed after each N-th step. If $N = 0$, the intermediate results are not printed at all. N is a number specified simultaneous with the input parameters. At the end of the computation of a case, there are always printed the maximum values of the aforementioned functions, i.e. max $M(1/2, \tau)$, max $y(1/2, \tau)$, max $R_1(\tau)$, max $R_2(\tau)$, and the values of τ at which the maxima occur.

To check the correctness of the programme we computed the first step of case No. 1 (Table 9.1) using the Ural 2 computer as well as a Cellatron-Mercedes electric desk calculator. In either case the computation was made by the method of Runge-Kutta-Nyström. As the results summarized in Table 9.2 suggest, the programme was set up correctly because the agreement between the two methods of computation is very good.

According to (9.49) we may expect sudden changes in the values of functions at the place $r_i(\xi)$ where some irregularity, e.g. impact of flat

*) The damping is neglected on the right-hand side of Eq. (5.6).

Table 9.2 Results of calculation of a check example
Case No. 1, $h_1 = 0{\cdot}001$

Calculating machine	$M(1/2, 0{\cdot}001)$	$y(1/2, 0{\cdot}001)$
Automatic computer Ural 2	−0·000 594 705 36	+0·000 000 566 883 88
Electric desk calculator	−0·000 594 705 337	+0·000 000 566 883 88

wheels, takes place. That is the reason for our introducing two values of integration step, h_1 and $h_2 (h_2 \leqq h_1)$ as input data. We will make use of the lesser value, h_2, only in the irregularity neighbourhood equal to double the irregularity length. The integration step h is therefore chosen as follows:

$$h = \begin{cases} h_2 & \text{for} \quad B_i + kA_i - h_1 < \xi_i \leqq B_i + kA_i + 2b_i\,; \\ & \qquad i = 1, 2, \quad k = 0, 1, 2, \ldots \\ h_1 & \text{in all other cases.} \end{cases} \tag{9.59}$$

The integration step h_1 was estimated using the method described in detail in [78]. In our case No. 30, h_1 turned out to be 0·001. The effect

Table 9.3 Effect of integration step length and number of equations
Case No. 30

h_1	0·001		0·0005	
h_2	0·0001		0·0001	
s	1	3	1	3
$\tau[\max M\,\frac{1}{2}, \tau)]$	1·571 0999	1·571 0999	1·571 0999	1·571 5999
$\max M\,\frac{1}{2}, \tau)$	3·558 3133	3·550 4918	3·558 3731	3·560 3028
$\tau[\max y(\frac{1}{2}, \tau)]$	1·572 0999	1·572 0999	1·572 5999	1·572 5999
$\max y(\frac{1}{2}, \tau)$	4·193 4544	4·201 1628	4·194 8825	4·203 8006
$\tau[\max R_1(\tau)]$	0·022 0000	0·022 0000	0·022 0000	0·022 0000
$\max R_1(\tau)$	1·138 5666	1·138 4909	1·138 5666	1·138 4909
$\tau[\max R_2(\tau)]$	0·020 9000	0·020 9000	0·020 9000	0·020 9000
$\max R_2(\tau)$	2·719 8116	2·719 8118	2·719 8118	2·719 8118

of the integration step length was examined for case No. 30 whose input parameters are summarized in Table 9.1. The most important results obtained for $h_1 = 0{\cdot}001$ and $h_1 = 0{\cdot}0005$ are reviewed in Table 9.3. Comparing them we see that they differ but very little; accordingly, integration step $h_1 = 0{\cdot}001$ is very suitable for case No. 30.

The other effect we followed was that of the number of Eqs. (9.52), i.e. the effect of number s in Eq. (9.54). We did so for case No. 30 and the integration steps $h_1 = 0{\cdot}001$ and $h_1 = 0{\cdot}0005$. The results of that examination are also reviewed in Table 9.3. The effect of s in case No. 30 was small enough and that is why we used $s = 3$ in most of our calculations.

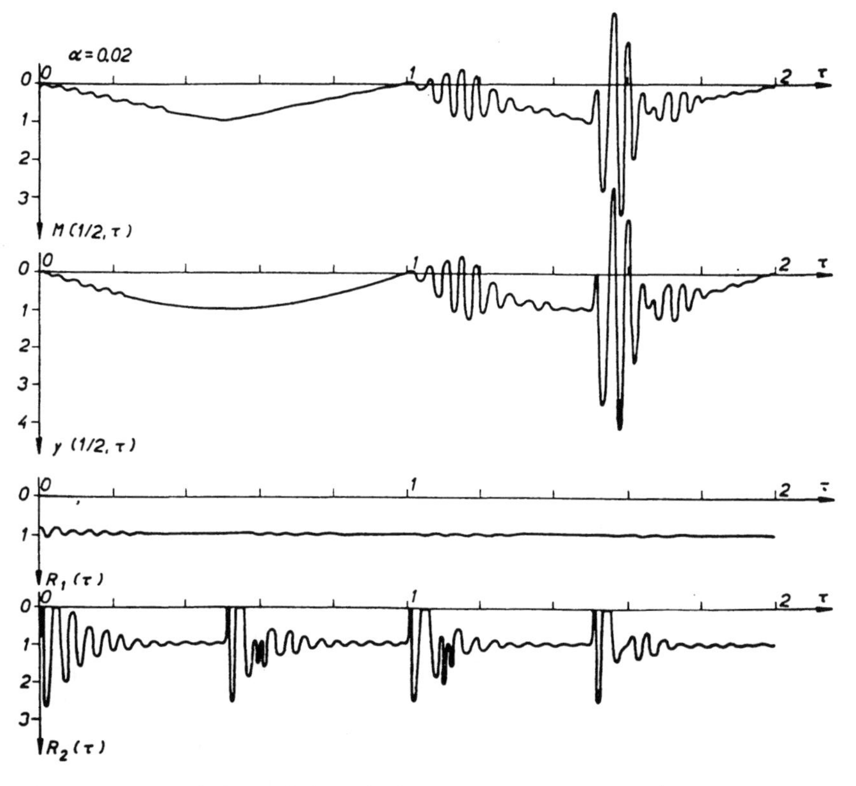

Fig. 9.5. Time variations of dimensionless bending moment $M(1/2, \tau)$ and deflection $y(1/2, \tau)$ at the centre of beam, and of forces $R_1(\tau)$, $R_2(\tau)$. Case No. 30, $a_1 = 0$, $a_2 = 20$, $\alpha = 0{\cdot}02$.

9.3 *The effect of various parameters*

The effect of some of the dimensionless parameters discussed in the foregoing was studied for our case No. 30 (Table 9.1). In the calculations always only one parameter was varied and the remaining were kept constant. Case No. 30 corresponds to the dynamic effect of a flat spot on the rear axle of a two-axle railway car traversing a bridge with span equal to the car axle base, $l = D$.

The results of the calculation are presented in Fig. 9.5, while Fig. 9.6 shows the same case but without irregularity of any kind.

The diagrams that follow plot only the maxima of functions $M(l/2, \tau)$, $y(l/2, \tau)$, $R_1(\tau)$ and $R_2(\tau)$ comprehensively designated by δ (dynamic coefficient) in relation to the parameters under examination.

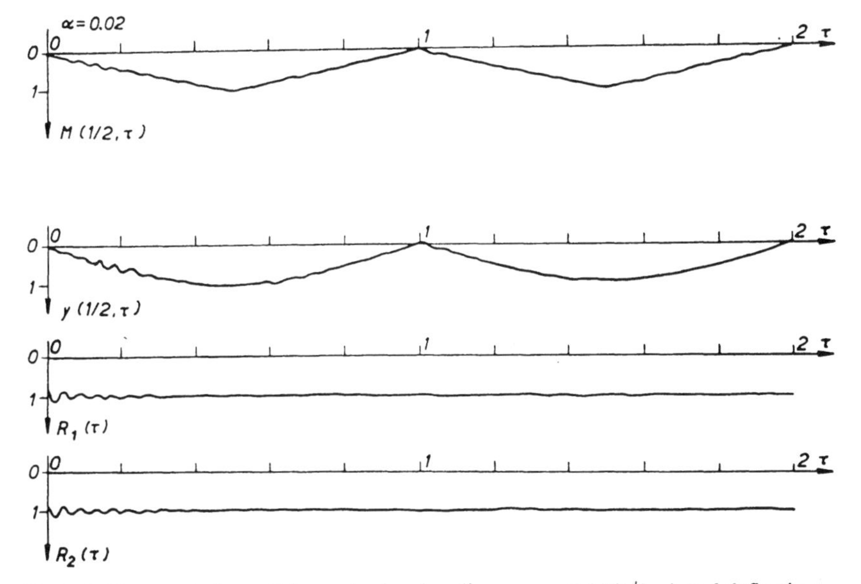

Fig. 9.6. Time variations of dimensionless bending moment $M(l/2, \tau)$ and deflection $y(l/2, \tau)$ at the centre of beam, and of forces $R_1(\tau)$, $R_2(\tau)$. Case No. 30, $a_1 = 0$, $a_2 = 0$, $\alpha = 0{\cdot}02$.

9.3.1 *The effect of speed*

The effect of speed parameter α (9.16) was studied in the range $0 \leqq \alpha \leqq$ $\leqq 0{\cdot}12$, for α ascending in steps of 0·005 (or smaller at lower α's). Fig. 9.7

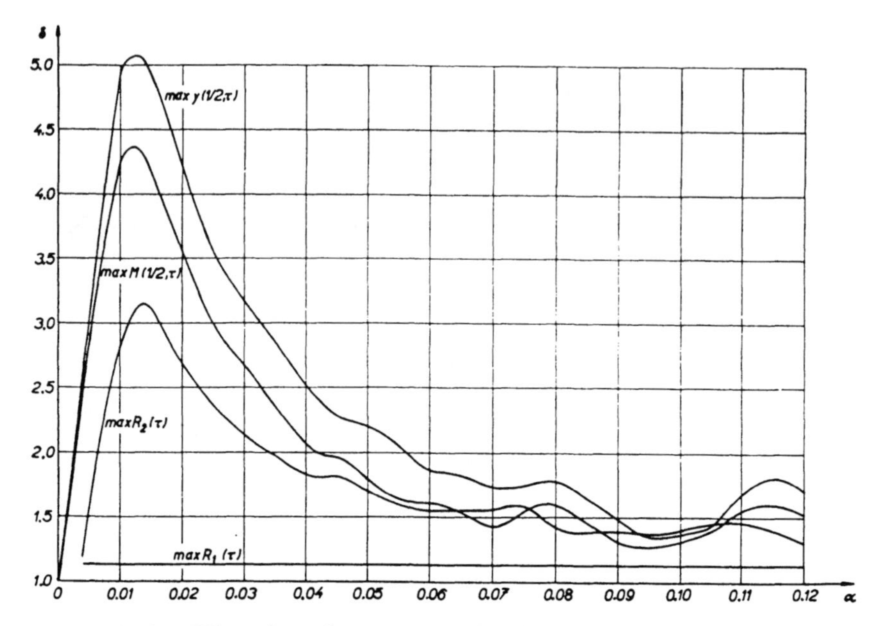

Fig. 9.7. Effect of speed parameter α. Case No. 30, $a_1 = 0$, $a_2 = 20$.

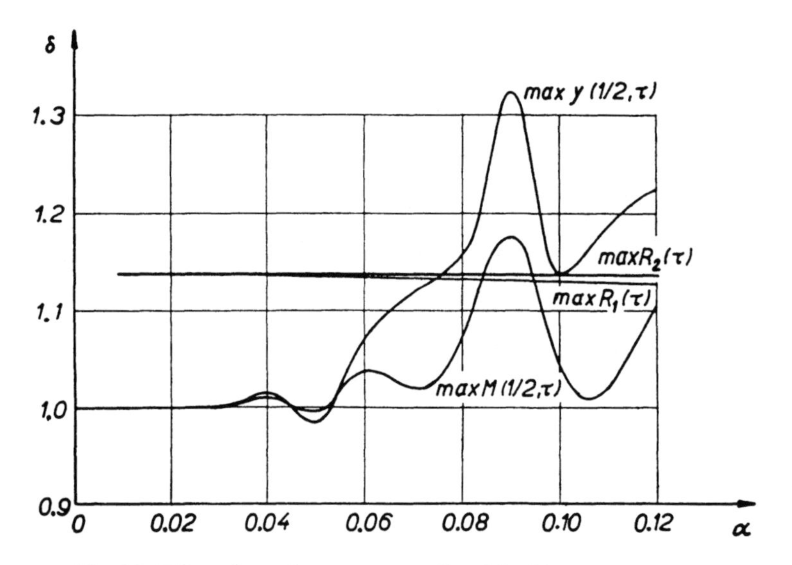

Fig. 9.8. Effect of speed parameter α. Case No. 30, $a_1 = 0$, $a_2 = 0$.

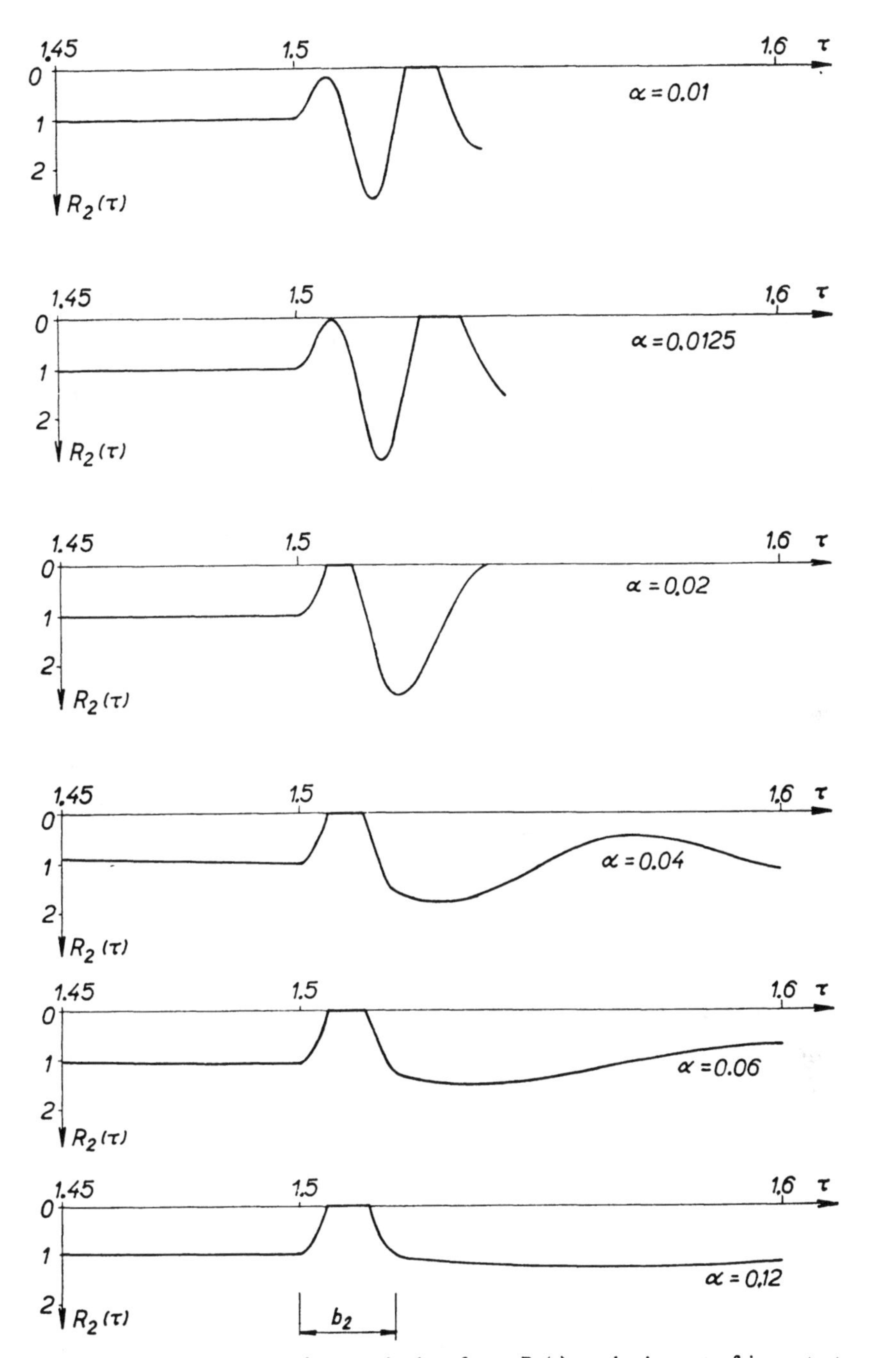

Fig. 9.9. Time variation of dimensionless force $R_2(\tau)$ at the instant of impact at various values of α. Case No. 30, $a_1 = 0$, $a_2 = 20$.

shows the dependence of δ on α for case No. 30 of Table 9.1. It is plain to see in the figure that the dynamic effects of isolated irregularities (flat wheel) reach their maximum at low speeds. For a wheel with no flat spot, the highest axle pressure, max $R_1(\tau)$, continues nearly unvaried for various α's.

Fig. 9.8 shows the same case but without any irregularity whatever. A comparison of Figs. 9.7 and 9.8 clearly shows the substantial effect of irregularities on beam stresses.

Fig. 9.9 is the time variation of force $R_2(\tau)$ for various α (case No. 30) drawn at an enlarged time scale. Consulting it we see that at very low speeds ($\alpha < 0{\cdot}0075$) no impact occurs at all and the contact force continues positive $[R_2(\tau) > 0]$. At speeds about $\alpha = 0{\cdot}0125$ the arrival of a flat wheel causes force $R_2(\tau)$ to ease off to a degree likely to bring it to zero. There is no impact at that time, but a secondary impact is apt to take place during the second pulse. That is the case when an irregularity leads to the highest possible values of bending moment and deflection. At still higher speeds ($\alpha \geqq 0{\cdot}02$), the wheel loses contact with the beam as soon as the flat spot arrives, and there follow one or even several impacts. But the corresponding maxima of bending moments and deflection already start to fall off. If the speed is increased yet further ($\alpha > 0{\cdot}05$), there always will occur unloading and primary impact – the latter, however, without any peak typical of impact phenomena. After impact, the force $R_2(\tau)$ follows nearly the same course as it had before the impact. So far as bending moments, and deflection are concerned, they are hardly affected by this kind of impact and therefore fall off considerably.

9.3.2 *The effect of the frequency parameter of unsprung mass*

For case No. 30 the relaiton between δ and the frequency parameter of unsprung mass, γ_i (9.17), is illustated in Fig. 9.10. The striking feature of this diagram is the rapid growth of dynamic stresses in the beam with growing parameter γ_i.

9.3.3 *The effect of the frequency parameter of sprung mass*

The effect of the frequency parameter of sprung mass, γ_{3i} (9.18), is clear to see in Fig. 9.11 which shows that the dynamic effects first increase, then slightly fall off with growing γ_{3i} (except for the region of highest values of γ_{3i} examined in the study).

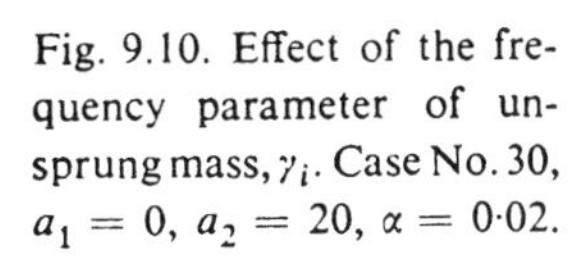

Fig. 9.10. Effect of the frequency parameter of unsprung mass, γ_i. Case No. 30, $a_1 = 0$, $a_2 = 20$, $\alpha = 0{\cdot}02$.

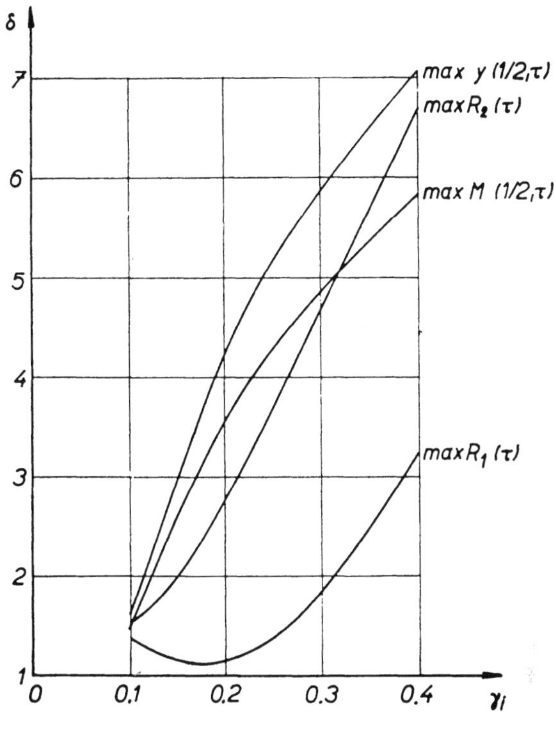

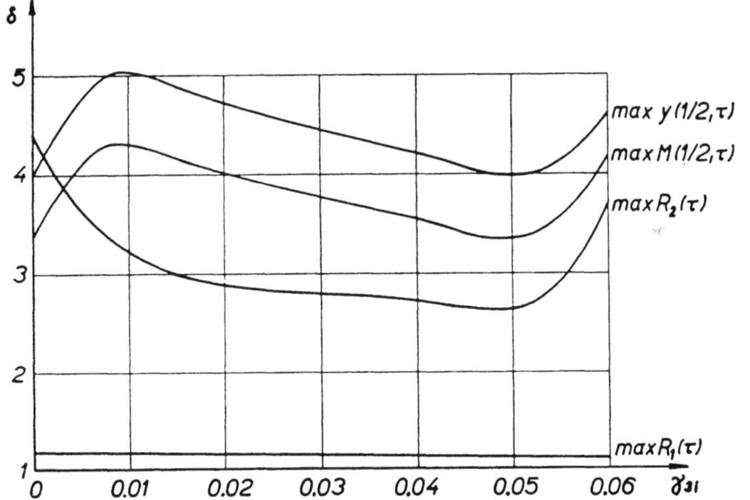

Fig. 9.11. Effect of the frequency parameter of sprung mass, γ_{3i}. Case No. 30, $a_1 = 0$, $a_2 = 20$, $\alpha = 0{\cdot}02$.

9.3.4 *The effect of the ratio between the weights of vehicle and beam*

The dependence of δ on the ratio between the weights of vehicle and beam – $\varkappa$ (9.19) – is recorded in Fig. 9.12 (case No. 30, Table 9.1). At constant α, the dynamic effects roughly increase with growing $\varkappa$ but the increase is somewhat slower after $\varkappa$ has reached a certain value.

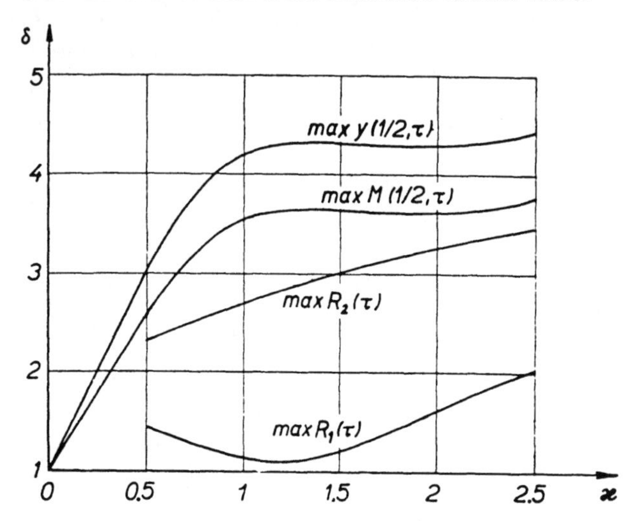

Fig. 9.12. Effect of the ratio between weights of vehicle and beam $\varkappa$. Case No. 30, $a_1 = 0$, $a_2 = 20$, $\alpha = 0{\cdot}02$.

9.3.5 *The effect of the ratio between the weights of unsprung parts and whole vehicle*

This effect is characterized by parameter $\varkappa_i$ (9.20). The relation between δ and parameter $\varkappa_i$ is analyzed in Fig. 9.13. As the figure suggests, in the range under examination the effect of unsprung masses first increases, and then the increase stops.

9.3.6 *The effect of the depth of irregularities*

The depth of an irregularity or a flat spot is described by parameter a_i (9.22). For case No. 30, the dependence of δ on depth a_2 is shown in Fig. 9.14. The dynamic effects grow nearly linear with the growing depth of irregularity. The growth slows down only at high values of a_2.

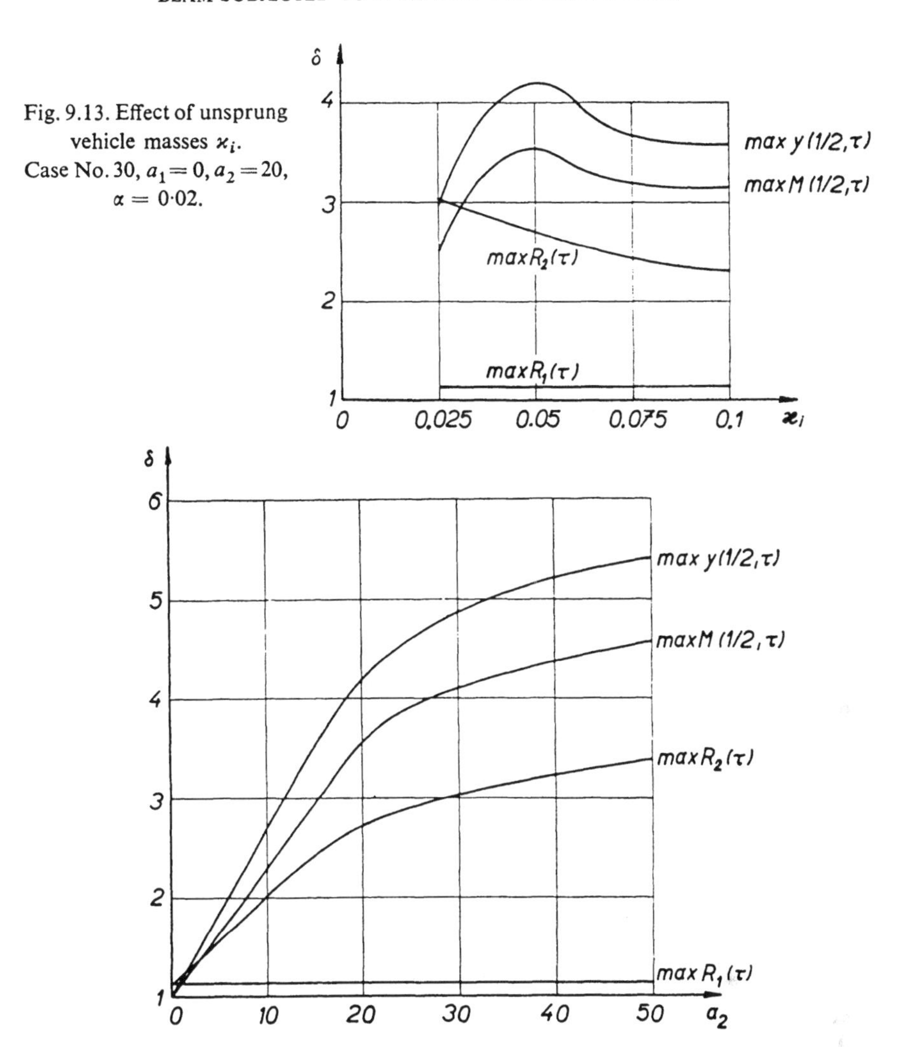

Fig. 9.13. Effect of unsprung vehicle masses $\varkappa_i$. Case No. 30, $a_1 = 0, a_2 = 20$, $\alpha = 0{\cdot}02$.

Fig. 9.14. Effect of the depth of track uneveness a_2. Case No. 30, $a_1 = 0, b_2 = 0{\cdot}02$, $\alpha = 0{\cdot}02$.

9.3.7 *The effect of the length of irregularities*

The length of an irregularity or a flat spot is expressed by parameter b_i (9.23). For case No. 30 (Table 9.1) the dependence of δ on length b_2 is in Fig. 9.15. The dynamic effects grow with the increasing length of irregularity up to a certain value after the attainment of which they slowly fall off. It goes without saying that in this as well as in the preceding figures, max $R_1(\tau)$ continues practically unvaried because we set $a_1 = 0$.

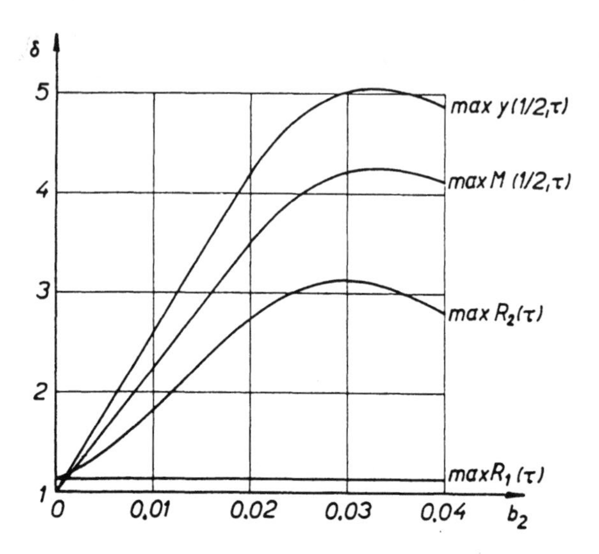

Fig. 9.15. Effect of the length of track uneveness b_2. Case No. 30, $a_1 = 0$, $a_2 = 20$, $\alpha = 0{\cdot}02$.

9.3.8 *The effect of other parameters*

One of the parameters examined in this part of our study, was parameter λ (9.21) expressing rotation of sprung parts. For case No. 30 (Table 9.1) and the range $0{\cdot}1 \leqq \lambda \leqq 0{\cdot}3$, the effect of λ on the maximum values of functions $M(1/2, \tau)$, $y(1/2, \tau)$, $R_1(\tau)$ and $R_2(\tau)$ was found to be almost negligible.

The effect investigated next was that of various values of initial conditions y_{10} (9.42). The calculation was made for case No. 30 in the range $8 \leqq y_{10} \leqq 16$, without consideration given to track irregularities, i.e. for $a_i = 0$. In this way it was possible to decide to what extent the effect of initial conditions is commensurable with that of track irregularities. It was found that the effect of track irregularities or flat wheels is greater by far than the effect of initial conditions.

9.4 *Application of the theory*

9.4.1 *Comparison of theory with experiments*

The theory expounded in the foregoing sections was experimentally verified on four railway bridges traversed by a two-axle car with flat wheels. Two of the bridges were of steel, two of reinforced concrete. One of the steel structures was a short-span bridge with direct fastened rails, the other a truss bridge with a fairly large span and a conventional roadway. Both reinforced concrete bridges were of the short-span kind, one was provided with a gravel bed, the other was without ballast.

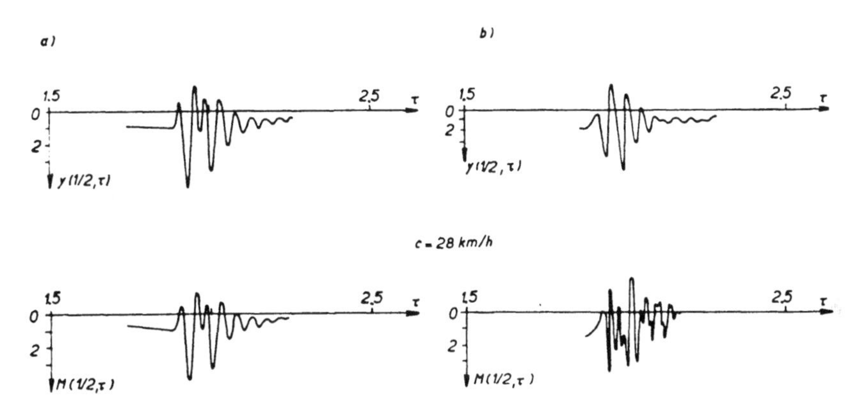

Fig. 9.16. Deflection $y(1/2, \tau)$ and bending moment $M(1/2, \tau)$ at the centre of span of a steel bridge, $l = 3{\cdot}6$ m, depth of flat spot $\bar{a}_2 = 6$ mm, speed $c = 28$ km/h, a) theory, b) experiments.

The experimental verification of the theory is described at length in [84]. Some of its typical results are presented in Fig. 9.16 drawing the computed and measured deflections and bending moments at mid-span of a 3·6 m-span steel bridge. The depth of the flat spot on the rear axle of a two-axle car was $\bar{a}_2 = 6$ mm. The dependence of the dynamic coefficient δ and increment Δ on speed is plotted in Fig. 9.17 (deflection) and Fig. 9.18 (stress). Coefficient δ and increment Δ were measured from the respective static value at the point where the flat wheel effect comes into play (for particulars see [84]). As evident from Figs. 9.16 to 9.18, the theory on the whole describes the dynamic effects of flat wheels on railway bridges to satisfaction.

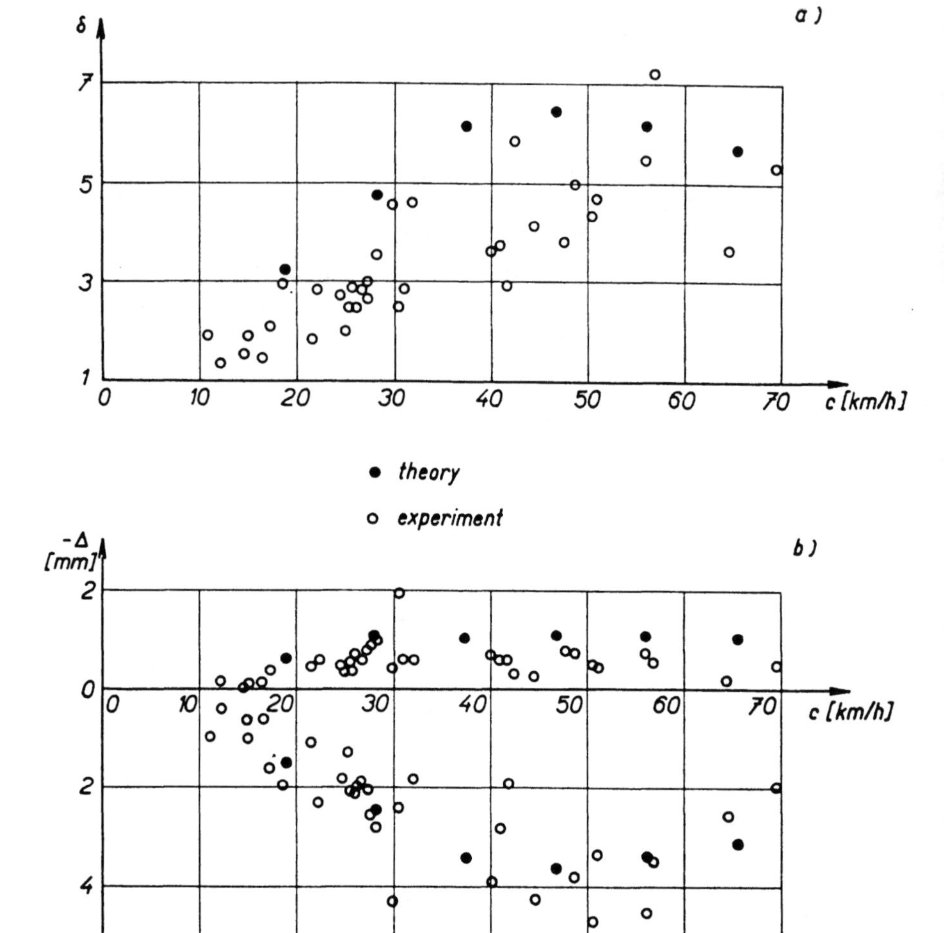

Fig. 9.17. Theoretical and experimental dependences of a) dynamic coefficient δ, b) dynamic increment Δ on speed c. Deflection at mid-span of a steel bridge, $l = 3{\cdot}6$ m, depth of flat spot $\bar{a}_2 = 6$ mm.

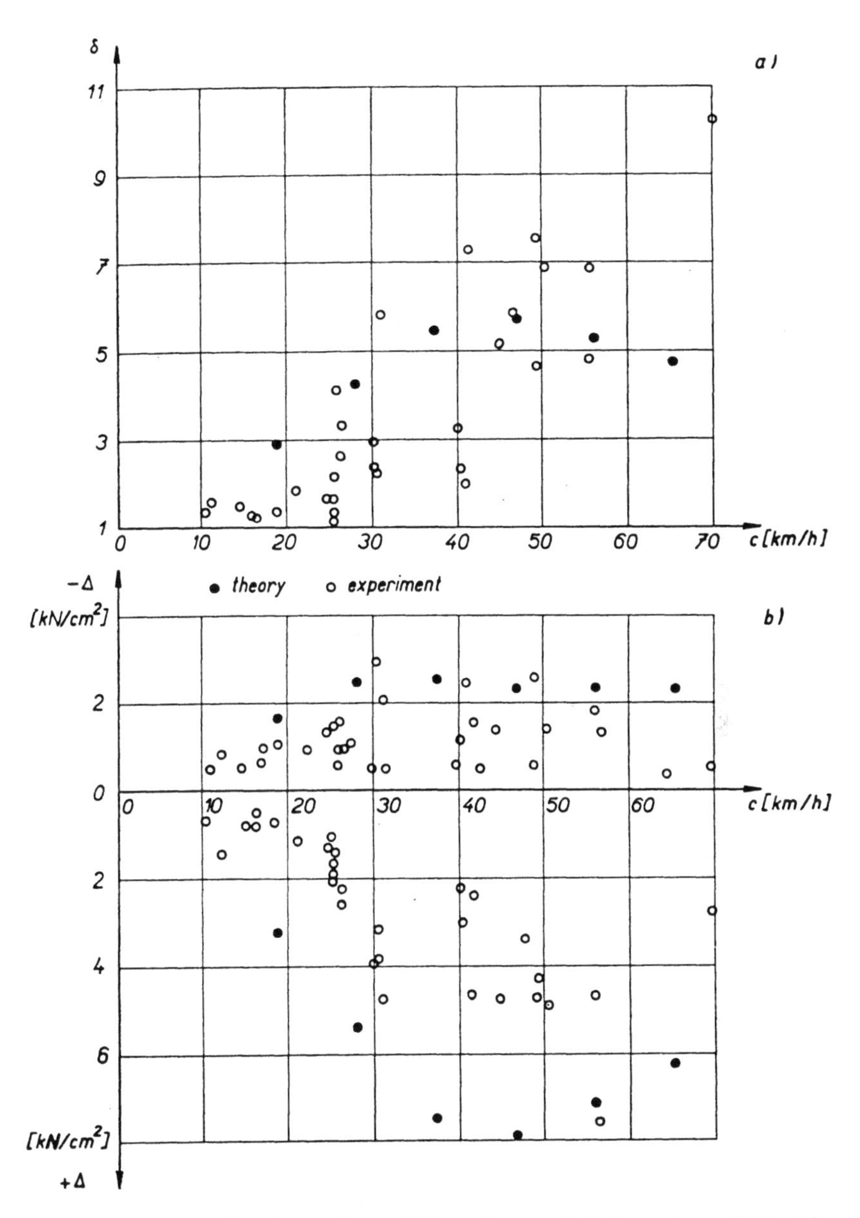

Fig. 9.18. Theoretical and experimental dependences of a) dynamic coefficient δ, b) dynamic increment Δ on speed c. Stress at mid-span of a steel bridge, $l = 3{\cdot}6$ m, depth of flat spot $\bar{a}_2 = 6$ mm.

9.4.2 *Dynamic stresses in short-span railway bridges*

It follows from the foregoing discussion that dynamic stresses in short-span railway bridges are first and foremost affected by impacts resulting from track or wheel irregularities (rail joints, flat wheels, etc.). An equally important factor is the effect of sprung and unsprung masses of vehicles that have been set to vibration ahead of the entrance to the bridge.

The dynamic effects of vehicles passing over short, isolated irregularities, are highest at low speeds (about 25 to 50 km/h for vehicles with flat wheels).

The dynamic effects in bridges roughly grow with growing vehicle weight, growing roadway stiffness and growing irregularity depth. On the whole, the unsprung masses of the vehicle exert but a small effect. Compared with the effects of irregularities, the effect of initial conditions of the vehicle as it enters the bridge, is small.

The theory works well for both railway and highway bridges built of steel, reinforced concrete or prestressed concrete. It is particularly well suited for bridges with spans of up to 20 and 30 m.

9.5 *Additional bibliography*

[24, 33, 52, 65, 69, 78, 84, 85, 93, 96, 121, 229, 230, 231, 242, 245, 246, 258, 271].

10

Beam subjected to a moving multi-axle system

Multi-axle vehicles are very complex mechanical systems with a number of degrees of freedom, linear and non-linear springs and damping. From the point of beam stress calculations they may be idealized sufficiently well by systems schematically shown in Fig. 10.1. However, those simplified models would not work to satisfaction if used in analysis of motions of all vehicle parts. In any case, the proposed idealization takes care of all the essential effects a multi-axle vehicle is likely to produce in a beam traversed by it. And that is just what we are primarily interested in – beam stresses and deflections.

10.1 *Formulation of the problem*

Similarly as in Chaps. 8 and 9 we shall start out from the following assumptions (next to assumptions 1, 3, 4 and 5 stated in Sect. 1.1):

1. The vehicle has $2n$ axles, and its total weight

$$P = P_1 + P_2 = mg \tag{10.1}$$

where $P_1 = m_1 g$ – weight of all unsprung parts of the vehicle,
$P_2 = m_2 g$ – total weight of sprung parts of the vehicle.

Each unsprung mass, $m_1/(2n)$, is free to perform vertical motion $v_{1i}(t)$. The sprung part of the vehicle can move in vertical direction – $v_2(t)$ – as well as rotate $-\bar{\varphi}(t)$. Quantities $v_{1i}(t)$ and $v_2(t)$ are measured from the equilibrium position in which springs C and K are compressed by the static action of the vehicle. Quantity $\bar{\varphi}(t)$ is measured from the horizontal in clockwise direction.

Each unsprung mass $m_1/(2n)$ is connected to the vehicle body by means of a spring, all springs having the same linear spring constant C with the coefficient of viscous damping C_b.

Unlike in Chap. 9 the vehicle is symmetric with respect to the vertical plane passing through the vehicle centroid O. The mass moment of inertia of the sprung part of the vehicle about the horizontal transverse axis is I. The distance between the i-th axle $(i = 1, 2, \ldots, 2n)$ and centroid O is $\bar{d}_i$. Because of symmetry

$$\bar{d}_{2i-1} = \bar{d}_{2i}\,, \quad i = 1, 2, \ldots, 2n\,. \tag{10.2}$$

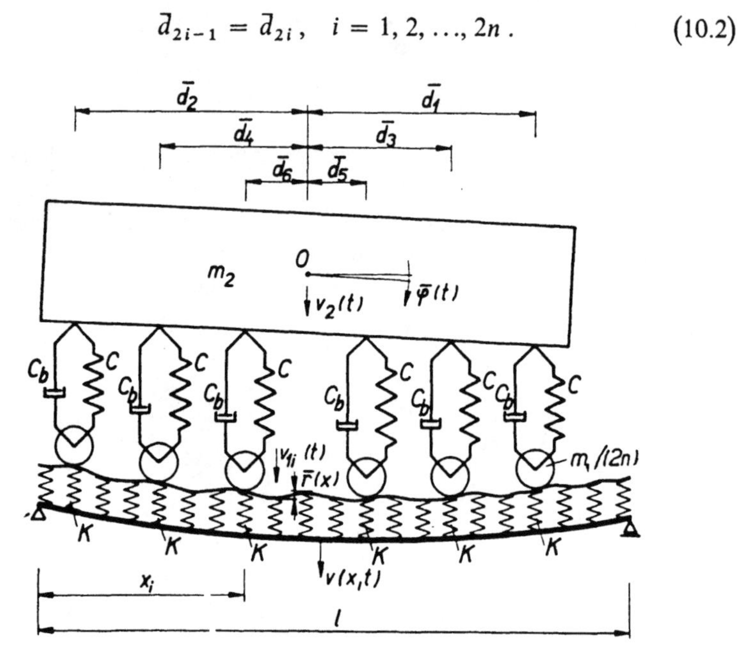

Fig. 10.1. Model of a beam with an elastic layer and irregularities, loaded with a moving multi-axle vehicle.

2. There is on the beam surface an elastic layer with spring constant K, or alternatively, with constant K underneath each axle (constant — in contradistinction to Chap. 8, and identical under all axles — in contradistinction to Chap. 9).

3. There are irregularities $\bar{r}(x)$ on the surface of the track.

4. The vehicle moves from left to right at uniform speed c along the beam; therefore the coordinates of the contact points of the i-th axle (Fig. 10.1) are

$$x_i = \begin{cases} ct - \bar{d}_1 + \bar{d}_i & \text{for} \quad i = 1, 3, 5, \ldots, 2n - 1 \\ ct - \bar{d}_1 - \bar{d}_{i-1} & \text{for} \quad i = 2, 4, 6, \ldots, 2n\,. \end{cases} \tag{10.3}$$

5. At the beginning of the event, neither the beam nor the vehicle possesses displacement or velocity in vertical direction. The event begins at the instant the first axle arrives on the beam.

Assuming the above we may express the problem by a system of $2n + 3$ differential equations successively describing rotation and vertical motion of the sprung mass, vertical motions of the unsprung masses and bending vibration of the beam. In the intervals $0 \leqq x \leqq l$, $0 \leqq t \leqq l/c + 2\bar{d}_1/c$ the equations take on the form

$$-I \frac{\mathrm{d}^2 \bar{\varphi}(t)}{\mathrm{d}t^2} + \sum_{i=1}^{n} \bar{d}_{2i-1}[Z_{2i}(t) - Z_{2i-1}(t) + Z_{b2i}(t) - Z_{b2i-1}(t)] = 0 \,, \tag{10.4}$$

$$-m_2 \frac{\mathrm{d}^2 v_2(t)}{\mathrm{d}t^2} - \sum_{i=1}^{2n} [Z_i(t) + Z_{bi}(t)] = 0 \,, \tag{10.5}$$

$$\frac{P_2}{2n} - \frac{m_1}{2n} \frac{\mathrm{d}^2 v_{1i}(t)}{\mathrm{d}t^2} + Z_i(t) + Z_{bi}(t) - \bar{R}_i(t) = 0 \,; \quad i = 1, 2, \ldots, 2n \,, \tag{10.6}$$

$$EJ \frac{\partial^4 v(x, t)}{\partial x^4} + \mu \frac{\partial^2 v(x, t)}{\partial t^2} + 2\mu\omega_b \frac{\partial v(x, t)}{\partial t} = \sum_{i=1}^{2n} \bar{\varepsilon}_i \, \delta(x - x_i) \left[\frac{P_1}{2n} + \bar{R}_i(t)\right]. \tag{10.7}$$

Similarly as in Sect. 9.1 the following notation was used in Eqs. (10.4) to (10.7):

$$Z_i(t) = C[v_{2i}(t) - v_{1i}(t)] \,, \tag{10.8}$$

$$Z_{bi}(t) = C_b[\dot{v}_{2i}(t) - \dot{v}_{1i}(t)] \,, \tag{10.9}$$

$$v_{2i}(t) = \begin{cases} v_2(t) + \bar{d}_i \, \bar{\varphi}(t) & \text{for} \quad i = 1, 3, 5, \ldots, 2n - 1 & (10.10) \\ v_2(t) - \bar{d}_i \, \bar{\varphi}(t) & \text{for} \quad i = 2, 4, 6, \ldots, 2n \,, & (10.11) \end{cases}$$

$$\bar{R}_i(t) = K[v_{1i}(t) - \bar{\varepsilon}_i \, v(x_i, t) - \bar{r}(x_i)] \,. \tag{10.12}$$

The remaining symbols are as those used in Chap. 9. To the system of equations (10.4) to (10.7) we shall specify the boundary conditions of a simply supported beam – (1.2) – and the initial conditions

$$v(x, 0) = 0 \,, \quad \left.\frac{\partial v(x, t)}{\partial t}\right|_{t=0} = 0 \,, \quad v_2(0) = 0 \,, \quad \left.\frac{\mathrm{d}v_2(t)}{\mathrm{d}t}\right|_{t=0} = 0 \,,$$

$$v_{1i}(0) = 0 \,, \quad \left.\frac{\mathrm{d}v_{1i}(t)}{\mathrm{d}t}\right|_{t=0} = 0 \,, \quad \bar{\varphi}(0) = 0 \,, \quad \left.\frac{\mathrm{d}\bar{\varphi}(t)}{\mathrm{d}t}\right|_{t=0} = 0 \,. \tag{10.13}$$

Now the contact force between the i-th axle and the beam is

$$\frac{P_1}{2n} + \bar{R}_i(t) \geqq 0 \,. \tag{10.14}$$

If in view of (10.12) expression (10.14) yields a negative value, we shall substitute in Eqs. (10.6) and (10.7)

$$\bar{R}_i(t) = -\frac{P_1}{2n} \,. \tag{10.15}$$

Note: If we consider the special case of $C \to \infty$ and divide the sprung mass into $2n$ parts, we get a sequence of $2n$ separate unsprung masses as shown in Fig. 10.2. In that case the now invalid equations (10.4) to (10.6) are replaced by the equations

$$-\frac{m_1}{2n}\frac{\mathrm{d}^2 v_{1i}(t)}{\mathrm{d}t^2} - \bar{R}_i(t) = 0, \quad i = 1, 2, \dots, 2n \,. \tag{10.16}$$

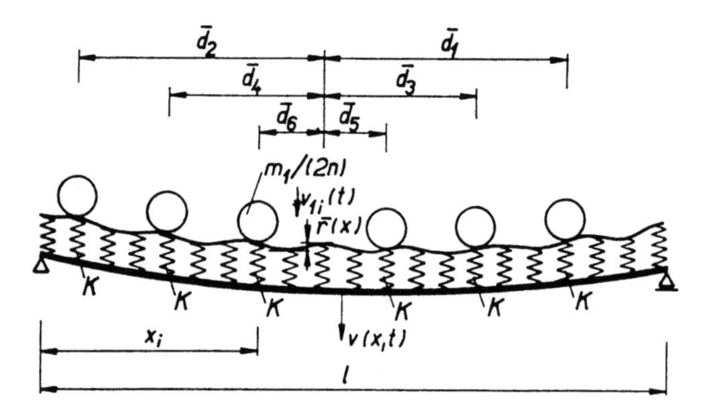

Fig. 10.2. Model of a beam with an elastic layer and irregularities, loaded with a moving series of masses.

10.2 *Solution of the problem*

10.2.1 *Dimensionless parameters*

Like in Chaps. 8 and 9 we shall carry out the numerical evaluation with the aid of the dimensionless parameters

$$\alpha = \frac{c}{2f_{(1)}l} \,, \tag{10.17}$$

$$\gamma_1 = f_1/f_{(1)}\,, \tag{10.18}$$

$$\gamma_2 = f_2/f_{(1)}\,, \tag{10.19}$$

$$\vartheta = \omega_b/f_{(1)}\,, \tag{10.20}$$

$$\vartheta_2 = C_b/(2mf_2)\,, \tag{10.21}$$

$$\varkappa = P/G\,, \tag{10.22}$$

$$\varkappa_1 = P_1/P\,, \quad \varkappa_2 = P_2/P = 1 - \varkappa_1\,, \tag{10.23}$$

$$\lambda = I/(ml^2)\,, \tag{10.24}$$

$$a = \bar{a}/v_0\,, \tag{10.25}$$

$$b = \bar{b}/l\,, \tag{10.26}$$

$$A = \bar{A}/l\,, \tag{10.27}$$

$$B = \bar{B}/l\,, \tag{10.28}$$

$$d_i = \bar{d}_i/l\,, \quad d_{2i} = d_{2i-1}\,, \quad i = 1, 3, 5, \ldots, 2n - 1\,. \tag{10.29}$$

In Eqs. (10.17) to (10.29)

$$v_0 = \frac{Pl^3}{48EJ}\,, \tag{10.30}$$

$$f_1 = \frac{1}{2\pi}\left(\frac{K}{m}\right)^{1/2}, \tag{10.31}$$

$$f_2 = \frac{1}{2\pi}\left(\frac{C}{m}\right)^{1/2}. \tag{10.32}$$

The remaining symbols are the same as in Chap. 9, and quantities a, b, A, B refer to track irregularities shown in Fig. 10.3.

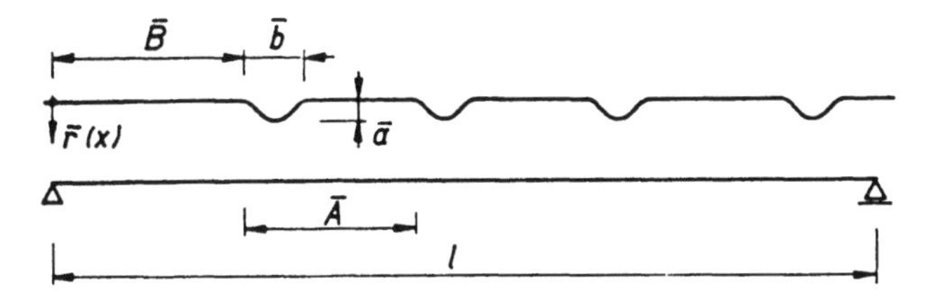

Fig. 10.3. Track irregularities.

The following parameters will also be considered among the input data:

number n – half the number of vehicle axles,

s – number of equations in (9.52) and (9.54), i.e. the number of harmonics,

h – integration step,

number N – following which the machine prints intermediate results.

10.2.2 *Reduction of the equations to the dimensionless form*

For the purposes of numerical evaluation we shall again introduce the independent variables (9.34) and the dependent variables

$$y(\xi, \tau) = v(x, t)/v_0\,, \quad y_{1i}(\tau) = v_{1i}(t)/v_0\,, \quad i = 1, 2, \ldots, 2n\,,$$
$$y_2(\tau) = v_2(t)/v_0\,, \quad \varphi(\tau) = \bar{\varphi}(t)\, l/v_0\,. \tag{10.33}$$

Substitution of (10.17) to (10.33) in (10.4) to (10.7) [with Eq. (10.7) solved by transformation (9.50) same as in paragraph 9.2.3] and rearrangement give

$$\ddot{\varphi}(\tau) = \frac{1}{\lambda\alpha^2} \sum_{i=1}^{n} d_{2i-1}\gamma_2\{\pi^2\gamma_2[z_{2i}(\tau) - z_{2i-1}(\tau)] +$$
$$+ \vartheta_2\alpha[\dot{z}_{2i}(\tau) - \dot{z}_{2i-1}(\tau)]\}\,, \tag{10.34}$$

$$\ddot{y}_2(\tau) = -\frac{1}{\varkappa_2\alpha^2} \sum_{i=1}^{2n} \gamma_2[\pi^2\gamma_2\, z_i(\tau) + \vartheta_2\alpha\, \dot{z}_i(\tau)]\,, \tag{10.35}$$

$$\ddot{y}_{1i}(\tau) = \frac{96n}{\pi^2\varkappa\varkappa_1\alpha^2}\left[\frac{\varkappa_2}{2n} + \frac{\pi^4}{48}\varkappa\gamma_2^2\, z_i(\tau) + \frac{\pi^2}{48}\vartheta_2\varkappa\alpha\gamma_2\, \dot{z}_i(\tau) - R_i(\tau)\right];$$
$$i = 1, 2, 3, \ldots, 2n\,, \tag{10.36}$$

$$\ddot{q}_{(j)}(\tau) = \sum_{i=1}^{2n} \frac{96}{\pi^2\alpha^2}\,\varepsilon_i\left[\frac{\varkappa_1}{2n} + R_i(\tau)\right]\sin j\pi\xi_i - j^4\frac{\pi^2}{\alpha^2}\,q_{(j)}(\tau) - \frac{\vartheta}{\alpha}\,\dot{q}_{(j)}(\tau)\,;$$
$$j = 1, 2, 3, \ldots, s \tag{10.37}$$

The initial conditions are

$$q_{(j)}(0) = 0\,; \quad \dot{q}_{(j)}(0) = 0\,,$$
$$y_{1i}(0) = 0\,; \quad \dot{y}_{1i}(0) = 0\,,$$
$$y_2(0) = 0\,; \quad \dot{y}_2(0) = 0\,,$$
$$\varphi(0) = 0\,; \quad \dot{\varphi}(0) = 0\,. \tag{10.38}$$

The integration range now is

$$0 \leqq \tau \leqq 1 + 2d_1 \tag{10.39}$$

and the coordinate of the contact point of the i-th axle

$$\xi_i = \begin{cases} \tau - d_1 + d_i & \text{for} \quad i = 1, 3, 5, \ldots, 2n - 1 \\ \tau - d_1 - d_{i-1} & \text{for} \quad i = 2, 4, 6, \ldots, 2n \,. \end{cases} \tag{10.40}$$

In Eqs. (10.34) to (10.37)

$$z_i(\tau) = y_2(\tau) - (-1)^i \, d_i \varphi(\tau) - y_{1i}(\tau) \,, \quad i = 1, 2, \ldots, 2n \,, \tag{10.41}$$

$$R_i(\tau) = \bar{R}_i(t)/P = \begin{cases} \dfrac{\pi^4}{48} \varkappa \gamma_1^2 [y_{1i}(\tau) - \varepsilon_i y(\xi_i, \tau) - r(\xi_i)] \\ \text{if} \quad \dfrac{\pi^4}{48} \varkappa \gamma_1^2 [y_{1i}(\tau) - \varepsilon_i y(\xi_i, \tau) - r(\xi_i)] \geqq - \dfrac{\varkappa_1}{2n} \,, \\ \text{otherwise} \quad - \dfrac{\varkappa_1}{2n} \,; \quad i = 1, 2, \ldots, 2n \,. \end{cases} \tag{10.42}$$

The meaning of ε_i, $\delta(\xi)$, $r(\xi)$ is the same as in Chap. 9, Eq. (9.49) is now without subscripts i.

Note: If $\gamma_2 \to \infty$, Eqs. (10.34) to (10.36) do not apply and in the sense of the Note at the end of Sect. 10.1 are replaced by the equation ($\varkappa_1 = 1$)

$$\ddot{y}_{1i}(\tau) = - \frac{96n}{\pi^2 \varkappa \varkappa_1 \alpha^2} R_i(\tau) \,. \tag{10.43}$$

10.2.3 *Numerical solution*

The dimensionless beam deflection is again computed from Eq. (9.54). Analogous to (9.56), the bending moment divided by bending moment M_0 of the centre of the beam loaded at point $x = l/2$ with force P,

$$M_0 = Pl/4 \,, \tag{10.44}$$

is

$$M(\xi, \tau) = \sum_{i=1}^{2n} M_{Ri}(\xi, \tau) + M_\mu(\xi, \tau) \tag{10.45}$$

where $M_\mu(\xi, \tau)$ has the same meaning and the same value as in (9.58), and in view of (10.42) and (10.44)

$$M_{Ri}(\xi, \tau) = \begin{cases} 4\varepsilon_i \left[\dfrac{\varkappa_1}{2n} + R_i(\tau)\right](1 - \xi_i)\,\xi & \text{for} \quad \xi_i \geqq \xi \\ 4\varepsilon_i \left[\dfrac{\varkappa_1}{2n} + R_i(\tau)\right](1 - \xi)\,\xi_i & \text{for} \quad \xi_i \leqq \xi . \end{cases} \tag{10.46}$$

The problem specified by the set of equations (10.34) to (10.37) or (10.43) with the initial conditions (10.38) was again solved by Runge-Kutta-Nyström's method according to [37]. The solution was programmed for a Leo 360 automatic computer. The computation starts with the printing of the input data, with the values of τ, $y(1/2, \tau)$ and $M(1/2, \tau)$ printed after each N-th step. The computation ends with the printing of the largest values of those functions, i.e. max $y(1/2, \tau)$, max $M(1/2, \tau)$ and the values of τ at which the maxima occur. The computer's output is either in the form of tabulated data or recorded on a punched tape which in turn serves as input to a automatic drawing device that draws directly the time variations of deflection and bending moment at the centre of the beam span.

It was found in the course of the computation that the accuracy of results is affected more substantially by a higher number of harmonics, s, than by the length of the integration step, h. A higher number of harmonics is favourable for yet another reason: it does not prolong the com-

Table 10.1 Effect of integration step length and number of equations

Model shown in Fig. ($n = 2$)	Fig. 10.2				Fig. 10.1	
s	1		3		3	5
h	0·001	0·0005	0·001	0·0005	0·001	
$\tau[\max v(\frac{1}{2}, \tau)]$	0·646	0·646	0·649	0·649 5	0·834	0·726
$\max v(\frac{1}{2}, \tau)$	0·940 9436	0·940 9123	0·947 3748	0·947 4257	0·822 2459	0·777 7311
$\tau[\max M(\frac{1}{2}, \tau)]$	0·548	0·548 5	0·573	0·573	0·704	0·699
$\max M(\frac{1}{2}, \tau)$	0·815 3738	0·815 5276	0·821 1936	0·821 1587	0·964 1011	0·732 3058
computing time [minute]	5	9	8	15	13	18

putation as does a shorter integration step. As proved in [78], a certain, optimum, number of steps is, of course, necessary. A comparison of the maximum values of $y(1/2, \tau)$ and $M(1/2, \tau)$ for different numbers of harmonics, s, and different lengths of the integration step, h, is presented in Table 10.1 for the two cases illustrated in Figs. 10.1 and 10.2. As a point of interest, the table also gives the computing time of the Leo 360 computer for the two cases.

10.3 *The effect of various parameters**)

10.3.1 *The effect of speed*

The effect of speed parameter α (10.17) is recorded in Fig. 10.4. The figure summarizes the results of our calculations relating to the model of a Swiss Oerlikon bridge tested at the University of Lausanne, to a German bridge across the river Paar and to an idealized 10-m span concrete plate bridge with ballast. The model of the Oerlikon bridge weighed 3050 N and was tested with and without a gravel bed (Fig. 10.4). The test locomotives were designated $P = 750$ N and Re 4/4. The 19·6-m span Paar bridge was tested with electric locomotives E 03 ($P = 1{\cdot}115$ MN) and E 10 ($P = 0{\cdot}85$ MN), respectively, up to the speed 200 km/h. The concrete plate bridge was calculated for traverse by an electric locomotive E 10; the track irregularities owing to the sleeper effect were assumed to have different depth parameters, i.e. $\bar{a} = 0, 0{\cdot}25, 0{\cdot}5$ mm.

Fig. 10.5 refers to the model of the Oerlikon bridge without a gravel bed, and shows the dependence of the strain of lower fibres, ΔA, on speed. Both the measured and the computed values of the maximum increment of strain of lower fibres at mid-span of the bridge are noted. According to measurement, the static values of this strain range between $A_0 = 419{\cdot}6 \times 10^{-6}$ to $440{\cdot}9 \times 10^{-6}$ while the computed $A_{0t} = 458 \times 10^{-6}$. The computed values of dynamic strain agree well with the measured data.

---

*) The author gratefully acknowledges the permission of the Office for Research and Experiments (ORE) of the International Union of Railways (UIC) to publish Sects. 10.3 and 10.4 and the experimental data in Figs. 10.5 to 10.8. The experimental data are taken over from the research question ORE D 23 "Determination of Dynamic Forces in Bridges", Report No. ORE D 23/RP 16 "Theoretical Study of Dynamic Forces in Bridges", Utrecht, 1970, in which the author collaborated, see [258].

The dependence of the dynamic coefficient on speed is also shown in Fig. 10.6, a review of the results of calculations of the 10 m span reinforced concrete plate bridge traversed by an 0·85 MN electric four-axle locomotive E 10 at different speeds.

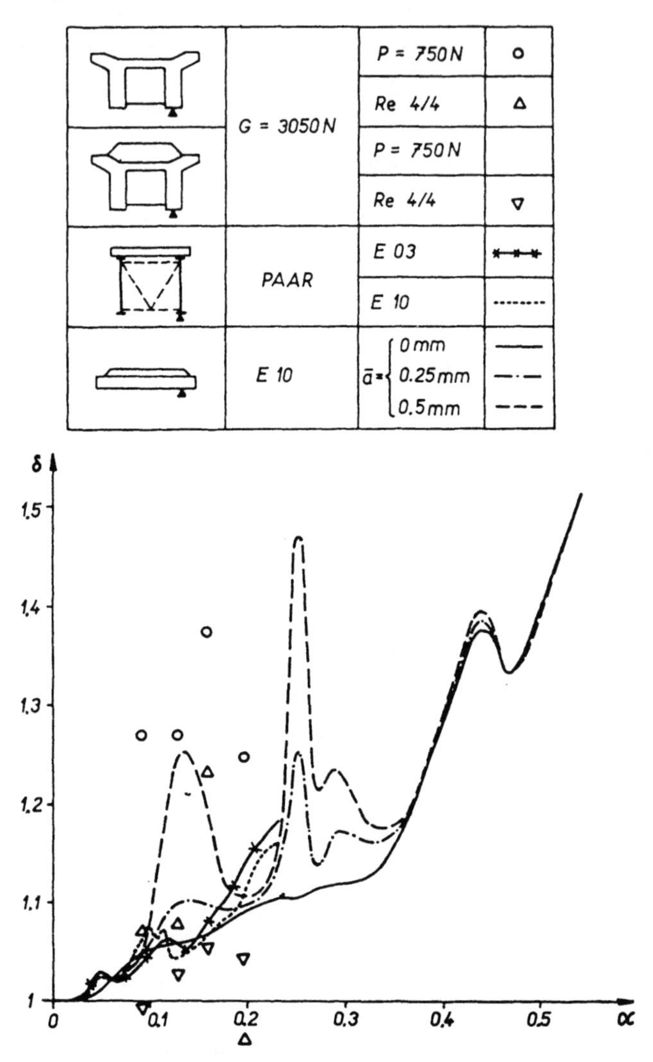

Fig. 10.4. Dependence of the dynamic coefficient δ of stresses on the dimensionless speed parameter α for the model of the Oerlikon bridge ($G = 3050$ kN), for the Paar river bridge and for a reinforced concrete plate.

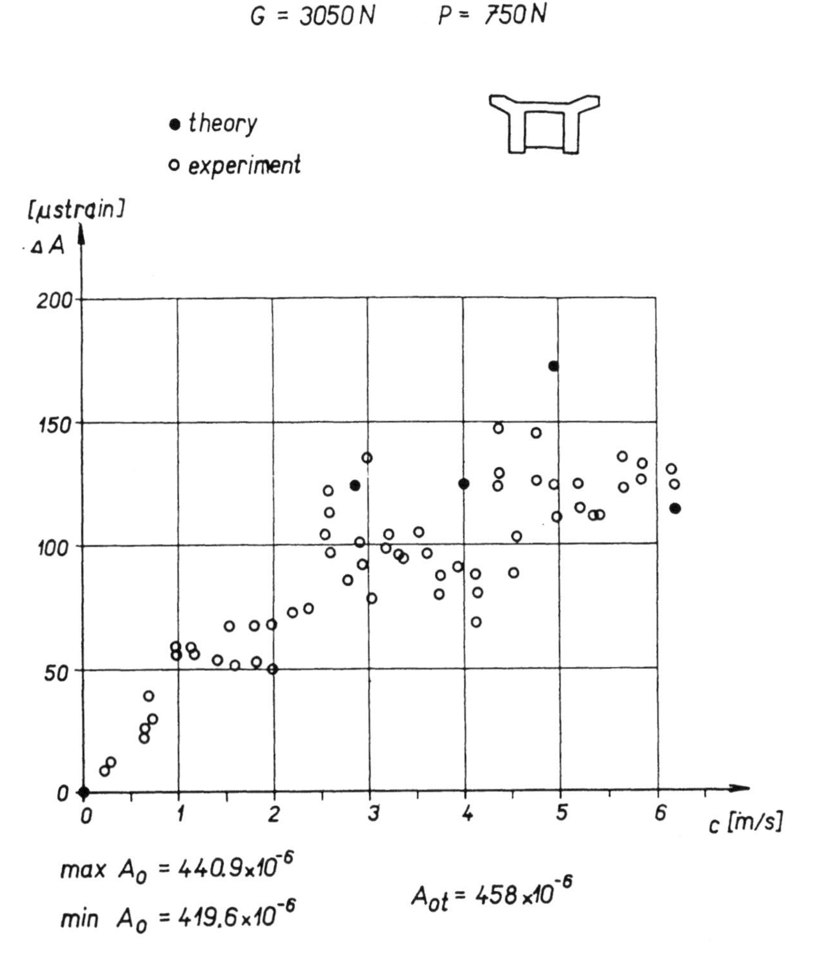

Fig. 10.5. Dependence of the dynamic increment of strain ΔA (in microstrains) on speed c for the model of the Oerlikon bridge.

Consulting Figs. 10.4 to 10.6 we see that the dependence of dynamic stresses in bridges on speed is highly complex. However, the dynamic effects of modern locomotives, the principal subject of our studies, may be said to roughly increase with growing speed.

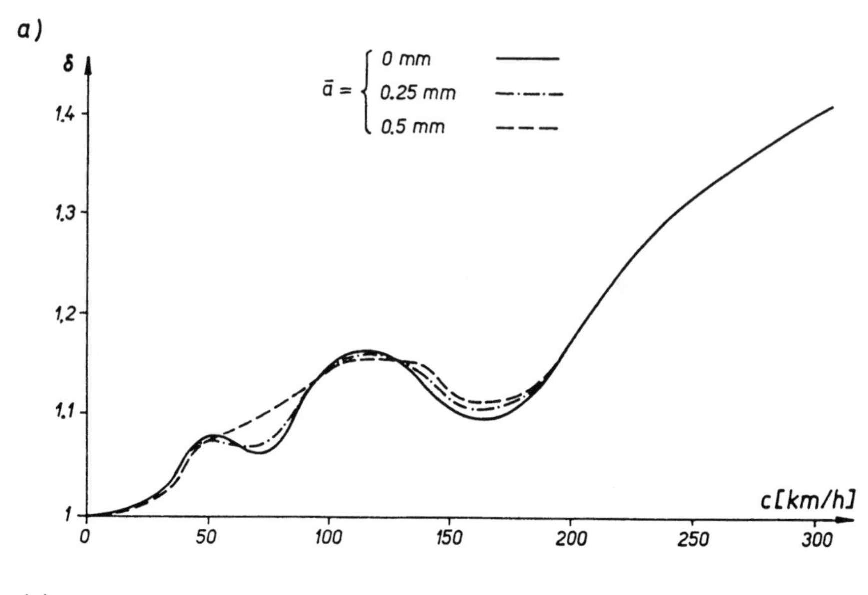

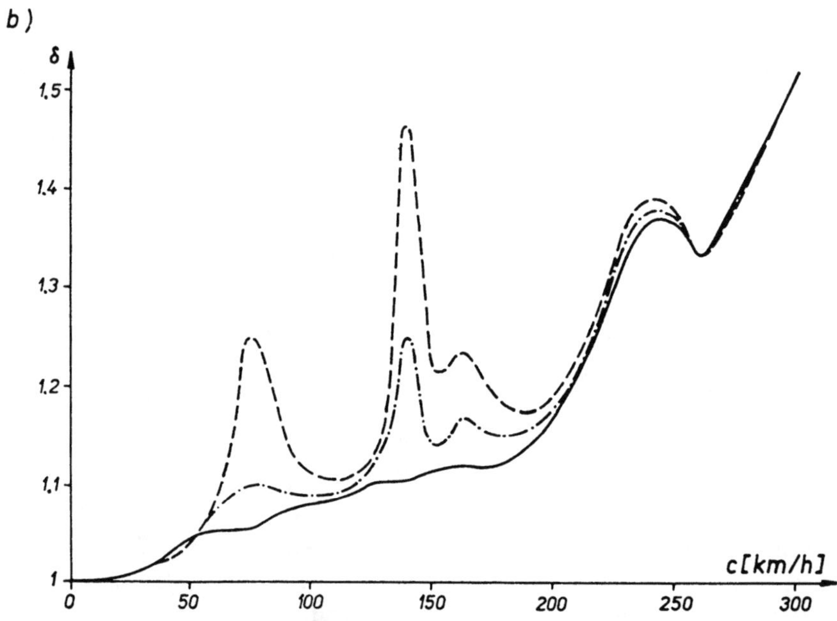

Fig. 10.6. Dependence of the dynamic coefficient δ on speed c for a reinforced concrete bridge with span $l = 10$ m traversed by an E 10 electric locomotive weighing 0·85 MN. Effect of the depth of uneveness: $\bar{a} = 0$, 0·25, 0·5 mm. a) Deflection, b) bending moment.

10.3.2 *The effect of other parameters*

As our theoretical studies suggest, the dynamic coefficient is considerably affected by parameter γ_1 (10.18) – characteristic of the elastic layer stiffness. Unfortunately, parameter γ_1 is very hard to determine even experimentally.

The other parameters of importance are $\varkappa$ (10.22), γ_2 (10.20) and a (10.25).

10.4 *Application of the theory*

10.4.1 *Comparison of theory with experiments*

Fig. 10.7 shows the computed and measured time variations of stresses in the lower fibres at mid-span of the Paar bridge, a 19·6 m span, single-track, slightly skew steel structure with two main girders and a roadway above. The bridge was subjected to tests by a 1·115 MN six-axle electric locomotive E 03 which attained speeds of up to 200 km/h during the tests. Judging by the diagrams in Fig. 10.7, the agreement between theory and experiments is very good; the minute deviations can readily be explained as caused by bridge torsion resulting from the skew support of the main girders, a factor that has not been considered in the theoretical analysis. The stress recorded after the departure of the locomotive from the bridge (Fig. 10.7, right) was produced by the measuring car hauled by the locomotive.

Fig. 10.8 shows the dependence of the dynamic coefficient of the Paar bridge on speed. The experiments conducted by German Federal Railways (DB) in 1965, 1967 and 1968 included measurements of stresses in the lower chord at mid-span of the bridge, at points 105 and 108 (denoted by letters MP in Fig. 10.8). In the measurements made at those points at different times there also appeared slightly different stresses from the static load A_0, which are likewise shown in Fig. 10.8. The test load consisted of the electric locomotives a) E 03 (P = 1·115 MN), b) E 10 (P = 0·85 MN) which travelled over the bridge at speeds of up to 200 km/h. The theoretical dependences obtained by the author were computed for speeds of up to 300 km/h. As the figure suggests, the ex-

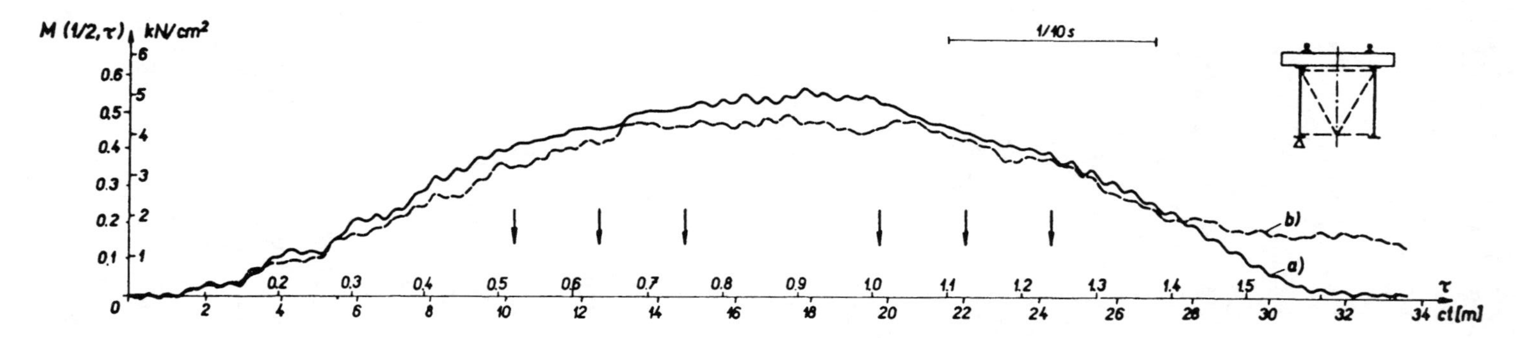

Fig. 10.7. Time variation of stresses at the centre of the Paar river bridge, $l = 19{\cdot}6$ m, electric locomotive E 03, $P = 1{\cdot}115$ MN, $c = 200$ km/h. a) Theory, b) experiment.

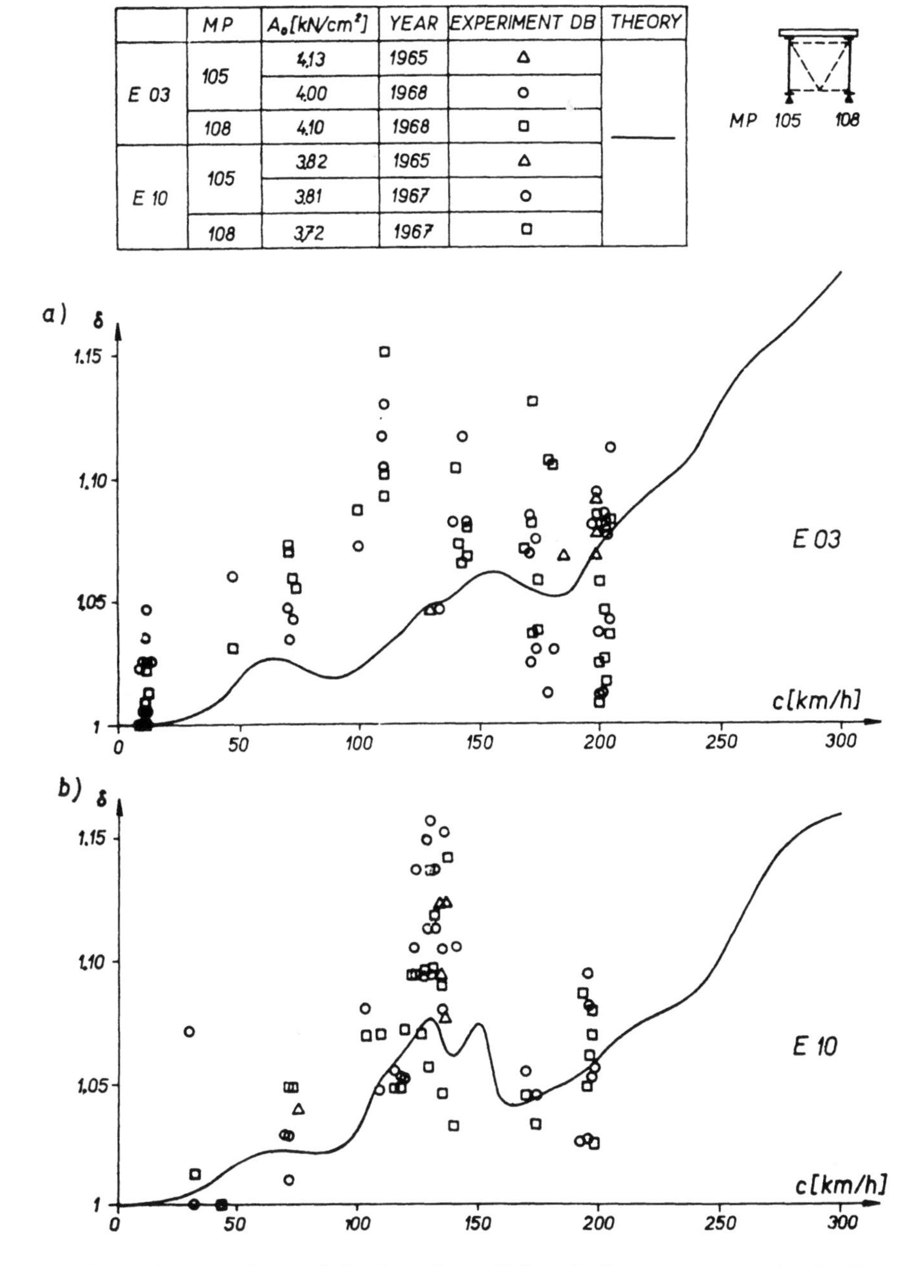

	MP	A_0 [kN/cm²]	YEAR	EXPERIMENT DB	THEORY
E 03	105	4.13	1965	△	
		4.00	1968	○	
	108	4.10	1968	□	——
E 10	105	3.82	1965	△	
		3.81	1967	○	
	108	3.72	1967	□	

Fig. 10.8. Dependence of the dynamic coefficient δ of stresses on speed c for the Paar river bridge. a) electric locomotive E 03, $P = 1{\cdot}115$ MN, b) electric locomotive E 10, $P = 0{\cdot}85$ MN. MP — points of measurement at mid-span of the bridge, A_0 — static values of stress in various years of experiments carried out by DB (German Federal Railways).

perimental results roughly follow the theoretical dependence, a considerable scatter notwithstanding.

It has been shown that the theoretical solution is sensitive to the change of some parameters. Therefore, the values presented in Figs. 10.4 to 10.8 are the results of a larger number of calculations.*)

The problem of the action of vehicles on bridges could further be refined by introducing nonlinear terms, by giving consideration to a larger number of degrees of freedom, to real damping, to other physical constants of the bridge and of the vehicle, etc. — all matters that could easily be mastered by help of large computers.

There are, however, limits to this process of refinement. For a solution of the type we have presented in Chapters 8 to 10 we need to know over 30 parameters whose determination by computation or experiment is very difficult. This makes of them the limiting factor in the solution of problems from the field discussed.

10.4.2 *Dynamic stresses in medium-span railway bridges at high speeds*

In the absence of flat wheels, rail joints, etc., dynamic stresses in bridges with spans less than about 30 m are primarily affected by irregularities arising owing to different deflections of rails above and between sleepers (the so-called sleeper effect). These periodic irregularities cause both the vehicle and the bridge structure to vibrate, with resonant vibration likely to set in at speeds between 100 and 200 km/h — see, for example, Fig. 10.6.

The dynamic effects in bridges grow roughly with growing speed, and are also sensitive to the stiffness of the roadway or of the gravel bed.

10.5 *Additional bibliography*

[24, 33, 52, 121, 135, 137, 200, 229, 231, 240, 242, 245, 246, 250, 251, 257, 258, 261, 272].

---

*) For further details see Report ORE D 23/RP 16 "Theoretical Study of Dynamic Forces in Bridges" mentioned in the footnote on page 165, see [258].

11

Systems of prismatic bars subjected to a moving load

Complicated systems of prismatic bars subjected to a moving load are usually solved by the application of the principle of superposition. Below we will give an outline of the procedure using the general frame system shown in Fig. 11.1 by way of an example and referring to that figure for actual directions of forces and moments acting on the system.

The actual state of bar g, h being traversed by force $P(t)$ (Fig. 11.1a) is resolved in two states (11.1b and 11.1c). In the first state (Fig. 11.1b) the ends of bar g, h – removed from the system – are immobile and the bar is traversed by force $P(t)$. The end moments and forces – considered to be the external load – are the same as in the case of a clamped bar. Therefore the deflection of bar g, h is computed from Eq. (6.25).

In the second state (Fig. 11.1c) the system is subjected only to joint moments and forces that act in the opposite direction than in the first state. Superposition of loads according to Figs. 11.1b and 11.1c gives the external load $P(t)$. Therefore the deformation of the whole system (Fig. 11.1a) is the sum of deformations indicated in Figs. 11.1b and 11.1c.

When force $P(t)$ passes from bar g, h to the neighbouring bar h, $h + 1$, the end deformations and velocities at the instant the force is at point h must be taken for the initial conditions of calculations relating to bar h, $h + 1$.

The above procedure makes it possible to reduce the given case to two partial problems:

1. Motion of concentrated force $P(t)$ along a single-span beam. This problem was solved in Chap. 6, Eq. (6.25). The bar need not always be considered clamped like in Fig. 11.1b but can be supported in any of the known ways – hinged at both ends, hinged at one and clamped at the other end, clamped at one and free at the other end, etc., depending on the specification.

2. Effects of joint forces and moments generally variable in time, acting on the whole system.

In the computation of the deflection of bar g, h subjected to forces $Y_{gh}(t)$, $Y_{hg}(t)$ and moments $M_{gh}(t)$ and $M_{hg}(t)$ at ends g and h, we now

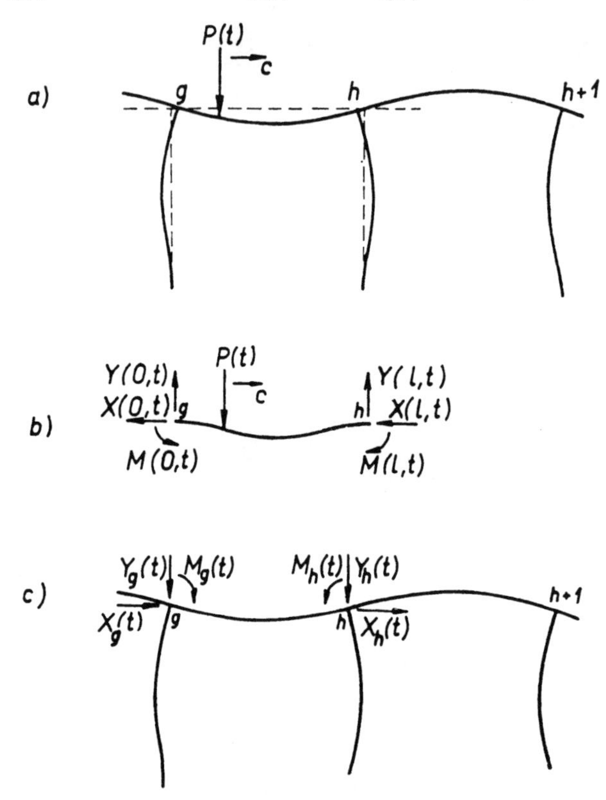

Fig. 11.1. Load and deformation of a frame system; a) actual state, b) a bar removed from the system, traversed by the load, c) load acting on joints of the frame system.

refer to Fig. 11.2 showing the positive direction of end forces and moments, and write the applicable equation

$$EJ \frac{\partial^4 v(x, t)}{\partial x^4} + \mu \frac{\partial^2 v(x, t)}{\partial t^2} = \delta(x)\, Y_{gh}(t) + \delta(x - l)\, Y_{hg}(t) -$$

$$- H_1(x)\, M_{gh}(t) - H_1(x - l)\, M_{hg}(t) \tag{11.1}$$

where $\delta(x)$ is the Dirac function, and $H_1(x)$ the second order impulse function, by means of which the effects of concentrated forces or concentrated bending moments are expressed. In writing the above we neglect the longitudinal vibration of the bar as one of lesser significance than the bending vibration.

Fig. 11.2. Action of end moments and forces generally variable in time.

The procedure of solving Eq. (11.1) is the same as that used in obtaining Eq. (6.25). First we make use of the generalized method of finite integral transformations (6.1), (6.2) and of the properties (4.20a) of the second order impulse function, then apply the Laplace-Carson transformation (1.15). Considering general boundary and initial conditions like in Chap. 6, we get

$$v(x, t) = \sum_{j=1}^{\infty} \left\{ \frac{1}{V_j \omega_{(j)}} v_{(j)}(x) \int_0^t \left[v_{(j)}(0)\, Y_{gh}(\tau) + v_{(j)}(l)\, Y_{hg}(\tau) + \right.\right.$$
$$\left. + v'_{(j)}(0)\, M_{gh}(\tau) + v'_{(j)}(l)\, M_{hg}(\tau) - EJ\, z(0, l, \tau) \right] .$$
$$\sin \omega_{(j)}(t - \tau)\, \mathrm{d}\tau + \frac{\mu}{V_j} G_1(j)\, v_{(j)}(x) \cos \omega_{(j)} t +$$
$$\left. + \frac{\mu}{V_j \omega_{(j)}} G_2(j)\, v_{(j)}(x) \sin \omega_{(j)} t \right\} . \tag{11.2}$$

The procedure makes it possible to solve motion along the system of not only a concentrated force but of the load mass as well. When the solution involves complicated systems, the procedure becomes unwieldy and time consuming even in the simplest case of a moving force of constant magnitude.

11.1 *Frame systems*

By way of an example of more complicated systems, we shall consider a frame, the horizontal bar of which is traversed by a concentrated constant force P at constant speed c (Fig. 11.3a). The bars of the system are of equal length l ($l_{0,1} = l_{1,2} = l_{2,3} = l$), have the same ratio μ/J ($\mu_{0,1}/J_{0,1} = \mu_{1,2}/J_{1,2} = \mu_{2,3}/J_{2,3} = \mu/J$) and the same λ ($\lambda_{0,1} = \lambda_{1,2} = = \lambda_{2,3} = \lambda$) where according to (6.5)

$$\lambda^4 = \omega_{(j)}^2 \frac{\mu l^4}{EJ} . \tag{11.3}$$

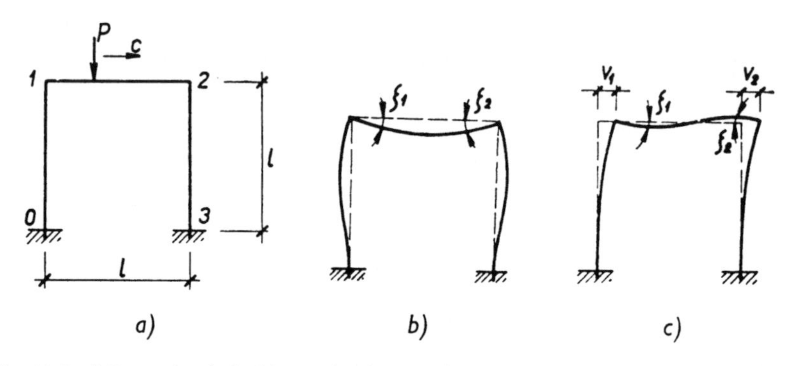

Fig. 11.3. a) Frame loaded with moving force P (actual state), b) symmetric vibration of the frame, c) antisymmetric vibration of the frame.

The speed of motion is assumed to be such that $\omega \neq \omega_{(j)}$ where ω is described by (6.26). When $\omega = \omega_{(j)}$, the computation would proceed as that outlined, for example, in Sect. 1.3, case 1.3.2.2.

11.1.1 *Free vibration*

An analysis of a frame subjected to a moving load starts with the computation of natural frequencies and normal modes of the structure. A method well suited for that purpose is the deformation method of Koloušek [130] according to which the moments and forces in free vibration must satisfy the equilibrium conditions. The conditions for the amplitudes of harmonic vibration are in the form

$$\begin{aligned} \sum_g M_{gh} &= 0 , \\ \sum_g X_{gh} &= 0 , \\ \sum_g Y_{gh} &= 0 . \end{aligned} \tag{11.4}$$

In the above the summation sign includes all the bars issuing from joint g. The conditions refer respectively to the sum of all end moments, horizontal and vertical forces of bars issuing from joint g.

In symmetric vibration (shown for our case in Fig. 11.3b) the joint displacements $v_0 = v_1 = v_2 = v_3 = 0$, and rotations $\zeta_0 = \zeta_3 = 0$, $\zeta_1 = -\zeta_2$. Then according to [130], the moments in joint 1

$$M_{10} = \frac{EJ}{l} F_2(\lambda)\,\zeta_1\,, \quad M_{12} = \frac{EJ}{l} F_2(\lambda)\,\zeta_1 - \frac{EJ}{l} F_1(\lambda)\,\zeta_1\,. \tag{11.5}$$

Functions $F_i(\lambda)$ are tabulated in [130]. Substitution of (11.5) in the first of conditions (11.4) results in the equation

$$2F_2(\lambda) - F_1(\lambda) = 0 \tag{11.6}$$

which in turn gives the values of λ for computing the natural frequencies according to (6.5). The first of λ's satisfying (11.6) turns out to be approximately $\lambda_{(1)} = 3{\cdot}56$.

The normal modes are computed with the aid of integration constants C_i, $i = 1, 2, 3, 4$, set forth in [130] for various cases of displacement and rotation of ends of bars supported in different ways. Constants A, B, C which we have introduced in Eq. (6.3) describing the normal mode of a bar, are bound with constants C_i by the following relations

$$A = \frac{C_3}{C_4}, \quad B = \frac{C_2}{C_4}, \quad C = \frac{C_1}{C_4}\,. \tag{11.7}$$

For bar 0,1

$$C_1 = -\frac{l}{2\lambda^2} F_1(\lambda)\,\zeta_1\,, \quad C_2 = \frac{l}{2\lambda^3} F_3(\lambda)\,\zeta_1\,, \quad C_3 = -C_1\,, \quad C_4 = -C_2\,, \tag{11.8}$$

$$A = -C = -\frac{\sinh\lambda - \sin\lambda}{\cosh\lambda - \cos\lambda}\,, \quad B = -1\,. \tag{11.9}$$

For bar 1,2

$$C_1 = -\frac{l}{2\lambda^2} F_2(\lambda)\,\zeta_1 + \frac{l}{2\lambda^2} F_1(\lambda)\,\zeta_1\,,$$

$$C_2 = -\frac{l}{2\lambda^3} F_4(\lambda)\,\zeta_1 + \frac{l}{2\lambda}\zeta_1 - \frac{l}{2\lambda^3} F_3(\lambda)\,\zeta_1\,,$$

$$C_3 = -C_1\,, \quad C_4 = \frac{l}{\lambda}\zeta_1 - C_2 \tag{11.10}$$

and

$$A = -C = \frac{\cos\lambda \sinh\lambda + \sinh\lambda - \sin\lambda(\cosh\lambda + 1)}{\sin\lambda \sinh\lambda + \cos\lambda + \cosh\lambda(\cos\lambda - 1) - 1},$$

$$B = \frac{-\sin\lambda \sinh\lambda + \cosh\lambda + \cos\lambda(\cosh\lambda - 1) - 1}{\sin\lambda \sinh\lambda + \cos\lambda + \cosh\lambda(\cos\lambda - 1) - 1}. \tag{11.11}$$

And analogously *for bar* 2,3

$$C_1 = -\frac{l}{2\lambda^2} F_2(\lambda)\,\zeta_2, \quad C_2 = \left(-\frac{l}{2\lambda^3} F_4(\lambda) + \frac{l}{2\lambda}\right)\zeta_2,$$

$$C_3 = -C_1, \qquad C_4 = \frac{l}{\lambda}\zeta_2 - C_2 \tag{11.12}$$

with constants A, B, C obtained from (11.7).

In antisymmetric vibration (Fig. 11.3c) $v_0 = v_3 = 0$, $v_1 = v_2$, $\zeta_0 = \zeta_3 = 0$, $\zeta_1 = \zeta_2$. According to [130], the end moments and forces in joint 1

$$M_{1,0} = \frac{EJ}{l} F_2(\lambda)\,\zeta_1 + \frac{EJ}{l^2} F_4(\lambda)\,v_1,$$

$$M_{1,2} = \frac{EJ}{l} F_2(\lambda)\,\zeta_1 + \frac{EJ}{l} F_1(\lambda)\,\zeta_1,$$

$$Y_{1,0} = \frac{EJ}{l^2} F_4(\lambda)\,\zeta_1 + \frac{EJ}{l^3} F_6(\lambda)\,v_1,$$

$$Y_{1,2} = -\frac{\mu l}{2}\omega^2 v_1 = -\frac{EJ}{2l^3}\lambda^4 v_1. \tag{11.13}$$

The horizontal force $Y_{1,2}$ includes half the inertia force of bar 1, 2 performing harmonic vibration with frequency ω and amplitude v_1.

Substitution of (11.13) in the first and third of conditions (11.4) gives

$$[2F_2(\lambda) + F_1(\lambda)]\,\zeta_1 + F_4(\lambda)\frac{v_1}{l} = 0,$$

$$F_4(\lambda)\,\zeta_1 + \left[F_6(\lambda) - \frac{\lambda^4}{2}\right]\frac{v_1}{l} = 0. \tag{11.14}$$

The condition that the determinant of the coefficients at the unknown deformations ζ_1 and v_1 must equal zero results in the equation for computing λ

$$[2\,F_2(\lambda) + F_1(\lambda)]\left[F_6(\lambda) - \frac{\lambda^4}{2}\right] - F_4^2(\lambda) = 0\,. \qquad (11.15)$$

The first of λ's approximately satisfying Eq. (11.15) is $\lambda_{(1)} = 1{\cdot}79$; with this value in, Eq. (6.5) gives the first natural frequency of antisymmetric vibration of the frame.

In the computation of normal modes, consideration must be given to the fact that deformations ζ_1 and v_1 are related via Eqs. (11.14). For example, from the first of those equations, it follows that

$$\frac{v_1}{l} = -\frac{2\,F_2(\lambda) + F_1(\lambda)}{F_4(\lambda)}\,\zeta_1\,. \qquad (11.16)$$

For *bar 0,1* the constants are

$$C_1 = -\frac{l}{2\lambda^2}\,F_1(\lambda)\,\zeta_1 + \frac{1}{2\lambda^2}\,F_3(\lambda)\,v_1\,,$$

$$C_2 = \frac{l}{2\lambda^3}\,F_3(\lambda)\,\zeta_1 + \frac{1}{2\lambda^3}\,F_5(\lambda)\,v_1\,,$$

$$C_3 = -C_1\,, \quad C_4 = -C_2\,. \qquad (11.17)$$

For *bar 1,2*

$$C_1 = -\frac{l}{2\lambda^2}\,F_2(\lambda)\,\zeta_1 - \frac{l}{2\lambda^2}\,F_1(\lambda)\,\zeta_1\,,$$

$$C_2 = -\frac{l}{2\lambda^3}\,F_4(\lambda)\,\zeta_1 + \frac{l}{2\lambda}\,\zeta_1 + \frac{l}{2\lambda^3}\,F_3(\lambda)\,\zeta_1\,,$$

$$C_3 = -C_1\,, \quad C_4 = \frac{l}{\lambda}\,\zeta_1 - C_2\,, \qquad (11.18)$$

and for *bar 2,3*

$$C_1 = -\frac{l}{2\lambda^2}\,F_2(\lambda)\,\zeta_1 + \left(\frac{1}{2\lambda^2}\,F_4(\lambda) + \frac{1}{2}\right)v_1\,,$$

$$C_2 = \left(-\frac{l}{2\lambda^3}\,F_4(\lambda) + \frac{l}{2\lambda}\right)\zeta_1 + \frac{1}{2\lambda^3}\,F_6(\lambda)\,v_1\,,$$

$$C_3 = v_1 - C_1\,, \quad C_4 = \frac{l}{\lambda}\,\zeta_1 - C_2\,. \qquad (11.19)$$

The constants of normal modes of all bars are obtained from (11.7) following substitution of (11.16) in Eqs. (11.17) to (11.19).

In what follows we shall also need — next to the computation of free vibration of the frame — a similar computation made for *bar* 1,2 assuming clamped ends of the bar, i.e.

$$v_{(j)} = 0\,, \quad v'_{(j)}(0) = 0, \quad v_{(j)}(l) = 0\,, \quad v'_{(j)}(l) = 0\,. \qquad (11.20)$$

Substitution of the normal mode in the form of (6.3) in boundary conditions (11.20) gives the equation for computing λ (and in turn the natural frequencies)

$$\cos\lambda \cosh\lambda = 1\,, \quad \lambda_{(1)} = 4{\cdot}73 \qquad (11.21)$$

and the constants

$$A = -C = -\frac{\sinh\lambda - \sin\lambda}{\cosh\lambda - \cos\lambda}\,, \quad B = -1\,. \qquad (11.22)$$

11.1.2 *Forced vibration*

Having obtained the natural frequencies and modes of vibration of our frame structure in the manner outlined, we can set out to compute the vibration brought about by motion of force P. In doing so we shall assume zero initial conditions.

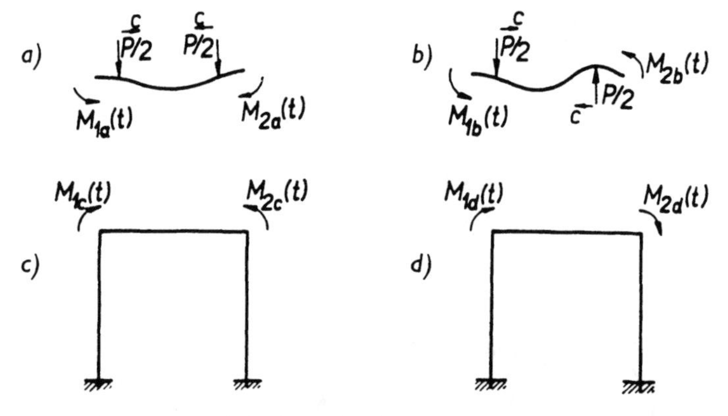

Fig. 11.4. Fictitious states of the frame load (figure represents the actual directions of forces and moments acting on the system). a) symmetric load on a bar removed from the system, b) antisymmetric load on a bar removed from the system, c) symmetric load acting on joints of the system, d) antisymmetric load acting on joints of the system.

First, we resolve the whole case of Fig. 11.3 to four states (Fig. 11.4):

a) Remove the horizontal bar from the frame structure and load it symmetrically with two forces $P/2$, one moving from left to right, the other from right to left, at constant speed c. The bar is further loaded with end moments such that the displacements and rotations produced by them are zero at both ends (Fig. 11.4a). The moments are therefore equal to bending moments at the ends of a clamped bar.

In this state a) the deflection of *bar 1,2*

$$v_{1,2a}(x, t) = \sum_{j=1}^{\infty} \frac{P\, v_{(j)}(x)}{2V_j} \left\{ \frac{1}{\omega_{(j)}^2 - \omega^2} \left[\left(\sin \omega t - \frac{\omega}{\omega_{(j)}} \sin \omega_{(j)} t \right) . \right. \right.$$
$$. (1 - \cos \lambda + A \sin \lambda) + (\cos \omega t - \cos \omega_{(j)} t) .$$
$$\left. . (A + \sin \lambda + A \cos \lambda) \right] + \frac{1}{\omega_{(j)}^2 + \omega^2} \left[\left(\sinh \omega t - \right. \right.$$
$$\left. - \frac{\omega}{\omega_{(j)}} \sin \omega_{(j)} t \right) (B - B \cosh \lambda - C \sinh \lambda) +$$
$$\left. \left. + (\cosh \omega t - \cos \omega_{(j)} t) (C + B \sinh \lambda + C \cosh \lambda) \right] \right\} =$$
$$= \sum_{j=1}^{\infty} v_{(j)}(x)\, q_{(j)a}(t) . \tag{11.23}$$

Expression (11.23) is the sum of Eqs. (6.27) and (6.29) with $P/2$ substituted for P.

Accordingly, the end moments – positive in the clockwise direction – are

$$M_{1a}(t) = \quad M_{1,2a}(0, t) = -EJ \sum_{j=1}^{\infty} v''_{(j)}(0)\, q_{(j)a}(t) ,$$
$$M_{2a}(t) = -M_{1,2a}(l, t) = \quad EJ \sum_{j=1}^{\infty} v''_{(j)}(l)\, q_{(j)a}(t) . \tag{11.24}$$

The natural frequencies and constants substituted in Eqs. (11.23) and (11.24) are those of a clamped bar, i.e. those described by Eqs. (11.21) and (11.22).

b) In the second state (Fig. 11.4b) bar 1,2, removed from the structure, is subjected to antisymmetric forces $P/2$, one moving from left to right, the other from right to left. Like in state a) the bar is also loaded with end moments producing zero displacements and rotations at both ends. In state b) the deflection of *bar 1,2* is

$$v_{1,2b}(x, t) = \sum_{j=1}^{\infty} \frac{P\, v_{(j)}(x)}{2V_j} \left\{ \frac{1}{\omega_{(j)}^2 - \omega^2} \left[\left(\sin \omega t - \frac{\omega}{\omega_{(j)}} \sin \omega_{(j)} t \right) . \right. \right.$$

$$. (1 + \cos \lambda - A \sin \lambda) + (\cos \omega t - \cos \omega_{(j)} t) .$$

$$\left. . (A - \sin \lambda - A \cos \lambda) \right] + \frac{1}{\omega_{(j)}^2 + \omega^2} \left[\left(\sinh \omega t - \right. \right.$$

$$\left. - \frac{\omega}{\omega_{(j)}} \sin \omega_{(j)} t \right) (B + B \cosh \lambda + C \sinh \lambda) +$$

$$\left. \left. + (\cosh \omega t - \cos \omega_{(j)} t) (C - B \sinh \lambda - C \cosh \lambda) \right] \right\} =$$

$$= \sum_{j=1}^{\infty} v_{(j)}(x)\, q_{(j)b}(t) . \tag{11.25}$$

Expression (11.25) is the difference between (6.27) and (6.29), again with $P/2$ substituted for P.

The end moments then are

$$M_{1b}(t) = \quad M_{1,2b}(0, t) = -EJ \sum_{j=1}^{\infty} v''_{(j)}(0)\, q_{(j)b}(t) ,$$

$$M_{2b}(t) = -M_{1,2b}(l, t) = \quad EJ \sum_{j=1}^{\infty} v''_{(j)}(l)\, q_{(j)b}(t) . \tag{11.26}$$

The constants substituted in (11.25) and (11.26) are those of a clamped bar, i.e. (11.21) and (11.22).

c) In the third state the frame is loaded with symmetric joint moments

$$M_{1c}(t) = -M_{1a}(t) , \quad M_{2c}(t) = -M_{2a}(t) \tag{11.27}$$

as indicated in Fig. 11.4c. In view of Eq. (11.2) the deflections of the three bars are respectively

$$v_{0,1c}(x, t) = \sum_{j=1}^{\infty} \frac{v_{(j)}(x)}{V_j \omega_{(j)}} \int_0^t v'_j(l)\, M_{1c}(\tau) \sin \omega_{(j)}(t - \tau)\, d\tau ,$$

$$v_{1,2c}(x, t) = \sum_{j=1}^{\infty} \frac{v_{(j)}(x)}{V_j \omega_{(j)}} \int_0^t [v'_j(0)\, M_{1c}(\tau) + v'_j(l)\, M_{2c}(\tau)] \sin \omega_{(j)}(t - \tau)\, d\tau ,$$

$$v_{2,3c}(x, t) = \sum_{j=1}^{\infty} \frac{v_{(j)}(x)}{V_j \omega_{(j)}} \int_0^t v'_j(0)\, M_{2c}(\tau) \sin \omega_{(j)}(t - \tau)\, d\tau . \tag{11.28}$$

The quantities to be substituted in (11.28) are the natural frequencies of symmetric vibration obtained from (11.6) and the respective constants of the bars, i.e. (11.8) to (11.12).

d) In the fourth state, d, the frame is loaded with antisymmetric joint moments (Fig. 11.4d)

$$M_{1d}(t) = -M_{1b}(t) \, ,$$
$$M_{2d}(t) = -M_{2b}(t) \, . \tag{11.29}$$

By (11.2) the deflections of the three bars are respectively

$$v_{0,1d}(x, t) = \sum_{j=1}^{\infty} \frac{v_{(j)}(x)}{V_j \omega_{(j)}} \int_0^t v_j'(l) \, M_{1d}(\tau) \sin \omega_{(j)}(t - \tau) \, d\tau \, ,$$

$$v_{1,2d}(x, t) = \sum_{j=1}^{\infty} \frac{v_{(j)}(x)}{V_j \omega_{(j)}} \int_0^t [v_j'(0) \, M_{1d}(\tau) + v_{(j)}'(l) \, M_{2d}(\tau)] \sin \omega_{(j)}(t - \omega) \, d\tau \, ,$$

$$v_{2,3d}(x, t) = \sum_{j=1}^{\infty} \frac{v_{(j)}(x)}{V_j \omega_{(j)}} \int_0^t v_j'(0) \, M_{2d}(\tau) \sin \omega_{(j)}(t - \tau) \, d\tau \, . \tag{11.30}$$

The quantities to be substituted in (11.30) are the natural frequencies of antisymmetric vibration obtained from (11.15), and the respective constants of the bars, i.e. (11.17) to (11.19).

The sum of the external forces and moments in states a, b, c and d (Fig. 11.4) gives load P which moves from left to right (Fig. 11.3a). The deflections of the individual bars will therefore be the sum of the deflections in the four states, i.e.

$$v_{0,1}(x, t) = v_{0,1c}(x, t) + v_{0,1d}(x, t) \, ,$$
$$v_{1,2}(x, t) = v_{1,2a}(x, t) + v_{1,2b}(x, t) + v_{1,2c}(x, t) + v_{1,2d}(x, t) \, ,$$
$$v_{2,3}(x, t) = v_{2,3c}(x, t) + v_{2,3d}(x, t) \, . \tag{11.31}$$

11.2 *Continuous beams*

Calculation of simple systems may be carried through to solutions in closed forms. We will show that it is so by solving the continuous beam with two equal spans traversed by a constant force P from left to right at speed c, illustrated in Fig. 11.5a.

Consider zero initial conditions at the instant of arrival of force P on the first span, and to simplify, a speed such that $\omega \neq \omega_{(j)}$, see (6.26). This problem was first solved by R. S. Ayre, G. Ford and L. S. Jacobsen [5]. In the next paragraphs we will evolve a different method of its solution.*)

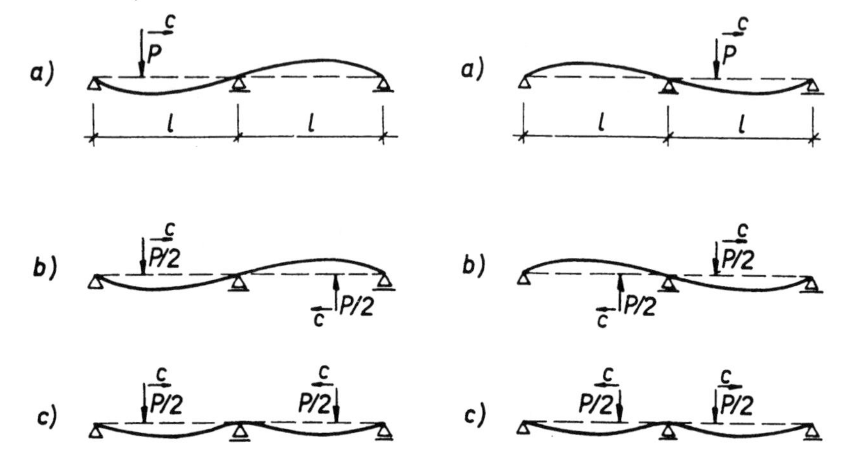

Fig. 11.5. Two-span continuous beam. a) force in the first span, b) antisymmetric load, c) symmetric load.

Fig. 11.6. Two-span continuous beam. a) force in the second span, b) antisymmetric load, c) symmetric load.

11.2.1 *Free vibration*

Like in the preceding case our first task is to establish the natural frequencies and normal modes of vibration of the continuous beam in question.

The antisymmetric vibration is very simple (same as for a simply supported beam). The natural frequencies are obtained from the condition

$$\sin \lambda = 0 \,, \quad \lambda_{(j)} = j\pi \tag{11.32}$$

and the normal modes of the first as well as of the second span ($A = B = C = 0$) from

$$v_{(j)}(x) = \sin \frac{\lambda}{l} x \,. \tag{11.33}$$

*) The dynamics of continuous beams subjected to a moving load is treated at great length by V. Koloušek in [130].

The symmetric vibration is the same as for a bar hinged at one and clamped at the other end. According to [8], the frequency equation is

$$\operatorname{tg} \lambda = \operatorname{tgh} \lambda\,, \quad \lambda_{(1)} = 3{\cdot}927\,. \tag{11.34}$$

The normal modes of the *first span* are obtained from the boundary conditions $v_{(j)}(0) = v''_{(j)}(0) = v_{(j)}(l) = v'_{(j)}(l) = 0$

$$A = C = 0\,, \quad B = -\frac{\sin \lambda}{\sinh \lambda} \tag{11.35}$$

in the form (6.3)

$$v_{(j)}(x) = \sin \frac{\lambda}{l} x + B \sinh \frac{\lambda}{l} x \tag{11.36}$$

and of the *second span* from the conditions $v_{(j)}(0) = v'_{(j)}(0) = v_{(j)}(l) = v''_{(j)}(l) = 0$

$$A = -C = -\frac{\sinh \lambda + \sin \lambda}{\cosh \lambda + \cos \lambda}\,, \quad B = -1 \tag{11.37}$$

in the form

$$v_{(j)}(x) = \sin \frac{\lambda}{l} x + A \cos \frac{\lambda}{l} x - \sinh \frac{\lambda}{l} x - A \cosh \frac{\lambda}{l} x\,. \tag{11.38}$$

The origin of the x-coordinates is assumed to be at the left-hand end of each span.

11.2.2 *Forced vibration*

In the calculation of a continuous beam we make distinction between the case where the force is in the first span (Fig. 11.5a) and the case where the force is in the second span (Fig. 11.6a). In either case the load is resolved into the antisymmetric and the symmetric component producing respectively antisymmetric (Figs. 11.5b and 11.6b) and symmetric (Figs. 11.5c and 11.6c) vibration. The sum of the two loads gives, of course, force P moving from left to right (Figs. 11.5a and 11.6a). In every case span 1 is examined separately from span 2. The origin of the t-coordinates is assumed to be the instant of arrival on the first and second span, respectively.

Force in the *first span* — The initial conditions at the instant the force arrives on the first span are assumed to be zero.

a) Antisymmetric vibration

Span 1: According to Fig. 11.5b the deflection is described by the equation (6.27)

$$v_{1a1}(x, t) = \sum_{j=1}^{\infty} \frac{P\, v_{(j)}(x)}{2V_j(\omega_{(j)}^2 - \omega^2)} \left(\sin \omega t - \frac{\omega}{\omega_{(j)}} \sin \omega_{(j)} t\right). \quad (11.39)$$

Span 2: The situation is analogous, force $-P/2$, motion from right to left (Fig. 11.5b), Eq. (6.29)

$$v_{1a2}(x, t) = \sum_{j=1}^{\infty} \frac{P\, v_{(j)}(x)}{2V_j(\omega_{(j)}^2 - \omega^2)} \left(\sin \omega t - \frac{\omega}{\omega_{(j)}} \sin \omega_{(j)} t\right) \cos \lambda\,. \quad (11.40)$$

Eqs. (11.32) and (11.33) are substituted in (11.39) and (11.40).

b) Symmetric vibration

Span 1: From Eqs. (6.27), (11.34) to (11.36) and Fig. 11.5c

$$v_{1b1}(x, t) = \sum_{j=1}^{\infty} \frac{P\, v_{(j)}(x)}{2V_j} \left[\frac{1}{\omega_{(j)}^2 - \omega^2} \left(\sin \omega t - \frac{\omega}{\omega_{(j)}} \sin \omega_{(j)} t\right) + \right.$$

$$\left. + \frac{B}{\omega_{(j)}^2 + \omega^2} \left(\sinh \omega t - \frac{\omega}{\omega_{(j)}} \sin \omega_{(j)} t\right)\right]. \quad (11.41)$$

Span 2: From Eqs. (6.29), (11.34), (11.37), (11.38) and Fig. 11.5c

$$v_{1b2}(x, t) = \sum_{j=1}^{\infty} \frac{P\, v_{(j)}(x)}{2V_j} \left\{\frac{1}{\omega_{(j)}^2 - \omega^2} \left[\left(\sin \omega t - \frac{\omega}{\omega_{(j)}} \sin \omega_{(j)} t\right) . \right.\right.$$

$$. \left(-\cos \lambda + A \sin \lambda\right) + \left(\cos \omega t - \cos \omega_{(j)} t\right) .$$

$$. \left.\left(\sin \lambda + A \cos \lambda\right)\right] + \frac{1}{\omega_{(j)}^2 + \omega^2} \left[\left(\sinh \omega t - \frac{\omega}{\omega_{(j)}} \sin \omega_{(j)} t\right) . \right.$$

$$. \left(\cosh \lambda + A \sinh \lambda\right) + \left(\cosh \omega t - \cos \omega_{(j)} t\right) .$$

$$. \left.\left.\left(-\sinh \lambda - A \cosh \lambda\right)\right]\right\}. \quad (11.42)$$

The initial conditions for the examination of the system when span 2 is being traversed are obtained from the end values of the traverse of span

1 $(t = l/c)$, i.e.

$$g_{1ik}(x) = v_{1ik}(x, l/c),$$
$$g_{2ik}(x) = \dot{v}_{1ik}(x, l/c), \quad i = a, b, \quad k = 1, 2. \tag{11.43}$$

These values are obtained by the substitution of $t = l/c$ in, and by the differentiation of Eqs. (11.39) to (11.42), respectively.

Force in *the second span* – The origin of the t-coordinates is now the instant the force arrives on span 2. The calculation proceeds as outlined above, the effect of the initial conditions being considered as it was in (6.25) by the last two terms.

a) Antisymmetric vibration

Span 1: From Eqs. (6.25), (6.29), (11.32), (11.33), (11.43) and Fig. 11.6b

$$v_{2a1}(x, t) = \sum_{j=1}^{\infty} \frac{P\, v_{(j)}(x)}{2V_j(\omega_{(j)}^2 - \omega^2)} \left[\sin \omega t \cos \lambda - \frac{\omega}{\omega_{(j)}} \sin \omega_{(j)} \left(t + \frac{l}{c}\right)\right]. \tag{11.44}$$

Span 2: From Eqs. (6.25), (6.27), (11.32), (11.33), (11.43) and Fig. 11.6b

$$v_{2a2}(x, t) = \sum_{j=1}^{\infty} \frac{P\, v_{(j)}(x)}{2V_j(\omega_{(j)}^2 - \omega^2)} \left[\sin \omega t - \frac{\omega}{\omega_{(j)}} \sin \omega_{(j)} \left(t + \frac{l}{c}\right) \cos \lambda\right]. \tag{11.45}$$

b) Symmetric vibration

Span 1: From Eqs. (6.25), (6.29), (11.34) to (11.36), (11.43) and Fig. 11.6c

$$v_{2b1}(x, t) = \sum_{j=1}^{\infty} \frac{P\, v_{(j)}(x)}{2V_j} \left\{\frac{1}{\omega_{(j)}^2 - \omega^2}\left[- \sin \omega \left(t - \frac{l}{c}\right) + \frac{\omega}{\omega_{(j)}} \cdot \left(2 \cos \lambda \sin \omega_{(j)} t - \sin \omega_{(j)} \left(t + \frac{l}{c}\right)\right)\right] + \right.$$
$$\left. + \frac{B}{\omega_{(j)}^2 + \omega^2}\left[- \sinh \omega \left(t - \frac{l}{c}\right) + \frac{\omega}{\omega_{(j)}} \cdot \left(2 \cosh \lambda \sin \omega_{(j)} t - \sin \omega_{(j)} \left(t + \frac{l}{c}\right)\right)\right]\right\}. \tag{11.46}$$

Span 2: From Eqs. (6.25), (6.27), (11.34), (11.37), (11.38), (11.43) and Fig. 11.6c

$$v_{2b2}(x, t) = \sum_{j=1}^{\infty} \frac{P\, v_{(j)}(x)}{2V_j} \left\{ \frac{1}{\omega_{(j)}^2 - \omega^2} \left[\sin \omega t + A \cos \omega t - 2 \frac{\omega}{\omega_{(j)}} \, . \right. \right.$$

$$. \sin \omega_{(j)} t - A \frac{\omega}{\omega_{(j)}} \sin \lambda \sin \omega_{(j)} \left(t + \frac{l}{c} \right) + \frac{\omega}{\omega_{(j)}} \cos \lambda \, .$$

$$. \sin \omega_{(j)} \left(t + \frac{l}{c} \right) - A \cos \lambda \cos \omega_{(j)} \left(t + \frac{l}{c} \right) - \sin \lambda \, .$$

$$. \cos \omega_{(j)} \left(t + \frac{l}{c} \right) \Bigg] + \frac{1}{\omega_{(j)}^2 + \omega^2} \Bigg[- \sinh \omega t - A \cosh \omega t +$$

$$+ 2 \frac{\omega}{\omega_{(j)}} \sin \omega_{(j)} t - \frac{\omega}{\omega_{(j)}} \cosh \lambda \sin \omega_{(j)} \left(t + \frac{l}{c} \right) -$$

$$- A \frac{\omega}{\omega_{(j)}} \sinh \lambda \sin \omega_{(j)} \left(t + \frac{l}{c} \right) + \sinh \lambda \cos \omega_{(j)} \left(t + \frac{l}{c} \right) +$$

$$\left. \left. + A \cosh \lambda \cos \omega_{(j)} \left(t + \frac{l}{c} \right) \right] \right\} . \tag{11.47}$$

The total deflection of the continuous beam then is

$$v(x, t) = \sum_{i=a}^{b} v_{hik}(x, t) , \quad h = 1, 2, \; i = a, b \, , \; k = 1, 2 \tag{11.48}$$

because one can add either the antisymmetric or the symmetric components of span deflection in equations that apply at the same time interval.

11.3 *Trussed bridges*

Since trussed bridges, too, are systems composed of prismatic bars, the calculation of their vibration arising due to the effect of a moving load may be carried out in the way indicated in the introduction to this chapter. However, such calculation is very laborious.

That is why the trusses are very often replaced by a plane girder of constant cross section and analyzed with the aid of the simple relations evolved in Chaps. 1 to 3 and 6. The approximate methods applicable to such analysis have been outlined earlier in the book (paragraph 1.4.4), see also paragraphs 11.4.1 and 11.4.2.

11.4 *Application of the theory*

In bridge engineering one is frequently called to dynamically examine continuous beams of equal spans traversed by a constant or a harmonic force.

Since the first normal mode of vertical vibration of continuous beams of equal spans is a sine curve, sin $\lambda x/l$, the dynamic coefficients can approximately be determined by the application of the simple beam theory, because the first normal mode of a simple beam is also a sine curve. The method has been used with success in two cases on which we will now report.

11.4.1 *Two-span continuous trussed bridge*

Our first theoretical and experimental study involved a continuous trussed steel railway bridge with two equal spans, $l = 43$ m, $E = 21$ MN/cm^2, mass per unit length $\mu = 2{\cdot}4$ t/m (see [75] for further details).

The moment of inertia was computed in two ways: using the areas of the upper and lower chords and formula (1.52) gave $J = 0{\cdot}382$ m^4, while the formula for the static deflection of a continuous beam resulted in $J = 0{\cdot}256$ m^4. Since both methods are but approximate, we took an average $J = 0{\cdot}319$ m^4.

This value in formula (6.5) gives the natural frequency

$$f_{(j)} = \frac{\lambda^2}{2\pi l^2} \left(\frac{EJ}{\mu}\right)^{1/2} = 0{\cdot}449\lambda^2 \, . \tag{11.49}$$

The value of λ is obtained from (11.32) for antisymmetric, and from (11.34) for symmetric vibration. A review of the measured and computed frequencies is in Table 11.1.

The computed first antisymmetric frequency agrees with the measured value.

The first natural frequency of the bridge loaded with a 524.1 steam locomotive was determined from the approximate formula (1.51) as $\bar{f}_{(1)} = 2{\cdot}60$ Hz. The measured value was $\bar{f}_{(1)} = 2{\cdot}42$ Hz, i.e. in good agreement with the theoretical one. The critical speeds obtained through the use of formula (2.16) are set out in Table 11.1. The dynamic coefficient computed from (2.12) for $c_{cr} = 37{\cdot}1$ km/h turned out to be $\delta = 1{\cdot}20$.

Table 11.1 Natural frequencies, damping, critical speeds and dynamic coefficient of a two-span continuous bridge

	Natural frequency of vertical vibration			Logarithmic decrement of damping — measured	Critical speed			Max. dynamic coefficient	
	measured	computed			measured	computed		measured	computed
		antisymmetric	symmetric			antisymmetric	symmetric		
j	$f_{(j)}$ [Hz]			ϑ [1]	c_{cr} [km/h]			δ [1]	
1	4·43	4·43	6·93	0·112	40·3	37·1	58·0	1·20	1·20
2		17·72	22·45			148·4	188·1		
3		39·9	46·8			334	392		

Only the first critical speed – about 40 km/h – was attained during the tests. The largest dynamic coefficient measured at 40·3 km/h was $\delta = 1{\cdot}20$. The agreement between the computed and measured values is good.

11.4.2 *Six-span continuous trussed bridge*

Next we made a theoretical and experimental study of a six-span continuous trussed steel railway bridge (cf. [79]). The four inner spans were each 62·7 m long, the end ones each 61·35 m.

Table 11.2 Natural frequencies of vertical vibration of a six-span continuous trussed bridge

j	Theory	Experiment
	$f_{(j)}$ [Hz]	
1	2·74	2·76
2	2·97	3·09
3	3·50	3·56
4	4·29	4·49
5	5·14	5·06
6	5·85	5·70

In the dynamic computations the bridge was assumed to have six equal spans, each 62·7 m long.

The computed natural frequencies are compared with the measured ones in Table 11.2. As we can see, they are in agreement.

In the dynamic experiments the bridge was traversed alternatively by a ČSD 434.2 steam locomotive and by the same locomotive hauling three Vsa cars. The computation was made using the approximate formulae of Sects. 2.3 and 3.4 (refer to [79] for particulars). The most important results of computation and measurement, i.e. the natural frequencies $\bar{f}_{(1)}$ of the loaded bridge, critical speeds c_{cr} and dynamic coefficients δ are summarized in Table 11.3. The agreement between theory and experiments is good.

Table 11.3 Forced vibration of a six-span continuous trussed bridge

Load	Symbol	Unit	Theory	Experiment
434.2	$\bar{f}_{(1)}$	Hz	2·55	2·58
	c_{cr}	km/h	37·8	39·8
	δ	1	1·17	1·16
434.2	$\bar{f}_{(1)}$	Hz	2·45	2·41
plus	c_{cr}	km/h	36·2	36·8
3 × Vsa	δ	1	1·09	1·11

11.5 *Additional bibliography*

[5, 6, 7, 56, 75, 79, 102, 103, 130, 131, 144, 149, 168, 189, 232, 262, 272].

12

Non-uniform beams and curved bars subjected to a moving load

Complex one-dimensional structures such as non-uniform beams or arches subjected to a moving load were first treated by V. Koloušek [130], [131] who solved most of the topical problems by the application of normal-mode analysis. Apart from his books the literature on the subject is very meagre, the only study worthy of mention being the solution of an arch subjected to a moving continuous load and to the effect of inertia (discussed for a simple beam in Chap. 3) presented by A. B. Morgaevskiĭ [158].

12.1 *Straight non-uniform bar*

A non-uniform bar has both its mass per unit length, $\mu(x)$, and the moment of inertia, $J(x)$, dependent on the length coordinate x. Hence its differential equation is in the form

$$\frac{\partial^2}{\partial x^2}\left[EJ(x)\frac{\partial^2 v(x,t)}{\partial x^2}\right] + \mu(x)\frac{\partial^2 v(x,t)}{\partial t^2} = p(x,t)\,. \tag{12.1}$$

Eq. (12.1) is a partial differential equation with variable coefficients amenable to direct integration only in some special cases. Otherwise it is solved numerically or by one of the approximate methods.

12.1.1 *The network method*

The network method ([186], p. 910) consists in covering the integration range $\langle 0, l\rangle$, $\langle 0, T\rangle$ with a network composed of a finite number of nodal points (Fig. 12.1). If there are I divisions of size h_1 in the direction of axis x, and K divisions of size h_2 in the direction of axis t, then the functional value $v(x, t)$ at the intersection of the i-th and the k-th division is

$$v_{i,k} = v(ih_1, kh_2) \tag{12.2}$$

and similarly so for the remaining functions in Eq. (12.1). At the nodal points the value of the derivative is replaced by the difference, the remainder – dependent on higher derivatives of the function under con-

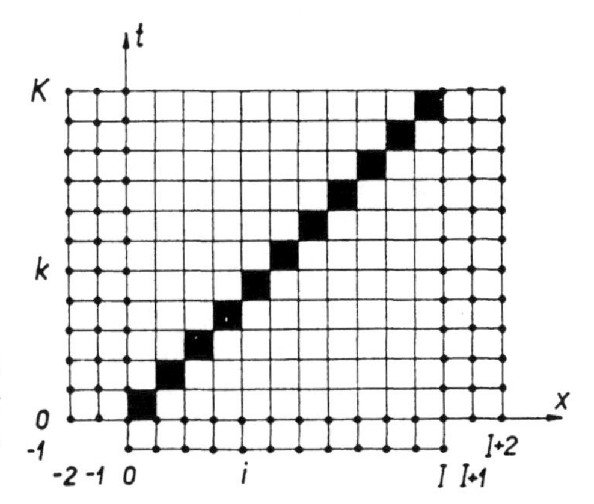

Fig. 12.1. The network in the integration range of a bar subjected to a moving load.

sideration – being neglected in the process. Difference formulae with an error of the order of h^2 are, for example

$$\frac{\partial v_{i,k}}{\partial x} = \frac{1}{2h_1}\left(v_{i+1,k} - v_{i-1,k}\right),$$

$$\frac{\partial^2 v_{i,k}}{\partial x^2} = \frac{1}{h_1^2}\left(v_{i+1,k} - 2v_{i,k} + v_{i-1,k}\right),$$

$$\frac{\partial^3 v_{i,k}}{\partial x^3} = \frac{1}{2h_1^3}\left(v_{i+2,k} - 2v_{i+1,k} + 2v_{i-1,k} - v_{i-2,k}\right),$$

$$\frac{\partial^4 v_{i,k}}{\partial x^4} = \frac{1}{h_1^4}\left(v_{i+2,k} - 4v_{i+1,k} + 6v_{i,k} - 4v_{i-1,k} + v_{i-2,k}\right),$$

$$\frac{\partial^2 v_{i,k}}{\partial x\,\partial t} = \frac{1}{4h_1h_2}\left(v_{i+1,k+1} - v_{i+1,k-1} - v_{i-1,k+1} + v_{i-1,k-1}\right). \tag{12.3}$$

With the above differences in, Eq. (12.1) takes on the form

$$E\frac{1}{h_1^4}\left[\left(J_{i+1} - 2J_i + J_{i-1}\right)\left(v_{i+1,k} - 2v_{i,k} + v_{i-1,k}\right) + \right.$$

$$+ \tfrac{1}{2}(J_{i+1} - J_{i-1})(v_{i+2,k} - 2v_{i+1,k} + 2v_{i-1,k} - v_{i-2,k}) +$$

$$+ J_i(v_{i+2,k} - 4v_{i+1,k} + 6v_{i,k} - 4v_{i-1,k} + v_{i-2,k})] +$$

$$+ \mu_i \frac{1}{h_2^2}(v_{i,k+1} - 2v_{i,k} + v_{i,k-1}) = p_{i,k} \,. \tag{12.4}$$

We have thus obtained in place of the original partial differential equation a set of algebraic equations with n unknowns, by means of which we can find the approximate values of the unknown function in n nodal points of the network. The set can be solved by any of the suitable numerical methods.

Before we start the calculation we need to know the initial conditions. Thus, for example, for zero initial conditions [the second relation is obtained from the first of Eqs. (12.3) written, of course, for $\partial v_{i,k}/\partial t$]

$$v_{i,0} = 0 \,, \quad v_{i,-1} = v_{i,1} \,. \tag{12.5}$$

The boundary conditions, too, have to be introduced in the calculation. For example, for a simply supported beam [the last two relations are obtained from the second of Eqs. (12.3)]

$$v_{0,k} = 0 \,, \quad v_{I,k} = 0 \,, \quad v_{-1,k} = -v_{1,k} \,,$$

$$v_{I+1,k} = -v_{I-1,k} \,. \tag{12.6}$$

It also follows from Eq. (12.4) that we also need to know the values $v_{-2,k}$ and $v_{I+2,k}$ outside the beam. We get them, for example, by linear extrapolation of (12.6),

$$v_{-2,k} = -2v_{1,k} \,, \quad v_{I+2,k} = -2v_{I-1,k} \,. \tag{12.7}$$

The values known from the boundary and initial conditions, or computed with their aid, are marked with circles in Fig. 12.1.

Only after (12.5) to (12.7) have been substituted in (12.4) can we start with the solution of the set of algebraic equations. The calculation is carried out in time rows and is best done on an automatic computer.

Load $p(x, t)$ is wholly arbitrary and its motion along the beam not necessarily at constant speed. We can even express the effect of its inertia action according to Eq. (3.3) by replacing the derivatives in (3.7) by differences (12.3).

In uniform motion of force $P(t)$ along the beam, when $p(x, t) =$ $= \delta(x - ct)\, P(t)$

$$p_{i,k} = \begin{cases} \dfrac{P_k}{h_1} & \text{for} \quad i = kch_2/h_1 \\ 0 & \text{for} \quad i \neq kch_2/h_1\,. \end{cases} \tag{12.8}$$

In Fig. 12.1 the squares in which this load acts are marked in black. As (12.8) implies, it is advantageous to choose the size of the divisions so as to have $ch_2 = h_1$.

12.1.2 *Galerkin's method*

Galerkin's method is one of the approximate methods best suited for solving diverse problems in dynamics of structures. The equation of motion of an element of structure solved with its aid is symbolically written in the form

$$L\,[v] - p = 0 \tag{12.9}$$

where L is the differential operator (not necessarily linear),
v – the structure displacement (even multi-dimensional),
p – the load acting on the structure.

A sequence of linearly independent functions $\varphi_j(x)$ satisfying the boundary conditions is chosen, and the approximate solution sought in the form

$$v_n = \sum_{j=1}^{n} q_j(t)\, \varphi_j(x)\,. \tag{12.10}$$

Function $q_j(t)$ is determined from the condition that expression (12.9) should be orthogonal to functions $\varphi_1, \ldots, \varphi_n$. In this way we get a set of ordinary differential equations

$$\int_0^l \{L[\sum_{j=1}^{n} q_j(t)\, \varphi_j(x)] - p\}\, \varphi_k(x)\, \mathrm{d}x = 0\,, \quad k = 1, 2, \ldots, n \tag{12.11}$$

from which we compute the unknown $q_j(t)$.

By way of an *example* let us solve a simply supported beam traversed at constant speed c by force P (Fig. 12.2). The beam has variable depth,

$h(x)$, mass, $\mu(x)$, as well as moment of inertia, $J(x)$:

$$h(x) = h_0(1 + \sin \pi x/l),$$
$$\mu(x) = \mu_0(1 + \sin \pi x/l),$$
$$J(x) = J_0(1 + \sin \pi x/l)^3. \tag{12.12}$$

We choose functions

$$\varphi_j(x) = \sin \frac{j\pi x}{l} \tag{12.13}$$

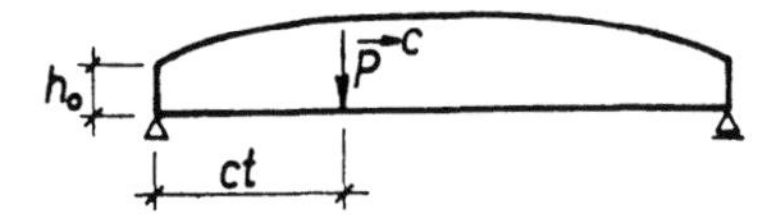

Fig. 12.2. Beam of variable cross section.

which satisfy the boundary conditions (1.2). Restricting our considerations to just one term $j = 1$, $n = 1$, we get by (12.1), (12.10) and (12.12) from Eq. (12.11)

$$\int_0^l \left\{\frac{\partial^2}{\partial x^2}\left[-\frac{\pi^2}{l^2} EJ_0\, q_1(t)\left(1 + \sin\frac{\pi x}{l}\right)^3 \sin\frac{\pi x}{l}\right] + \mu_0\left(1 + \sin\frac{\pi x}{l}\right)\cdot\right.$$
$$\left. \cdot\, \ddot{q}_1(t) \sin\frac{\pi x}{l} - \delta(x - ct)\, P\right\} \sin\frac{\pi x}{l}\, \mathrm{d}x = 0\,.$$

The derivative of the first term of this equation and integration give

$$\ddot{q}_1(t) + \omega_1^2\, q_1(t) = \frac{2P}{\mu_0 l}\, \frac{1}{1 + 8/(3\pi)} \sin \omega t \tag{12.14}$$

where by (1.13) $\omega = \pi c/l$ and the first natural frequency of a non-uniform beam is approximately

$$\omega_1^2 = \frac{195\pi + 608}{20(3\pi + 8)}\, \frac{\pi^4}{l^4}\, \frac{EJ}{\mu_0} = \frac{3{\cdot}5\pi^4}{l^4}\, \frac{EJ}{\mu_0}\,. \tag{12.15}$$

Eq. (12.14) is readily solved by the Laplace-Carson integral transformation using relations (27.4) (27.18) and (27.32). For zero initial conditions and $\omega_1 \neq \omega$, (12.10) thus results in the approximate solution

$$v_1(x, t) = \frac{2P}{\mu_0 l}\, \frac{1}{1 + 8/(3\pi)}\, \frac{1}{\omega_1^2 - \omega^2}\left(\sin \omega t - \frac{\omega}{\omega_1} \sin \omega_1 t\right) \sin \frac{\pi x}{l}$$

analogous to expression (1.31) in Chap. 1.

12.2 *Curved bars*

The equation of motion of a non-uniform bar of spatial curvature can be expressed by a single vector equation, namely

$$\mu(s)\frac{\partial^2 \mathbf{r}(s,t)}{\partial t^2} + 2\mu(s)\,\omega_b\frac{\partial \mathbf{r}(s,t)}{\partial t} - \mathbf{R}(s,t) = \mathbf{p}(s,t) \qquad (12.16)$$

first derived by V. Koloušek [130]. Eq. (12.16) is the vector sum of forces acting on unit length of bar (Fig. 12.3). The independent variable s is measured along the bar centreline. In (12.16) denotes:

$\mathbf{r}(s,t)$ – displacement vector with components $u(s,t)$, $v(s,t)$, $w(s,t)$,
$\mathbf{R}(s,t)$ – vector of restoring elastic force which depends on the type of structure in question and has components $X(s,t)$, $Y(s,t)$, $Z(s,t)$,
$\mathbf{p}(s,t)$ – load vector,
$\mu(s)$ – mass per unit length of curved bar,
ω_b – circular frequency of viscous damping.

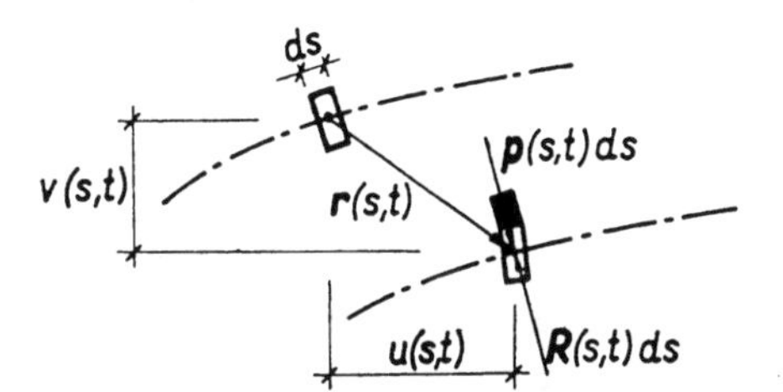

Fig. 12.3. Forces acting on an element of a curved bar.

The system performs harmonic vibration with natural frequencies $\omega_{(j)}$ in infinitely many modes $\mathbf{r}_{(j)}(s)$ whose components are $u_{(j)}(s)$, $v_{(j)}(s)$ and $w_{(j)}(s)$. The force acting on a length element is $\mathbf{R}_{(j)}(s)$. By assumption it therefore holds in free vibration that

$$\mathbf{r}(s,t) = \mathbf{r}_{(j)}(s)\sin\omega_{(j)}t\,, \quad \mathbf{R}(s,t) = \mathbf{R}_{(j)}(s)\sin\omega_{(j)}t\,. \qquad (12.17)$$

The free vibration of such a system is computed from Eq. (12.16) without the terms expressing damping and external load. Substitution of (12.17) in such simplified Eq. (12.16) gives

$$\mathbf{R}_{(j)}(s) = -\omega_{(j)}^2\,\mu(s)\,\mathbf{r}_{(j)}(s) \qquad (12.18)$$

with $\omega_{(j)}$ computed as the characteristic number (eigenvalue) of differential equation (12.18) expressed vectorially.

12.2.1 *Normal-mode analysis*

Knowing the natural frequencies $\omega_{(j)}$ and normal modes $\mathbf{r}_{(j)}(s)$ we can set out to calculate the forced vibration. We expand both the displacement and the external load in normal modes

$$\mathbf{r}(s, t) = \sum_{j=1}^{\infty} q_{(j)}(t)\, \mathbf{r}_{(j)}(s)\,, \tag{12.19}$$

$$\mathbf{p}(s, t) = \sum_{j=1}^{\infty} \mu(s)\, Q_{(j)}(t)\, \mathbf{r}_{(j)}(s) \tag{12.20}$$

where $q_{(j)}(t)$ is the (hitherto unknown) generalized coordinate of displacement, and $Q_{(j)}(t)$ the analogous coordinate of load, both variable in time. The expressions on the right-hand side of Eqs. (12.19) and (12.20) are thought of as vector sums.

$Q_{(j)}(t)$ is computed from (12.20) using the orthogonal properties of normal modes

$$\int \mu(s)\, \mathbf{r}_{(j)}(s)\, \mathbf{r}_{(k)}(s) = \begin{cases} 0 & \text{for } j \neq k \\ R_j & \text{for } j = k \end{cases} \tag{12.21}$$

where

$$R_j = \int \mu(s)\, r^2_{(j)}(s)\, \mathrm{d}s\,. \tag{12.22}$$

The product of the vectors in (12.21) is a scalar product, and the integrals in (12.21) and (12.22) refer to all the bars of the system. $r_{(j)}(s)$ is the modulus of vector $\mathbf{r}_{(j)}(s)$, $r_{(j)}(s) = |\mathbf{r}_{(j)}(s)|$. In view of (12.21), scalar multiplication of both sides of Eq. (12.20) by vector $\mathbf{r}_{(k)}(s)$ and integration over all bars thus give the generalized coordinate of the load

$$Q_{(j)}(t) = \frac{1}{R_j} \int \mathbf{p}(s, t)\, \mathbf{r}_{(j)}(s)\, \mathrm{d}s\,. \tag{12.23}$$

With $Q_{(j)}(t)$ thus obtained, expressions (12.19) and (12.20) are substituted in (12.16). If the latter equation is to apply to all s's, it follows from our procedure that $q_{(j)}(t)$ must satisfy the equation

$$\ddot{q}_{(j)}(t) + 2\omega_b\, \dot{q}_{(j)}(t) + \omega^2_{(j)}\, q_{(j)}(t) = Q_{(j)}(t)\,. \tag{12.24}$$

By solving the above we have also solved our problem, because the resultant displacement is implied by (12.19).

If, for example, an arch is traversed by a concentrated force $\mathbf{P}(t)$, the point of action of which moves according to the law $s = f(t)$, i.e. $\mathbf{p}(s, t) = \delta[s - f(t)]\, \mathbf{P}(t)$, the generalized load (12.23) is

$$Q_{(j)}(t) = \frac{1}{R_j} \mathbf{P}(t)\, \mathbf{r}_{(j)}[f(t)] \,. \tag{12.25}$$

In flat, plane arches subjected to vertical force $|\mathbf{P}(t)| = P(t)$ we may neglect component $u_{(j)}(s)$ of vector $\mathbf{r}_{(j)}(s)$ and get

$$Q_{(j)}(t) = \frac{1}{R_j} P(t)\, v_{(j)}[f(t)] \tag{12.26}$$

where

$$R_{(j)} = \int \mu(s)\, v_{(j)}^2(s)\, \mathrm{d}s \,.$$

The above method of computing the forced vibration of arches is simple enough; what is difficult is the determination of free vibration and normal modes (for details refer to [130]).

12.2.2 *Circular arch of constant cross section*

We shall now restrict our considerations to the solution of a plane circular arch of constant cross section and an incompressible centreline*). With reference to Fig. 12.4 we shall introduce polar coordinates with the origin at the centre of the arch, and denote by r the radius of the arch and by φ the variable angle so that the arch length is $s = \varphi r$ and $\mathrm{d}s = r\, \mathrm{d}\varphi$.

We take out an element of the arch (Fig. 12.4) and write for it the conditions of equilibrium in the tangential and normal directions as well as the moment condition. Neglecting the effect of shear, element rotation and infinitely small quantities of higher orders we get the following equations

$$\frac{\partial T(\varphi, t)}{\partial \varphi} + N(\varphi, t) - \mu r \frac{\partial^2 v(\varphi, t)}{\partial t^2} + r\, p_n(\varphi, t) = 0 \,,$$

*) Deformations of thin curved bars are almost exclusively produced by bending. Compared to them, the elongation of fibres in the bar axis (i.e. in the centreline) as well as the effect of shear are negligibly small.

$$\frac{\partial N(\varphi, t)}{\partial \varphi} - T(\varphi, t) - \mu r \frac{\partial^2 u(\varphi, t)}{\partial t^2} + r\, p_t(\varphi, t) = 0\,,$$

$$\frac{\partial M(\varphi, t)}{\partial \varphi} - r\, T(\varphi, t) + r\, p_m(\varphi, t) = 0 \tag{12.27}$$

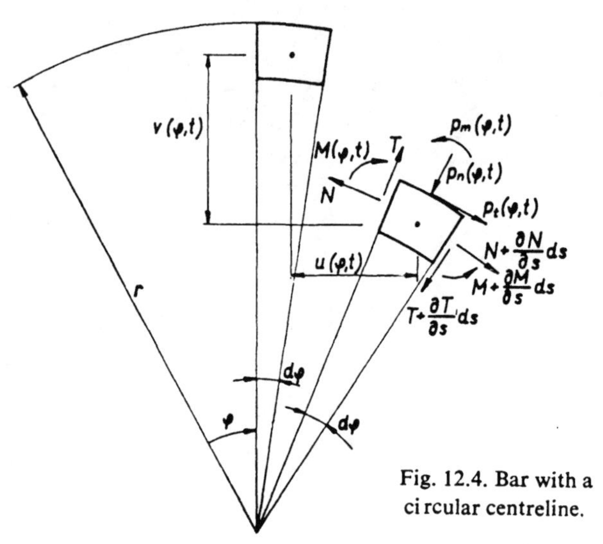

Fig. 12.4. Bar with a circular centreline.

where $u(\varphi, t)$ — tangential displacement,
$v(\varphi, t)$ — radial displacement,
$T(\varphi, t)$ — shear force,
$N(\varphi, t)$ — normal force,
$M(\varphi, t)$ — bending moment,
$p_n(\varphi, t)$ — normal load per unit length of arch,
$p_t(\varphi, t)$ — tangential load per unit length of arch,
$p_m(\varphi, t)$ — bending load per unit length of arch,
μ — constant mass per unit length of arch.

Further, for arches with incompressible centrelines and large radii of curvature (cf. [63], [218])

$$M(\varphi, t) = -\frac{EJ}{r^2}\left[\frac{\partial^2 v(\varphi, t)}{\partial \varphi^2} + \frac{\partial u(\varphi, t)}{\partial \varphi}\right],$$

$$v(\varphi, t) = \frac{\partial u(\varphi, t)}{\partial \varphi}\,. \tag{12.28}$$

By eliminating all unknowns except $u(\varphi, t)$ we can readily reduce Eqs. (12.27) and (12.28) to a single partial differential equation, viz.

$$\frac{\partial^6 u(\varphi, t)}{\partial \varphi^6} + 2\frac{\partial^4 u(\varphi, t)}{\partial \varphi^4} + \frac{\partial^2 u(\varphi, t)}{\partial \varphi^2} + \frac{\mu r^4}{EJ}\left[\frac{\partial^4 u(\varphi, t)}{\partial \varphi^2\, \partial t^2} - \frac{\partial^2 u(\varphi, t)}{\partial t^2}\right] =$$

$$= \frac{r^3}{EJ}\left[r\,\frac{\partial p_n(\varphi, t)}{\partial \varphi} - r\,p_t(\varphi, t) + \frac{\partial^2 p_m(\varphi, t)}{\partial \varphi^2} + p_m(\varphi, t)\right]. \quad (12.29)$$

The normal modes of the arch are obtained from the respective homogeneous equation, i.e. from Eq. (12.29) without the right-hand side, on the assumption that $u(\varphi, t) = u_{(j)}(\varphi) \sin \omega_{(j)} t$. The homogeneous equation ensuing from this is in the form

$$u_{(j)}^{(6)}(\varphi) + 2u_{(j)}^{(4)}(\varphi) + (1 - \lambda^4)\, u_{(j)}^{(2)}(\varphi) + \lambda^4\, u_{(j)}(\varphi) = 0 \quad (12.30)$$

where

$$\lambda^4 = \omega_{(j)}^2 \frac{\mu r^4}{EJ}.$$

The normal modes are considered in the form $u_{(j)}(\varphi) = e^{in\varphi}$ where n are the roots of the characteristic equation

$$n^6 - 2n^4 + (1 - \lambda^4)\, n^2 - \lambda^4 = 0. \quad (12.31)$$

Since the roots of the above equation are $n = \pm n_1, \pm n_2, \pm n_3$, we may also write the normal modes in the form

$$u_{(j)}(\varphi) = A_1 \cos n_1\varphi + B_1 \sin n_1\varphi + A_2 \cos n_2\varphi + B_2 \sin n_2\varphi +$$
$$+ A_3 \cos n_3\varphi + B_3 \sin n_3\varphi. \quad (12.32)$$

The natural frequency $\omega_{(j)}$ and the ratios of constants A and B are computed from the system of homogeneous equations resulting from six boundary conditions of the arch.

Not until we have gone through all the steps just outlined can we tackle the solution of the non-homogeneous equation (12.29). The procedure, though possible, is very time consuming and laborious.

In what follows we shall therefore show an approximate computation by Galerkin's method. Consider a two-hinged arch with central angle Φ and radius r, with the hinges at equal heights (Fig. 12.5a). Along a horizontal roadway masslessly connected to the arch, or directly along the arch, moves a vertical load, $p(x, t)$, whose components in polar coordinates are

$$\begin{aligned} p_n(\varphi, t) &= p(x, t) \cos(\varphi - \Phi/2), \\ p_t(\varphi, t) &= p(x, t) \sin(\varphi - \Phi/2), \\ p_m(\varphi, t) &= 0. \end{aligned} \tag{12.33}$$

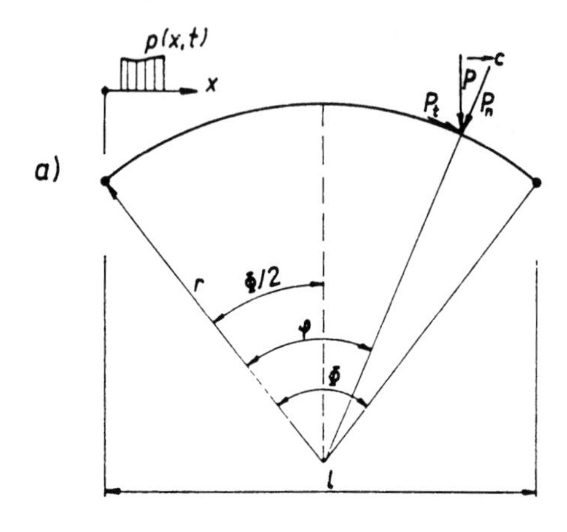

c)

Fig. 12.5. Two-hinged circular arch.
a) Motion of a vertical force,
b) bending vibration,
c) radial vibration.

For vertical force P moving horizontally at uniform speed c

$$p(x, t) = \delta(x - ct) P \approx \frac{P}{r} \delta(\varphi - \Omega t) \tag{12.34}$$

where

$$\Omega = \frac{c}{r} \approx \frac{c\Phi}{l}.$$

If the arch is very flat, we can set $\cos(\varphi - \Phi/2) \approx 1$ and $\sin(\varphi - \Phi/2) \approx \approx \varphi - \Phi/2$, and from (12.33) and (12.34) [see also (4.20a)] get for the moving force

$$p_n(\varphi, t) = \frac{P}{r}\,\delta(\varphi - \Omega t)\,; \quad \frac{\partial p_n(\varphi, t)}{\partial \varphi} = \frac{P}{r}\,H_1(\varphi - \Omega t)\,,$$

$$p_t(\varphi, t) = \frac{P}{r}\left(\varphi - \frac{\Phi}{2}\right)\delta(\varphi - \Omega t)\,,$$

$$p_m(\varphi, t) = 0\,. \tag{12.35}$$

As it is well known, e.g. [131], in two-hinged arches one differentiates between bending vibration (Fig. 12.5b) and radial vibration (Fig. 12.5c)*). It is usually the bending vibration that has the lowest natural frequency, and that is why we shall treat only that vibration by Galerkin's method. In harmony with the second of Eqs. (12.28) and Fig. 12.5b, we will choose for the first approximation the following functions

$$u(\varphi, t) = q(t)\,\frac{\Phi}{2\pi}\left(1 - \cos\frac{2\pi\varphi}{\Phi}\right),$$

$$v(\varphi, t) = q(t)\sin\frac{2\pi\varphi}{\Phi} \tag{12.36}$$

which also satisfy the boundary conditions of a two-hinged bar

$$v(0, t) = 0\,, \quad v(\Phi, t) = 0\,, \quad u(0, t) = 0\,, \quad u(\Phi, t) = 0\,,$$
$$M(0, t) = 0\,, \quad M(\Phi, t) = 0\,.$$

Once we have established expressions (12.35) and (12.36) we are in a position to solve Eq. (12.29) by Galerkin's method. By (12.11)

$$\int_0^{\Phi}\left[q(t)\left(\frac{2\pi}{\Phi}\right)^5 \cos\frac{2\pi\varphi}{\Phi} - 2q(t)\left(\frac{2\pi}{\Phi}\right)^3 \cos\frac{2\pi\varphi}{\Phi} + q(t)\,\frac{2\pi}{\Phi}\cos\frac{2\pi\varphi}{\Phi} + \right.$$
$$+ \frac{\mu r^4}{EJ}\,\ddot{q}(t)\,\frac{2\pi}{\Phi}\cos\frac{2\pi\varphi}{\Phi} - \frac{\mu r^4}{EJ}\,\ddot{q}(t)\,\frac{\Phi}{2\pi}\left(1 - \cos\frac{2\pi\varphi}{\Phi}\right) -$$
$$\left. - \frac{r^4}{EJ}\,\frac{P}{r}\,H_1(\varphi - \Omega t) + \frac{r^4}{EJ}\,\frac{P}{r}\left(\varphi - \frac{\Phi}{2}\right)\delta(\varphi - \Omega t)\right].$$
$$\cdot\,\frac{\Phi}{2\pi}\left(1 - \cos\frac{2\pi\varphi}{\Phi}\right)\mathrm{d}\varphi = 0\,.$$

*) Radial vibration does not satisfy the assumption concerning the incompressibility of the centreline. However, in large-span, thin, curved bars the elongation of the centreline is very small compared with the effect of bending.

Integration of the above results in the ordinary differential equation

$$\ddot{q}(t) + \omega_1^2\, q(t) = \frac{1}{1 + 3\Phi^2/(4\pi^2)} \left[\frac{2P}{\mu r \Phi} \sin \omega t + \frac{P}{\mu r \pi} \cdot \left(\Omega t - \frac{\Phi}{2} \right) (1 - \cos \omega t) \right] \tag{12.37}$$

where

$$\omega = \frac{2\pi\Omega}{\Phi}, \qquad \omega_1^2 = \frac{(4\pi^2 - \Phi^2)^2}{1 + 3\Phi^2/(4\pi^2)} \frac{1}{\Phi^4 r^4} \frac{EJ}{\mu}. \tag{12.38}$$

ω_1 is the first approximation to the first natural frequency of bending vibration of the arch.

Eq. (12.37) is readily solved by the Laplace-Carson integral transformation. For zero initial conditions and $\omega_1 \neq \omega$, relations (27.18), (27.37), (27.32), (27.30), (27.4), (27.17), (27.15), (27.33), (27.65) will give the following result:

$$\begin{aligned} q(t) = {} & \frac{1}{1 + 3\Phi^2/(4\pi^2)} \left\{ \frac{2P}{\mu r \Phi} \frac{1}{\omega_1^2 - \omega^2} \left(\sin \omega t - \frac{\omega}{\omega_1} \sin \omega_1 t \right) + \right. \\ & + \frac{P}{\mu r \pi \omega_1^2} \left[\Omega t - \frac{\Omega}{\omega_1} \sin \omega_1 t - \frac{\Phi}{2} (1 - \cos \omega_1 t) + \frac{\Phi}{2} \frac{\omega_1^2}{\omega_1^2 - \omega^2} \cdot \right. \\ & \cdot (\cos \omega t - \cos \omega_1 t) - \frac{\omega_1^2}{\omega_1^2 - \omega^2} \Omega t \cos \omega t - \\ & \left. \left. - \frac{2\omega\Omega\omega_1^2}{(\omega_1^2 - \omega^2)^2} \sin \omega t + \frac{(\omega_1^2 + \omega^2)\, \Omega \omega_1}{(\omega_1^2 - \omega^2)^2} \sin \omega_1 t \right] \right\}. \end{aligned} \tag{12.39}$$

$q(t)$ is then substituted in (12.36) to get the resultant displacements. In (12.39) the first expression represents the solution due to the radial action, the second, with the square brackets, that due to the tangential component of force P.

If the arch radius is increased, $r \to \infty$, $\Phi \to 0$, $\Phi r \to l$, (12.38) gives the second natural frequency and (12.36) and (12.39) a result analogous to (1.31), Chap. 1, for the second normal mode of a simply supported beam.

12.3 *Application of the theory*

The theory expounded in the foregoing is used in dynamic calculations of large-span bridges whose supporting structure is usually formed by a non-uniform beam or arch. Because of the large ratio between the masses of bridge structure and vehicle, one can merely take for the load a moving constant or harmonic force. Example calculations and experimental results relating to problems of this sort are presented in [130] and [131].

12.4 *Additional bibliography*

[14, 24, 128, 129, 130, 131, 158].

13

Infinite beam on elastic foundation

The problem of a force moving along an infinite beam on an elastic foundation is of great theoretical and practical significance. It was first solved by S. P. Timoshenko [217], and for the case of a constant force theoretically refined by J. Dörr [54] as to the transient phenomenon, and by J. T. Kenney [122]. The motion of a harmonic force was considered by P. M. Mathews [152], that of a sprung mass with two harmonic forces by E. G. Goloskokov and A. P. Filippov [98]. The case of an infinite beam traversed by a force generally variable in time was approximately solved by the author [68].

In this chapter we shall present a detailed solution of the problem of a constant force moving along an infinite beam on an elastic foundation, including in it all possible speeds and values of viscous damping.

13.1 *Formulation of the problem*

We will formulate the problem as follows: an infinite beam on an elastic foundation is traversed by a constant force P moving from infinity to infinity at constant speed c. The elastic foundation is assumed to be of the Winkler type, i.e. one with the foundation reaction directly proportional to beam deflection.*)

On these assumptions the transverse vibration of the beam is described by the differential equation

$$EJ\,\frac{\partial^4 v(x,t)}{\partial x^4} + \mu\,\frac{\partial^2 v(x,t)}{\partial t^2} + 2\mu\omega_b\,\frac{\partial v(x,t)}{\partial t} + kv(x,t) = P\,\delta(x-ct) \quad (13.1)$$

*) The case of a non-Winkler type of foundation (i.e. when the non-loaded parts of the foundation also deflect) was solved by W. J. van der Eb and A. D. de Pater: Calcul des barres de longeur infinie sur appuis élastiques avec observation de l'affaissement que subit un appui, en chargeant un autre appui. Bull. mens. Assoc. Intern. Congrès Chemins de fer, 28 (1951), No 8, 511—559.

where k – the coefficient of Winkler foundation [force per length squared], and the remaining symbols are as explained in Chap. 1.

For the present we shall specify the boundary and initial conditions but verbally, stipulating that at infinite distance to the right as well as to the left of force P, the deflection, the slope of the deflection line, the bending moment and the shear force should be zero at each instant.

For the problem formulated as above there comes into existence the so-called quasi-stationary state in which the beam is at rest relative to the moving coordinate system. This state comes into being after a sufficiently long time of load travel if the moving load – otherwise no longer dependent on time – depends only on the distance from the origin of the coordinate axes (which also moves uniformly). For the purpose of further discussion it is convenient to introduce the new independent variable

$$s = \lambda(x - ct) \tag{13.2}$$

which in dimensionless form expresses the fact that its origin moves together with the load at uniform speed c. By convention

$$\lambda = \left(\frac{k}{4EJ}\right)^{1/4}. \tag{13.3}$$

For the quasi-stationary state we shall then assume that the solution $v(x, t)$ of the equation (13.1) will be in the form

$$v(x, t) = v_0 v(s) \tag{13.4}$$

where $v(s)$ – the dimensionless deflection of beam,

$$v_0 = \frac{P}{8\lambda^3 EJ} = \frac{P\lambda}{2k} \tag{13.5}$$

– the static deflection of beam, $v(0) = v_0$, underneath immobile load P – cf. [68], [72], [83].

According to (1.8) the dimensionless form of the Dirac function $\delta(x)$ on the right-hand side of (13.1) is

$$\delta(s) = \frac{1}{\lambda}\,\delta(x) \tag{13.6}$$

with $\int_{-\infty}^{\infty} \delta(x)\,\mathrm{d}x = 1$

$$\int_{-\infty}^{\infty} \bar{\delta}(s)\,\mathrm{d}s = \int_{-\infty}^{\infty} \frac{1}{\lambda}\,\delta(x)\,\lambda\,\mathrm{d}x = 1\,. \tag{13.7}$$

Differentiation of expressions (13.2) and (13.4) gives

$$\frac{\partial s}{\partial x} = \lambda\,, \quad \frac{\partial s}{\partial t} = -\lambda c\,,$$

$$\frac{\partial^4 v(x,t)}{\partial x^4} = \lambda^4 v_0 \frac{\mathrm{d}^4 v(s)}{\mathrm{d}s^4}\,, \quad \frac{\partial v(x,t)}{\partial t} = -\lambda c v_0 \frac{\mathrm{d}v(s)}{\mathrm{d}s}\,,$$

$$\frac{\partial^2 v(x,t)}{\partial t^2} = \lambda^2 c^2 v_0 \frac{\mathrm{d}^2 v(s)}{\mathrm{d}s^2}\,. \tag{13.8}$$

Let us further introduce the dimensionless parameters

$$\alpha = \frac{c}{c_{cr}} = \frac{c}{2\lambda}\left(\frac{\mu}{EJ}\right)^{1/2}, \tag{13.9}$$

$$\beta = \left(\frac{\mu}{k}\right)^{1/2} \omega_b\,, \tag{13.10}$$

where

$$c_{cr} = 2\lambda\left(\frac{EJ}{\mu}\right)^{1/2}, \tag{13.11}$$

to express the effect of speed — parameter α, and the effect of damping — parameter β.

Substitution of expressions (13.6), (13.8) to (13.10) in the primary equation (13.1) and rearrangement give the ordinary differential equation of the fourth order

$$\frac{\mathrm{d}^4 v(s)}{\mathrm{d}s^4} + 4\alpha^2 \frac{\mathrm{d}^2 v(s)}{\mathrm{d}s^2} - 8\alpha\beta \frac{\mathrm{d}v(s)}{\mathrm{d}s} + 4v(s) = 8\bar{\delta}(s)\,. \tag{13.12}$$

Solution $v(s)$ of Eq. (13.12) is the particular integral of Eq. (13.1) and merely expresses the steady-state vibration; although it fails to satisfy the initial conditions, it may be shown (see [54]) that the quasi-stationary state sets in with adequate accuracy in a comparatively short time interval after the moving force starts to act.

We are now in a position to formulate the boundary conditions of Eq. (13.12) even mathematically, namely

$$\text{for } s \to +\infty,\ s \to -\infty: \quad v(s) = v'(s) = v''(s) = v'''(s) = 0\,. \tag{13.13}$$

We will solve Eq. (13.12) with its boundary conditions (13.13) by the method of Fourier integral transformations using the following fundamental relations

$$v(s) = \frac{1}{2\pi}\int_{-\infty}^{\infty} V(q)\,\mathrm{e}^{\mathrm{i}sq}\,\mathrm{d}q\,,$$

$$V(q) = \int_{-\infty}^{\infty} v(s)\,\mathrm{e}^{-\mathrm{i}qs}\,\mathrm{d}s \tag{13.14}$$

where q is a variable in the complex plane, and $V(q)$ the transform of function $v(s)$ – see (27.78).

In view of boundary conditions (13.13) and expressions (27.79) to (27.84) transforming Eq. (13.12) in the above way gives

$$q^4\,V(q) - 4\alpha^2 q^2\,V(q) - \mathrm{i}\,8\alpha\beta q\,V(q) + 4V(q) = 8 \tag{13.15}$$

from which it follows that

$$V(q) = \frac{8}{q^4 - 4\alpha^2 q^2 - \mathrm{i}\,8\alpha\beta q + 4}\,. \tag{13.16}$$

According to (13.14) the solution of (13.12) was thus obtained in the form of the Fourier integral of the function of a complex variable

$$v(s) = \frac{4}{\pi}\int_{-\infty}^{\infty} \frac{\mathrm{e}^{\mathrm{i}sq}}{q^4 - 4\alpha^2 q^2 - \mathrm{i}\,8\alpha\beta q + 4}\,\mathrm{d}q\,. \tag{13.17}$$

13.2 *Poles of the function of a complex variable*

The poles of the function of the complex variable in the integrand of (13.17) will be assumed to be in the form (Fig. 13.1a)

$$A_1 = a_1 + \mathrm{i}b\,,\quad A_2 = -a_1 + \mathrm{i}b\,,$$

$$A_3 = a_2 - \mathrm{i}b\,,\quad A_4 = -a_2 - \mathrm{i}b\,. \tag{13.18}$$

As a matter of fact, in this case the poles are determined by the roots of the denominator $Q(q)$ in (13.17) and for light damping may be assumed to have the form of (13.18). The values of a_1, a_2, b are computed from the condition

$$Q(q) = q^4 - 4\alpha^2 q^2 - i\,8\alpha\beta q + 4 = \\ = (q - A_1)(q - A_2)(q - A_3)(q - A_4). \tag{13.19}$$

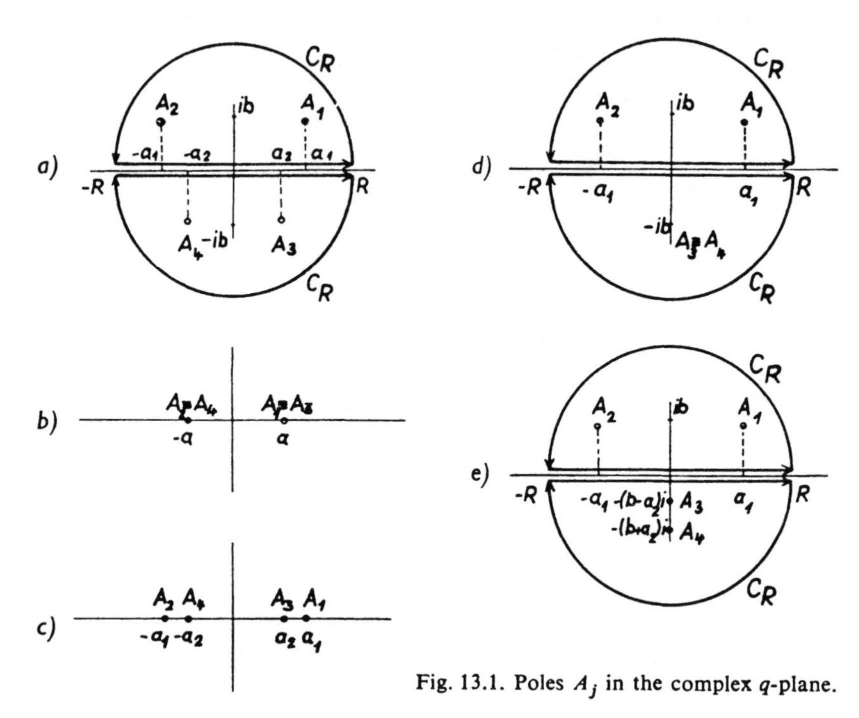

Fig. 13.1. Poles A_j in the complex q-plane.

On substituting (13.18) in (13.19) we find that it is necessary to satisfy the following relations:

$$2b^2 - a_1^2 - a_2^2 = -4\alpha^2, \\ 2bi(a_2^2 - a_1^2) = -8\alpha\beta i, \\ (a_1^2 + b^2)(a_2^2 + b^2) = 4. \tag{13.20}$$

From the first two of Eqs. (13.20) we get

$$a_1^2 = 2\alpha^2 + b^2 + 2\alpha\beta/b, \tag{13.21}$$

$$a_2^2 = 2\alpha^2 + b^2 - 2\alpha\beta/b, \tag{13.22}$$

and after substitution in the third of (13.20)

$$b^6 + 2\alpha^2 b^4 + (\alpha^4 - 1) b^2 - \alpha^2\beta^2 = 0\,. \tag{13.23}$$

So as not to lose the sense of expressions (13.18) we shall take only the positive one of the six roots of Eq. (13.23). In fact, according to Descartes' rule of signs ([186], sect. 31.2, p. 962), Eq. (13.23) always has a positive root at $\alpha \geqq 0$, $\beta \geqq 0$. It is readily solved as an equation of the third degree in b^2 ([186], Sect. 1.20, p. 67) and a_1, a_2 are then obtained from (13.21) and (13.22).

We will now take up some special cases of parameters α and β, and some approximate solutions of Eqs. (13.21) to (13.23).

13.2.1 *Static case* ($\alpha = 0$)

$$b = 1\,, \quad a_1 = 1\,, \quad a_2 = 1 \tag{13.24}$$

13.2.2 *Case with no damping* ($\beta = 0$)

13.2.2.1 $\alpha < 1\,,\quad \beta = 0$

$$b = (1 - \alpha^2)^{1/2}\,, \quad a = a_1 = a_2 = (1 + \alpha^2)^{1/2} \tag{13.25}$$

13.2.2.2 $\alpha = 1\,,\quad \beta = 0$ (Fig. 13.1b)

$$b = 0\,, \quad a = a_1 = a_2 = 2^{1/2} \tag{13.26}$$

13.2.2.3 $\alpha > 1\,,\quad \beta = 0$ (Fig. 13.1c)

$$b = 0\,,$$

$$a_{1,2} = \{2[\alpha^2 \pm (\alpha^4 - 1)^{1/2}]\}^{1/2} = (\alpha^2 + 1)^{1/2} \pm (\alpha^2 - 1)^{1/2} \tag{13.27}$$

13.2.3 *Light damping* ($\beta \ll 1$) – *approximate solutions*

13.2.3.1 $\alpha < 1,\quad \beta \ll 1$

$$b \approx (1 - \alpha^2)^{1/2}\,, \quad a_{1,2} \approx \left[1 + \alpha^2 \pm \frac{2\alpha\beta}{(1 - \alpha^2)^{1/2}}\right]^{1/2} \tag{13.28}$$

13.2.3.2 $\alpha = 1\,,\quad \beta \ll 1$

$$b \approx 2^{-1/4}\beta^{1/2}\,, \quad a \approx a_1 \approx a_2 \approx 2^{1/2}(1 + 2^{-3/4}\beta^{1/2}) \tag{13.29}$$

13.2.3.3 $\alpha > 1\,, \quad \beta \ll 1$

$$b \approx \frac{\alpha\beta}{(\alpha^4 - 1)^{1/2}}\,, \quad a_{1.2} \approx (\alpha^2 + 1)^{1/2} \pm (\alpha^2 - 1)^{1/2} \tag{13.30}$$

13.2.4 *Critical damping* ($\beta = \beta_{cr}$)

Critical damping occurs whenever $a_2 = 0$ (Fig. 13.1d). Then from the condition $2\alpha\beta_{cr}/b = 2\alpha^2 + b^2$, which follows from Eq. (13.22), it is

$$\alpha\beta_{cr} = \tfrac{1}{2}b(2\alpha^2 + b^2)\,. \tag{13.31}$$

If (13.31) is substituted in (13.23) the solution of the latter takes on the form

$$b^2 = \tfrac{2}{3}[-\alpha^2 + (\alpha^4 + 3)^{1/2}]\,. \tag{13.32}$$

β_{cr} is obtained from (13.31) with (13.32) substituted for b

$$\beta_{cr} = \frac{2^{1/2}}{3^{3/2}}[-\alpha^2 + (\alpha^4 + 3)^{1/2}]^{1/2}\left[2\alpha + \frac{1}{\alpha}(\alpha^4 + 3)^{1/2}\right]. \tag{13.33}$$

It is plain to see from Eq. (13.33) that the value of critical damping depends on the speed of the load motion, i.e. on parameter α. Analogously, in vibration of systems with one degree of freedom the critical damping depends on the natural frequency of the system $\omega_b = \omega_0$ ([130], Vol. I, p. 42). Eq. (13.33) describing the dependence of critical damping β_{cr} on speed α is graphically represented in Fig. 13.2.

By (13.21) and (13.22) Eqs. (13.31) and (13.32) then yield

$$a_1^2 = \tfrac{4}{3}[2\alpha^2 + (\alpha^4 + 3)^{1/2}]\,, \quad a_2 = 0\,. \tag{13.34}$$

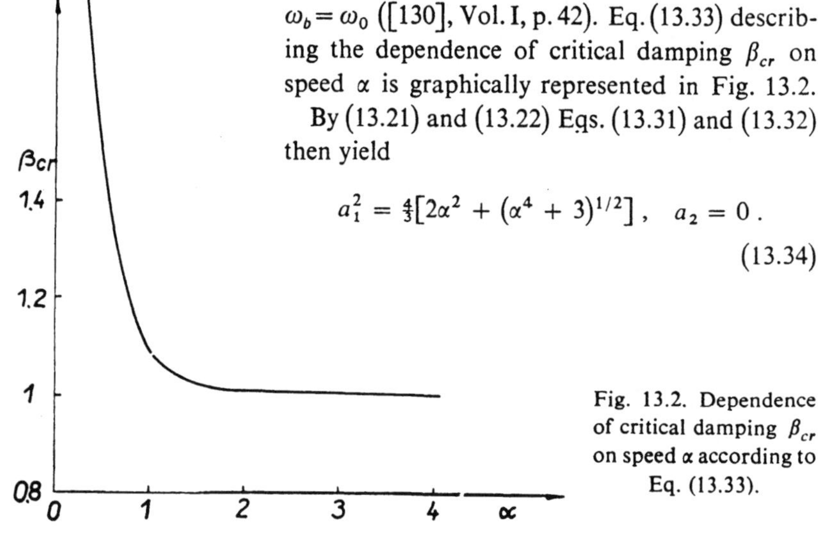

Fig. 13.2. Dependence of critical damping β_{cr} on speed α according to Eq. (13.33).

13.2.5 *Supercritical damping* ($\beta > \beta_{cr}$)

For supercritical damping, expression (13.22) becomes negative. Let us, therefore, designate the poles in this case as (Fig. 13.1e)

$$A_1 = \quad a_1 + ib\,, \quad A_3 = -(b - \alpha_2)\,\mathrm{i}\,,$$
$$A_2 = -a_1 + ib\,, \quad A_4 = -(b + a_2)\,\mathrm{i}\,. \tag{13.35}$$

Using a procedure analogous to that of deriving Eqs. (13.19) to (13.22) we get

$$a_2^2 = 2\alpha\beta/b - 2\alpha^2 - b^2\,. \tag{13.36}$$

The values of a_1 and b continue to be defined by (13.21) and (13.23). In the case of supercritical damping it is always $a_2 > 0$ (as we have assumed), and $a_2 < b$ [as can be proved by Eq. (13.36)]. Therefore, the poles are always in the position indicated in Fig. 13.1e.

13.3 *Solution of various cases*

If we know poles A_j (13.18) of the integrand in Eq. (13.17) we can evaluate the integral along the real axis by the application of the methods of the function of a complex variable. We express the integral in Eq. (13.17) $v(s) = (4/\pi)\int_{-\infty}^{\infty} F(q)\,\mathrm{d}q$ as the limit (see Fig. 13.1a)

$$\int_{-\infty}^{\infty} F(q)\,\mathrm{d}q = \int_{-\infty}^{\infty} \frac{\mathrm{e}^{\mathrm{i}sq}}{Q(q)}\,\mathrm{d}q = \lim_{R\to\infty}\int_{-R}^{R} \frac{\mathrm{e}^{\mathrm{i}sq}}{Q(q)}\,\mathrm{d}q \tag{13.37}$$

where R is the radius of semicircle C_R.

According to Cauchy's residue theorem ([186], Sect. 20.5, p. 795), the integral in the counter-clockwise direction around the closed curve C consisting of segments $-R$, $+R$ and semicircle C_R at $\lim R = \infty$ is

$$\oint_C \frac{\mathrm{e}^{\mathrm{i}sq}}{Q(q)}\,\mathrm{d}q = \lim_{R\to\infty}\left[\int_{-R}^{+R} \frac{\mathrm{e}^{\mathrm{i}sq}}{Q(q)}\,\mathrm{d}q + \int_{C_R} \frac{\mathrm{e}^{\mathrm{i}sq}}{Q(q)}\,\mathrm{d}q\right] =$$
$$= 2\pi\mathrm{i}\sum_{j=1}^{n} \operatorname{res} F(q)\big|_{q=A_j} \tag{13.38}$$

where C_R is the semicircle that passes around all poles in a half-plane, and $\operatorname{res} F(q)|_{q=A_j}$ is the residue of function $F(q)$ in the pole A_j.

Integral (13.38) converges because $F(q)$ is a regular function in the upper half-plane and on the real axis except for a finite number of poles that lie in the upper half-plane, and at $q \to \infty$, $q\,F(q)$ uniformly tends to zero in the same region. Function $Q(q)$ is namely a polynomial of the fourth degree, and

$$\text{for } s > 0\,, \quad \left|e^{isq}\right| = \left|e^{is(\xi + i\eta)}\right| = \left|e^{-s\eta + is\xi}\right| = e^{-s\eta} \leqq 1 \quad \text{for} \quad \eta > 0\,,$$

and (13.39)

$$\text{for } s < 0\,, \quad \left|e^{isq}\right| = \left|e^{-i(-s)q}\right| = \left|e^{-i(-s)(\xi + i\eta)}\right| = \left|e^{(-s)\eta - i(-s)\xi}\right| = $$
$$= e^{(-s)\eta} \leqq 1 \quad \text{for} \quad \eta < 0\,.$$

In the above the complex variable is expressed as $q = \xi + i\eta$. That is why in the evaluation of integrals (13.38) the semicircle C_R is drawn in the upper half-plane ($\eta > 0$) for $s > 0$, and in the lower half-plane ($\eta < 0$) for $s < 0$ (Fig. 13.1a).

There remains the evaluation of the second of the integrals in the square brackets of (13.38). $Q(q)$ is a polynomial of the fourth degree [generally $Q(q) = q^n + a_1 q^{n-1} + \ldots + a_n$, $n > 1$] and therefore $|Q(q)| \geqq R^n$ on semicircle C_R. Hence, with the aid of (13.39) the absolute value of function $F(q)$ is

$$\left|F(q)\right| = \left|\frac{e^{isq}}{Q(q)}\right| = \frac{\left|e^{isq}\right|}{\left|Q(q)\right|} \leqq \frac{1}{R^n} \tag{13.40}$$

and by the theorem on integral estimate (also by the Jordan theorem — see [201], Vol. III/2, p. 224) the integral around C_R is

$$\lim_{R \to \infty} \left| \int_{C_R} F(q)\, dq \right| \leqq \frac{1}{R^n} \pi R = \frac{\pi}{R^{n-1}} = 0\,. \tag{13.41}$$

Therefore, by (13.37) to (13.41) the solution is

$$\int_{-\infty}^{\infty} F(q)\, dq = \pm 2\pi i \sum_{j=1}^{n} \operatorname{res} F(q)\big|_{q = A_j} \tag{13.42}$$

where the plus sign is taken for $s > 0$ (integration in the upper half-plane), and the minus sign for $s < 0$ (integration in the lower half-plane, carried out in the opposite direction).

The residue of function $F(q)$ for a pole of order m is computed from the familiar relation ([186], Sect. 20.5, p. 795) which in our case of the pole of the first order gives

$$\operatorname{res} F(q)|_{q=A_j} = F(q)\,(q - A_j)|_{q=A_j},$$

e.g. for $j = 2$

$$\operatorname{res} F(q)|_{q=A_2} = \frac{\mathrm{e}^{\mathrm{i}sA_2}}{(A_2 - A_1)(A_2 - A_3)(A_2 - A_4)}, \tag{13.43}$$

and for the pole of the second order, e.g. at $Q(q) = (q - A_j)^2 \cdot (q - A_1) \cdot$ $\cdot (q - A_2)$ gives

$$\operatorname{res} F(q)|_{q=A_j} = \frac{\mathrm{e}^{\mathrm{i}sA_j}}{(A_j - A_1)(A_j - A_2)}\left[\mathrm{i}s - \frac{2A_j - A_1 - A_2}{(A_j - A_1)(A_j - A_2)}\right]. \tag{13.44}$$

On substituting (13.42) to (13.44) and (13.18) in (13.17) and writing simply

$$D_1 = a_1 b, \qquad D_3 = a_2 b,$$
$$D_2 = b^2 - \tfrac{1}{4}(a_1^2 - a_2^2), \quad D_4 = b^2 + \tfrac{1}{4}(a_1^2 - a_2^2) \tag{13.45}$$

we get after some handling the solution of Eq. (13.12), i.e. the dimensionless deflection with the poles situated as indicated in Fig. 13.1a*)

$$v(s) = \frac{1}{a_1(D_1^2 + D_2^2)}\left[(D_1 - D_2\mathrm{i})\,\mathrm{e}^{\mathrm{i}A_1 s} + (D_1 + D_2\mathrm{i})\,\mathrm{e}^{\mathrm{i}A_2 s}\right] =$$
$$= \frac{2}{a_1(D_1^2 + D_2^2)}\,\mathrm{e}^{-bs}(D_1 \cos a_1 s + D_2 \sin a_1 s) \quad \text{for} \quad s > 0, \tag{13.46}$$

$$v(s) = \frac{1}{a_2(D_3^2 + D_4^2)}\left[(D_3 + D_4\mathrm{i})\,\mathrm{e}^{\mathrm{i}A_3 s} + (D_3 - D_4\mathrm{i})\,\mathrm{e}^{\mathrm{i}A_4 s}\right] =$$
$$= \frac{2}{a_2(D_3^2 + D_4^2)}\,\mathrm{e}^{bs}(D_3 \cos a_2 s - D_4 \sin a_2 s) \quad \text{for} \quad s < 0. \tag{13.47}$$

In the dimensionless form, the bending moment $M(x, t) = -EJ \,.$ $. \, v''(x, t) = M_0\, M(s)$ and the shear force $T(x, t) = -EJ\, v'''(x, t) =$

*) Expressions (13.46), (13.47) and (13.58) describing the deflection were evolved earlier by J. T. Kenney [122] using a different method.

$= T_0\, T(s)$, where $[M_0 = M(0)$ is the static bending moment underneath load P – cf. [68], [83]]

$$M_0 = \frac{P}{4\lambda}, \quad T_0 = P, \tag{13.48}$$

will turn out to be

$$M(s) = -\frac{1}{2}\,v''(s) = \frac{1}{a_1(D_1^2 + D_2^2)}\,\mathrm{e}^{-bs}\,[(a_1^2 D_1 + 2a_1 b D_2 - b^2 D_1)\,.$$
$$.\cos a_1 s + (a_1^2 D_2 - 2a_1 b D_1 - b^2 D_2)\sin a_1 s] \quad \text{for} \quad s > 0\,, \tag{13.49}$$

$$M(s) = \frac{1}{a_2(D_3^2 + D_4^2)}\,\mathrm{e}^{bs}[(a_2^2 D_3 + 2a_2 b D_4 - b^2 D_3)\cos a_2 s -$$
$$-(a_2^2 D_4 - 2a_2 b D_3 - b^2 D_4)\sin a_2 s] \quad \text{for} \quad s < 0\,, \tag{13.50}$$

$$T(s) = -\frac{1}{8}\,v'''(s) = \frac{1}{4a_1(D_1^2 + D_2^2)}\,\mathrm{e}^{-bs}[(a_1^3 D_2 - 3a_1^2 b D_1 - 3a_1\,.$$
$$.\,b^2 D_2 + b^3 D_1)\cos a_1 s - (a_1^3 D_1 + 3a_1^2 b D_2 - 3a_1 b^2 D_1 -$$
$$- b^3 D_2)\sin a_1 s] \quad \text{for} \quad s > 0\,, \tag{13.51}$$

$$T(s) = -\frac{1}{4a_2(D_3^2 + D_4^2)}\,\mathrm{e}^{bs}[(a_2^3 D_4 - 3a_2^2 b D_3 - 3a_2 b^2 D_4 + b^3 D_3)\,.$$
$$.\cos a_2 s + (a_2^3 D_3 + 3a_2^2 b D_4 - 3a_2 b^2 D_3 - b^3 D_4)\sin a_2 s]$$
$$\text{for} \quad s < 0\,. \tag{13.52}$$

In the next paragraphs we will discuss some simplifications of Eqs. (13.46) to (13.52) obtainable in special cases.

13.3.1 *Static case* ($\alpha = 0$)

For an infinite beam on elastic foundation subjected to static force P, substitution of (13.24) in (13.45) to (13.52) results in

$$\begin{aligned} v(s) &= \mathrm{e}^{-|s|}(\cos s + \sin |s|)\,, \\ M(s) &= \mathrm{e}^{-|s|}(\cos s - \sin |s|)\,, \\ T(s) &= -\operatorname{sign} s\,.\,\tfrac{1}{2}\mathrm{e}^{-|s|}\cos s\,. \end{aligned} \tag{13.53}$$

Functions (13.53) are plotted in Fig. 13.3a.

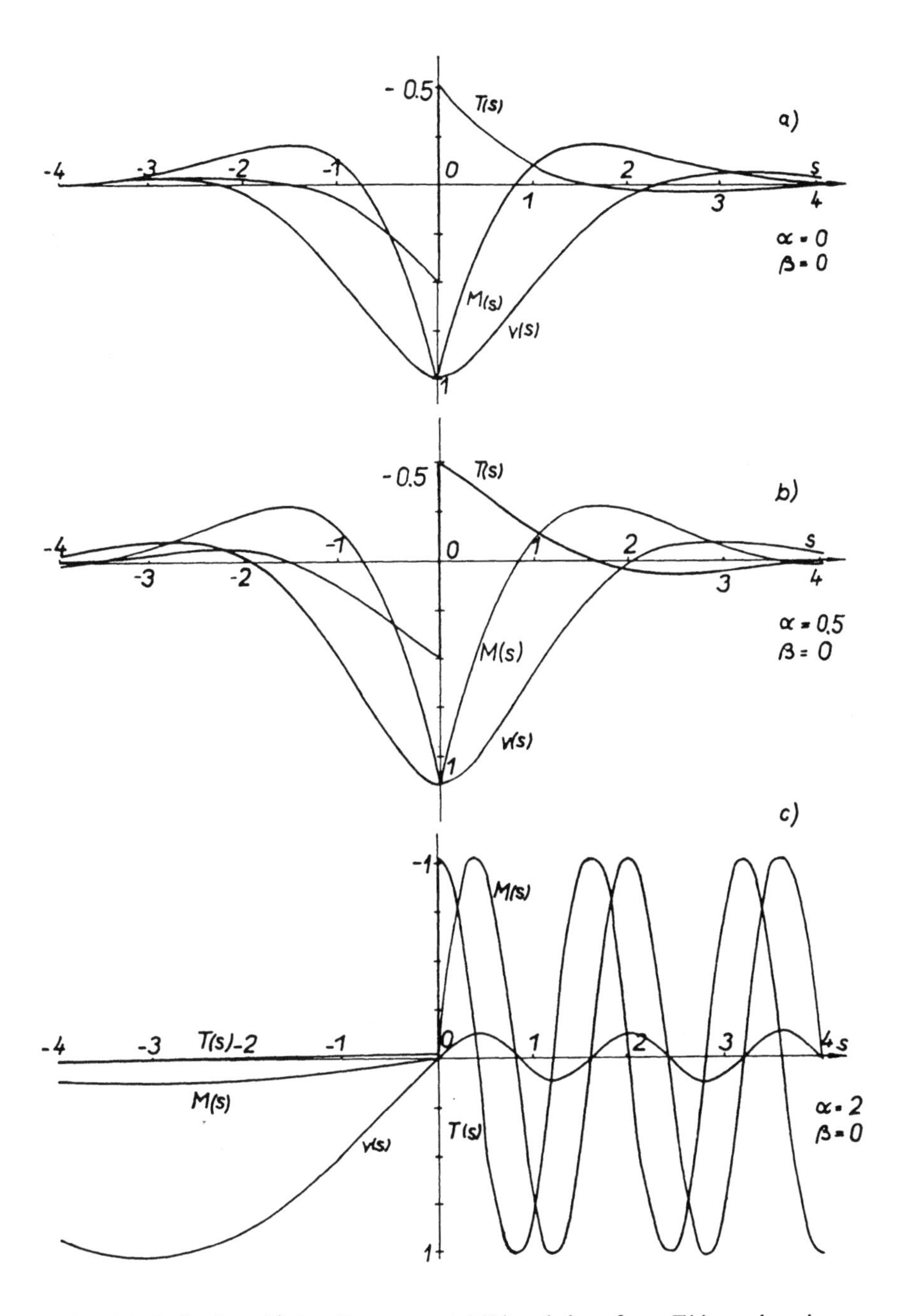

Fig. 13.3. Deflection $v(s)$, bending moment $M(s)$ and shear force $T(s)$, no damping. a) $\alpha = 0$, $\beta = 0$, b) $\alpha = 0{\cdot}5$, $\beta = 0$, c) $\alpha = 2$, $\beta = 0$.

13.3.2 *Case with no damping* ($\beta = 0$)

13.3.2.1 $\alpha < 1,\ \beta = 0$

With (13.25) in, Eqs. (13.45) to (13.52) give

$$v(s) = \frac{1}{ab}\,e^{-b|s|}(a\cos as + b\sin a|s|)\,,$$

$$M(s) = \frac{1}{ab}\,e^{-b|s|}(a\cos as - b\sin a|s|)\,,$$

$$T(s) = -\frac{1}{2ab}\,e^{-b|s|}(\operatorname{sign} s\,.\,ab\cos as + \alpha^2\sin as)\,. \qquad (13.54)$$

Functions (13.54) are illustrated in Fig. 13.3b for $\alpha = 0{\cdot}5$, $\beta = 0$.

13.3.2.2 $\alpha = 1,\ \beta = 0$

In this case, speed c [Eq. (13.9)] will attain its critical value, c_{cr}. The solution of Eq. (13.12) is not defined like that of compressed bars under critical load. Actually, Eqs. (13.1) and (13.12) are formally the same as those for a beam on elastic foundation subjected to axially-applied compressive force $S = \mu c^2$. Since for such a beam the critical force $S_{cr} = 4EJ\lambda^2$, [109], according to (13.9) this value, $S = S_{cr}$, is only reached for $\alpha = 1$. We can suppose that at this speed of load motion the beam will lose stability.

13.3.2.3 $\alpha > 1,\ \beta = 0$

At speeds higher than the critical one, as b approaches zero [Eq. (13.27)], expressions (13.46) to (13.52) approach the limits

$$v(s) = -\frac{2}{a_1(\alpha^4 - 1)^{1/2}}\sin a_1 s \quad \text{for} \quad s > 0\,,$$

$$v(s) = -\frac{2}{a_2(\alpha^4 - 1)^{1/2}}\sin a_2 s \quad \text{for} \quad s < 0\,,$$

$$M(s) = -\frac{a_1}{(\alpha^4 - 1)^{1/2}}\sin a_1 s \quad \text{for} \quad s > 0\,,$$

$$M(s) = -\frac{a_2}{(\alpha^4 - 1)^{1/2}} \sin a_2 s \quad \text{for} \quad s < 0\,,$$

$$T(s) = -\frac{a_1^2}{4(\alpha^4 - 1)^{1/2}} \cos a_1 s \quad \text{for} \quad s > 0\,,$$

$$T(s) = -\frac{a_2^2}{4(\alpha^4 - 1)^{1/2}} \cos a_2 s \quad \text{for} \quad s < 0\,. \qquad (13.55)$$

Functions (13.55) are represented in Fig. 13.3c for $\alpha = 2$, $\beta = 0$. As the figure shows, the solution has a wave character; waves with frequency a_1 form ahead of, waves with frequency a_2 beyond the load. Solution (13.55) naturally fails to satisfy the boundary conditions (13.13); it was J. Dörr [54] who evolved the expressions of the fronts and rears of the waves, which would also satisfy the boundary conditions.

13.3.3 *Light damping* ($\beta \ll 1$)

In the case of light damping, a_1, a_2, b are approximately computed from formulae (13.28) to (13.30) and the solution is found in the form of (13.45) to (13.52). Figs. 13.4a, b, c are the graphical representations of the case in question for damping $\beta = 0{\cdot}1$ and speeds $\alpha = 0{\cdot}5$, 1, 2.

13.3.4 *Critical damping* ($\beta = \beta_{cr}$)

In the case of critical damping (13.33), the coordinates of poles A_j are computed from Eqs. (13.32) and (13.34) (Fig. 13.1d). For $s > 0$ the integration is carried out in the upper half-plane and the solutions thus obtained are the same as (13.46), (13.49) and (13.51). For $s < 0$, semi-circle C_R passes around the double pole $A_3 \equiv A_4$ and (13.42) is computed using relation (13.44). This gives for $s < 0$

$$v(s) = \frac{2}{\alpha^4 + 3}\, e^{bs}[b - (\alpha^4 + 3)^{1/2}\, s]\,,$$

$$M(s) = -\frac{b}{\alpha^4 + 3}\, e^{bs}[b^2 - 2(\alpha^4 + 3)^{1/2} - b(\alpha^4 + 3)^{1/2}\, s]\,,$$

$$T(s) = -\frac{b^2}{4(\alpha^4 + 3)}\, e^{bs}[b^2 - 3(\alpha^4 + 3)^{1/2} - b(\alpha^4 + 3)^{1/2}\, s]\,. \qquad (13.56)$$

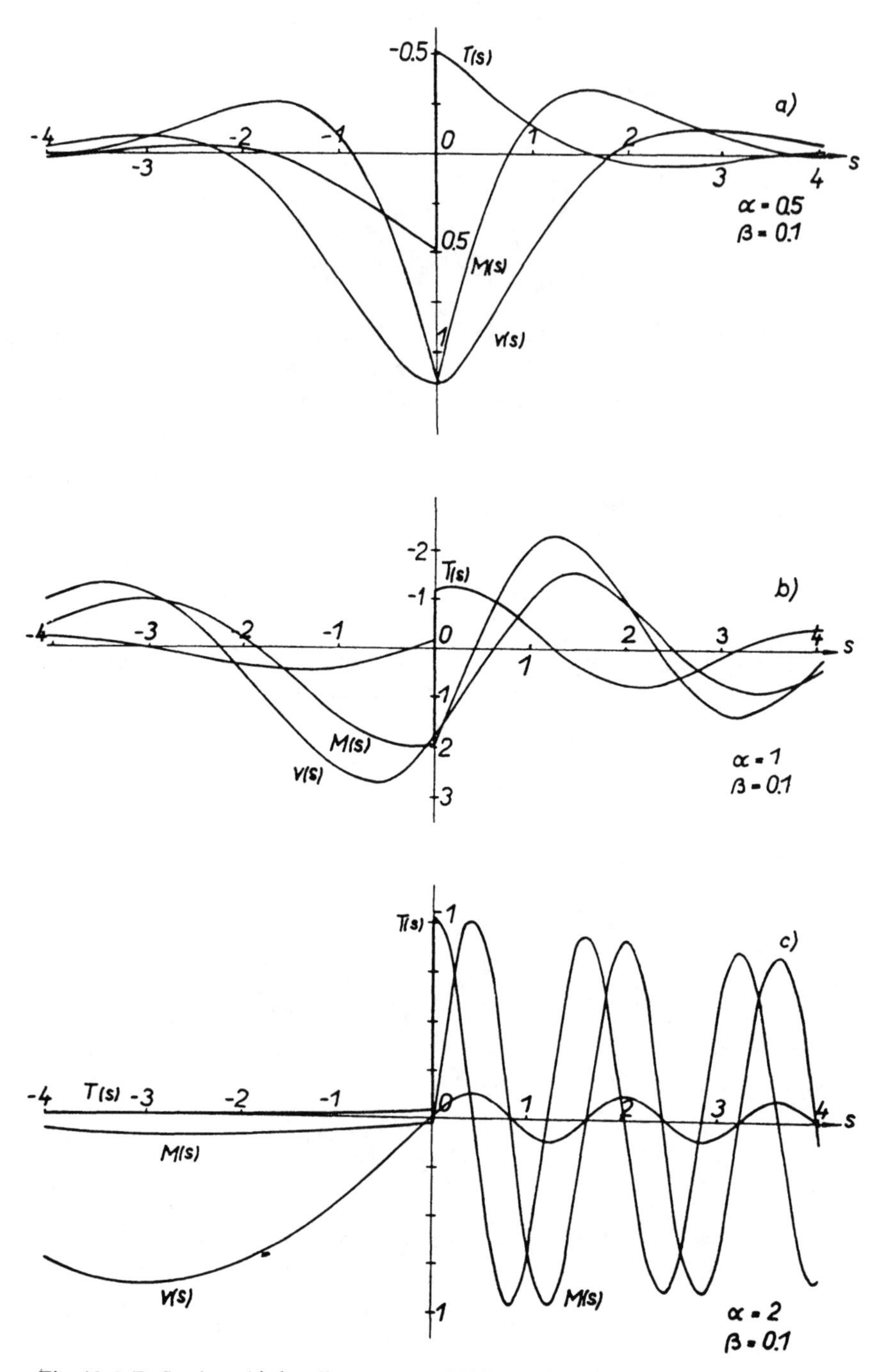

Fig. 13.4. Deflection $v(s)$, bending moment $M(s)$ and shear force $T(s)$, light damping. a) $\alpha = 0{\cdot}5$, $\beta = 0{\cdot}1$, b) $\alpha = 1$, $\beta = 0{\cdot}1$, c) $\alpha = 2$, $\beta = 0{\cdot}1$.

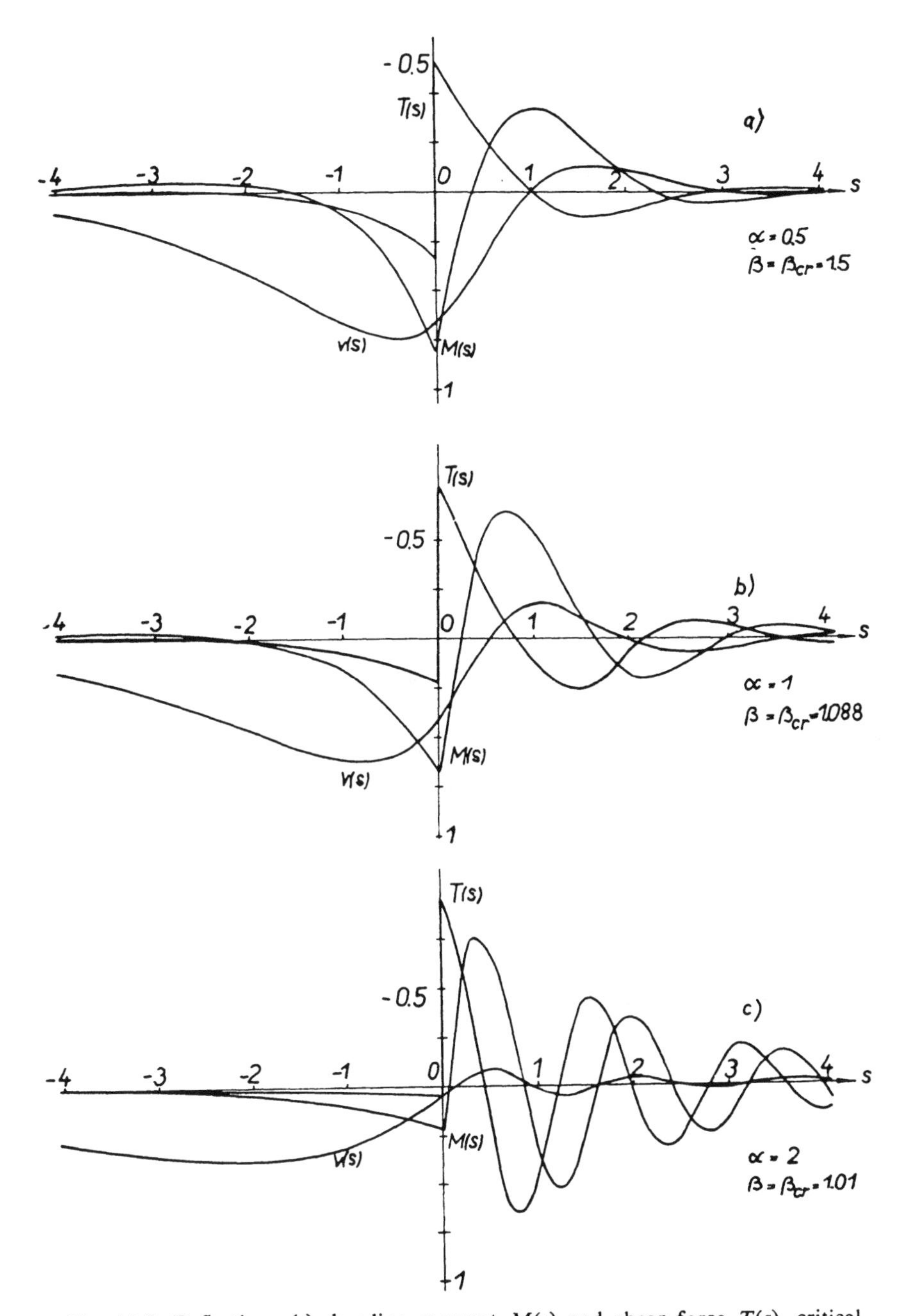

Fig. 13.5. Deflection $v(s)$, bending moment $M(s)$ and shear force $T(s)$, critical damping. a) $\alpha = 0{\cdot}5$, $\beta = \beta_{cr} = 1{\cdot}5$, b) $\alpha = 1$, $\beta = \beta_{cr} = 1{\cdot}088$, c) $\alpha = 2$, $\beta = \beta_{cr} = 1{\cdot}01$.

Eqs. (13.46), (13.49), (13.51) and (13.56) are represented in Figs. 13.5a, b,c for the following three pairs of α and β

$$\alpha = 0{\cdot}5\,, \quad 1\,, \qquad 2\,,$$
$$\beta = \beta_{cr} = 1{\cdot}5\,, \quad 1{\cdot}088\,, \quad 1{\cdot}01\,.$$

13.3.5 *Supercritical damping* ($\beta > \beta_{cr}$)

At supercritical damping the coordinates of poles (13.35) are computed from (13.21), (13.23) and (13.36).

Writing

$$\begin{aligned} D_1 &= a_1 b\,, & D_3 &= a_2 b\,, \\ D_2 &= b^2 - \tfrac{1}{4}(a_1^2 + a_2^2)\,, & D_4 &= b^2 + \tfrac{1}{4}(a_1^2 + a_2^2)\,, \end{aligned} \tag{13.57}$$

we get for $s > 0$ expressions wholly identical with (13.46), (13.49) and (13.51) because the poles in the upper half-plane of Fig. 13.1e are situated in the same way as those in Fig. 13.1a.

For $s < 0$ the integration is carried out in the lower half-plane of Fig. 13.1e, and Eqs. (13.42) and (13.43) give

$$\begin{aligned} v(s) &= \frac{1}{a_2(D_4^2 - D_3^2)}\left[(D_3 + D_4)\,e^{(b-a_2)s} - (D_4 - D_3)\,e^{(b+a_2)s}\right], \\ M(s) &= -\frac{1}{2a_2(D_4^2 - D_3^2)}\left[(D_3 + D_4)(b - a_2)^2\,e^{(b-a_2)s} - (D_4 - D_3)\,.\right. \\ &\qquad \left. .\,(b + a_2)^2\,e^{(b+a_2)s}\right], \\ T(s) &= -\frac{1}{8a_2(D_4^2 - D_3^2)}\left[(D_3 + D_4)(b - a_2)^3\,e^{(b-a_2)s} - (D_4 - D_3)\,.\right. \\ &\qquad \left. .\,(b + a_2)^3\,e^{(b+a_2)s}\right]. \end{aligned} \tag{13.58}$$

Functions (13.46), (13.49), (13.51) and (13.58) are drawn in Figs. 13.6a,b,c for damping $\beta = 2$ and speeds $\alpha = 0{\cdot}5$, 1, 2.

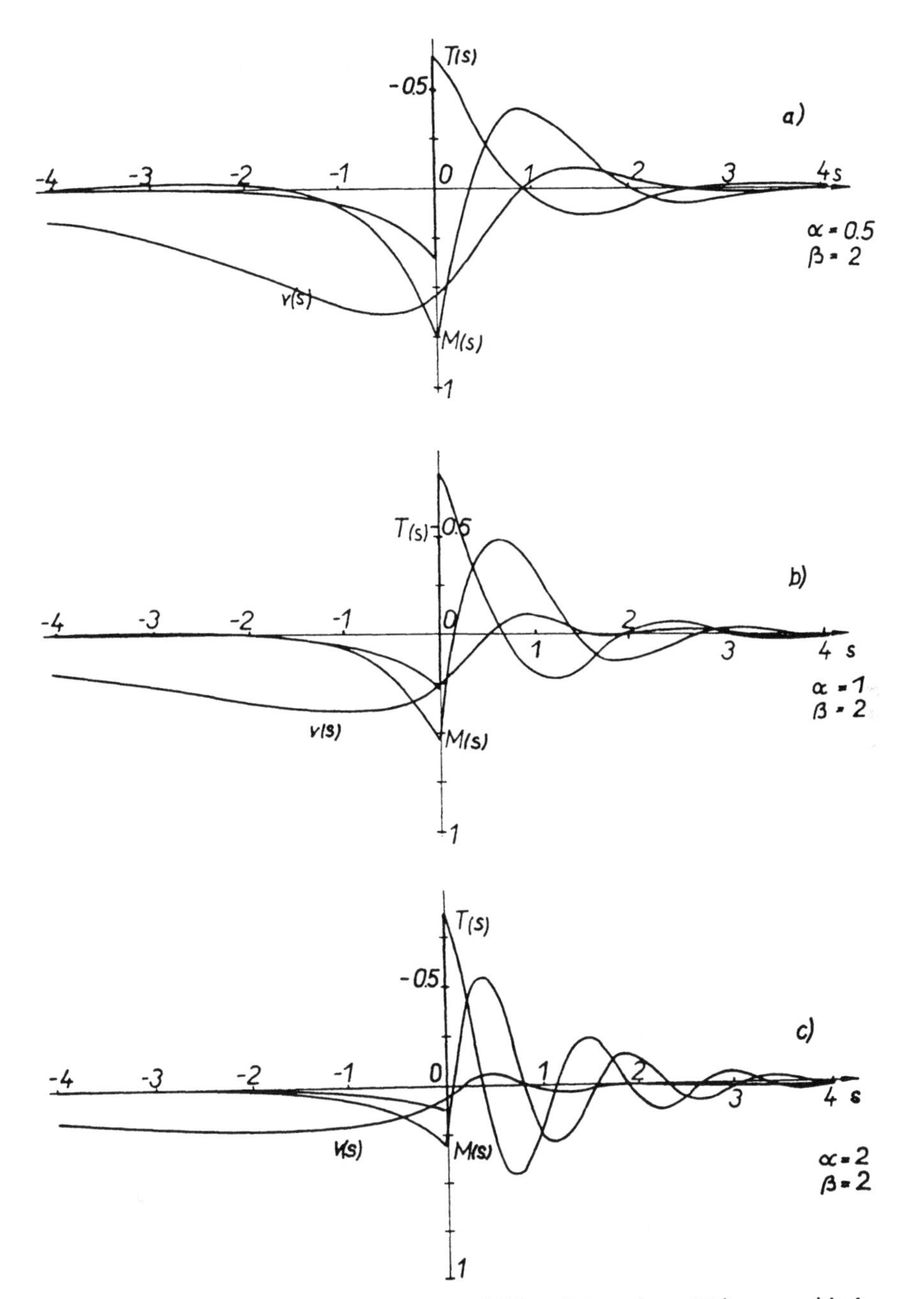

Fig. 13.6. Deflection $v(s)$, bending moment $M(s)$ and shear force $T(s)$, supercritical damping. a) $\alpha = 0{\cdot}5$, $\beta = 2$, b) $\alpha = 1$, $\beta = 2$, c) $\alpha = 2$, $\beta = 2$.

13.4 *Deflection, bending moment and shear force at the point of load action*

At point $s = 0$, i.e. underneath the moving load, the dimensionless deflection, bending moment and shear force $[T(+0) = \lim_{s \to 0_+} T(s), T(-0) = = \lim_{s \to 0_-} T(s)]$ attain the values that are computed from Eqs. (13.45) to (13.52) after the substitution $s = 0$

$$v(0) = \frac{2b}{3b^4 + 4\alpha^2 b^2 + \alpha^4 - 1},$$

$$M(0) = \frac{2b(\alpha^2 + b^2)}{3b^4 + 4\alpha^2 b^2 + \alpha^4 - 1},$$

$$T(\pm 0) = \mp \frac{1}{2}\left[1 \pm \frac{\alpha\beta(\alpha^2 + b^2)}{b(3b^4 + 4\alpha^2 b^2 + \alpha^4 - 1)}\right]. \tag{13.59}$$

In the last expression either the upper or the lower signs apply throughout.

For our special cases we then get:

13.4.1 *Static case* ($\alpha = 0$)

$$v(0) = M(0) = 1, \quad T(\pm 0) = \mp\tfrac{1}{2}. \tag{13.60}$$

13.4.2 *Case with no damping* ($\beta = 0$)

13.4.2.1 $\alpha < 1$, $\beta = 0$

$$v(0) = M(0) = (1 - \alpha^2)^{-1/2}, \quad T(\pm 0) = \mp\tfrac{1}{2}. \tag{13.61}$$

13.4.2.2 $\alpha = 1$, $\beta = 0$

In the case of critical speed there are obtained two values for every quantity at the limit $\alpha = 1$ from the left or from the right [from the limits of (13.61) or (13.63) for $\alpha \to 1$]:

$$\alpha \to 1_- \quad v(0) = M(0) = \infty, \quad T(\pm 0) = \mp\tfrac{1}{2},$$
$$\alpha \to 1_+ \quad v(0) = M(0) = 0, \quad T(\pm 0) = -\infty. \tag{13.62}$$

13.4.2.3 $\alpha > 1,\ \beta = 0$

$$v(0) = M(0) = 0\,, \quad T(\pm 0) = \mp \frac{1}{2}\left[1 \pm \frac{\alpha^2}{(\alpha^4 - 1)^{1/2}}\right]. \qquad (13.63)$$

13.4.3 *Light damping* ($\beta \ll 1$)

13.4.3.1 $\alpha < 1,\ \beta \ll 1$

$$v(0) \quad = M(0) = \frac{1}{(1 - \alpha^2)^{1/2}\,[1 + \frac{1}{2}\alpha^2\beta^2/(1 - \alpha^2)^2]}\,,$$

$$T(\pm 0) = \mp \frac{1}{2}\left[1 \pm \frac{\alpha\beta}{2(1 - \alpha^2)^{3/2} + \alpha^2\beta^2(1 - \alpha^2)^{-1/2}}\right]. \qquad (13.64)$$

13.4.3.2 $\alpha = 1,\ \beta \ll 1$

$$v(0) \quad = \frac{2^{3/4}}{\beta^{1/2}(\frac{3}{2}\beta + 2^{3/2})}\,,$$

$$M(0) \quad = \frac{2^{3/4}(1 + 2^{-1/2}\beta)}{\beta^{1/2}(\frac{3}{2}\beta + 2^{3/2})}\,,$$

$$T(\pm 0) = \mp \frac{1}{2}\left[1 \pm \frac{2^{1/4}(1 + 2^{-1/2}\beta)}{\beta^{1/2}(\frac{3}{2}\beta + 2^{3/2})}\right]. \qquad (13.65)$$

13.4.3.3 $\alpha > 1,\ \beta \ll 1$

$$v(0) \quad = \frac{2\alpha\beta}{(\alpha^4 - 1)^{3/2}\,[1 + 4\alpha^4\beta^2/(\alpha^4 - 1)^2 + 3\alpha^4\beta^4/(\alpha^4 - 1)^3]}\,,$$

$$M(0) \quad = \frac{2\alpha\beta[\alpha^2 + \alpha^2\beta^2/(\alpha^4 - 1)]}{(\alpha^4 - 1)^{3/2}\,[1 + 4\alpha^4\beta^2(\alpha^4 - 1)^2 + 3\alpha^4\beta^4/(\alpha^4 - 1)^3]}\,,$$

$$T(\pm 0) = \mp \frac{1}{2}\left\{1 \pm \frac{\alpha^2 + \alpha^2\beta^2/(\alpha^4 - 1)}{(\alpha^4 - 1)^{1/2}[1 + 4\alpha^4\beta^2/(\alpha^4 - 1)^2 + 3\alpha^4\beta^4/(\alpha^4 - 1)^3]}\right\}. \qquad (13.66)$$

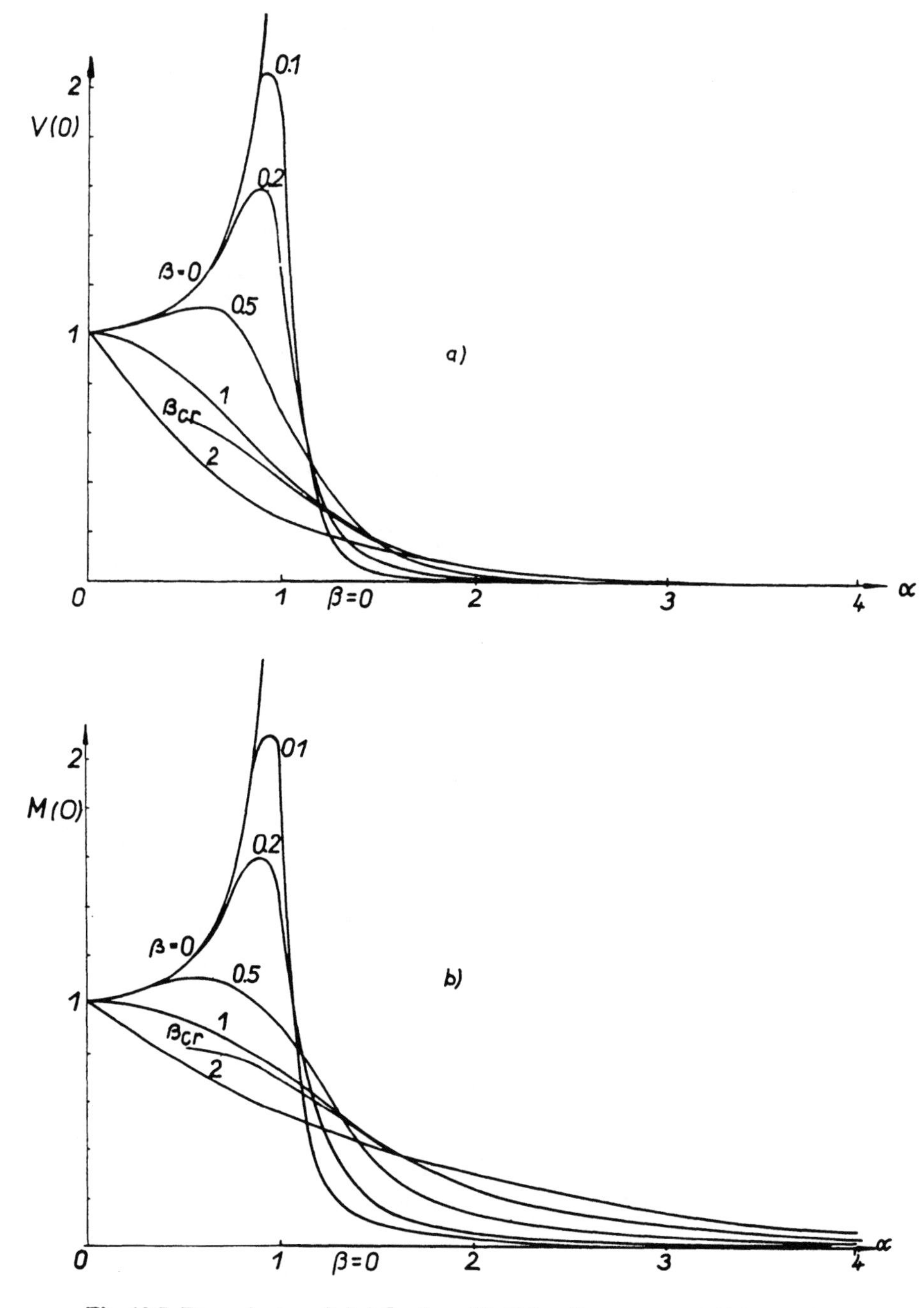

Fig. 13.7. Dependences of a) deflection $v(0)$, b) bending moment $M(0)$, and c) shear force $T(\pm 0)$ at point $s = 0$, on speed α; damping $\beta = 0$, 0·1, 0·2, 0·5, 1, β_{cr}, 2.

13.4.4 *Critical damping* ($\beta = \beta_{cr}$)

$$v(0) = \frac{2^{3/2}[-\alpha^2 + (\alpha^4 + 3)^{1/2}]^{1/2}}{3^{1/2}(\alpha^4 + 3)},$$

$$M(0) = \frac{2^{3/2}[\alpha^2 + 2(\alpha^4 + 3)^{1/2}]}{3^{3/2}(\alpha^4 + 3)}[-\alpha^2 + (\alpha^4 + 3)^{1/2}]^{1/2},$$

$$T(\pm 0) = \mp \frac{1}{2}\left[1 \pm \frac{4\alpha^4 + 5\alpha^2(\alpha^4 + 3)^{1/2} + 6}{9(\alpha^4 + 3)}\right]. \qquad (13.67)$$

When dealing with the case of supercritical damping, we would proceed similarly as above.

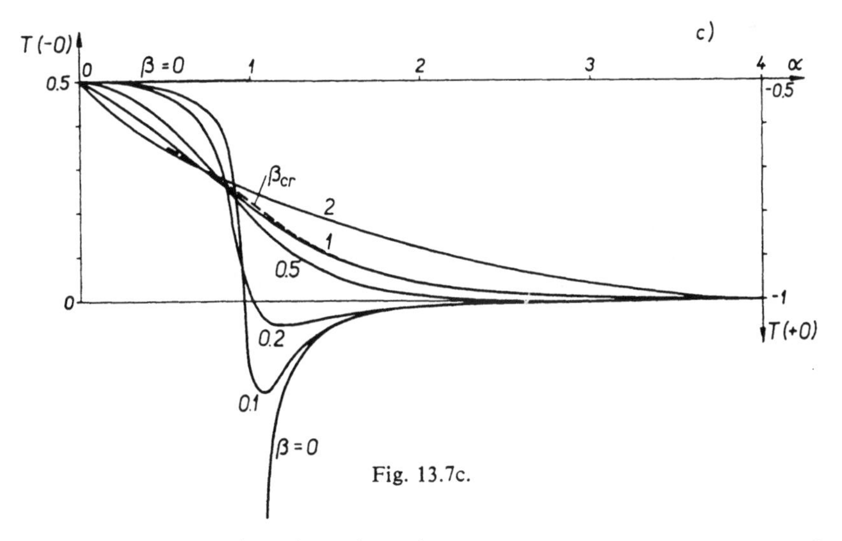

Fig. 13.7c.

From formulae (13.59) to (13.67) we computed the dependences of $v(0)$, $M(0)$ and $T(\pm 0)$ on speed in the range $0 \leqq \alpha \leqq 4$ for damping $\beta = 0$, 0·1, 0·2, 0·5, 1, β_{cr}, 2, and plotted them in Figs. 13.7a,b,c. As the curves – analogous to the resonance curves of systems with n degrees of freedom – suggest, at $\alpha > 1$, the effect of damping is wholly opposite to that in the latter systems. The values of $v(0)$ and $M(0)$ are higher at heavy than at light damping. It should be noted, however, that the maximum values, max $v(s)$ and max $M(s)$ are likely to lie elsewhere, too, (for $s \neq 0$) as evident from Figs. 13.3 to 13.6.

13.5 *Application of the theory*

13.5.1 *The effect of moving mass*

The theory presented in the foregoing sections is widely used in computations relating to structures founded on soil and traversed by machines. Typical examples of such structures are rails and longitudinal railway sleepers. Their treatment by the theory presented is to be found in [12], [83], [217]. The theory was further elaborated to include complex effects of dynamic forces on infinite beams on elastic foundations, for example, for the computation of impacts produced by flat wheels, etc., see [68], [69], [72].

In practical applications, inertia effects deserve equal considerations as the force effects of load P. If we take them into account we write Eq. (13.1) in the form

$$p(x, t) = \delta(x - ct)\left[P - m\,\frac{\mathrm{d}^2 v(ct, t)}{\mathrm{d}t^2}\right] \tag{13.68}$$

where m is the load mass ($P = mg$).

The derivative in (13.68) is

$$\frac{\mathrm{d}^2 v(ct, t)}{\mathrm{d}t^2} = \left[c^2\,\frac{\partial^2 v(x, t)}{\partial x^2} + 2c\,\frac{\partial^2 v(x, t)}{\partial x\,\partial t} + \frac{\partial^2 v(x, t)}{\partial t^2}\right]_{x = ct}. \tag{13.69}$$

Using Eqs. (13.8) it holds for the quasi-stationary state defined by (13.4) and (13.2) that

$$\begin{aligned}
\frac{\partial^2 v(x, t)}{\partial x^2} &= \lambda^2 v_0\,\frac{\mathrm{d}^2 v(s)}{\mathrm{d}s^2}\,,\\
\frac{\partial^2 v(x, t)}{\partial x\,\partial t} &= -\lambda^2 c v_0\,\frac{\mathrm{d}^2 v(s)}{\mathrm{d}s^2}\,,\\
\frac{\partial^2 v(x, t)}{\partial t^2} &= \lambda^2 c^2 v_0\,\frac{\mathrm{d}^2 v(s)}{\mathrm{d}s^2}\,.
\end{aligned} \tag{13.70}$$

For the case considered, substitution of (13.70) in (13.69) gives

$$\frac{\mathrm{d}^2 v(ct, t)}{\mathrm{d}t^2} = 0\,. \tag{13.71}$$

From this we may draw the important conclusion that in the quasi-stationary state a load exerts no inertia effects. This goes, for example, even for the case when an infinitely long beam is traversed by a system with two degrees of freedom (sprung and unsprung masses), etc.

If the load does not satisfy the conditions of the quasi-stationary state, e.g. if there acts next to it also a harmonic force, etc., one must consider Eq. (13.68) including (13.69) (cf. [98], [152]).

13.5.2 *The effect of speed*

The critical speeds computed from (13.11) are actually so high, that till now they have never been attained in railway transport. The critical speeds appertaining to a superstructure with $J = 1862\ \mathrm{cm}^4$, $E = 21\ \mathrm{MN/cm}^2$, $\mu = 0{\cdot}125\ \mathrm{t/m}$ and different values of foundation coefficient k are shown in Fig. 13.8.

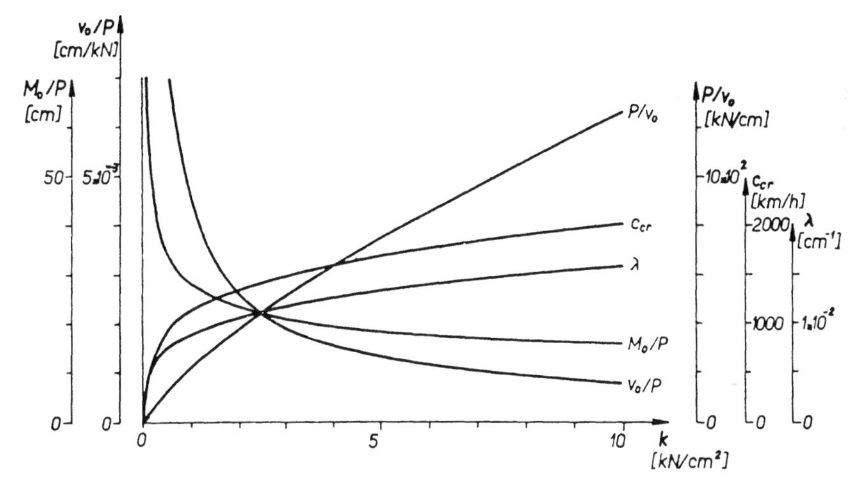

Fig. 13.8. Dependences of deflection v_0/P and P/v_0, bending moment M_0/P, critical speed c_{cr} and coefficient λ on foundation coefficient k.

At speeds common today and for some time to come, the highest values of $v(s)$ and $M(s)$ are at the point of load application ($s = 0$), as suggested by Figs. 13.3 to 13.7, and differ but slightly from the static ones. This, of course, holds on the assumption of an ideally straight beam without irregularities, and an ideal vehicle. However, in high-speed transport envisaged for the future, solutions for $\alpha = 1$ and $\alpha > 1$, too, are bound to be of great importance.

13.5.3 *The effect of foundation*

The accuracy of results is substantially affected by the coefficient of elastic foundation, k, a quantity very hard to determine. The dependence of λ, v_0/P, P/v_0, M_0/P and c_{cr} on k is shown in Fig. 13.8. As one can see in the figure, for k ranging from 0·5 to about 4 kN/cm^2, the inaccuracy of k determination bears strongly on v_0/P and P/v_0. The calculation of bending moment, on the other hand, is not so sensitive to errors in k. A survey of experimental results is in [83].

13.6 *Additional bibliography*

[12, 18, 54, 59, 62, 68, 69, 72, 83, 87, 98, 122, 136, 148, 152, 163, 178, 193, 198, 214, 217, 223, 240, 241, 253, 263, 265, 267, 270].

14

String subjected to a moving load

Modern structures make frequent use of one-dimensional elements resistant to tension but not to bending, such as strings, fibres, ropes, cables, chains, etc. Vibration studies of these elements subjected to a moving load have started but recently [105], [202].

14.1 *String with a moving force*

According to [131] the differential equation describing vibration of a mass string is

$$-N \frac{\partial^2 v(x, t)}{\partial x^2} + \mu \frac{\partial^2 v(x, t)}{\partial t^2} = p(x, t) \qquad (14.1)$$

where – with reference to Fig. 14.1 –

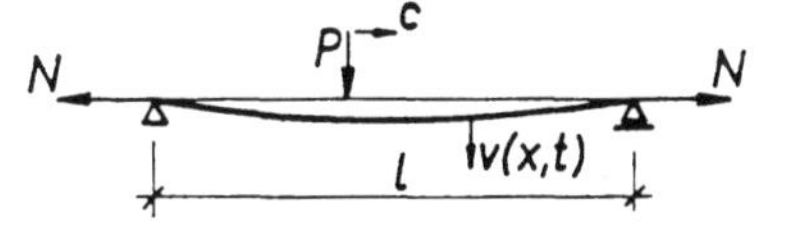

Fig. 14.1. String subjected to a moving force P.

$v(x, t)$ – deflection of string at point x and time t, measured from the state under static load. The string is considered to be in horizontal position, and its deflections are assumed to be very small against l,

μ – mass per unit length of string,

N – horizontal force stretching the string (the force is so large as to continue virtually unvaried during string deformation),*)

$p(x, t)$ – external load per unit length.

*) It has been found necessary to judge each particular case as to whether or not this assumption is justified. Thus, e.g. in an infinitely long continuous cable of equal spans such as that used in cable belt conveyors, the axial force continues approximately constant. In cables with stretching weights common in cabin cableways, on the other hand, the axial force in the cable is likely to vary, sometimes even within wide limits.

The boundary conditions of a string of length l are

$$v(0, t) = 0\,, \quad v(l, t) = 0 \tag{14.2}$$

while the initial conditions are usually zero, i.e. (1.3).

By way of an example we will solve the motion of a constant force P moving along a string at uniform speed c; in that case the load

$$p(x, t) = \delta(x - ct)\, P\,. \tag{14.3}$$

In the solution we shall use the Fourier finite sine transformation (1.9) which in view of (14.2) and (27.68) will turn Eq. (14.1) into the ordinary differential equation

$$\ddot{V}(j, t) + \omega_{(j)}^2\, V(j, t) = \frac{P}{\mu} \sin \omega t \tag{14.4}$$

where

$$\omega_{(j)}^2 = \frac{j^2 \pi^2}{l^2} \frac{N}{\mu} \tag{14.5}$$

is the natural circular frequency of the string, and

$$\omega = \frac{j \pi c}{l}\,. \tag{14.6}$$

Following the Laplace-Carson transformation (1.15) we get — with the initial conditions (1.3) and with (27.4) and (27.18) —

$$V^*(j, p) = \frac{P}{\mu} \frac{\omega p}{(p^2 + \omega_{(j)}^2)(p^2 + \omega^2)} \tag{14.7}$$

and after the inverse transformation (27.32) and (1.9) the sought-for solution (for $\omega_{(j)} \neq \omega$; if $\omega_{(j)} = \omega$, the procedure would be that used in Sect. 1.3, case 1.3.2.2):

$$v(x, t) = \sum_{j=1}^{\infty} \frac{2P}{\mu l} \frac{1}{\omega_{(j)}^2 - \omega^2} \left(\sin \omega t - \frac{\omega}{\omega_{(j)}} \sin \omega_{(j)} t \right) \sin \frac{j \pi x}{l}\,. \tag{14.8}$$

It is clear to see that the solution is wholly analogous to that of a simply supported beam traversed by force P [Eq. (1.31)]. This means that so far as the deflection of a mass string is concerned, we are free to make use of all the results derived in Chaps. 1 to 4. The similarity stops at the natural frequency of the string (14.5) which differs from the analogous value of a simply supported beam, (1.11). Since the former is proportional to j^2 (and the latter to j^4), the convergence of all series figuring in the computations relating to a string, will be slower than that of the analogous series for beam deflection. As will be made clear in Sect. 14.2, the deflection of a mass string is actually wholly analogous to the bending moment of a simple beam.

Since the homogeneous equation appertaining to (14.1) is also satisfied by functions $f_1(x - ct)$ and $f_2(x + ct)$, one can even derive the solution in a form descriptive of the wave character (see [202] for further details).

14.2 *Motion of a mass along a massless string*

We shall now turn to the other extreme case, that of the motion of a mass along a massless string (Fig. 14.2). We shall again start out from the integro-differential equation (5.5) and considering $\mu \to 0$, write it in the form

$$v(x, t) = \int_0^l G(x, s)\, p(s, t)\, \mathrm{d}s\,. \tag{14.9}$$

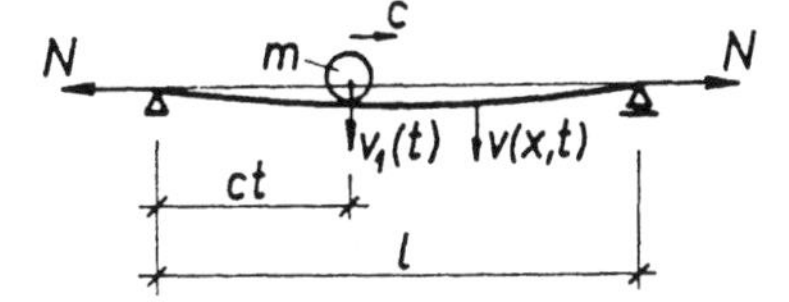

Fig. 14.2. Massless string subjected to moving mass m.

Taking into account also the inertia effect of mass m of load P, the equation of load $p(x, t)$ turns out to be

$$p(x, t) = \delta(x - ct)\left[P - m\,\frac{\mathrm{d}^2 v(ct, t)}{\mathrm{d}t^2}\right]. \tag{14.10}$$

$G(x, s)$ is the so-called influence function of string deflection or Green's function. It is obtained from the static solution of Eq. (14.1) for $\mu = 0$, $p(x, t) = \delta(x - s)$, i.e. for the string loaded with force $P = 1$ at point $x = s$, namely

$$G(x, s) = \begin{cases} \dfrac{1}{N}\left(1 - \dfrac{s}{l}\right) x & \text{for} \quad x \leqq s \\[2ex] \dfrac{1}{N}\left(1 - \dfrac{x}{l}\right) s & \text{for} \quad x \geqq s\,. \end{cases} \tag{14.11}$$

Comparing Eqs. (14.11) and (5.9) we again see perfect analogy between the string deflection and the bending moment of a simple beam.

According to Fig. 14.2, the vertical displacement of mass m, $v_1(t)$, is equal to the string deflection at point $x = ct$ and time t, i.e.

$$v_1(t) = v(ct, t) . \tag{14.12}$$

From Eqs. (14.9), (14.10) and (14.12) then follows the ordinary differential equation

$$v_1(t) = \left[P - m \frac{\mathrm{d}^2 v_1(t)}{\mathrm{d}t^2}\right] G(ct, ct) \tag{14.13}$$

where

$$G(ct, ct) = \frac{1}{N}\left(1 - \frac{ct}{l}\right) ct$$

with zero initial conditions (7.4).

We shall now introduce the dimensionless dependent variable and the dimensionless independent variable

$$y(\tau) = v_1(t)/v_0 , \quad \tau = ct/l \tag{14.14}$$

where

$$v_0 = \frac{Pl}{4N} \tag{14.15}$$

is the deflection at mid-span of the string subjected to force P placed at $x = l/2$.

Using the new variables we write Eq. (14.13) as

$$\tau(1 - \tau)\, \ddot{y}(\tau) + 2\alpha'\, y(\tau) = 8\alpha'\, \tau(1 - \tau) . \tag{14.16}$$

The initial conditions are (7.10) and the solution is sought in the interval $0 \leqq \tau \leqq 1$. Symbol

$$\alpha' = \frac{Nl}{2mc^2} \tag{14.17}$$

used in (14.16) characterizes all the data necessary for the solution of the problem.

Case $\alpha' \neq 1$

Eq. (14.16) is an ordinary differential equation of the second order with variable coefficients. In the range $\langle 0, 1 \rangle$ its singular points are $\tau = 0$ and $\tau = 1$.

It is solved by the substitution $y(\tau) = \tau(1 - \tau)\, u(\tau)$. After a bit of handling we get

$$\tau(1 - \tau)\, \ddot{u}(\tau) + (2 - 4\tau)\, \dot{u}(\tau) - 2(1 - \alpha')\, u(\tau) = 8\alpha' . \quad (14.18)$$

The homogeneous equation appertaining to (14.18) can be reduced to the hypergeometric (Gauss') differential equation [120], [186]

$$\tau(1 - \tau)\, \ddot{u}(\tau) + [c - (a + b + 1)\, \tau]\, \dot{u}(\tau) - ab\, u(\tau) = 0 \quad (14.19)$$

if we set

$a + b + 1 = 4$, $ab = 2(1 - \alpha')$, $c = 2$. From this follows that

$$a = \tfrac{1}{2}[3 \mp (1 + 8\alpha')^{1/2}] , \quad b = \tfrac{1}{2}[3 \pm (1 + 8\alpha')^{1/2}] , \quad c = 2 .$$

According to [100] (Eq. 9.153.3) the fundamental system of Eq. (14.19) is formed by the expressions

$$u_1(\tau) = \mathrm{F}(a, b, 2, \tau) ,$$

$$u_2(\tau) = \mathrm{F}(a, b, 2, \tau) \ln \tau + \sum_{k=1}^{\infty} \frac{(a)_k (b)_k}{(2)_k} [h(k) - h(0)]\, \tau^k + $$

$$+ \frac{1}{(1 - a)(1 - b)\, \tau} \quad (14.20)$$

where the hypergeometric series

$$\mathrm{F}(a, b, c, \tau) = 1 + \sum_{k=1}^{\infty} \frac{(a)_k (b)_k}{(c)_k} \frac{\tau^k}{k!} ,$$

$$(a)_k = a(a + 1) \dots (a + k - 1) ,$$

$$h(k) = \psi(a + k) + \psi(b + k) - \psi(2 + k) - \psi(1 + k) ,$$

$$\psi(x) = -C - \frac{1}{x} + x \sum_{n=1}^{\infty} \frac{1}{n(x + n)} \quad \text{Euler's psi function.}$$

For $\alpha' \neq 1$ the particular solution to the non-homogeneous equation (14.18) was readily obtained as

$$u_p(\tau) = \frac{4\alpha'}{\alpha' - 1} . \tag{14.21}$$

With (14.20) and (14.21) the general solution of Eq. (14.16) thus is

$$y(\tau) = [A_1 u_1(\tau) + A_2 u_2(\tau) + u_p(\tau)] \tau (1 - \tau) .$$

Constants A_1, A_2 are established from the initial conditions (7.10) with the above solution substituted in. At the limit, for τ tending to zero, $A_2 = 0$, $A_1 = -4\alpha'/(\alpha' - 1)$. Therefore, the sought solution that satisfies the initial conditions, too, is

$$y(\tau) = \frac{4\alpha'}{\alpha' - 1} \tau (1 - \tau) [1 - F(a, b, 2, \tau)] . \tag{14.22}$$

Solution (14.20) applies for $|\tau| < 1$ and is a distribution of solutions for the neighbourhood of point $\tau = 0$. For the neighbourhood of $\tau = 1$ there exists another system of fundamental solutions (cf. [165], Vol. I, p. 776); the case is analyzed at length in [202].

Case $\alpha' = 1$

If the parameter (14.17) equals one, $\alpha' = 1$, the case may be solved in closed form. Eq. (14.16) will then take on the form

$$\tau(1 - \tau) \ddot{y}(\tau) + 2y(\tau) = 8\tau(1 - \tau) . \tag{14.23}$$

First, we shall solve the homogeneous equation appertaining to (14.23). Note that the left-hand side of (14.23) is the so-called exact equation since it is the total differential of function $\tau(1 - \tau) \dot{y}(\tau) - (1 - 2\tau) y(\tau)$. We may, therefore, reduce the order of the homogeneous equation (14.23) to the first and consider the equation (except for the integration constants which will be introduced in the calculation at the end) in the form

$$\frac{dy}{y} = \frac{1 - 2\tau}{\tau(1 - \tau)} d\tau . \tag{14.24}$$

The last equation can readily be integrated by the substitution $\tau(1 - \tau) = z$, $dz = (1 - 2\tau)\, d\tau$, which gives

$$\ln |y| = \ln |\tau(1 - \tau)| .$$

From there follows the desired solution (again except for the integration constant) in the range $\langle 0, 1 \rangle$, namely

$$y_1(\tau) = \tau(1 - \tau) . \tag{14.25}$$

If we know one solution, (14.25), of the homogeneous equation (14.23), we may obtain the other one by the substitution

$$y_2(\tau) = y_1(\tau)\, u(\tau) = \tau(1 - \tau)\, u(\tau) . \tag{14.26}$$

With (14.26) in (14.23) we get

$$\tau(1 - \tau)\, \ddot{u}(\tau) + 2(1 - 2\tau)\, \dot{u}(\tau) = 0 ;$$

by further substitution, $\dot{u}(\tau) = u_1(\tau)$, it is

$$\tau(1 - \tau)\, \dot{u}_1(\tau) + 2(1 - 2\tau)\, u_1(\tau) = 0$$

and from there

$$\frac{du_1}{u_1} = -2 \frac{1 - 2\tau}{\tau(1 - \tau)}\, d\tau .$$

Like (14.24), the last equation is easily integrated by the substitution $\tau(1 - \tau) = z$, $dz = (1 - 2\tau)\, d\tau$ with the result (except for the integration constant)

$$u_1(\tau) = [\tau(1 - \tau)]^{-2} .$$

Integration of the above equation (cf. [186], Sect. 13.5, Eq. 19) gives

$$u(\tau) = \frac{1}{1 - \tau} - \frac{1}{\tau} - 2 \ln \frac{1 - \tau}{\tau} .$$

Hence, according to (14.26) the second solution of the homogeneous equation is

$$y_2(\tau) = 2\tau - 1 - 2\tau(1 - \tau) \ln \frac{1 - \tau}{\tau} . \tag{14.27}$$

Expressions (14.25) and (14.27) form the fundamental system of the homogeneous equation, because within the whole of interval $\langle 0, 1 \rangle$ the Wronskian is different from zero,

$$W = y_1 \dot{y}_2 - \dot{y}_1 y_2 = 1 . \tag{14.28}$$

Using the method of variation of parameters we get the particular solution of non-homogeneous equation (14.16) by mere quadratures

$$y_p(\tau) = A_1(\tau)\, y_1(\tau) + A_2(\tau)\, y_2(\tau) \tag{14.29}$$

with functions $A_1(\tau)$ and $A_2(\tau)$ satisfying the equations

$$\begin{aligned} \dot{A}_1(\tau)\, y_1(\tau) + \dot{A}_2(\tau)\, y_2(\tau) &= 0 , \\ \dot{A}_1(\tau)\, \dot{y}_1(\tau) + \dot{A}_2(\tau)\, \dot{y}_2(\tau) &= 8 . \end{aligned} \tag{14.30}$$

The substitution of (14.25), (14.27) and (14.28) in (14.30) gives $\dot{A}_1$ and $\dot{A}_2$, and their integration A_1 and A_2. After all the operations have been carried out, the particular solution (14.29) turns out to be

$$\begin{aligned} y_p(\tau) = \frac{4}{3}\tau^2(3 - 2\tau)\left[2\tau - 1 - 2\tau(1 - \tau)\ln\frac{1 - \tau}{\tau}\right] - 8\tau(1 - \tau) \,. \\ .\left\{\tau^2 - \tau^2 \ln\frac{1 - \tau}{\tau} + \ln(1 - \tau) + \frac{2}{3}\left[\tau^3 \ln\frac{1 - \tau}{\tau} - \frac{1}{2} \,. \right.\right. \\ \left.\left. .\,(\tau + 2)\,\tau + \ln\frac{1}{1 - \tau}\right]\right\} . \end{aligned} \tag{14.31}$$

With (14.25), (14.27) and (14.31) the general integral of non-homogeneous Eq. (14.23) thus is

$$y(\tau) = A_1\, y_1(\tau) + A_2\, y_2(\tau) + y_p(\tau) \tag{14.32}$$

where A_1 and A_2 are the integration constants. For zero initial conditions (7.10), $A_1 = A_2 = 0$ so that

$$y(\tau) = y_p(\tau) . \tag{14.33}$$

Case $\alpha' = 0$

For $\alpha' = 0$, i.e. $c \to \infty$, and zero initial conditions (7.10), Eq. (14.16) gives

$$y(\tau) = 0 . \tag{14.34}$$

Case $\alpha' \to \infty$

For $\alpha' \to \infty$, i.e. for the static case with $c = 0$, (14.16) yields

$$y(\tau) = 4\tau(1 - \tau) \tag{14.35}$$

which corresponds to the trajectory of mass m, i.e. $v_1(t) = P\,G(ct, ct)$.

14.3 *String with ends suspended at unequal heights*

In practice one often comes across the case of a string with ends A, B (Fig. 14.3a) suspended at points with a height difference $h = l \sin \Phi$,

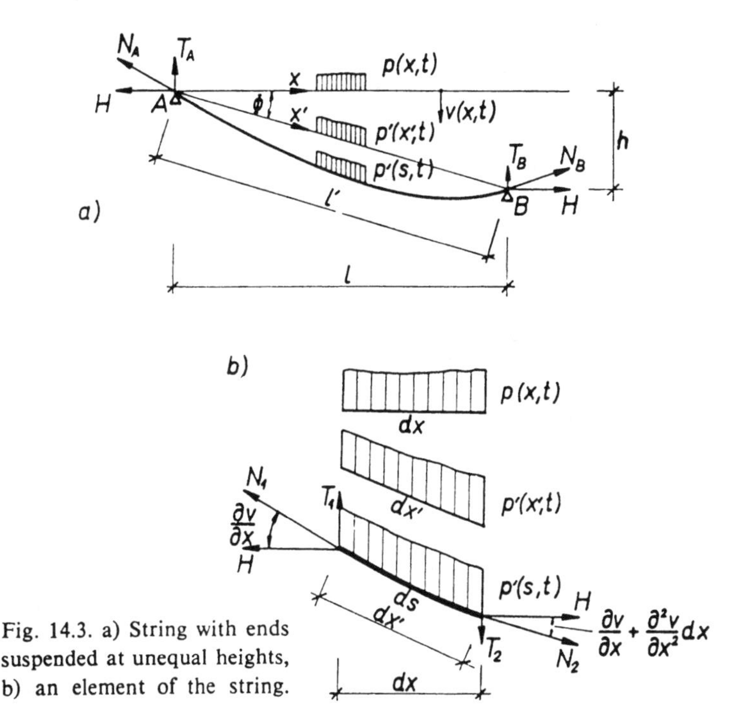

Fig. 14.3. a) String with ends suspended at unequal heights, b) an element of the string.

where Φ is the angle of chord $\overline{AB}$ from the horizontal. Let us derive the differential equation of motion of such a string. Draw the horizontal axis x through point A as the origin, the oblique axis x' also through the origin but in the direction of chord $\overline{AB}$. The string deflection $v(x, t)$ is measured from axis x in the downward direction. The string is subjected to load $p(x, t)$ per unit length of axis x, load $p'(x, t)$ per unit length of axis x', and to load $p'(s, t)$ per unit length of the curved string. All the loads act in the vertical direction.

As it is well known from the theory of catenaries, there acts at each point of the string a constant horizontal force H. We shall accept this assumption for our case, too, and suppose that the force is large enough to be virtually constant even under a load varying in time. The string can only be stretched by the normal force N which resolves into the constant horizontal force H and the vertical force T (Figs. 14.3a,b).

In the derivation of the respective differential equation, we first take from the string a length element $\mathrm{d}s$ and replace the action of the other parts of the string at its ends by normal forces N_1 and N_2 (Fig. 14.3b). The respective vertical components T_1 and T_2 are computed from the rotation and its increment

$$T_1 = H\,\frac{\partial v(x, t)}{\partial x}\,, \quad T_2 = H\left[\frac{\partial v(x, t)}{\partial x} + \frac{\partial^2 v(x, t)}{\partial x^2}\,\mathrm{d}x\right]. \tag{14.36}$$

Next we write the equation of equilibrium of all vertical forces acting on element $\mathrm{d}s$ from Fig. 14.3b:

$$-T_1 + T_2 + p(x, t)\,\mathrm{d}x + p'(x', t)\,\mathrm{d}x' + p'(s, t)\,\mathrm{d}s - \\ - \mu'\,\frac{\partial^2 v(x, t)}{\partial t^2}\,\mathrm{d}s = 0\,. \tag{14.37}$$

In the above, the last component is the inertia force exerted by mass μ' uniformly distributed along the length of the deflected string. On substituting (14.36) and factoring out $\mathrm{d}s$ we get from (14.37) the differential equation of the string

$$-H\,\frac{\mathrm{d}x}{\mathrm{d}s}\,\frac{\partial^2 v(x, t)}{\partial x^2} + \mu'\,\frac{\partial^2 v(x, t)}{\partial t^2} = p(x, t)\,\frac{\mathrm{d}x}{\mathrm{d}s} + p'(x', t)\,\frac{\mathrm{d}x'}{\mathrm{d}s} + p'(s, t)\,. \tag{14.38}$$

The general theory of catenaries counts with a curved coordinate of the string, i.e. $\mathrm{d}s = \mathrm{d}x[1 + (\partial v/\partial x)^2]^{1/2}$. Our considerations are confined to the so-called parabolic theory of strings according to which

$$\mathrm{d}s \approx \mathrm{d}x'\,, \quad \mathrm{d}x' \cos\Phi = \mathrm{d}x\,, \quad \mathrm{d}x/\mathrm{d}s \approx \mathrm{d}x/\mathrm{d}x' = \cos\Phi\,.$$

With these values in, Eq. (14.38) takes on the form [$p'(x', t)$ now merges into one with $p'(s, t)$]:

$$-H \cos \Phi \frac{\partial^2 v(x, t)}{\partial x^2} + \mu' \frac{\partial^2 v(x, t)}{\partial t^2} = p(x, t) \cos \Phi + p'(x', t) . \quad (14.39)$$

14.3.1 *Static deflection of a string produced by dead weight and by a concentrated force*

To illustrate we will first solve the static [i.e. without the effect of the second term of Eq. (14.39)] deflection of a string subjected to vertical load q' uniformly distributed along axis x' (that is, to the dead weight of the string $q' = \mu' g$) and to concentrated force P acting at point $s = s' \cos \Phi$ on axis x, or at point s' on axis x'. Thus:

$$p'(x', t) = q' ; \quad p(x, t) = \delta(x - s) P , \quad \text{or possibly} , \quad p'(x', t) = \\ = \delta(x' - s') P = \cos \Phi \, \delta(x - s) P . \quad (14.40)$$

The solution of Eq. (14.39) with (14.40) on its right-hand side is

$$v(x, s) = A_1 + A_2 x - \frac{q' x^2}{2H \cos \Phi} + P \, G(x, s) . \quad (14.41)$$

The first two terms on the right-hand side of this equation are the general solution of the homogeneuos equation, the last two, the particular solutions for the given load [$G(x, s)$ is defined by Eq. (14.11) with N replaced by H]. The integration constants A_1, A_2 are obtained from the boundary conditions $v(0) = 0$, $v(l) = h$ in the form $A_1 = 0$, $A_2 = h/l + q' l/(2H \cos \Phi)$. With them the solution (14.41) takes on the form

$$v(x, s) = \frac{h}{l} x + \frac{q' x(l - x)}{2H \cos \Phi} + P \, G(x, s) . \quad (14.42)$$

14.3.2 *Motion of a force along a string*

For the second illustrative example we choose the analysis of the motion of force P on axis x' at speed c' ($x' \cos \Phi = x$, $c' \cos \Phi = c$). In view of

(1.8) the load in Eq. (14.39) becomes

$$p'(x', t) = \delta(x' - c't)\,P = \cos\Phi\,\delta(x - ct)\,P \quad \text{or}$$

$$p(x, t) = \delta(x - ct)\,P\,. \tag{14.43}$$

The solution of Eq. (14.39) with load (14.43) is wholly analogous to that shown in Sect. 14.1. At $\omega_{(j)} \neq \omega$ the result is also analogous to Eq. (14.8):

$$v(x, t) = \sum_{j=1}^{\infty} \frac{2P\cos\Phi}{\mu' l} \frac{1}{\omega_{(j)}^2 - \omega^2} \left(\sin\omega t - \frac{\omega}{\omega_{(j)}} \sin\omega_{(j)} t\right) \sin\frac{j\pi x}{l}\,. \tag{14.44}$$

The circular frequency of a string suspended at unequal heights is

$$\omega_{(j)}^2 = \frac{j^2\pi^2}{l^2} \frac{H\cos\Phi}{\mu'}\,. \tag{14.45}$$

The string deflection (14.44) is now of course added to the deflection from the dead-weight load, i.e. to the first two terms on the right-hand side of Eq. (14.42).

14.4 *Application of the theory*

The theory outlined in the preceding sections can be used in calculations of supporting cables of cableways, cable cranes, and other structures of that sort in which a moving load comes into play. Even though the latter — because of the low speeds of its motion — bears a lesser effect than some other kinds of dynamic loading (e.g. sudden emptying or fall of the truck, sudden load applied to the truck, etc.), it should nevertheless be taken into account. It has been observed in many instances that as the moving truck or car suddenly stops, the cable starts to vibrate intensively — a phenomenon especially unpleasant in passenger cableways.*) The theory can also be applied to textile engineering, e.g. to the analysis of shuttle motion over textile fibres.

*) Because the phenomena accompanying stopping and starting of cableways are more complicated still, the theory explained in the foregoing sections cannot be applied directly to the case of sudden stopping of a cableway car.

As to cables and ropes: the calculation should include both the effect of their mass and the effect produced by the inertia force of load P, i.e. load (14.10). An exact calculation is again quite complicated (cf. [202]). In view of the very low speeds common in cableways or cable cranes, and the large deflections due to the cable dead-weight, the second derivative in (14.10) may approximately be computed from the static deflection (14.42) of a string loaded with dead weight q' and force P. Eq. (14.42) [$s = ct$ also because of (14.11) where $N = H$] yields

$$v(ct, t) = \frac{h}{l} ct + \frac{q' ct(l - ct)}{2H \cos \Phi} + P\, G(s, s),$$

$$\frac{\mathrm{d}^2 v(ct, t)}{\mathrm{d}t^2} = - \frac{q'c^2}{H \cos \Phi} - \frac{2Pc^2}{Hl} \tag{14.46}$$

so that the concentrated force in (14.10) turns out approximately

$$P' = P - m \frac{\mathrm{d}^2 v(ct, t)}{\mathrm{d}t^2} = P \left[1 + \frac{c^2}{gH} \left(\frac{q'}{\cos \Phi} + \frac{2P}{l} \right) \right]. \tag{14.47}$$

If P' according to (14.47) is substituted for P in (14.44), the string deflection thus obtained will include even an approximate expression of the effect of inertia of the moving load.

14.5 *Additional bibliography*

[42, 83, 105, 202, 210].

Part III

TWO-DIMENSIONAL SOLIDS

Part III

15

Plates subjected to a moving load

Vibration of plates under the action of a moving load has so far received but scant attention. The problem was treated by W. Nowacki [170] and K. Piszczek [180].

The method of plate vibration analysis is based on the following assumptions:

1. The small elastic strains arising in the body are within the scope of Hooke's law.

2. There exists in the plate the so-called neutral surface. The distances between points lying on that surface do not vary with plate deflection. The surface contains the coordinate plane xy, axis z points downward (Fig. 15.1).

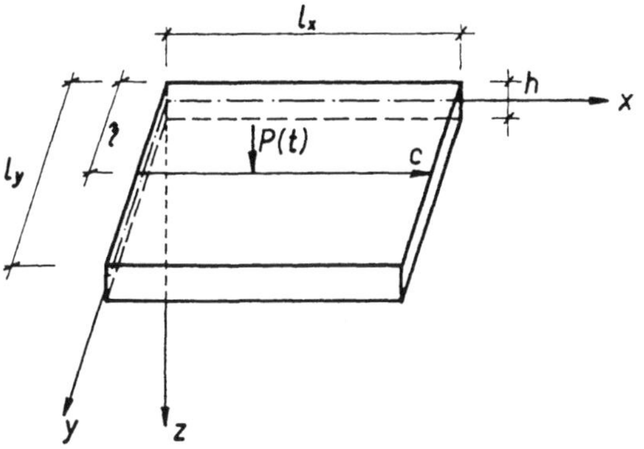

Fig. 15.1. Rectangular plate subjected to load $P(t)$ moving parallel to axis x.

3. Mass particles lying on the normal line to the neutral surface continue to lie on it even after the plate has been deformed.

The differential equation of plate vibration written on the above assumptions is in the form

$$D\left[\frac{\partial^4 w(x, y, t)}{\partial x^4} + 2\,\frac{\partial^4 w(x, y, t)}{\partial x^2\,\partial y^2} + \frac{\partial^4 w(x, y, t)}{\partial y^4}\right] + \mu\,\frac{\partial^2 w(x, y, t)}{\partial t^2} = p(x, y, t) \qquad (15.1)$$

where

$w(x, y, t)$	– vertical deflection of the plate at point with coordinates x, y, and time t,
$D = \dfrac{Eh^3}{12(1 - \nu)}$	– bending rigidity of the plate,
E	– Young's modulus of the plate,
h	– thickness of the plate,
ν	– Poisson's ratio ($\nu < 1$),
μ	– mass per unit area of the plate,
$p(x, y, t)$	– external load per unit area of the plate.

In case the inertia effect of mass $\mu_p(x, y, t)$ is considered too, the external load takes on the complicated form

$$p(x, y, t) - \mu_p(x, y, t) \frac{d^2 w}{dt^2} \tag{15.2}$$

where the acceleration $d^2 w/dt^2$ is computed from the total differential of the second order assuming that coordinates x and y that describe the load motion on the plate, are functions of time

$$\frac{d^2 w}{dt^2} = \frac{\partial^2 w}{\partial x^2}\left(\frac{dx}{dt}\right)^2 + \frac{\partial^2 w}{\partial y^2}\left(\frac{dy}{dt}\right)^2 + \frac{\partial^2 w}{\partial t^2} + 2\frac{\partial^2 w}{\partial x\, \partial y}\frac{dx}{dt}\frac{dy}{dt} +$$

$$+ 2\frac{\partial^2 w}{\partial x\, \partial t}\frac{dx}{dt} + 2\frac{\partial^2 w}{\partial y\, \partial t}\frac{dy}{dt} + \frac{\partial w}{\partial x}\frac{d^2 x}{dt^2} + \frac{\partial w}{\partial y}\frac{d^2 y}{dt^2}\,. \tag{15.3}$$

Since the solution is very difficult to manage, we will restrict our considerations to the force action of the external load and neglect its inertia effect described by (15.2).

Next we shall write the boundary conditions of plates (for details see, e.g. [8], [170]). Thus for example, for rectangular plates the boundary conditions on edges parallel to axis y are for a simply supported edge

$$w = 0\,, \quad \frac{\partial^2 w}{\partial x^2} + \nu \frac{\partial^2 w}{\partial y^2} = 0\,, \tag{15.4}$$

for a clamped edge

$$w = 0\,, \quad \frac{\partial w}{\partial x} = 0\,, \tag{15.5}$$

and for a free edge

$$\frac{\partial^2 w}{\partial x^2} + \nu \frac{\partial^2 w}{\partial y^2} = 0\,, \quad \frac{\partial^3 w}{\partial x^3} + (2 - \nu) \frac{\partial^3 w}{\partial x\, \partial y^2} = 0\,. \tag{15.6}$$

The initial conditions are

$$w(x, y, t)\big|_{t=0} = g_1(x, y)\,, \quad \left.\frac{\partial w(x, y, t)}{\partial t}\right|_{t=0} = g_2(x, y)\,. \tag{15.7}$$

15.1 *Simply supported rectangular plate*

The rectangular plate considered has spans l_x in the direction of axis x, and l_y in the direction of axis y (Fig. 15.1). According to (15.4) the boundary conditions for $x = 0$, $x = l_x$ are

$$w = 0\,, \quad \frac{\partial^2 w}{\partial x^2} + \nu \frac{\partial^2 w}{\partial y^2} = 0\,, \tag{15.8}$$

and for $y = 0$, $y = l_y$

$$w = 0\,, \quad \frac{\partial^2 w}{\partial y^2} + \nu \frac{\partial^2 w}{\partial x^2} = 0\,. \tag{15.9}$$

For boundary conditions of the above kind it is convenient to use the two-dimensional Fourier finite sine integral transformations [204] defined by the relations

$$W(i, j, t) = \int_0^{l_x} \int_0^{l_y} w(x, y, t) \sin \frac{i\pi x}{l_x} \sin \frac{j\pi y}{l_y}\, \mathrm{d}x\, \mathrm{d}y\,, \tag{15.10}$$

$$w(x, y, t) = \sum_{i=1}^{\infty} \sum_{j=1}^{\infty} \frac{4}{l_x l_y} W(i, j, t) \sin \frac{i\pi x}{l_x} \sin \frac{j\pi y}{l_y}\,. \tag{15.11}$$

In view of the boundary conditions (15.8) and (15.9), (15.10) results in the following transformations of the derivatives

$$\int_0^{l_x} \int_0^{l_y} \frac{\partial^4 w}{\partial x^4} \sin \frac{i\pi x}{l_x} \sin \frac{j\pi x}{l_y}\, \mathrm{d}x\, \mathrm{d}y = \frac{i^4 \pi^4}{l_x^4} W(i, j, t)\,,$$

$$\int_0^{l_x} \int_0^{l_y} \frac{\partial^4 w}{\partial x^2\, \partial y^2} \sin \frac{i\pi x}{l_x} \sin \frac{j\pi y}{l_y}\, \mathrm{d}x\, \mathrm{d}y = \frac{i^2 j^2 \pi^4}{l_x^2 l_y^2} W(i, j, t)\,,$$

$$\int_0^{l_x} \int_0^{l_y} \frac{\partial^4 w}{\partial y^4} \sin \frac{i\pi x}{l_x} \sin \frac{j\pi y}{l_y}\, \mathrm{d}x\, \mathrm{d}y = \frac{j^4 \pi^4}{l_y^4} W(i, j, t)\,. \tag{15.12}$$

15.1.1 *Force variable in time moving parallel to x-axis*

We are now in a position to undertake an analysis of the case in which concentrated force $P(t)$ moves on a plate at uniform speed c. Since the force moves along the straight line $y = \eta$ parallel to x-axis (Fig. 15.1), the load is

$$p(x, y, t) = \delta(x - ct)\, \delta(y - \eta)\, P(t)\,. \tag{15.13}$$

Eq. (15.1) with (15.13) on its right-hand side is transformed according to (15.10). In view of (15.12) we thus get the ordinary differential equation

$$\ddot{W}(i, j, t) + \omega_{(i,j)}^2\, W(i, j, t) = \frac{P(t)}{\mu} \sin \omega_x t \sin \frac{j\pi\eta}{l_y} \tag{15.14}$$

where

$$\omega_{(i,j)}^2 = \left(\frac{i^4\pi^4}{l_x^4} + 2\, \frac{i^2 j^2 \pi^4}{l_x^2 l_y^2} + \frac{j^4\pi^4}{l_y^4} \right) \frac{D}{\mu} \tag{15.15}$$

is the natural circular frequency of a simply supported rectangular plate, and

$$\omega_x = \frac{i\pi c}{l_x}\,. \tag{15.16}$$

To simplify we shall assume zero initial conditions (15.7), i.e. $g_1(x, y) = 0$, $g_2(x, y) = 0$. In view of (27.4) and (27.11), Eq. (15.14) solved by the Laplace-Carson transformation will then give

$$W^*(i, j, p) = \frac{\sin j\pi\eta/l_y}{\mu}\, \frac{1}{p^2 + \omega_{(i,j)}^2}\, \frac{p}{2i} \left[\frac{P^*(p - i\omega_x)}{p - i\omega_x} - \frac{P^*(p + i\omega_x)}{p + i\omega_x} \right] \tag{15.17}$$

where $W^*(i, j, p)$ and $P^*(p)$ are the Laplace-Carson transformations of deflection $W(i, j, t)$ and force $P(t)$, respectively. Following the inverse transformations (27.5), (27.18) and (15.11) we get the result

$$w(x, y, t) = \sum_{i=1}^{\infty} \sum_{j=1}^{\infty} \frac{4}{\mu l_x l_y \omega_{(i,j)}} \sin \frac{i\pi x}{l_x} \sin \frac{j\pi y}{l_y} \sin \frac{j\pi\eta}{l_y}\,.$$

$$\cdot \int_0^t P(\tau) \sin \omega_x \tau \sin \omega_{(i,j)}(t - \tau)\, d\tau \tag{15.18}$$

valid in the range $0 \leqq t \leqq l_x/c$.

15.1.2 *Motion of a constant force*

If the force considered is of constant magnitude, $P(t) = P$, (15.17) will give

$$W^*(i, j, p) = \frac{P \sin j\pi\eta/l_y}{\mu} \frac{\omega_x p}{(p^2 + \omega_{(i,j)}^2)(p^2 + \omega_x^2)}$$

and following the inverse transformations (27.32) and (15.11), from there

$$w(x, y, t) = \sum_{i=1}^{\infty} \sum_{j=1}^{\infty} \frac{4P}{\mu l_x l_y} \frac{1}{\omega_{(i,j)}^2 - \omega_x^2} \sin \frac{i\pi x}{l_x} \sin \frac{j\pi y}{l_y} \, .$$

$$. \sin \frac{j\pi\eta}{l_y} \left(\sin \omega_x t - \frac{\omega_x}{\omega_{(i,j)}} \sin \omega_{(i,j)} t \right) . \tag{15.19}$$

Eq. (15.19) applies in the interval $0 \leqq t \leqq l_x/c$ at $\omega_x \neq \omega_{(i,j)}$. For $\omega_x = \omega_{(i,j)}$ one could deduce a solution analogous to that of paragraph 1.3.2, case 1.3.2.2.

15.1.3 *Motion of a force along a straight line*

Consider now the case where a simply supported plate is traversed by a constant force P at uniform speed c (Fig. 15.2), the motion of the force

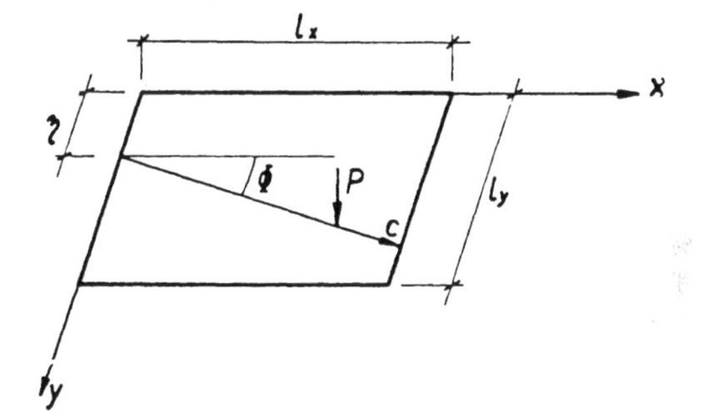

Fig. 15.2.
Rectangular plate; force P moves along a straight line.

being effected along a section of a straight line issuing from point $y = \eta$ on axis y and forming angle Φ with axis x. Hence the components of speed c on axes x and y

$$c_x = c \cos \Phi \, , \quad c_y = c \sin \Phi \, .$$

In this case the load is

$$p(x, y, t) = \delta(x - c_x t)\, \delta(y - \eta - c_y t)\, P\,. \tag{15.20}$$

The solution of Eq. (15.1) with (15.20) for its right-hand side is completely analogous to those discussed in paragraphs 15.1.1 and 15.1.2. Following the transformation (15.10) we get

$$\ddot{W}(i, j, t) + \omega_{(i,j)}^2 W(i, j, t) = \frac{P}{\mu} \sin \frac{i\pi c_x t}{l_x} \sin \frac{j\pi(\eta + c_y t)}{l_y}\,. \tag{15.21}$$

Writing now

$$\omega_x = \frac{i\pi c_x}{l_x}\,, \qquad \omega_y = \frac{j\pi c_y}{l_y}\,, \tag{15.22}$$

$$r_1 = \omega_x + \omega_y\,, \quad r_2 = \omega_x - \omega_y \tag{15.23}$$

and transforming Eq. (15.21) with respect to time, we get — in view of (27.4), (27.24) and (27.25) —

$$W^*(i, j, p) = \frac{P}{\mu} \frac{1}{p^2 + \omega_{(i,j)}^2} \left[\sin \frac{j\pi\eta}{l_y} \frac{p}{2} \left(\frac{r_2}{p^2 + r_2^2} + \frac{r_1}{p^2 + r_1^2}\right) + \right.$$
$$\left. + \cos \frac{j\pi\eta}{l_y} \frac{p^2}{2} \left(\frac{1}{p^2 + r_2^2} - \frac{1}{p^2 + r_1^2}\right)\right]. \tag{15.24}$$

The result valid for $\omega_{(i,j)} \neq r_1, r_2$ so long as the force continues to act on the plane, is obtained following the inverse trasnformations carried out with the aid of (27.32), (27.33) and (15.11). It is

$$w(x, y, t) = \sum_{i=1}^{\infty} \sum_{j=1}^{\infty} \frac{2P}{\mu l_x l_y} \sin \frac{i\pi x}{l_x} \sin \frac{j\pi y}{l_y} \left\{\frac{1}{\omega_{(i,j)}^2 - r_2^2} \cdot \right.$$
$$\cdot \left[\cos\left(\frac{j\pi\eta}{l_y} - r_2 t\right) - \cos \frac{j\pi\eta}{l_y} \cos \omega_{(i,j)} t - \frac{r_2}{\omega_{(i,j)}} \sin \frac{j\pi\eta}{l_y} \cdot \right.$$
$$\left. . \sin \omega_{(i,j)} t\right] - \frac{1}{\omega_{(i,j)}^2 - r_1^2} \left[\cos\left(\frac{j\pi\eta}{l_y} + r_1 t\right) - \cos \frac{j\pi\eta}{l_y} \cdot \right.$$
$$\left.\left. . \cos \omega_{(i,j)} t + \frac{r_1}{\omega_{(i,j)}} \sin \frac{j\pi\eta}{l_y} \sin \omega_{(i,j)} t\right]\right\}. \tag{15.25}$$

15.2 *Rectangular plate simply supported on opposite edges*

Dynamic analysis of a rectangular plate is straightforward only in the case two of its opposite edges are simply supported. The other two opposite edges may be supported at will. In what follows we will consider the plate to be simply supported on edges $x = 0$, $x = l_x$.

In the solution we shall make use of the generalized two-dimensional integral transformation arrived at through generalization and analogy of the results obtained in Sects. 15.1 and 6.1. The transformation is defined by the relations

$$W(i, j, t) = \int_0^{l_x} \int_0^{l_y} w(x, y, t) \sin \frac{i\pi x}{l_x} w_{(j)}(y) \, dx \, dy \,, \tag{15.26}$$

$$w(x, y, t) = \sum_{i=1}^{\infty} \sum_{j=1}^{\infty} \frac{2}{l_x} \frac{\mu}{W_j} W(i, j, t) \sin \frac{i\pi x}{l_x} w_{(j)}(y) \tag{15.27}$$

where $w_{(j)}(y)$ is a function that satisfies the boundary conditions on edges $y = 0$, $y = l_y$ as well as the equation of free vibration of the plate [i.e. Eq. (15.1) without the right-hand side]. It is calculated from the ordinary differential equation

$$D\left[\frac{i^4\pi^4}{l_x^4} w_{(j)}(y) - 2\frac{i^2\pi^2}{l_x^2} w''_{(j)}(y) + w^{IV}_{(j)}(y)\right] - \mu\omega^2_{(i,j)} w_{(j)}(y) = 0 \tag{15.28}$$

which results from the substitution $w(x, y, t) = \sin(i\pi x/l_x)\, w_{(j)}(y)\,.$ $. \sin \omega_{(i,j)}t$ in (15.1).

Further, in (15.27)

$$W_j = \int_0^{l_y} \mu w^2_{(j)}(y) \, dy \,. \tag{15.29}$$

We have proved the validity of relations (15.26) and (15.27) by making use of the orthogonal properties of normal modes

$$\int_0^{l_x} \int_0^{l_y} \mu \sin \frac{i\pi x}{l_x} \sin \frac{i'\pi x}{l_x} w_{(j)}(y)\, w_{(j')}(y) \, dx \, dy =$$

$$= \begin{cases} 0 & \text{for} \quad i \neq i',\ j \neq j' \\ W_j \dfrac{l_x}{2} & \text{for} \quad i = i',\ j = j' \,. \end{cases} \tag{15.30}$$

The first relation, (15.26), is proved after substituting (15.27) for $w(x, y, t)$

$$W(i, j, t) = \int_0^{l_x} \int_0^{l_y} \sum_{i'=1}^{\infty} \sum_{j'=1}^{\infty} \frac{2}{l_x} \frac{\mu}{W_{j'}} W(i', j', t) \sin \frac{i' \pi x}{l_x} .$$

$$. \, w_{(j')}(y) \sin \frac{i \pi x}{l_x} w_{(j)}(y) \, dx \, dy = \frac{2}{l_x} \frac{1}{W_j} W(i, j, t) \, W_j \frac{l_x}{2} = W(i, j, t) .$$

The above was obtained on the assumption that series (15.27) is uniformly convergent in the interval $0 \leqq x \leqq l_x$, $0 \leqq y \leqq l_y$ and therefore the integration and the summation may be exchanged in sequence, and further, that (15.29) and (15.30) apply. Relation (15.27) may be proved in a similar way.

15.2.1 *Force variable in time moving parallel to x-axis under arbitrary initial conditions*

Proceeding on the lines established in paragraph 15.1.1 we will now find the deflection of a plate simply supported on two opposite edges with the other two edges supported in any of the usual ways (Fig. 15.1). Force $P(t)$ moves parallel to x-axis at a distance of $y = \eta$ with speed c. The initial conditions are described by (15.7). Functions $g_1(x, y)$ and $g_2(x, y)$ are assumed to be expressible in terms of relations (15.26) and (15.27) as follows:

$$G_1(i, j) = \int_0^{l_x} \int_0^{l_y} g_1(x, y) \sin \frac{i \pi x}{l_x} w_{(j)}(y) \, dx \, dy ,$$

$$G_2(i, j) = \int_0^{l_x} \int_0^{l_y} g_2(x, y) \sin \frac{i \pi x}{l_x} w_{(j)}(y) \, dx \, dy ,$$

$$g_1(x, y) = \sum_{i=1}^{\infty} \sum_{j=1}^{\infty} \frac{2}{l_x} \frac{\mu}{W_j} G_1(i, j) \sin \frac{i \pi x}{l_x} w_{(j)}(y) ,$$

$$g_2(x, y) = \sum_{i=1}^{\infty} \sum_{j=1}^{\infty} \frac{2}{l_x} \frac{\mu}{W_j} G_2(i, j) \sin \frac{i \pi x}{l_x} w_{(j)}(y) . \qquad (15.31)$$

To be able to transform Eq. (15.1) we also need to know the transforms of the derivatives. In view of the boundary conditions of simply supported edges $x = 0$ and $x = l_x$ [Eq. (15.8)] integration by parts will give

$$\int_0^{l_x}\int_0^{l_y} \frac{\partial^4 w}{\partial x^4} \sin\frac{i\pi x}{l_x} w_{(j)}(y)\, dx\, dy = \frac{i^4\pi^4}{l_x^4} W(i, j, t),$$

$$\int_0^{l_x}\int_0^{l_y} \frac{\partial^4 w}{\partial x^2\, \partial y^2} \sin\frac{i\pi x}{l_x} w_{(j)}(y)\, dx\, dy = -\frac{i^2\pi^2}{l_x^2}\left\{\int_0^{l_x}\left[\frac{\partial w}{\partial y} w_{(j)}(y) - \right.\right.$$

$$\left.\left. - w\, w'_{(j)}(y)\right]_0^{l_y} \sin\frac{i\pi x}{l_x} dx + \int_0^{l_x}\int_0^{l_y} w \sin\frac{i\pi x}{l_x} w''_{(j)}(y)\, dx\, dy\right\},$$

$$\int_0^{l_x}\int_0^{l_y} \frac{\partial^4 w}{\partial y^4} \sin\frac{i\pi x}{l_x} w_{(j)}(y)\, dx\, dy = \int_0^{l_x}\left[\frac{\partial^3 w}{\partial y^3} w_{(j)}(y) - \frac{\partial^2 w}{\partial y^2}\, .\right.$$

$$\left. . w'_{(j)}(y) - \frac{\partial w}{\partial y} w''_{(j)}(y) - w\, w'''_{(j)}(y)\right]_0^{l_y} \sin\frac{i\pi x}{l_x} dx + \int_0^{l_x}\int_0^{l_y} w \sin\frac{i\pi x}{l_x}\, .$$

$$. w^{IV}_{(j)}(y)\, dx\, dy\, . \tag{15.32}$$

In the above, symbol $[\ ]_0^{l_y}$ denotes the substitution of the limits from the evaluation of the definite integral with respect to variable y. In the calculation we will also use the relation

$$D\left[\frac{i^4\pi^4}{l_x^4}\int_0^{l_x}\int_0^{l_y} w \sin\frac{i\pi x}{l_x} w_{(j)}(y)\, dx\, dy - 2\frac{i^2\pi^2}{l_x^2}\int_0^{l_x}\int_0^{l_y} w \sin\frac{i\pi x}{l_x}\, .\right.$$

$$\left. . w''_{(j)}(y)\, dx\, dy + \int_0^{l_x}\int_0^{l_y} w \sin\frac{i\pi x}{l_x} w^{IV}_{(j)}(y)\, dx\, dy\right] = \mu\omega^2_{(i,j)} W(i, j, t) \tag{15.33}$$

which can be obtained from (15.28) on multiplication by $w \sin(i\pi\, x/l_x)$ and integration with respect to x and y between the limits 0, l_x and 0, l_y, respectively.

We are now ready to transform Eq. (15.1) with (15.13) on its right-hand side, according to (15.26). In view of (1.8) we get

$$\ddot{W}(i, j, t) + \omega^2_{(i,j)} W(i, j, t) = \frac{P(t)}{\mu} \sin\omega_x t\, w_{(j)}(\eta) - \frac{D}{\mu} z(0, l, t)\, . \tag{15.34}$$

In obtaining the above we have used notation (15.16) and relations (15.32) and (15.33). Function $z(0, l, t)$ expresses the effect of the boundary conditions for $y = 0$ and $y = l_y$

$$z(0, l, t) = \int_0^{l_x} \left[\frac{\partial^3 w}{\partial y^3} w_{(j)}(y) - 2 \frac{i^2 \pi^2}{l_x^2} \frac{\partial w}{\partial y} w_{(j)}(y) - \frac{\partial^2 w}{\partial y^2} w'_{(j)}(y) + \right.$$

$$\left. + \frac{\partial w}{\partial y} w''_{(j)}(y) - w\, w'''_{(j)}(y) + 2 \frac{i^2 \pi^2}{l_x^2} w\, w'_{(j)}(y) \right]_0^{l_y} \sin \frac{i\pi x}{l_x} \mathrm{d}x\,. \quad (15.35)$$

For the usual boundary conditions (15.4) to (15.6) written, of course, for edges $y = 0$ and $y = l_y$ parallel to axis x, the function $z(0, l, t)$ is zero.

Eq. (15.34) is further solved by the Laplace-Carson transformation. In view of (15.7), (15.31)

$$W^*(i, j, p) = \frac{w_{(j)}(\eta)}{\mu} \frac{P^*(p)}{p^2 + \omega^2_{(i,j)}} - \frac{D}{\mu} \frac{Z^*(p)}{p^2 + \omega^2_{(i,j)}} + \frac{p^2\, G_1(i, j) + p\, G_2(i, j)}{p^2 + \omega^2_{(i,j)}}$$

where $P^*(p)$ and $Z^*(p)$ are the transformations of functions $P(t) \sin \omega_x t$ and $z(0, l, t)$, respectively. After the inverse transformations we get the result

$$w(x, y, t) = \sum_{i=1}^{\infty} \sum_{j=1}^{\infty} \frac{2}{l_x} \frac{\mu}{W_j} \sin \frac{i\pi x}{l_x} w_{(j)}(y) \left\{ \frac{1}{\mu \omega_{(i,j)}} \cdot \right.$$

$$\cdot \int_0^t [P(\tau) \sin \omega_x \tau\, w_{(j)}(\eta) - D\, z(0, l, \tau)] \sin \omega_{(i,j)}(t - \tau)\, \mathrm{d}\tau +$$

$$\left. + G_1(i, j) \cos \omega_{(i,j)} t + \frac{1}{\omega_{(i,j)}} G_2(i, j) \sin \omega_{(i,j)} t \right\} \quad (15.36)$$

valid between the limits $0 \leqq t \leqq l_x/c$.

15.2.2 *Motion of a constant force*

If the force is constant, $P(t) = P$, $z(0, l, t) = 0$ and the initial conditions (15.7) are zero, Eq. (15.26) simplifies to

$$w(x, y, t) = \sum_{i=1}^{\infty} \sum_{j=1}^{\infty} \frac{2P}{l_x W_j} \frac{1}{\omega^2_{(i,j)} - \omega_x^2} \sin \frac{i\pi x}{l_x} w_{(j)}(y)\, w_{(j)}(\eta)\,.$$

$$\cdot \left(\sin \omega_x t - \frac{\omega_x}{\omega_{(i,j)}} \sin \omega_{(i,j)} t \right). \quad (15.37)$$

15.3 *Application of the theory*

The theory may be applied to plate structures of bridges, whose two opposite edges are usually simply supported and the other edges are free.

The calculation of normal modes $w_{(j)}(y)$ in the direction of y-axis*) makes use of Eq. (15.28) written in the form

$$\frac{\mathrm{d}^4 w_{(j)}(y)}{\mathrm{d}y^4} - 2\,\frac{i^2\pi^2}{l_x^2}\,\frac{\mathrm{d}^2 w_{(j)}(y)}{\mathrm{d}y^2} + \left(\frac{i^4\pi^4}{l_x^4} - \frac{\mu\omega_{(i,j)}^2}{D}\right)\omega_{(j)}(y) = 0\,. \qquad (15.38)$$

For $y = 0$, $y = l_y$, the boundary conditions according to (15.6) now are

$$w''_{(j)} - \nu\,\frac{i^2\pi^2}{l_x^2}\,w_j = 0\,, \quad w'''_{(j)} - (2-\nu)\,\frac{i^2\pi^2}{l_x^2}\,w'_{(j)} = 0\,. \qquad (15.39)$$

The solution to the homogeneous equation (15.38) is $w_{(j)}(y) = \mathrm{e}^{\lambda y/l_y}$ where λ must satisfy the characteristic equation

$$\frac{\lambda^4}{l_y^4} - 2\,\frac{i^2\pi^2}{l_x^2}\,\frac{\lambda^2}{l_y^2} + \left(\frac{i^4\pi^4}{l_x^4} - \frac{\mu\omega_{(i,j)}^2}{D}\right) = 0\,.$$

There are four roots:

$$\lambda_1 = -\lambda_3 = l_y\left[\frac{i^2\pi^2}{l_x^2} - \left(\frac{\mu\omega_{(i,j)}^2}{D}\right)^{1/2}\right]^{1/2}, \qquad (15.40)$$

$$\lambda_2 = -\lambda_4 = l_y\left[\frac{i^2\pi^2}{l_x^2} + \left(\frac{\mu\omega_{(i,j)}^2}{D}\right)^{1/2}\right]^{1/2}. \qquad (15.41)$$

Case 1. – If $i^2\pi^2/l_x^2 > (\mu\omega_{(i,j)}^2/D)^{1/2}$, all the roots, (15.40) and (15.41), are real and the general solution of Eq. (15.38) is in the form

$$w_{(j)}(y) = \sinh\frac{\lambda_1 y}{l_y} + A\cosh\frac{\lambda_1 y}{l_y} + B\sinh\frac{\lambda_2 y}{l_y} + C\cosh\frac{\lambda_2 y}{l_y}\,. \qquad (15.42)$$

The integration constants A, B, C as well as the frequency equation

*) The solution of normal modes of plates supported in various ways is discussed in detail in several of the references (e.g. [131]).

are obtained after (15.42) has been substituted in four boundary conditions (15.39):

$$A = \frac{\sinh \lambda_1 - \dfrac{\lambda_1 D_1^2}{\lambda_2 D_2^2} \sinh \lambda_2}{\cosh \lambda_2 - \cosh \lambda_1},$$

$$B = \frac{\lambda_1 D_1}{\lambda_2 D_2},$$

$$C = \frac{\dfrac{D_2}{D_1} \sinh \lambda_1 - \dfrac{\lambda_1 D_1}{\lambda_2 D_2} \sinh \lambda_2}{\cosh \lambda_2 - \cosh \lambda_1}, \tag{15.43}$$

$$2\lambda_1 \lambda_2 D_1^2 D_2^2 (\cosh \lambda_1 \cosh \lambda_2 - 1) - (\lambda_1^2 D_1^4 + \lambda_2^2 D_2^4) \sinh \lambda_1 \sinh \lambda_2 = 0$$

where

$$D_1 = l_y^2 \left[\quad (1 - \nu) \frac{i^2 \pi^2}{l_x^2} + \left(\frac{\mu \omega_{(i,j)}^2}{D} \right)^{1/2} \right],$$

$$D_2 = l_y^2 \left[- (1 - \nu) \frac{i^2 \pi^2}{l_x^2} + \left(\frac{\mu \omega_{(i,j)}^2}{D} \right)^{1/2} \right]. \tag{15.44}$$

The natural frequency of the plate, $\omega_{(i,j)}$, is computed from Eq. (15.44).

Case 2. — If $i^2\pi^2/l_x < (\mu\omega_{(i,j)}^2/D)^{1/2}$

$$\lambda_1 = -\lambda_3 = l_y \left[\left(\frac{\mu \omega_{(i,j)}^2}{D} \right)^{1/2} - \frac{i^2 \pi^2}{l_x^2} \right]^{1/2} \tag{15.45}$$

while λ_2 and λ_4 are given by (15.41). The solution of Eq. (15.38) now is in the form

$$w_{(j)}(y) = \sin \frac{\lambda_1 y}{l_y} + A \cos \frac{\lambda_1 y}{l_y} + B \sinh \frac{\lambda_2 y}{l_y} + C \cosh \frac{\lambda_2 y}{l_y} \tag{15.46}$$

and the integration constants and the frequency equation are

$$A = \frac{\sin \lambda_1 - \dfrac{\lambda_1 D_1^2}{\lambda_2 D_2^2} \sinh \lambda_2}{\cosh \lambda_2 - \cos \lambda_1},$$

$$B = \frac{\lambda_1 D_1}{\lambda_2 D_2},$$

$$C = \frac{\dfrac{D_2}{D_1} \sin \lambda_1 - \dfrac{\lambda_1 D_1}{\lambda_2 D_2} \sinh \lambda_2}{\cosh \lambda_2 - \cos \lambda_1}, \tag{15·47}$$

$$2\lambda_1 \lambda_2 D_1^2 D_2^2 (\cos \lambda_1 \cosh \lambda_2 - 1) + (\lambda_1^2 D_1^4 - \lambda_2^2 D_2^4) \sin \lambda_1 \sinh \lambda_2 = 0 .$$

Case 3. – If $i^2\pi^2/l_x^2 = (\mu\omega_{(i,j)}^2/D)^{1/2}$,

$\lambda_1 = \lambda_3 = 0$, and the solution of Eq. (15.38) is in the form

$$w_{(j)}(y) = 1 + Ay + B \sin \frac{\lambda_2 y}{l_y} + C \cosh \frac{\lambda_2 y}{l_y} \tag{15.48}$$

where

$$A = \frac{v^2}{(2 - v)^2} \frac{\lambda_2}{l_y} \frac{1 - \cosh \lambda_2}{\sinh \lambda_2 - \lambda_2},$$

$$B = \frac{v}{2 - v} \frac{1 - \cosh \lambda_2}{\sinh \lambda_2 - \lambda_2},$$

$$C = \frac{v}{2 - v},$$

and the frequency equation is

$$2(\cosh \lambda_2 - 1) - \lambda_2 \sinh \lambda_2 = 0 . \tag{15.49}$$

The solution of the frequency equations of plates is laborious, because $\omega_{(i,j)}$ must be examined for two values of subscripts i and j. Naturally, if the normal modes of the plate considered are known, the solution is no longer difficult because the function of time by which the normal modes are multiplied is the same as for other structures discussed at length in Part II of the book.

15.4 *Additional bibliography*

[44, 56, 99, 112, 124, 161, 167, 170, 179, 180, 184, 234].

16

Infinite plate on elastic foundation

In many instances, the plate considered may be idealized by an infinite plate on elastic foundation whose mass is approximately neglected. The problem has been dealt with by B. G. Korenev [133] and P. Ferrari [60].

The examination starts out from the assumption that the constant force P moves from infinity to infinity at uniform speed c along a straight line passing through the origin (Fig. 16.1). The x- and y-components

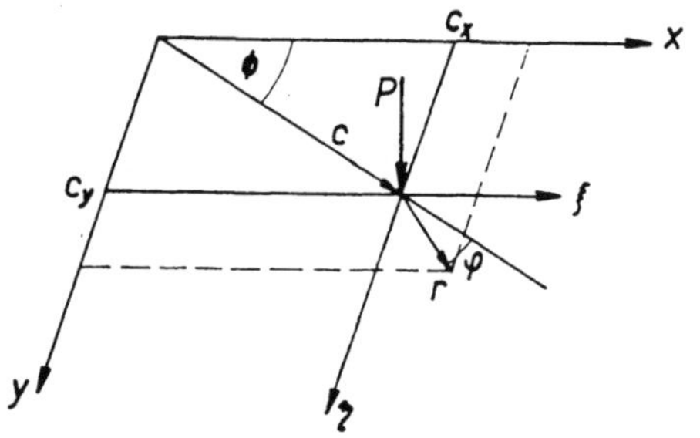

Fig. 16.1. Infinite plate on elastic foundation, subjected to moving force P.

of speed c are c_x and c_y, so that $c^2 = c_x^2 + c_y^2$. On these assumptions the plate deflection can be described by the differential equation

$$D\left(\frac{\partial^2}{\partial x^2} + \frac{\partial^2}{\partial y^2}\right)^2 w(x, y, t) + \mu \frac{\partial^2 w(x, y, t)}{\partial t^2} + k\, w(x, y, t) =$$

$$= \delta(x - c_x t)\, \delta(y - c_y t)\, P \tag{16.1}$$

where the meaning of the symbols is the same as in Chap. 15, and k [force per length cubed] is the coefficient of elastic Winkler foundation.

For the time being we will specify the boundary and initial conditions but verbally: at each instant the deflection and its first to third derivatives with respect to variables x and y at an infinite distance from load P are equal to zero.

16.1 *Steady-state vibration*

Similarly as in Chap. 13 we will merely follow the so-called quasi-stationary state in which the plate is at rest relative to the mobile coordinate system. We shall therefore introduce the new independent variables according to Fig. 16.1

$$\xi = x - c_x t\,, \quad \eta = y - c_y t \tag{16.2}$$

and consider the plate deflection to be in the form

$$w(x, y, t) = w(\xi, \eta)\,. \tag{16.3}$$

Substitution of (16.2) and (16.3) in (16.1) gives

$$D\left(\frac{\partial^2}{\partial \xi^2} + \frac{\partial^2}{\partial \eta^2}\right)^2 w(\xi, \eta) + \mu\left(c_x \frac{\partial}{\partial \xi} + c_y \frac{\partial}{\partial \eta}\right)^2 w(\xi, \eta) + $$

$$+ k\, w(\xi, \eta) = \delta(\xi)\, \delta(\eta)\, P\,. \tag{16.4}$$

To the above equation we shall prescribe the following boundary conditions:

$$\text{for} \quad \xi \to \infty,\ \xi \to -\infty: \quad w(\xi, \eta) = \partial w(\xi, \eta)/\partial \xi = \partial^2 w(\xi, \eta)/\partial \xi^2 = $$
$$= \partial^3 w(\xi, \eta)/\partial \xi^3 = 0\,,$$

$$\text{for} \quad \eta \to \infty,\ \eta \to -\infty: \quad w(\xi, \eta) = \partial w(\xi, \eta)/\partial \eta = \partial^2 w(\xi, \eta)/\partial \eta^2 = $$
$$= \partial^3 w(\xi, \eta)/\partial \eta^3 = 0\,. \tag{16.5}$$

Eq. (16.4) with boundary conditions (16.5) may be solved by the double Fourier integral transformations defined by the relations

$$W(q_1, q_2) = \int_{-\infty}^{\infty} \int_{-\infty}^{\infty} w(\xi, \eta)\, e^{-i(q_1 \xi + q_2 \eta)}\, d\xi\, d\eta\,,$$

$$w(\xi, \eta) = \frac{1}{4\pi^2} \int_{-\infty}^{\infty} \int_{-\infty}^{\infty} W(q_1, q_2)\, e^{i(\xi q_1 + \eta q_2)}\, dq_1\, dq_2\,. \tag{16.6}$$

Transformation of Eq. (16.4) according to (16.6) will give — in view of (27.80), (27.82) and (27.84) —

$$D(q_1^4 + 2q_1^2 q_2^2 + q_2^4)\, W(q_1, q_2) - \mu(c_x^2 q_1^2 + 2c_x c_y q_1 q_2 + c_y^2 q_2^2)\,.$$
$$.\, W(q_1, q_2) + k\, W(q_1, q_2) = P\,. \tag{16.7}$$

With the notation

$$\lambda^4 = \frac{k}{D}, \tag{16.8}$$

$$\alpha = \frac{c}{c_{cr}}; \quad c_{cr}^2 = \frac{2\lambda^2 D}{\mu} \tag{16.9}$$

the transformed solution may be written as

$$W(q_1, q_2) = \frac{P}{D} \frac{1}{(q_1^2 + q_2^2)^2 - \frac{2\lambda^2\alpha^2}{c^2}(c_x q_1 + c_y q_2)^2 + \lambda^4}. \tag{16.10}$$

In accordance with the second of relations (16.6) we get the solution for the plate deflection in the form

$$w(\xi, \eta) = \frac{P}{D} \frac{1}{4\pi^2} \int_{-\infty}^{\infty} \int_{-\infty}^{\infty} \frac{e^{i(\xi q_1 + \eta q_2)}}{(q_1^2 + q_2^2)^2 - \frac{2\lambda^2\alpha^2}{c^2}(c_x q_1 + c_y q_2)^2 + \lambda^4} \, . \, dq_1 \, dq_2 \, . \tag{16.11}$$

Similarly as in the one-dimensional case in Chap. 13, the integrals of Eq. (16.11) could be evaluated by the method of the function of two complex variables. We will not attempt to do so because the calculation is too difficult to manage.

16.2 *Solution in polar coordinates*

Another way of solving the problem of an infinite plate traversed by a moving load is to use polar coordinates. We will introduce in Eq. (16.1) variables r, φ, τ and $w(r, \varphi, \tau)$ in place of variables x, y, t and $w(x, y, t)$, and assume that the centre of the polar coordinates will move on the plate simultaneously with the load. According to Fig. 16.1 the relation between the old and the new independent variables ($\operatorname{tg} \Phi = c_y/c_x$) is

$$\begin{aligned} x &= c_x t + r \cos(\varphi + \Phi), \\ y &= c_y t + r \sin(\varphi + \Phi), \\ t &= \tau . \end{aligned} \tag{16.12}$$

With the new variables in, Eq. (16.1) takes on the form

$$D\left(\frac{\partial^2}{\partial r^2}+\frac{1}{r}\frac{\partial}{\partial r}+\frac{1}{r^2}\frac{\partial^2}{\partial \varphi^2}\right)^2 w(r,\varphi,\tau)+\mu\left(\frac{\partial^2}{\partial \tau^2}+c^2\cos^2\varphi\frac{\partial^2}{\partial r^2}+\right.$$
$$+\frac{1}{r^2}c^2\sin^2\varphi\frac{\partial^2}{\partial \varphi^2}-\frac{2}{r}c^2\sin\varphi\cos\varphi\frac{\partial^2}{\partial r\,\partial\varphi}-2c\cos\varphi\frac{\partial^2}{\partial r\,\partial\tau}+$$
$$\left.+\frac{2}{r}c\sin\varphi\frac{\partial^2}{\partial\varphi\,\partial\tau}+\frac{1}{r}c^2\sin^2\varphi\frac{\partial}{\partial r}+\frac{1}{r^2}2c^2\sin\varphi\cos\varphi\frac{\partial}{\partial\varphi}\right)\cdot$$
$$\cdot\, w(r,\varphi,\tau)+k\,w(r,\varphi,\tau)=\frac{P\,\delta(r)}{2\pi r} \tag{16.13}$$

where according to [204] the right-hand side expresses in polar coordinates a concentrated load placed at the origin. Eq. (16.13) is amenable to solution for several special cases.

16.2.1 *Static solution*

In the static case, $c = 0$, the solution of (16.13) will clearly be independent of φ and τ, and the equation may be written in the form

$$D\left(\frac{\partial^2}{\partial r^2}+\frac{1}{r}\frac{\partial}{\partial r}\right)^2 w(r)+k\,w(r)=\frac{P\,\delta(r)}{2\pi r}\,. \tag{16.14}$$

For $r \to \infty$ the boundary conditions will be specified as follows

$$w(r)=\mathrm{d}w(r)/\mathrm{d}r=\mathrm{d}^2w(r)/\mathrm{d}r^2=\mathrm{d}^3w(r)/\mathrm{d}r^3=0\,;$$

while for $r=0$ and $r\to\infty$ the deflection is assumed to satisfy the conditions

$$r\,w(r)=0\,,\quad r\,\mathrm{d}w(r)/\mathrm{d}r=0\,. \tag{16.15}$$

Eq. (16.14) will be solved by the Hankel integral transformation (27.91) defined by the relations

$$W(q)=\int_0^\infty w(r)\,r\,\mathrm{J}_0(qr)\,\mathrm{d}r\,,$$
$$w(r)=\int_0^\infty W(q)\,q\,\mathrm{J}_0(rq)\,\mathrm{d}q \tag{16.16}$$

where $J_0(x)$ denotes the Bessel function of the first kind of the zero-th order.

In view of (27.92) and (27.93) transformation of Eq. (16.14) will give

$$Dq^4 W(q) + k W(q) = P/(2\pi) \tag{16.17}$$

because $J_0(0) = 1$. Hence the transformed solution is

$$W(q) = \frac{P}{2\pi D} \frac{1}{q^4 + \lambda^4}. \tag{16.18}$$

According to (27.94) (see [100], Eq. 6.537) the inverse transformation will give

$$w(r) = \frac{P}{2\pi D} \int_0^\infty \frac{J_0(rq)\, q\, dq}{q^4 + \lambda^4} = -\frac{P}{2\pi\lambda^2 D} \operatorname{kei}(\lambda r) \tag{16.19}$$

where kei (x) is the so-called Thomson function of the zero-th order, [113]. From the last equation (see also [100], Eq. 3.241.2) the deflection underneath the load at $r = 0$ is given by

$$w_0 = w(0) = \frac{P}{8\lambda^2 D} \tag{16.20}$$

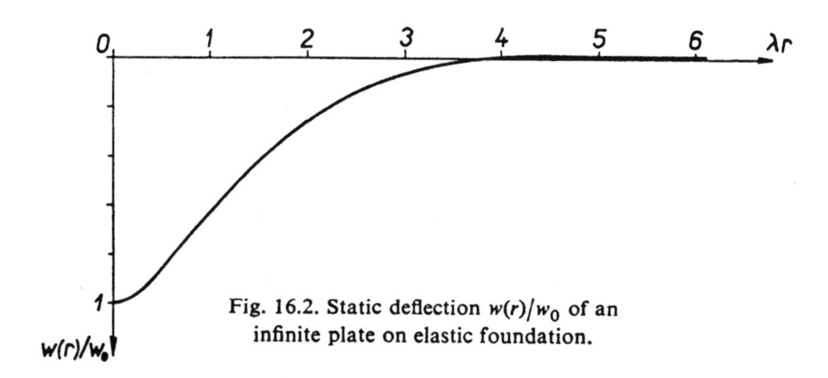

Fig. 16.2. Static deflection $w(r)/w_0$ of an infinite plate on elastic foundation.

so that the deflection equation (16.19) may be expressed as

$$w(r) = -w_0 \frac{4}{\pi} \operatorname{kei}(\lambda r). \tag{16.21}$$

Expression $w(r)/w_0 = -(4/\pi) \operatorname{kei}(\lambda r)$ is graphically represented in Fig. 16.2.

16.2.2 *Approximate solution for subcritical speed*

At subcritical speed $c < c_{cr}$ the solution of Eq. (16.13) may approximately be assumed to depend merely on r, i.e. $w(r, \varphi, \tau) = w(r)$. Eq. (16.13) then takes on the form

$$D\left(\frac{\partial^2}{\partial r^2} + \frac{1}{r}\frac{\partial}{\partial r}\right)^2 w(r) + \mu c^2\left(\frac{\partial^2}{\partial r^2} + \frac{1}{r}\frac{\partial}{\partial r}\right) w(r) + k\,w(r) = \frac{P\,\delta(r)}{2\pi r}\,. \tag{16.22}$$

This equation is again solved by the Hankel transformation (16.16). In view of (27.92), (27.93), (16.8) and (16.9) we get

$$Dq^4\,W(q) - \mu c^2 q^2\,W(q) + k\,W(q) = P/(2\pi)$$

and following the inverse transformation procedure the resultant deflection is

$$w(r) = \frac{P}{2\pi D}\int_0^\infty \frac{\mathrm{J}_0(rq)\,q\,\mathrm{d}q}{q^4 - 2\alpha^2\lambda^2 q^2 + \lambda^4}\,. \tag{16.23}$$

The same result will approximately be obtained also from Eq. (16.11) with the substitution $\xi = r\cos\varphi$, $\eta = r\sin\varphi$, $q_1 = q\cos\Theta$, $q_2 = q\sin\Theta$, $\Theta = \Phi$.

The integral in Eq. (16.23) may be evaluated by the method of the function of a complex variable, or numerically. The deflection underneath the load [for $r = 0$, i.e. $w(0)$] which is of primary interest can be expressed quite simply. According to [100] we can evaluate the integral of the type $\int x\,\mathrm{d}x/(a + bx^2 + cx^4)$ to which it is possible to reduce $\lim_{r\to 0} w(r)$ in Eq. (16.23) because $\mathrm{J}_0(0) = 1$. Following the evaluation of that integral and some manipulation, we get in view of (16.20)

$$w(0) = w_0\,\frac{1}{(1-\alpha^4)^{1/2}}\left[1 + \frac{2}{\pi}\,\mathrm{arctg}\,\frac{\alpha^2}{(1-\alpha^4)^{1/2}}\right]. \tag{16.24}$$

Eq. (16.24) applies only at subcritical speed, $\alpha < 1$. The procedure does not work at supercritical speed $\alpha > 1$, for then the solution of (16.13) depends on angle φ, too. This follows from the analogy between Fig. 13.3c and the deflection surface of the plate. In this case one must therefore take account of damping, too.

The dynamic deflection of the plate underneath a moving load, $w(0)/w_0$, Eq. (16.24), versus speed α, Eq. (16.9), is shown in Fig. 16.3.

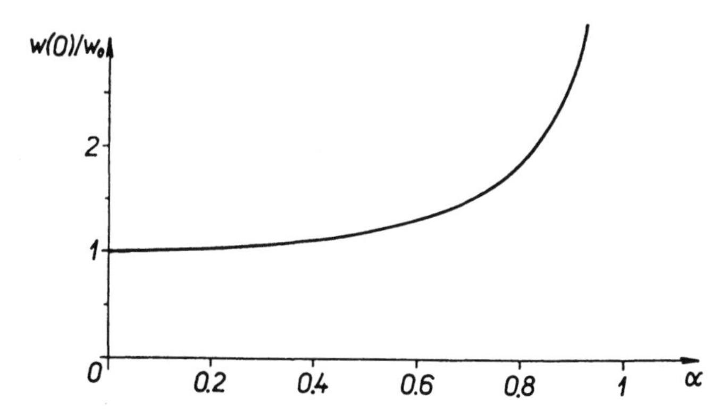

Fig. 16.3. Dynamic deflection $w(0)/w_0$ of an infinite plate on elastic foundation underneath the moving force P, versus speed parameter α.

16.3 *Application of the theory*

The theory expounded in the foregoing can be used in calculations of plates of roadways or runways. The plates, concrete or reinforced concrete for the most, have large areas and rest on variously prepared foundations that can very roughly be approximated by the Winkler foundation. The calculation carried through in the preceding sections reproduces but the basic effect produced by a moving constant force whereas the highest dynamic effects derive from track irregularities, sprung and unsprung masses of vehicles, etc.

16.4 *Additional bibliography*

[4, 60, 124, 133, 134, 164, 180, 255, 256].

Part IV

THREE-DIMENSIONAL SOLIDS

Part IV

17

Elastic space with a moving force

Solutions of problems concerned with dynamic effects in three-dimensional elastic solids are of comparatively recent date. Interestingly enough, it was exactly the problem of the effects of moving forces that first attracted the attention of scientists in the field of dynamic theory of elasticity once the classic questions of the theory of elastic wave propagation had been answered to satisfaction. Credit for this should be given to L. A. Galin [89], and particularly to I. N. Sneddon [204] who with his associates [55] solved a number of topical problems in that area.

Our discussion will be confined to a single subject – the effects of a force that moves at right angle to its direction in an elastic space. The problem was first solved by G. Eason, J. Fulton and I. N. Sneddon [55] using quadruple transformation; we shall present here a somewhat simpler treatment basing on the three-dimensional Fourier integral transformation, and as the first, derive the solutions at transonic and supersonic speeds. The static alternative of our case was dealt with, for example, by A. J. Lur'e [147].

As it is well known [130], the equations of motion for an elastic, isotropic solid are

$$(\lambda + G)\frac{\partial \Theta}{\partial x_i} + G\,\nabla^2 u_i + X_i = \varrho\,\frac{\partial^2 u_i}{\partial t^2}, \quad i = 1, 2, 3 \tag{17.1}$$

where

$u_i = u_i(x_1, x_2, x_3, t)$ – displacement in the direction of axis x_i,

ϱ – constant density of the solid,

$X_i = X_i(x_1, x_2, x_3, t)$ – force per unit volume in the direction of axis x_i,

$\nabla^2 = \sum_{j=1}^{3} \frac{\partial^2}{\partial x_j^2}$ – Laplace operator,

$\Theta = \sum_{j=1}^{3} \frac{\partial u_j}{\partial x_j}$ — relative change of volume,

$\lambda = \frac{\nu E}{(1 + \nu)(1 - 2\nu)}$ — Lamé's constant,

$G = \frac{E}{2(1 + \nu)}$ — modulus of elasticity in shear,

$\nu < 1$ — Poisson's ratio,

E — Young's modulus.

From displacement u_i one then computes the components of normal stress

$$\sigma_i = 2G\varepsilon_i + \lambda\Theta \tag{17.2}$$

and of tangential stress

$$\tau_{ij} = G\gamma_{ij}\,. \tag{17.3}$$

In the above, ε_i is the strain in the direction of axis x_i

$$\varepsilon_i = \frac{\partial u_i}{\partial x_i} \tag{17.4}$$

and γ_{ij} the shear strain in plane $x_i x_j$ $(i = 1, 2, 3,\ j = 1, 2, 3,\ i \neq j)$

$$\gamma_{ij} = \frac{\partial u_j}{\partial x_i} + \frac{\partial u_i}{\partial x_j}\,. \tag{17.5}$$

There are two more constants of utmost significance for the dynamic theory of elasticity, namely the velocity of propagation of longitudinal waves, c_1, and the velocity of propagation of transverse waves, c_2, $(c_2 < c_1)$, defined by the relations

$$c_1 = \left(\frac{\lambda + 2G}{\varrho}\right)^{1/2}, \quad c_2 = \left(\frac{G}{\varrho}\right)^{1/2}. \tag{17.6}$$

17.1 *Quasi-stationary motion of a force in elastic space*

Suppose that a constant force, P, moves at constant speed c in an elastic space from infinity to infinity (Fig. 17.1). Orient the coordinate system so that force P will act in the direction of axis x_3 and move in the direction of axis x_1. Then the volume forces are of the form

$$X_1 = 0\,, \quad X_2 = 0\,, \quad X_3 = P\,\delta(x_1 - ct)\,\delta(x_2)\,\delta(x_3)\,. \tag{17.7}$$

Introduce in place of coordinate system x_i the system

$$x = x_1 - ct\,, \quad x_2 = y\,, \quad x_3 = z \tag{17.8}$$

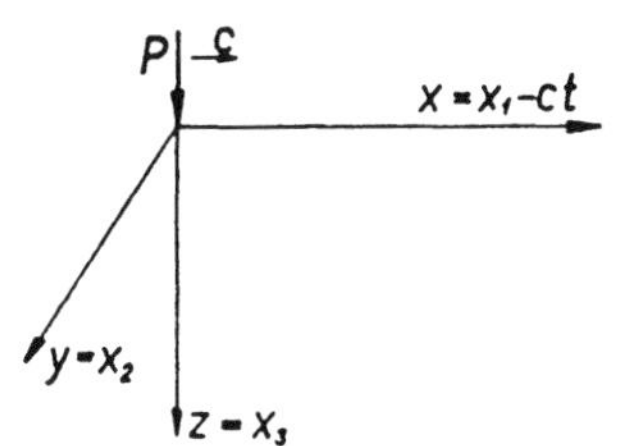

Fig. 17.1. Motion of a force in an elastic space.

that moves simultaneously with force P (Fig. 17.1). After a sufficiently long time of force action, the displacement components will have the steady-state form

$$u_1(x_1, x_2, x_3, t) = u(x, y, z)\,, \quad u_2(x_1, x_2, x_3, t) = v(x, y, z)\,,$$
$$u_3(x_1, x_2, x_3, t) = w(x, y, z)\,. \tag{17.9}$$

Our task thus consists in finding the particular solution of the problem, represented by functions u, v, w. Substitution of (17.7) to (17.9) in (17.1) gives the following system of partial differential equations

$$(\lambda + G)\frac{\partial}{\partial x}\left(\frac{\partial u}{\partial x} + \frac{\partial v}{\partial y} + \frac{\partial w}{\partial z}\right) + G\left(\frac{\partial^2}{\partial x^2} + \frac{\partial^2}{\partial y^2} + \frac{\partial^2}{\partial z^2}\right) u = c^2 \varrho \frac{\partial^2 u}{\partial x^2}\,,$$

$$(\lambda + G)\frac{\partial}{\partial y}\left(\frac{\partial u}{\partial x} + \frac{\partial v}{\partial y} + \frac{\partial w}{\partial z}\right) + G\left(\frac{\partial^2}{\partial x^2} + \frac{\partial^2}{\partial y^2} + \frac{\partial^2}{\partial z^2}\right) v = c^2 \varrho \frac{\partial^2 v}{\partial x^2}\,,$$

$$(\lambda + G)\frac{\partial}{\partial z}\left(\frac{\partial u}{\partial x} + \frac{\partial v}{\partial y} + \frac{\partial w}{\partial z}\right) + G\left(\frac{\partial^2}{\partial x^2} + \frac{\partial^2}{\partial y^2} + \frac{\partial^2}{\partial z^2}\right) w +$$
$$+ P\,\delta(x)\,\delta(y)\,\delta(z) = c^2 \varrho \frac{\partial^2 w}{\partial x^2}\,. \tag{17.10}$$

We will assume about the boundary conditions that at infinite distance from the load the displacements as well as their first derivatives are zero, i.e. for $x \to \pm\infty$, $y \to \pm\infty$, $z \to \pm\infty$

$$u = 0\,, \quad v = 0\,, \quad w = 0\,, \quad \partial u/\partial x = 0\,, \quad \partial v/\partial y = 0\,,$$
$$\partial w/\partial z = 0\,. \tag{17.11}$$

System (17.10) is conveniently solved by the method of triple Fourier integral transformation defined by the relations

$$U(q_1, q_2, q_3) = \int_{-\infty}^{\infty}\int_{-\infty}^{\infty}\int_{-\infty}^{\infty} u(x, y, z)\, \mathrm{e}^{-\mathrm{i}(q_1x+q_2y+q_3z)}\, \mathrm{d}x\, \mathrm{d}y\, \mathrm{d}z\,,$$
$$u(x, y, z) = \frac{1}{(2\pi)^3}\int_{-\infty}^{\infty}\int_{-\infty}^{\infty}\int_{-\infty}^{\infty} U(q_1, q_2, q_3)\, \mathrm{e}^{\mathrm{i}(xq_1+yq_2+zq_3)}\, \mathrm{d}q_1\,.$$
$$.\, \mathrm{d}q_2\, \mathrm{d}q_3\,. \tag{17.12}$$

Similar relations hold true between the other functions, $v(x, y, z)$ and $V(q_1, q_2, q_3)$, and $w(x, y, z)$ and $W(q_1, q_2, q_3)$, respectively; hereinafter we shall leave out the arguments of functions u, v, w, U, V and W.

Transforming now the system (17.10) in accordance with (17.12) we get — in view of (17.11), (27.80), (17.7) and (27.84) — the following system of algebraic equations

$$-(\lambda + G)(q_1^2U + q_1q_2V + q_1q_3W) - G(q_1^2 + q_2^2 + q_3^2)\,U = -c^2\varrho q_1^2 U\,,$$
$$-(\lambda + G)(q_1q_2U + q_2^2V + q_2q_3W) - G(q_1^2 + q_2^2 + q_3^2)\,V = -c^2\varrho q_1^2 V\,,$$
$$-(\lambda + G)(q_1q_3U + q_2q_3V + q_3^2W) - G(q_1^2 + q_2^2 + q_3^2)\,W + P =$$
$$= -c^2\varrho q_1^2 W\,. \tag{17.13}$$

With the notation [see (17.6)]

$$\alpha_1 = c/c_1\,, \quad \alpha_2 = c/c_2\,,$$
$$b^2 = \frac{\alpha_2^2}{\alpha_1^2} = \frac{c_1^2}{c_2^2} = \frac{\lambda + 2G}{G} = \frac{2(1 - \nu)}{1 - 2\nu}\,, \tag{17.14}$$

the unknowns U, V and W may be obtained from (17.13). After some formal manipulations

$$U = -\frac{(b^2 - 1)\,P}{b^2G}\,\frac{q_1q_3}{[(1 - \alpha_1^2)\,q_1^2 + q_2^2 + q_3^2]\,[(1 - \alpha_2^2)\,q_1^2 + q_2^2 + q_3^2]}\,,$$

$$V = -\frac{(b^2-1)P}{b^2G}\frac{q_2q_3}{[(1-\alpha_1^2)q_1^2+q_2^2+q_3^2][(1-\alpha_2^2)q_1^2+q_2^2+q_3^2]},$$

$$W = \frac{P}{b^2G}\left\{\frac{b^2}{(1-\alpha_2^2)q_1^2+q_2^2+q_3^2} - \right.$$

$$\left. -(b^2-1)\frac{q_3^2}{[(1-\alpha_1^2)q_1^2+q_2^2+q_3^2][(1-\alpha_2^2)q_1^2+q_2^2+q_3^2]}\right\}. \tag{17.15}$$

The inverse transformation is made in accordance with the second of Eqs. (17.12). Before we apply it, let us note that the character of Eqs. (17.15) is such that their inverse transformations can be expressed by just two integrals

$$I_1 = \int_{-\infty}^{\infty}\int_{-\infty}^{\infty}\int_{-\infty}^{\infty} \cdot$$

$$\cdot \frac{e^{i(xq_1+yq_2+zq_3)}}{[(1-\alpha_1^2)q_1^2+q_2^2+q_3^2][(1-\alpha_2^2)q_1^2+q_2^2+q_3^2]}\,dq_1\,dq_2\,dq_3,$$

$$I_2 = \int_{-\infty}^{\infty}\int_{-\infty}^{\infty}\int_{-\infty}^{\infty}\frac{e^{i(xq_1+yq_2+zq_3)}}{(1-\alpha_2^2)q_1^2+q_2^2+q_3^2}\,dq_1\,dq_2\,dq_3. \tag{17.16}$$

With the above notation the inverse transformation of (17.15) in accordance with (17.12) may be given the form

$$u = \frac{(b^2-1)P}{8\pi^3Gb^2}\frac{\partial^2 I_1}{\partial x\,\partial z}, \qquad v = \frac{(b^2-1)P}{8\pi^3Gb^2}\cdot\frac{\partial^2 I_1}{\partial y\,\partial z},$$

$$w = \frac{P}{8\pi^3Gb^2}\left[b^2I_2 + (b^2-1)\frac{\partial^2 I_1}{\partial z^2}\right]. \tag{17.17}$$

17.1.1 *Subsonic speed* $c < c_2 < c_1$

What remains to be done is to evalutae integrals (17.16). We will start with the second one, and introduce the following substitutions:

$$q_2 = \varrho\cos\varphi, \quad q_3 = \varrho\sin\varphi, \quad y = r\cos\Phi, \quad z = r\sin\Phi,$$

$$\varrho^2 = q_2^2 + q_3^2, \quad r^2 = y^2 + z^2,$$

$$\xi_i = \frac{x}{a_i}, \quad \varrho_i = \frac{\varrho}{a_i}, \quad a_i^2 = 1 - \alpha_i^2, \quad i = 1, 2. \tag{17.18}$$

With regard to (17.14) $\alpha_1 < \alpha_2 < 1$ at subsonic speeds. Following the substitutions stated above,

$$I_2 = \frac{1}{a_2^2} \int_{-\infty}^{\infty} \int_0^{\infty} \int_0^{2\pi} \frac{\varrho e^{i[xq_1 + r\varrho \cos(\Phi - \varphi)]}}{q_1^2 + \varrho_2^2} dq_1 \, d\varrho \, d\varphi \, .$$

In view of the integral expression of Bessel functions ([100], Eq. 8.411.1) integration with respect to φ will give

$$I_2 = \frac{2\pi}{a_2^2} \int_{-\infty}^{\infty} \int_0^{\infty} \frac{\varrho \, J_0(r\varrho) \, e^{ixq_1}}{q_1^2 + \varrho_2^2} dq_1 \, d\varrho$$

where $J_0(x)$ is the Bessel function of the first kind of the zero-th order. The simple poles of the function in the integrand in variable q_1 are $\pm i\varrho_2$. We will, therefore, use the residue theorem [186] in the integration with respect to q_1 in the complex plane and get

$$I_2 = \frac{2\pi^2}{a_2} \int_0^{\infty} e^{-|\xi_2 \varrho|} \, J_0(r\varrho) \, d\varrho \, .$$

This integral is already easy to evaluate (see [100], formula 6.611.1)

$$I_2 = \frac{2\pi^2}{a_2(r^2 + \xi_2^2)^{1/2}} = \frac{2\pi^2}{R_2} \, . \tag{17.19}$$

In the last expression we have used the notation

$$R_i^2 = x^2 + a_i^2 r^2 = a_i^2(\xi_i^2 + r^2) \, , \quad i = 1, 2 \, . \tag{17.20}$$

A similar procedure is adopted for evaluating the integral I_1 (17.16). Following the substitutions (17.18) we get

$$I_1 = \frac{1}{a_1^2 a_2^2} \int_{-\infty}^{\infty} \int_0^{\infty} \int_0^{2\pi} \frac{\varrho e^{i[xq_1 + r\varrho \cos(\Phi - \varphi)]}}{(q_1^2 + \varrho_1^2)(q_1^2 + \varrho_2^2)} dq_1 \, d\varrho \, d\varphi \, ;$$

the integration with respect to φ gives

$$I_1 = \frac{2\pi}{a_1^2 a_2^2} \int_{-\infty}^{\infty} \int_0^{\infty} \frac{\varrho \, J_0(r\varrho) \, e^{ixq_1}}{(q_1^2 + \varrho_1^2)(q_1^2 + \varrho_2^2)} dq_1 \, d\varrho \, .$$

The simple poles of the function in the integrand now are $\pm i\varrho_1$, and $\pm i\varrho_2$.

In the integration with respect to q_1 we will again use the residue theorem, and after some manipulation get

$$I_1 = \frac{2\pi^2}{\alpha_1^2(b^2 - 1)} \int_0^\infty \left(a_1 e^{-|\xi_1 \varrho|} - a_2 e^{-|\xi_2 \varrho|}\right) \frac{J_0(r\varrho)}{\varrho^2} \, d\varrho \, .$$

This expression is no longer integrable in elementary functions. We can, however, find its derivative with respect to r. Having regard to the fact that $dJ_0(x)/dx = -J_1(x)$, and to formula 6.623.3 in [100] we may write

$$\frac{\partial I_1}{\partial r} = \frac{2\pi^2}{\alpha_1^2(b^2 - 1)} \int_0^\infty \left(a_2 e^{-|\xi_2 \varrho|} - a_1 e^{-|\xi_1 \varrho|}\right) \frac{J_1(r\varrho)}{\varrho} \, d\varrho =$$

$$= \frac{2\pi^2}{\alpha_1^2(b^2 - 1)\, r} (R_2 - R_1) \, . \tag{17.21}$$

The last expression is all we need to compute the solution (17.17), for the required derivatives of integral I_1 will merely be functions of $\partial I_1/\partial r$ because

$$\frac{\partial I_1}{\partial z} = \frac{z}{r} \frac{\partial I_1}{\partial r} \, .$$

Therefore,

$$\frac{\partial^2 I_1}{\partial x \, \partial z} = \frac{z}{r} \frac{\partial^2 I_1}{\partial x \, \partial r} \, ,$$

$$\frac{\partial^2 I_1}{\partial y \, \partial z} = -\frac{yz}{r^3} \frac{\partial I_1}{\partial r} + \frac{yz}{r^2} \frac{\partial^2 I_1}{\partial r^2} \, ,$$

$$\frac{\partial^2 I_1}{\partial z^2} = \frac{y^2}{r^3} \frac{\partial I_1}{\partial r} + \frac{z^2}{r^2} \frac{\partial^2 I_1}{\partial r^2} \, . \tag{17.22}$$

Differentiating (17.21) with respect to x and r

$$\frac{\partial^2 I_1}{\partial x \, \partial r} = \frac{2\pi^2}{\alpha_1^2(b^2 - 1)} \frac{x}{r} \left(\frac{1}{R_2} - \frac{1}{R_1}\right),$$

$$\frac{\partial^2 I_1}{\partial r^2} = \frac{2\pi^2}{\alpha_1^2(b^2 - 1)} \frac{x^2}{r^2} \left(\frac{1}{R_1} - \frac{1}{R_2}\right), \tag{17.23}$$

and substituting (17.22) and (17.23) in (17.17) will eventually give the solution

$$u = \frac{P}{4\pi G\alpha_2^2} \frac{xz}{r^2}\left(\frac{1}{R_2} - \frac{1}{R_1}\right),$$

$$v = -\frac{P}{4\pi G\alpha_2^2} \frac{yz}{r^4}\left[R_2 - R_1 + x^2\left(\frac{1}{R_2} - \frac{1}{R_1}\right)\right],$$

$$w = \frac{P}{4\pi G\alpha_2^2}\left[\frac{\alpha_2^2}{R_2} + \frac{y^2}{r^4}(R_2 - R_1) - \frac{x^2 z^2}{r^4}\left(\frac{1}{R_2} - \frac{1}{R_1}\right)\right] \qquad (17.24)$$

where

$$x = x_1 - ct\,, \quad r^2 = y^2 + z^2\,, \quad R_i = x^2 + a_i^2 r^2\,, \quad \alpha_i = c/c_i\,, \quad i = 1, 2\,.$$

Solution (17.24) applies at motion speeds lesser than the velocity of propagation of transverse waves, $c < c_2 < c_1$ $(\alpha_1 < \alpha_2 < 1)$.

17.1.2 *Transonic speed* $c_2 < c < c_1$

If the motion speed is higher than the velocity of propagation of transverse waves but lower than the velocity of propagation of longitudinal waves, $c_2 < c < c_1$, $\alpha_1 < 1 < \alpha_2$, the solution is obtained in an analogous manner. We shall introduce the notation

$$a_1^2 = 1 - \alpha_1^2\,, \quad a_2^2 = \alpha_2^2 - 1\,, \quad R_1^2 = x^2 + a_1^2 r^2\,, \quad R_2^2 = x^2 - a_2^2 r^2 \qquad (17.25)$$

next to the symbols used in (17.18). With these symbols the expression I_2 of (17.16) may be given the form

$$I_2 = -\frac{1}{a_2^2}\int_{-\infty}^{\infty}\int_{-\infty}^{\infty}\int_{-\infty}^{\infty} \frac{e^{i(xq_1 + yq_2 + zq_3)}}{q_1^2 - (q_2^2 + q_3^2)/a_2^2}\, dq_1\, dq_2\, dq_3\,.$$

We again introduce in it the new variables (17.18)

$$I_2 = -\frac{1}{a_2^2}\int_{-\infty}^{\infty}\int_0^{\infty}\int_0^{2\pi} \frac{\varrho e^{i[xq_1 + r\varrho\cos(\Phi - \varphi)]}}{q_1^2 - \varrho_2^2}\, dq_1\, d\varrho\, d\varphi\,,$$

and integrate the above first with respect to φ

$$I_2 = -\frac{2\pi}{a_2^2}\int_{-\infty}^{\infty}\int_0^{\infty}\frac{\varrho\,\mathrm{J}_0(r\varrho)\,\mathrm{e}^{\mathrm{i}xq_1}}{q_1^2-\varrho_2^2}\,\mathrm{d}q_1\,\mathrm{d}\varrho\,,$$

then with respect to q_1 in the complex plane with poles $\pm\varrho_2$

$$I_2 = \frac{2\pi^2}{a_2}\int_0^{\infty}\mathrm{J}_0(r\varrho)\sin(\xi_2\varrho)\,\mathrm{d}\varrho\,,$$

and finally with respect to ϱ using formula 6.671.1 in [100]

$$I_2 = \frac{2\pi^2}{R_2}H(x-a_2r)\,. \tag{17.26}$$

Here $H(x)$ is the Heaviside unit function (3.23).

Integral I_1 of (17.16) is handled and evaluated in a similar way:

$$I_1 = -\frac{1}{a_1^2a_2^2}\int_{-\infty}^{\infty}\int_{-\infty}^{\infty}\int_{-\infty}^{\infty}\frac{\mathrm{e}^{\mathrm{i}(xq_1+yq_2+zq_3)}}{[q_1^2+(q_2^2+q_3^2)/a_1^2]\,[q_1^2-(q_2^2+q_3^2)/a_2^2]}\,.$$

$$.\,\mathrm{d}q_1\,\mathrm{d}q_2\,\mathrm{d}q_3 =$$

$$= -\frac{1}{a_1^2a_2^2}\int_{-\infty}^{\infty}\int_0^{\infty}\int_0^{2\pi}\frac{\varrho\mathrm{e}^{\mathrm{i}[xq_1+r\varrho\cos(\Phi-\varphi)]}}{(q_1^2+\varrho_1^2)(q_1^2-\varrho_2^2)}\,\mathrm{d}q_1\,\mathrm{d}\varrho\,\mathrm{d}\varphi\,.$$

It is first integrated with respect to φ

$$I_1 = -\frac{2\pi}{a_1^2a_2^2}\int_{-\infty}^{\infty}\int_0^{\infty}\frac{\varrho\,\mathrm{J}_0(r\varrho)\,\mathrm{e}^{\mathrm{i}xq_1}}{(q_1^2+\varrho_1^2)(q_1^2-\varrho_2^2)}\,\mathrm{d}q_1\,\mathrm{d}\varrho\,,$$

then with respect to q_1 in the complex plane with poles $\pm\mathrm{i}\varrho_1$, $\pm\varrho_2$

$$I_1 = \frac{2\pi^2}{\alpha_1^2(b^2-1)}\int_0^{\infty}\left(a_1\mathrm{e}^{-|\xi_1\varrho|}+a_2\sin\xi_2\varrho\right)\frac{\mathrm{J}_0(r\varrho)}{\varrho^2}\,\mathrm{d}\varrho\,.$$

Since we need but the derivative of the last expression with respect to r to compute the displacement we get

$$\frac{\partial I_1}{\partial r} = -\frac{2\pi^2}{\alpha_1^2(b^2-1)}\int_0^{\infty}\left(a_1\mathrm{e}^{-|\xi_1\varrho|}+a_2\sin\xi_2\varrho\right)\frac{\mathrm{J}_1(r\varrho)}{\varrho}\,\mathrm{d}\varrho\,.$$

This integral can be computed with the aid of formulae 6.623.3 and 6.963.1 in [100] and rearranged to the form

$$\frac{\partial I_1}{\partial r} = \frac{2\pi^2}{\alpha_1^2(b^2 - 1)\, r}\,[-R_1 + R_2\, H(x - a_2 r)]\,. \tag{17.27}$$

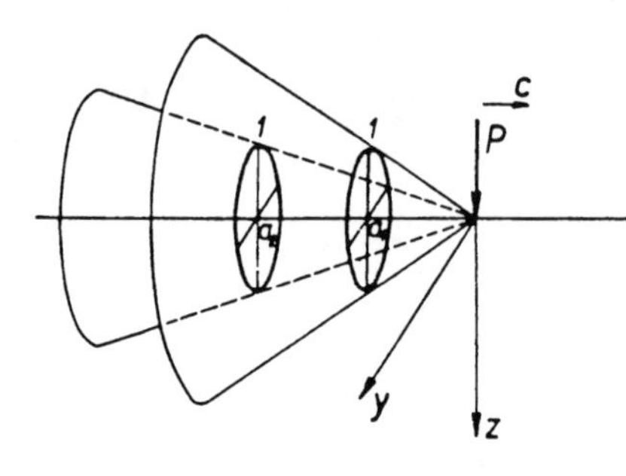

Fig. 17.2. Singularities on the lateral area of two cones of revolution at transonic and supersonic speeds.

The derivatives of (17.27) with respect to x and r are

$$\frac{\partial^2 I_1}{\partial x\, \partial r} = \frac{2\pi^2}{\alpha_1^2(b^2 - 1)\, r}\left[-\frac{x}{R_1} + \frac{x}{R_2}\, H(x - a_2 r) + R_2\, \delta(x - a_2 r)\right],$$

$$\frac{\partial^2 I_1}{\partial r^2} = \frac{2\pi^2}{\alpha_1^2(b^2 - 1)\, r^2}\left[\frac{x^2}{R_1} - \frac{x^2}{R_2}\, H(x - a_2 r) - a_2 r R_2\, \delta(x - a_2 r)\right] \tag{17.28}$$

where $\delta(x)$ is the Dirac function [cf. (1.4)].

Following the substitution of (17.27) and (17.28) in (17.22), (17.17) will give the following solution

$$u = \frac{P}{4\pi G\alpha_2^2}\,\frac{xz}{r^2}\left[-\frac{1}{R_1} + \frac{1}{R_2}\, H(x - a_2 r) + \frac{R_2}{x}\, \delta(x - a_2 r)\right],$$

$$v = -\frac{P}{4\pi G\alpha_2^2}\,\frac{yz}{r^4}\left[-R_1 + R_2\, H(x - a_2 r) - \frac{x^2}{R_1} + \frac{x^2}{R_2}\, H(x - a_2 r) + \right.$$
$$\left. + a_2 r R_2\, \delta(x - a_2 r)\right],$$

$$w = \frac{P}{4\pi G\alpha_2^2}\left\{\frac{\alpha_2^2}{R_2}\, H(x - a_2 r) + \frac{y^2}{r^4}\,[-R_1 + R_2\, H(x - a_2 r)] - \right.$$
$$\left. - \frac{x^2 z^2}{r^4}\left[-\frac{1}{R_1} + \frac{1}{R_2}\, H(x - a_2 r) + \frac{a_2 r R_2}{x^2}\, \delta(x - a_2 r)\right]\right\}. \tag{17.29}$$

The displacements obtained above have singularities at points $x = x_1 - ct = a_2 r$ on the lateral area of the cone of revolution $x^2/a_2^2 = y^2 + z^2$ with the vertex at the origin and the circular base of radius 1 in plane $x = -a_2$ (Fig. 17.2).

17.1.3 *Supersonic speed* $c_2 < c_1 < c$

The case of motion speed c higher than the velocities of propagation of both transverse and longitudinal waves, $c_2 < c_1 < c$, $1 < \alpha_1 < \alpha_2$, is solved analogously to the former. With the notation

$$a_i^2 = \alpha_i^2 - 1\,, \quad R_i^2 = x^2 - a_i^2 r^2\,, \quad i = 1, 2\,, \tag{17.30}$$

expression I_2 of (17.16) turns out to be the same as in the preceding section, i.e. Eq. (17.26) applies. Having regard to notation (17.30), we shall write (17.16) for the purposes of computing I_1 in the form

$$I_1 = \frac{1}{a_1^2 a_2^2} \int_{-\infty}^{\infty} \int_{-\infty}^{\infty} \int_{-\infty}^{\infty} \frac{e^{i(xq_1 + yq_2 + zq_3)}}{[q_1^2 - (q_2^2 + q_3^2)/a_1^2]\,[q_1^2 - (q_2^2 + q_3^2)/a_2^2]} \,.$$
$$. \, dq_1\, dq_2\, dq_3\,,$$

and introduce the new variables (17.18)

$$I_1 = \frac{1}{a_1^2 a_2^2} \int_{-\infty}^{\infty} \int_0^{\infty} \int_0^{2\pi} \frac{\varrho e^{i[xq_1 + r\varrho \cos(\Phi - \varphi)]}}{(q_1^2 - \varrho_1^2)(q_1^2 - \varrho_2^2)}\, dq_1\, d\varrho\, d\varphi\,.$$

The above is integrated first with respect to φ

$$I_1 = \frac{2\pi}{a_1^2 a_2^2} \int_{-\infty}^{\infty} \int_0^{\infty} \frac{\varrho\, J_0(r\varrho)\, e^{ixq_1}}{(q_1^2 - \varrho_1^2)(q_1^2 - \varrho_2^2)}\, dq_1\, d\varrho\,,$$

then with respect to q_1 in the complex plane with poles $\pm\varrho_1$, $\pm\varrho_2$

$$I_1 = \frac{2\pi^2}{\alpha_1^2(b^2 - 1)} \int_0^{\infty} (-a_1 \sin \xi_1 \varrho + a_2 \sin \xi_2 \varrho)\, \frac{J_0(r\varrho)}{\varrho^2}\, d\varrho\,.$$

All we need for the subsequent computation is the derivative of the last expression with respect to r

$$\frac{\partial I_1}{\partial r} = \frac{2\pi^2}{\alpha_1^2(b^2-1)} \int_0^\infty (a_1 \sin \xi_1 \varrho - a_2 \sin \xi_2 \varrho) \frac{\mathrm{J}_1(r\varrho)}{\varrho} \, \mathrm{d}\varrho \, .$$

This integral can be evalutead using formula 6.693.1 in [100]; after some handling we get

$$\frac{\partial I_1}{\partial r} = \frac{2\pi^2}{\alpha_1^2(b^2-1)\, r} \left[-R_1 \, H(x - a_1 r) + R_2 \, H(x - a_2 r) \right] . \quad (17.31)$$

Expression (17.31) is further differentiated with respect to x and r

$$\frac{\partial^2 I_1}{\partial x \, \partial r} = \frac{2\pi^2}{\alpha_1^2(b^2-1)\, r} \left[-\frac{x}{R_1} H(x - a_1 r) - R_1 \, \delta(x - a_1 r) + \right.$$

$$\left. + \frac{x}{R_2} H(x - a_2 r) + R_2 \, \delta(x - a_2 r) \right] ,$$

$$\frac{\partial^2 I_1}{\partial r^2} = \frac{2\pi^2}{\alpha_1^2(b^2-1)} \frac{x^2}{r^2} \left\{ \frac{1}{R_1} H(x - a_1 r) - \frac{1}{R_2} H(x - a_2 r) + \right.$$

$$\left. + \frac{r}{x^2} \left[a_1 R_1 \, \delta(x - a_1 r) - a_2 R_2 \, \delta(x - a_2 r) \right] \right\} . \quad (17.32)$$

With expressions (17.31) and (17.32) in (17.17) the solution for the displacements becomes

$$u = \frac{P}{4\pi G \alpha_2^2} \frac{xz}{r^2} \left[-\frac{1}{R_1} H(x - a_1 r) - \frac{R_1}{x} \delta(x - a_1 r) + \right.$$

$$\left. + \frac{1}{R_2} H(x - a_2 r) + \frac{R_2}{x} \delta(x - a_2 r) \right] ,$$

$$v = -\frac{P}{4\pi G \alpha_2^2} \frac{yz}{r^4} \left\{ -R_1 \, H(x - a_1 r) + R_2 \, H(x - a_2 r) - x^2 \, . \right.$$

$$\cdot \left[\frac{1}{R_1} H(x - a_1 r) - \frac{1}{R_2} H(x - a_2 r) + \frac{r}{x^2} (a_1 R_1 \, \delta(x - a_1 r) - \right.$$

$$\left. \left. - a_2 R_2 \, \delta(x - a_2 r)) \right] \right\} ,$$

$$w = \frac{P}{4\pi G\alpha_2^2}\left\{\frac{\alpha_2^2}{R_2} H(x - a_2 r) + \frac{y^2}{r^4}\left[-R_1 H(x - a_1 r) + R_2 \,.\right.\right.$$

$$\left.. H(x - a_2 r)\right] - \frac{x^2 z^2}{r^4}\left[-\frac{1}{R_1} H(x - a_1 r) + \frac{1}{R_2} H(x - a_2 r) - \right.$$

$$\left.\left. - \frac{r}{x^2}\left(a_1 R_1\, \delta(x - a_1 r) - a_2 R_2\, \delta(x - a_2 r)\right)\right]\right\}. \tag{17.33}$$

The displacements have singularities at points $x = x_1 - ct = a_1 r$ and $x = x_1 - ct = a_2 r$ on the lateral area of two cones of revolution described by the equations $x^2/a_i^2 = y^2 + z^2$, $i = 1, 2$, whose vertices lie in the origin and circular bases with radius 1 in planes $x = -a_1$ and $x = -a_2$ (Fig. 17.2).

17.2 *Stresses in elastic space*

Once we know the displacement components (17.24), (17.29) or (17.33) we can compute any stress component according to (17.2) and (17.3) by mere differentiation. By way of an example we will determine here the component of normal stress σ_z in the vertical direction defined by (17.2) as

$$\sigma_z = 2G\varepsilon_z + \lambda(\varepsilon_x + \varepsilon_y + \varepsilon_z)$$

where

$$\varepsilon_x = \frac{\partial u}{\partial x}, \quad \varepsilon_y = \frac{\partial v}{\partial y}, \quad \varepsilon_z = \frac{\partial w}{\partial z}.$$

For the case of subsonic speed, $\alpha_1 < \alpha_2 < 1$, the differentiation as indicated will give

$$\sigma_z = \frac{P}{2\pi\alpha_2^2}\left\{(1 - \alpha_2^2)\frac{z}{R_2^3}\left[\frac{1}{r^4}(y^2 R_2^2 - x^2 z^2) - \alpha_2^2\right] - (1 - \alpha_1^2)\,.\right.$$

$$. \frac{z}{R_1^3 r^4}(y^2 R_1^2 + x^2 z^2) - \frac{4y^2 z}{r^6}(R_2 - R_1) - \frac{2}{r^6} x^2 z\,.$$

$$\left.. (y^2 - z^2)\left(\frac{1}{R_2} - \frac{1}{R_1}\right)\right\} + \frac{\lambda P}{4\pi G\alpha_2^2}\left\{\frac{z}{r^2}\left(\frac{1}{R_2} - \frac{1}{R_1}\right) + \frac{x^2 z}{r^2}\,.\right.$$

$$\cdot\left(\frac{1}{R_1^3} - \frac{1}{R_2^3}\right) - \frac{z}{r^6}(r^2 - 4y^2)\left[R_2 - R_1 + x^2\left(\frac{1}{R_2} - \frac{1}{R_1}\right)\right] -$$

$$- \frac{y^2 z}{r^2}\left[\frac{(1 - \alpha_2^2)^2}{R_2^3} - \frac{(1 - \alpha_1^2)^2}{R_1^3}\right] + (1 - \alpha_2^2)\frac{z}{R_2^3} \cdot$$

$$\cdot\left[\frac{1}{r^4}(y^2 R_2^2 - x^2 z^2) - \alpha_2^2\right] - (1 - \alpha_1^2)\frac{z}{R_1^3 r^4} \cdot$$

$$\cdot (y^2 R_1^2 + x^2 z^2) - \frac{4y^2 z}{r^6}(R_2 - R_1) - \frac{2}{r^6}x^2 z(y^2 - z^2) \cdot$$

$$\cdot\left(\frac{1}{R_2} - \frac{1}{R_1}\right)\bigg\} \cdot \tag{17.34}$$

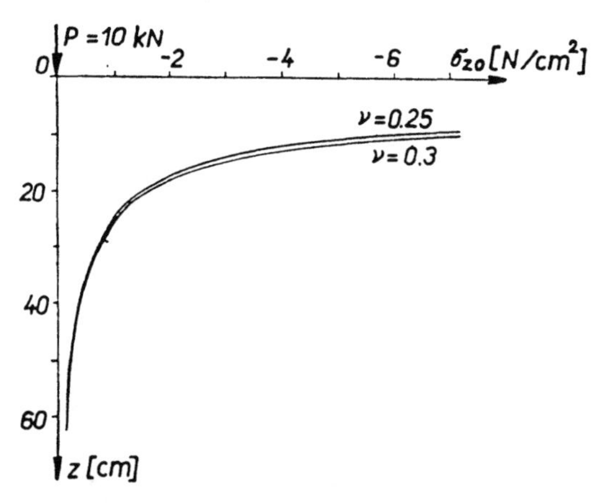

Fig. 17.3. Vertical static component of stress, σ_{z0}, in dependence on depth z underneath load $P = 10$ kN.

The normal stress in the vertical direction will clearly attain its maximum underneath the moving load. Hence the substitution $x = 0$, $y = 0$ in Eq. (17.34) will give

$$\sigma_z(0, 0, z) = \sigma_{z0}\delta \tag{17.35}$$

where σ_{z0} is the component of vertical normal stress from the static (immobile) load P (see [147])

$$\sigma_{z0} = -\frac{P}{4\pi z^2}\frac{2 - \nu}{1 - \nu} \tag{17.36}$$

and δ the dynamic coefficient of stress resulting from the effect of moving force P

$$\delta = \frac{1 - \nu}{2 - \nu}\left[\frac{2}{(1 - \alpha_2^2)^{1/2}} + \frac{\nu}{1 - \nu}\frac{1}{(1 - \alpha_1^2)^{1/2}}\right]. \qquad (17.37)$$

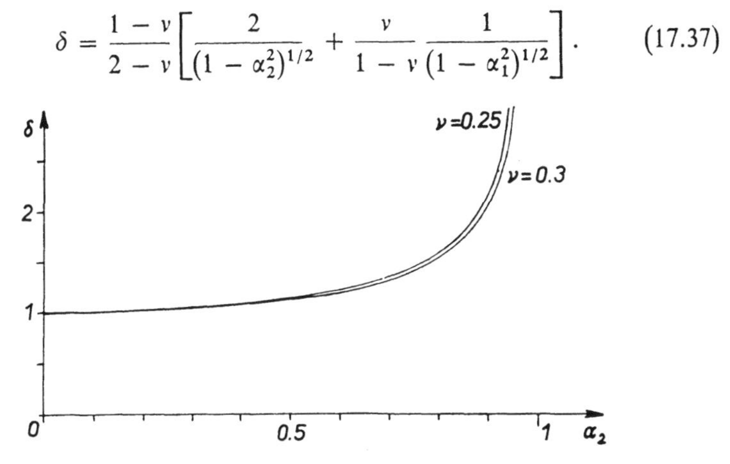

Fig. 17.4. Dynamic coefficient of stress, δ, resulting from the effect of a moving force, versus the force speed parameter α_2.

The distribution of vertical static stress σ_{z0} along a straight line underneath load $P = 10$ kN, computed from formula (17.36), is graphically represented in Fig. 17.3. As the figure reveals, the stress reaches very high values in the narrow neighbourhood of the force point of action. This is due to the load being concentrated on a very small area. At greater depths underneath the load the stress falls off rapidly and eventually reaches usable values. The effect of Poisson's ratio ν on the process is negligible.

We have also evaluated formula (17.37) and plotted the result – the dependence of δ on the force speed of motion α_2 – in Fig. 17.4. The expression of α_1 which was substituted in (17.37) follows from (17.14) and from the relations between λ and G

$$\alpha_1^2 = \frac{\alpha_2^2}{2}\frac{1 - 2\nu}{1 - \nu}. \qquad (17.38)$$

It is clear to see in Fig. 17.4 that at low speeds stress $\sigma_z(0, 0, z)$ has nearly a static character. It does not grow until at speeds approaching the velocity of propagation of transverse waves, $c \to c_2$.

At transonic and supersonic speeds the stress components possess singularities arising by differentiation of the generalized functions $H(x)$

and $\delta(x)$. To obtain results that are more like those met with in practical cases of this sort, we would also have to give consideration to the damping in the elastic space.

17.3 *Application of the theory*

The theory finds use in calculations of stresses produced in elastic solids by bodies traversing them, e.g. of stresses produced in soil by the motion of vehicles in underground tunnels. The case that we have just solved is suggestive of the motion of space ships in the cosmos, of aircrafts in air or submarines in water. Even though the media just mentioned are governed by other laws than those of the theory of elasticity, and stream past the bodies in a way not considered in our exposition, Fig. 17.2 can nevertheless be regarded as a very simple representation of a disturbance (e.g. sound) that propagates from a body moving at trans- or supersonic speed.

17.4 *Additional bibliography*

[30, 55, 57, 89, 170, 204].

18

Force moving on elastic half-space

Problems involving motion of a force on an elastic half-space are of great importance for engineering. They were first solved by I. N. Sneddon [203], J. Cole, J. Huth [35], and A. P. Filippov [62]. We shall here discuss only two fundamental problems, the motion of a concentrated force and that of a line load on an elastic half-space. The three methods which we will make use of in the solution are, however, applicable to other cases, too. Lately, attention has also been accorded to yet more complicated problems, such as the motion of a force along an infinitely long beam or an infinite plate resting on an elastic half-space, etc., see [62], [4].

The equation of motion of an elastic half-space is the same as Eq. (17.1) for an elastic space. Provided there exist no internal forces, it contains no terms expressing the volume forces, i.e.

$$X_i = 0\,, \quad i = 1, 2, 3\,. \tag{18.1}$$

The external force that moves on the plane bounding the half-space, is included in the boundary conditions.

Assuming quasi-stationary motion of force P executed at speed c in the direction of axis x_1, we can again introduce transformation (17.8) in Eq. (17.1) and get the particular solution in the form of (17.9). Following the substitutions (18.1), (17.8) and (17.9) in (17.1) we may write the system of equations describing the motion of a half-space (and of an elastic space in general) as

$$(\lambda + G)\frac{\partial \Theta}{\partial x} + G\,\nabla^2 u = c^2 \varrho\,\frac{\partial^2 u}{\partial x^2}\,,$$

$$(\lambda + G)\frac{\partial \Theta}{\partial y} + G\,\nabla^2 v = c^2 \varrho\,\frac{\partial^2 v}{\partial x^2}\,,$$

$$(\lambda + G)\frac{\partial \Theta}{\partial z} + G\,\nabla^2 w = c^2 \varrho\,\frac{\partial^2 w}{\partial x^2}\,. \tag{18.2}$$

Next to those familiar from Chap. 17 [Eq. (17.1)] we have used the following symbols in (18.2):

$$\Theta = \frac{\partial u}{\partial x} + \frac{\partial v}{\partial y} + \frac{\partial w}{\partial z} \tag{18.3}$$

denoting the relative volume change, and

$$\nabla^2 = \frac{\partial^2}{\partial x^2} + \frac{\partial^2}{\partial y^2} + \frac{\partial^2}{\partial z^2} \tag{18.4}$$

denoting the Laplace operator.

18.1 *Motion of a concentrated force on elastic half-space*

Consider a half-space bounded by plane $z = 0$ on which moves a concentrated force of constant magnitude. In the moving system of coordinates x, y, z, Fig. 18.1 shows the case of force P moving from infinity to infinity at uniform speed c in the positive direction of axis x.

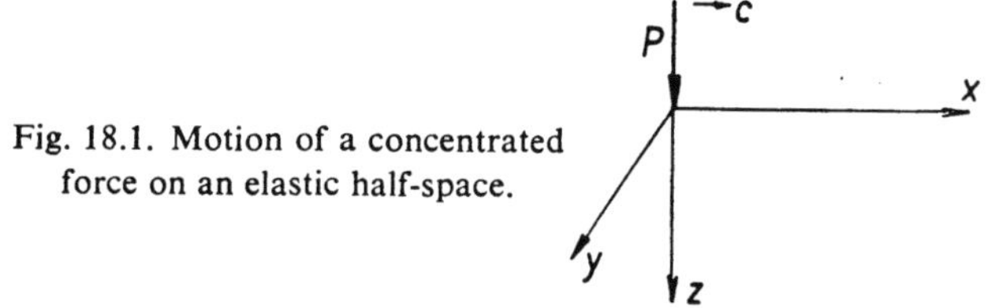

Fig. 18.1. Motion of a concentrated force on an elastic half-space.

In plane $z = 0$ we shall specify the boundary conditions as follows: except at the origin $x = 0$, $y = 0$, where they have singular values, the normal stresses in the direction of axis z are everywhere zero, and so are the components of tangential stress in that plane. Expressed mathematically the boundary conditions for $z = 0$ are

$$\sigma_z = -P\,\delta(x)\,\delta(y)\,, \quad \tau_{xz} = 0\,, \quad \tau_{yz} = 0\,. \tag{18.5}$$

Further, the boundary conditions for $x = \pm\infty$, $y = \pm\infty$, $z = +\infty$ are represented by

$$u = 0\,, \quad v = 0\,, \quad w = 0\,, \quad \partial u/\partial x = 0\,, \quad \partial v/\partial y = 0\,, \quad \partial w/\partial z = 0\,. \tag{18.6}$$

The solution of the system of equations (18.2) is assumed to be in the form

$$u = \frac{\partial f}{\partial x} + g_1 ,$$

$$v = \frac{\partial f}{\partial y} + g_2 ,$$

$$w = \frac{\partial f}{\partial z} + g_3 , \tag{18.7}$$

the unknown functions $f(x, y, z)$ and $g_i(x, y, z)$, $i = 1, 2, 3$, to be determined from Eq. (18.2) and from boundary conditions (18.5) and (18.6).

On substituting expressions (18.7) in (18.2) we can readily see that, if they are to be the solution of our problem according to (18.7), functions f and g_i must satisfy the conditions

$$\left(\nabla^2 - \alpha_1^2 \frac{\partial^2}{\partial x^2}\right) f = 0 , \tag{18.8}$$

$$\left(\nabla^2 - \alpha_2^2 \frac{\partial^2}{\partial x^2}\right) g_i = 0 , \quad i = 1, 2, 3 , \tag{18.9}$$

$$\Theta_i = \frac{\partial g_1}{\partial x} + \frac{\partial g_2}{\partial y} + \frac{\partial g_3}{\partial z} = 0 . \tag{18.10}$$

In the above, α_1, α_2 are again the dimensionless speeds (17.14), (17.16) and ∇^2 is the Laplace operator (18.4).

Eqs. (18.8) and (18.9) are conveniently solved by the method of two-dimensional Fourier integral transformation in coordinates x and y

$$F = \int_{-\infty}^{\infty} \int_{-\infty}^{\infty} f \, \mathrm{e}^{-\mathrm{i}(xq_1 + yq_2)} \, \mathrm{d}x \, \mathrm{d}y ,$$

$$f = \frac{1}{4\pi^2} \int_{-\infty}^{\infty} \int_{-\infty}^{\infty} F \mathrm{e}^{\mathrm{i}(xq_1 + yq_2)} \, \mathrm{d}q_1 \, \mathrm{d}q_2 ,$$

$$G_i = \int_{-\infty}^{\infty} \int_{-\infty}^{\infty} g_i \mathrm{e}^{-\mathrm{i}(xq_1 + yq_2)} \, \mathrm{d}x \, \mathrm{d}y , \quad i = 1, 2, 3 ,$$

$$g_i = \frac{1}{4\pi^2} \int_{-\infty}^{\infty} \int_{-\infty}^{\infty} G_i \mathrm{e}^{\mathrm{i}(xq_1 + yq_2)} \, \mathrm{d}q_1 \, \mathrm{d}q_2 . \tag{18.11}$$

Since in view of (18.5) the boundary conditions for coordinate z are not as advantageous as those for coordinates x and y (18.6), the transforms $F(q_1, q_2, z)$ and $G_i(q_1, q_2, z)$ continue to be functions of z.

To simplify we introduce the symbols

$$D = \frac{\partial}{\partial z}, \quad D^2 = \frac{\partial^2}{\partial z^2} \tag{18.12}$$

and transform Eqs. (18.8) and (18.9) in accordance with (18.11). In view of the boundary conditions (18.6) and of relation (27.80) we get the ordinary differential equations

$$\begin{aligned} &(D^2 - n_1^2)\, F = 0\,, \\ &(D^2 - n_2^2)\, G_i = 0\,, \quad i = 1, 2, 3 \end{aligned} \tag{18.13}$$

where

$$\begin{aligned} n_1^2 &= (1 - \alpha_1^2)\, q_1^2 + q_2^2\,, \\ n_2^2 &= (1 - \alpha_2^2)\, q_1^2 + q_2^2\,. \end{aligned} \tag{18.14}$$

18.1.1 *Subsonic speed*

At subsonic speeds, $c < c_2 < c_1$, $\alpha_1 < \alpha_2 < 1$ and hence also $n_i^2 > 0$, $i = 1, 2$. The solution of Eqs. (18.13) may therefore be assumed to be in the form

$$\begin{aligned} F &= A_4 \mathrm{e}^{-n_1 z}\,, \\ G_i &= A_i \mathrm{e}^{-n_2 z}\,, \quad i = 1, 2, 3 \end{aligned} \tag{18.15}$$

where the integration constants A_i, A_4 are naturally functions of q_1 and q_2. In expressions (18.15) we have already considered the boundary conditions (18.6) for $z = +\infty$ because we had omitted the general solutions in the form of $\mathrm{e}^{n_1 z}$ and $\mathrm{e}^{n_2 z}$. So long as the conditions (18.6) are to be satisfied for $z = +\infty$ the integration constants of the latter expressions are necessarily zero.

For the four integration constants A_1 to A_4 we have available three boundary conditions (18.5) and the hitherto unused condition (18.10). Of course, all of those conditions must be considered in the transformed state. Following transformations (18.11) and in view of (17.2) to (17.5) and (27.80), the left-hand sides of Eqs. (18.5) and (18.10) will successively give

$$
\begin{aligned}
\bar{\sigma}_z &= 2G(n_1^2 A_4 e^{-n_1 z} - n_2 A_3 e^{-n_2 z}) - \lambda \alpha_1^2 q_1^2 A_4 e^{-n_1 z}, \\
\bar{\tau}_{xz} &= G(-2iq_1 n_1 A_4 e^{-n_1 z} - n_2 A_1 e^{-n_1 z} + iq_1 A_3 e^{-n_2 z}), \\
\bar{\tau}_{yz} &= G(-2iq_2 n_1 A_4 e^{-n_1 z} - n_2 A_2 e^{-n_2 z} + iq_2 A_3 e^{-n_2 z}), \\
\bar{\Theta}_i &= iq_1 G_1 + iq_2 G_2 + DG_3 = 0 .
\end{aligned}
\tag{18.16}
$$

Symbols $\bar{\sigma}_z$, $\bar{\tau}_{xz}$, $\bar{\tau}_{yz}$ and $\bar{\Theta}_i$ denote the transformations of the respective functions σ_z, τ_{xz}, τ_{yz} and Θ_i according to (18.11). In the expression of $\bar{\sigma}_z$ [Eq. (17.2)], use was also made of the property of functions Θ (18.3), Θ_i (18.10) and (18.8) because

$$\Theta = \nabla^2 f = \alpha_1^2 \frac{\partial^2 f}{\partial x^2} .$$

After transformations of the right-hand sides of (18.5) we get for $z = 0$ [in view of (27.84)]

$$\bar{\sigma}_z = -P , \quad \bar{\tau}_{xz} = 0 , \quad \bar{\tau}_{yz} = 0 . \tag{18.17}$$

Substitution of (18.17) in the first three of Eqs. (18.16) for $z = 0$, and of the expression of G_i from (18.15) in the fourth, together with some manipulation will result in the system of algebraic equations

$$
\begin{aligned}
- 2n_2 A_3 + (q_1^2 + q_2^2 + n_2^2) A_4 &= -P/G , \\
-n_2 A_1 + iq_1 A_3 - 2iq_1 n_1 A_4 &= 0 , \\
-n_2 A_2 + iq_2 A_3 - 2iq_2 n_1 A_4 &= 0 , \\
q_1 A_1 + q_2 A_2 + in_2 A_3 &= 0 .
\end{aligned}
\tag{18.18}
$$

System (18.18) is easy to solve and results in

$$
\begin{aligned}
A_1 &= \frac{P}{G} \frac{2iq_1 n_1 n_2}{B} , & A_2 &= \frac{P}{G} \frac{2iq_2 n_1 n_2}{B} , \\
A_3 &= -\frac{P}{G} \frac{2n_1(q_1^2 + q_2^2)}{B} , & A_4 &= -\frac{P}{G} \frac{q_1^2 + q_2^2 + n_2^2}{B}
\end{aligned}
\tag{18.19}
$$

where $B = (q_1^2 + q_2^2 + n_2^2)^2 - 4n_1 n_2 (q_1^2 + q_2^2)$.

The integration constants A_1 to A_4 are substituted in (18.7) so that

$$U = \mathrm{i}q_1 A_4 \mathrm{e}^{-n_1 z} + A_1 \mathrm{e}^{-n_2 z},$$
$$V = \mathrm{i}q_2 A_4 \mathrm{e}^{-n_1 z} + A_2 \mathrm{e}^{-n_2 z},$$
$$W = -n_1 A_4 \mathrm{e}^{-n_1 z} + A_3 \mathrm{e}^{-n_2 z},$$

and after inverse transformations (18.11) the displacement components are obtained in the form

$$u = \frac{P\mathrm{i}}{4\pi^2 G} \int_{-\infty}^{\infty} \int_{-\infty}^{\infty} \frac{1}{B} [-q_1(q_1^2 + q_2^2 + n_2^2)\,\mathrm{e}^{-n_1 z} + 2q_1 n_1 n_2 \mathrm{e}^{-n_2 z}] \,. \\ . \,\mathrm{e}^{\mathrm{i}(xq_1 + yq_2)}\,\mathrm{d}q_1\,\mathrm{d}q_2\,,$$

$$v = \frac{P\mathrm{i}}{4\pi^2 G} \int_{-\infty}^{\infty} \int_{-\infty}^{\infty} \frac{1}{B} [-q_2(q_1^2 + q_2^2 + n_2^2)\,\mathrm{e}^{-n_1 z} + 2q_2 n_1 n_2 \mathrm{e}^{-n_2 z}] \,. \\ . \,\mathrm{e}^{\mathrm{i}(xq_1 + yq_2)}\,\mathrm{d}q_1\,\mathrm{d}q_2\,,$$

$$w = \frac{P}{4\pi^2 G} \int_{-\infty}^{\infty} \int_{-\infty}^{\infty} \frac{1}{B} [n_1(q_1^2 + q_2^2 + n_2^2)\,\mathrm{e}^{-n_1 z} - 2n_1(q_1^2 + q_2^2)\,\mathrm{e}^{-n_2 z}] \,. \\ . \,\mathrm{e}^{\mathrm{i}(xq_1 + yq_2)}\,\mathrm{d}q_1\,\mathrm{d}q_2\,. \tag{18.20}$$

The stress components according to (17.2) to (17.5) are

$$\sigma_x = \frac{P}{4\pi^2} \int_{-\infty}^{\infty} \int_{-\infty}^{\infty} \frac{q_1^2}{B} [(2 - 2\alpha_1^2 + \alpha_2^2)(q_1^2 + q_2^2 + n_2^2)\,\mathrm{e}^{-n_1 z} - \\ - 4n_1 n_2 \mathrm{e}^{-n_2 z}]\,\mathrm{e}^{\mathrm{i}(xq_1 + yq_2)}\,\mathrm{d}q_1\,\mathrm{d}q_2\,,$$

$$\sigma_y = \frac{P}{4\pi^2} \int_{-\infty}^{\infty} \int_{-\infty}^{\infty} \frac{1}{B} [(2q_2^2 - 2\alpha_1^2 q_1^2 + \alpha_2^2 q_1^2)(q_1^2 + q_2^2 + n_2^2)\,\mathrm{e}^{-n_1 z} - \\ - 4q_2^2 n_1 n_2 \mathrm{e}^{-n_2 z}]\,\mathrm{e}^{\mathrm{i}(xq_1 + yq_2)}\,\mathrm{d}q_1\,\mathrm{d}q_2\,,$$

$$\sigma_z = -\frac{P}{4\pi^2} \int_{-\infty}^{\infty} \int_{-\infty}^{\infty} \frac{1}{B} [(q_1^2 + q_2^2 + n_2^2)^2\,\mathrm{e}^{-n_1 z} + \\ + 4n_1 n_2 (q_1^2 + q_2^2)\,\mathrm{e}^{-n_2 z}]\,\mathrm{e}^{\mathrm{i}(xq_1 + yq_2)}\,\mathrm{d}q_1\,\mathrm{d}q_2\,,$$

$$\tau_{xy} = \frac{P}{4\pi^2} \int_{-\infty}^{\infty} \int_{-\infty}^{\infty} \frac{2q_1 q_2}{B} [(q_1^2 + q_2^2 + n_2^2)\,\mathrm{e}^{-n_1 z} - 2n_1 n_2 \mathrm{e}^{-n_2 z}] \,. \\ . \,\mathrm{e}^{\mathrm{i}(xq_1 + yq_2)}\,\mathrm{d}q_1\,\mathrm{d}q_2\,,$$

$$\tau_{xz} = \frac{Pi}{4\pi^2} \int_{-\infty}^{\infty} \int_{-\infty}^{\infty} \frac{2q_1 n_1}{B} \left(q_1^2 + q_2^2 + n_2^2\right) \left(e^{-n_1 z} - e^{-n_2 z}\right) . \\ . \, e^{i(xq_1 + yq_2)} \, dq_1 \, dq_2 \, ,$$

$$\tau_{yz} = \frac{Pi}{4\pi^2} \int_{-\infty}^{\infty} \int_{-\infty}^{\infty} \frac{2q_2 n_1}{B} \left(q_1^2 + q_2^2 + n_2^2\right) \left(e^{-n_1 z} - e^{-n_2 z}\right) . \\ . \, e^{i(xq_1 + yq_2)} \, dq_1 \, dq_2 \, . \tag{18.21}$$

So far we have not succeeded in evaluating the integrals in Eqs. (18.20) and (18.21). To get particular results we shall no doubt have to resort to automatic computers.

18.1.2 *Transonic speed* $c_2 < c < c_1$

At transonic speed, $c_2 < c < c_1$, $\alpha_1 < 1 < \alpha_2$, and in place of (18.14) we shall write

$$n_1^2 = \left(1 - \alpha_1^2\right) q_1^2 + q_2^2 \, , \\ n_2^2 = \left(\alpha_2^2 - 1\right) q_1^2 - q_2^2 \, . \tag{18.22}$$

Eqs. (18.13) will then take on the form

$$\left(D^2 - n_1^2\right) F = 0 \, , \\ \left(D^2 + n_2^2\right) G_i = 0 \, , \quad i = 1, 2, 3 \, . \tag{18.23}$$

At transonic speed there is always $n_1^2 > 0$. As to n_2^2, there can arise the following two cases:

1. for $\left(\alpha_2^2 - 1\right) q_1^2 < q_2^2$, $n_2^2 < 0$ and the solution of Eqs. (18.23) may be assumed to be of the form of (18.15).

2. for $\left(\alpha_2^2 - 1\right) q_1^2 > q_2^2$, $n_2^2 > 0$, and the solution of (18.23) is considered to be in the form

$$F = A_4 e^{-n_1 z} \, , \\ G_i = A_i e^{i n_2 z} \, , \quad i = 1, 2, 3 \, . \tag{18.24}$$

In this case the equation defining G_i is already of a wave character and unless damping is taken into account, the boundary conditions (18.6) for $z = +\infty$ can be satisfied no longer. Subsequent computation would be analogous to that outlined in paragraph 18.1.1.

18.1.3 *Supersonic speed* $c_2 < c_1 < c$

At supersonic speed $c_2 < c_1 < c$, $1 < \alpha_1 < \alpha_2$, and in place of expressions (18.14) we now write

$$n_1^2 = (\alpha_1^2 - 1)\, q_1^2 - q_2^2 ,$$
$$n_2^2 = (\alpha_2^2 - 1)\, q_1^2 - q_2^2 . \qquad (18.25)$$

Eqs. (18.13) will then take on the form

$$(D^2 + n_1^2)\, F = 0 ,$$
$$(D^2 + n_2^2)\, G_i = 0 , \quad i = 1, 2, 3 . \qquad (18.26)$$

Altogether four cases may arise at this speed:

1. for $(\alpha_i^2 - 1)\, q_1^2 < q_2^2$, $n_i^2 < 0$, $i = 1, 2$, and the solution of Eq. (18.26) is in the form of (18.15),

2. for $(\alpha_1^2 - 1)\, q_1^2 < q_2^2$, $(\alpha_2^2 - 1)\, q_1^2 > q_2^2$, $n_1^2 < 0$, $n_2^2 > 0$, and the solution of Eqs. (18.26) is in the form of (18.24),

3. for $(\alpha_1^2 - 1)\, q_1^2 > q_2^2$, $(\alpha_2^2 - 1)\, q_1^2 < q_2^2$, $n_1^2 > 0$, $n_2^2 < 0$, and the solution of Eqs. (18.26) is in the form

$$F = A_4 e^{in_1 z} ,$$
$$G_i = A_i e^{-n_2 z} , \quad i = 1, 2, 3 , \qquad (18.27)$$

4. for $(\alpha_i^2 - 1)\, q_1^2 > q_2^2$, $n_i^2 > 0$, $i = 1, 2$, and the solution of Eqs. (18.26) takes on the form

$$F = A_4 e^{in_1 z} ,$$
$$G_i = A_i e^{in_2 z} , \quad i = 1, 2, 3 . \qquad (18.28)$$

The evaluation of constants A_i, A_4 would proceed as outlined in paragraph 18.1.1 but would have to be made separately for each of the four cases. Because of the four cases of integration path, the resultant integration of equations analogous to (18.20) and (18.21) would, too, be highly complicated.

18.2 *Motion of a line load on elastic half-space*

In case of plane strains or plane stresses, our problem can be solved all the way to resultant simple formulae. In an elastic half-space, plane strains arise whenever the surface of the half-space is traversed by a continuous line load (Fig. 18.2). It is again assumed that the line load of magnitude P per unit length [force per length] moves from infinity to infinity at constant speed c in the direction of positive axis $x_1 \equiv x$.

Equations describing the motion of a half-space under plane strain start out from the fundamental equations of an elastic space (17.1) in which one substitutes the conditions of plane strain, i.e. the conditions specifying that the displacements in the direction of axis $x_3 \equiv z$ (according to the orientation of axes indicated in Fig. 18.2) and all their derivatives with respect to that coordinate should be zero [57]

$$u_3 = 0\,, \quad \frac{\partial u_3}{\partial x_i} = 0\,, \quad \frac{\partial^2 u_3}{\partial x_i^2} = 0\,, \quad \frac{\partial}{\partial x_3} = 0\,, \quad \frac{\partial^2}{\partial x_3^2} = 0\,. \tag{18.29}$$

When load P performs quasi-stationary motion at speed c in the direction of axis x_1, one can again introduce transformation (17.8) in Eq. (17.1) and get the particular solution in the form of (17.9). After the substitutions (17.8), (17.9), (18.1) and (18.29) the system of equations that describe the motion of a half-space (or of an elastic space in general) under plane strain turns out to be

$$(\lambda + G)\frac{\partial \Theta}{\partial x} + G\,\nabla^2 u = c^2 \varrho \frac{\partial^2 u}{\partial x^2}\,,$$

$$(\lambda + G)\frac{\partial \Theta}{\partial y} + G\,\nabla^2 v = c^2 \varrho \frac{\partial^2 v}{\partial x^2}\,. \tag{18.30}$$

The symbols are the same as those in Chap. 17, and further,

$$\Theta = \frac{\partial u}{\partial x} + \frac{\partial v}{\partial y}\,, \tag{18.31}$$

$$\nabla^2 = \frac{\partial^2}{\partial x^2} + \frac{\partial^2}{\partial y^2}\,. \tag{18.32}$$

Furthermore, it is assumed that the half-space is bounded by plane $y = 0$ on which the load moves. In the mobile system of coordinates

x, y, z this is represented in Fig. 18.2. Hence the boundary conditions will be

for $y = 0$
$$\sigma_y = -P\,\delta(x)\,, \quad \tau_{xy} = 0\,, \tag{18.33}$$

and for $x = \pm\infty$, $y = +\infty$
$$u = 0\,, \quad v = 0\,, \quad \partial u/\partial x = 0\,, \quad \partial v/\partial y = 0\,. \tag{18.34}$$

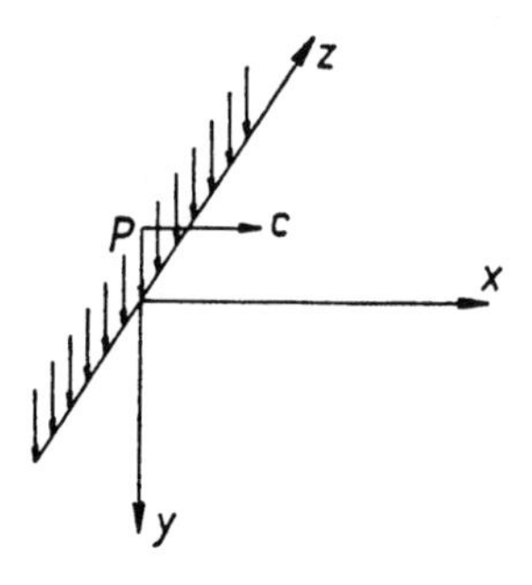

Fig. 18.2. Motion of a line load on an elastic half-space.

The boundary conditions for $x = \pm\infty$ make it possible to use the method of Fourier integral transformation with respect to x. Since the boundary conditions in the direction of axis y are more complicated and the method not applicable in that direction, we shall use the one-dimensional transformation

$$U = \int_{-\infty}^{\infty} u e^{-ixq}\,dq\,, \quad u = \frac{1}{2\pi}\int_{-\infty}^{\infty} U e^{ixq}\,dq\,,$$
$$V = \int_{-\infty}^{\infty} v e^{-ixq}\,dq\,, \quad v = \frac{1}{2\pi}\int_{-\infty}^{\infty} V e^{ixq}\,dq\,, \tag{18.35}$$

with the transforms $U(q, y)$, $V(q, y)$ of functions $u(x, y)$, $v(x, y)$ continuing to be functions of variable y.

With the use of the symbols
$$D = \frac{\partial}{\partial y}\,, \quad D^2 = \frac{\partial^2}{\partial y^2}\,,$$

Eq. (18.30) is transformed in accordance with (18.35). In view of the boundary conditions (18.34) and of (27.79) and (27.80)

$$(\lambda + G)\,iq(iqU + DV) + G(-q^2 + D^2)\,U = -c^2\varrho q^2 U\,,$$
$$(\lambda + G)\,D(iqU + DV) + G(-q^2 + D^2)\,V = -c^2\varrho q^2 V\,.$$

This system of ordinary differential equations in variable y may be rearranged with the aid of substitutions for α_1, α_2, b [see (17.14)] to

$$[D^2 - (b^2 - \alpha_2^2) q^2] U + i(b^2 - 1) q DV = 0 ,$$
$$i(b^2 - 1) q DU + [b^2 D^2 - (1 - \alpha_2^2) q^2] V = 0 . \qquad (18.36)$$

The system (18.36) will be satisfied if the determinant of the system, computed wholly formally as if operators D and D^2 were algebraic expressions, will be equal to zero, i.e.

$$b^2(D^2 - n_1^2)(D^2 - n_2^2)(U, V) = 0 . \qquad (18.37)$$

In the above

$$n_i^2 = a_i^2 q^2 , \quad a_i^2 = 1 - \alpha_i^2 , \quad i = 1, 2 . \qquad (18.38)$$

18.2.1 *Subsonic speed $c < c_2 < c_1$*

At subsonic speed $c < c_2 < c_1$, $\alpha_1 < \alpha_2 < 1$ and therefore also $a_i^2 > 0$ and $n_i^2 > 0$. With regard to the form of expression (18.37) the general solution of system (18.36) will therefore be

$$U = A_1 e^{-n_1 y} + A_2 e^{-n_2 y} ,$$
$$V = B_1 e^{-n_1 y} + B_2 e^{-n_2 y} \qquad (18.39)$$

where $A_{1,2}$ and $B_{1,2}$ are integration constants dependent, of course, on q. When writing (18.39) we have left out of the general solution (18.37) the terms containing $e^{n_1 y}$ and $e^{n_2 y}$ because – in view of the boundary conditions (18.34) for $y = +\infty$ – the integration constants of those terms must be zero.

So far as the determination of the four integration constants is concerned, we have available two boundary conditions (18.33) and two equations (18.36) which must be identically satisfied by expressions (18.39). In view of (17.2) to (17.5) and (18.29) the transformations of functions σ_y and τ_{xy} according to (18.38) denoted $\bar{\sigma}_y$ and $\bar{\tau}_{xy}$, respectively, give

$$\bar{\sigma}_y = -(\lambda + 2G)(n_1 B_1 e^{-n_1 y} + n_2 B_2 e^{-n_2 y}) + i\lambda q(A_1 e^{-n_1 y} + A_2 e^{-n_2 y}) ,$$
$$\bar{\tau}_{xy} = G[-n_1 A_1 e^{-n_1 y} - n_2 A_2 e^{-n_2 y} + iq(B_1 e^{-n_1 y} + B_2 e^{-n_2 y})] .$$

By (18.33), for $y = 0$, $\bar{\sigma}_y = -P$, $\bar{\tau}_{xy} = 0$. With the use of (17.14) and (18.38) and some manipulation the last equations will give

$$-q^2(2 - \alpha_2^2) B_1 - 2n_1 n_2 B_2 = -\frac{P}{G} n_1 ,$$

$$2B_1 + (2 - \alpha_2^2) B_2 = 0 .$$

The other two relations will be obtained by substituting (18.39) in (18.36)

$$n_1 A_1 = -\mathrm{i} q B_1 , \quad q A_2 = -\mathrm{i} n_2 B_2 .$$

From those four equations we determine the integration constants

$$A_1 = -\mathrm{i}\frac{PC_1}{Gq}, \quad A_2 = \mathrm{i}\frac{PC_2 a_2}{Gq},$$

$$B_1 = \frac{PC_1 a_1}{Gq}, \quad B_2 = -\frac{PC_2}{Gq}, \tag{18.40}$$

where in shorthand notation

$$C_1 = \frac{2 - \alpha_2^2}{(2 - \alpha_2^2)^2 - 4a_1 a_2}, \quad C_2 = \frac{2a_1}{(2 - \alpha_2^2)^2 - 4a_1 a_2} . \tag{18.41}$$

What remains is to substitute (18.41) and (18.40) in (18.39) and effect the inverse transformation according to (18.35). In the calculation we shall need the integrals of the type

$$\int_{-\infty}^{\infty} \mathrm{e}^{(-y+\mathrm{i}x)q}\, dq = \frac{2y}{x^2 + y^2}, \quad \mathrm{i}\int_{-\infty}^{\infty} \mathrm{e}^{(-y+\mathrm{i}x)q}\, dq = -\frac{2x}{x^2 + y^2} . \tag{18.42}$$

First, we shall calculate all the stress components. Using the integrals (18.42) we get from (17.2) and (17.5), (18.29) and (18.35) after some handling

$$\sigma_x = \frac{P}{\pi}\left[C_1(\alpha_2^2 - 2\alpha_1^2 + 2)\frac{a_1 y}{r_1^2} - 2C_2 a_2 \frac{a_2 y}{r_2^2}\right],$$

$$\sigma_y = -\frac{P}{\pi}\left[C_1(2 - \alpha_2^2)\frac{a_1 y}{r_1^2} - 2C_2 a_2 \frac{a_2 y}{r_2^2}\right],$$

$$\tau_{xy} = -\frac{P}{\pi} 2C_1 a_1 x \left(\frac{1}{r_1^2} - \frac{1}{r_2^2}\right) \tag{18.43}$$

where $r_i^2 = x^2 + a_i^2 y^2$, $i = 1, 2$.

Unlike the stress components, the displacement component u is not specified uniquely in this case. However, we can prove that our problem will be satisfied by the solution

$$u = \frac{P}{\pi G}\left[C_1(\pi - \varphi_1) - C_2 a_2(\pi - \varphi_2)\right],$$

$$v = \frac{P}{\pi G}\left(C_2 \ln|r_2| - C_1 a_1 \ln|r_1|\right) \tag{18.44}$$

where $\varphi_i = \operatorname{arctg} a_i y/x$, $0 \leqq \varphi_i \leqq \pi$.

18.2.2 *Transonic speed* $c_2 < c < c_1$

At transonic speed $c_2 < c < c_1$, $\alpha_1 < 1 < \alpha_2$, so that in place of (18.38) we write

$$n_i^2 = a_i^2 q^2, \quad i = 1, 2, \quad a_1^2 = 1 - \alpha_1^2 > 0,$$
$$a_2^2 = \alpha_2^2 - 1 > 0. \tag{18.45}$$

This case, too, could be solved by the method described in paragraph 18.2.1. Since its use would eventually lead to difficulties in the integration, we will resort to another method.

Assume the solution of the system (18.30) to be in the form

$$u = \frac{\partial f}{\partial x} - \frac{\partial g}{\partial y},$$

$$v = \frac{\partial f}{\partial y} + \frac{\partial g}{\partial x} \tag{18.46}$$

where $f(x, y)$ and $g(x, y)$ are new unknown functions. On substituting (18.46) in (18.30) we find that these functions must satisfy the equations

$$\left(\nabla^2 - \alpha_1^2 \frac{\partial^2}{\partial x^2}\right) f = 0,$$

$$\left(\nabla^2 - \alpha_2^2 \frac{\partial^2}{\partial x^2}\right) g = 0 \tag{18.47}$$

if they are to be – through the intermediary of (18.46) – the solutions of system (18.30).

Using notation (18.45) we may also write Eqs. (18.47) in the form

$$a_1^2 \frac{\partial^2 f}{\partial x^2} + \frac{\partial^2 f}{\partial y^2} = 0\,,$$

$$a_2^2 \frac{\partial^2 g}{\partial x^2} - \frac{\partial^2 g}{\partial y^2} = 0\,. \tag{18.48}$$

The first of the partial differential equations is of the elliptic type satisfied by the real part of any analytic function of the complex variable z_1 (see [201], Vol. IV, Sect. 159)

$$f = \operatorname{Re} f(z_1) + \mathrm{i} \operatorname{Im} f(z_1)\,, \quad z_1 = x + \mathrm{i}a_1 y\,, \tag{18.49}$$

for its derivatives are

$$\frac{\partial f}{\partial x} = \operatorname{Re} f'(z_1) + \mathrm{i} \operatorname{Im} f'(z_1)\,, \quad \frac{\partial^2 f}{\partial x^2} = \operatorname{Re} f''(z_1) + \mathrm{i} \operatorname{Im} f''(z_1)\,,$$

$$\frac{\partial f}{\partial y} = -a_1 \operatorname{Im} f'(z_1) + \mathrm{i}a_1 \operatorname{Re} f'(z_1)\,,$$

$$\frac{\partial^2 f}{\partial y^2} = -a_1^2 \operatorname{Re} f''(z_1) - \mathrm{i}a_1^2 \operatorname{Im} f''(z_1)\,,$$

$$\frac{\partial^2 f}{\partial x\,\partial y} = -a_1 \operatorname{Im} f''(z_1) + \mathrm{i}a_1 \operatorname{Re} f''(z_1) \tag{18.50}$$

where the primes denote the derivatives of function $f(z_1)$ with respect to z_1. In all the above expressions the real parts stand in the first place on the right-hand side. As the substitution of the expressions of $\partial^2 f/\partial x^2$ and $\partial^2 f/\partial y^2$ in the first of Eqs. (18.48) will prove, the latter is really satisfied by any function $f(z_1)$.

The second of Eqs. (18.48) is of the hyperbolic type (wave equation) satisfied by any two real functions $g(x \pm a_2 y)$ of the argument $x \pm a_2 y$. Having regard to the character of our problem we shall take for the solution only the function

$$g = g(x + a_2 y)\,. \tag{18.51}$$

This function describes the waves that move to the left from the load (with the force moving to the right — Fig. 18.3). The reason for our not taking the function $g(x - a_2 y)$ for the solution is that in the immobile system of coordinates x_1, y_1, z_1 it describes waves propagating at a velocity higher than the velocity of propagation of transverse waves, which is clearly impossible.

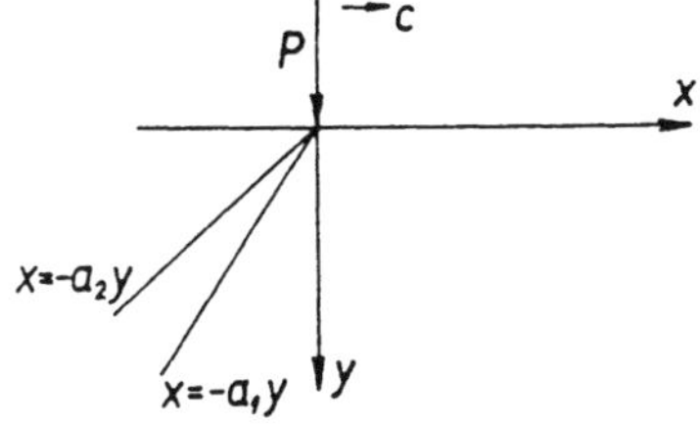

Fig. 18.3. Singularities obtaining as a line load moves on a half-space at transonic and supersonic speeds.

The form of functions f and g will be determined from the boundary conditions. Following the substitution of (18.45) in (17.2) to (17.5) the boundary conditions (18.33) become

for $y = 0$

$$\sigma_y = G\left[(\alpha_2^2 - 2)\frac{\partial^2 f}{\partial x^2} + 2\frac{\partial^2 g}{\partial x\,\partial y}\right] = -P\,\delta(x)\,,$$

$$\tau_{xy} = G\left[2\frac{\partial^2 f}{\partial x\,\partial y} + (2 - \alpha_2^2)\frac{\partial^2 g}{\partial x^2}\right] = 0\,. \tag{18.52}$$

Eqs. (18.52) may be simplified by integration with respect to x. Since they can affect the displacement constants at most, the integration constants are assumed to be zero in the operation. The integration and substitution of (18.49) and (18.51) then give

for $y = 0$

$$(2 - \alpha_2^2)\,\mathrm{Re}\,f' - 2a_2 g' = \frac{P}{G}\,H(x)\,,$$

$$-2a_1\,\mathrm{Im}\,f' + (2 - \alpha_2^2)\,g' = 0 \tag{18.53}$$

where $H(x)$ is the Heaviside function (1.4), (3.23), and g' denotes the derivative of function (18.51) with respect to the argument $x + a_2 y$.

According to [35], conditions (18.53) are satisfied by the function

$$f'(z_1) = \frac{P}{\pi G}(C_3 + \mathrm{i}C_4)(\ln z_1 - \mathrm{i}\pi) = \frac{P}{\pi G}\{C_3 \ln|r_1| +$$
$$+ C_4(\pi - \varphi_1) + \mathrm{i}[C_4 \ln|r_1| - C_3(\pi - \varphi_1)]\}\,,$$

$$g'(x + a_2 y) = \frac{P}{\pi G}\frac{2a_1}{2 - \alpha_2^2}[C_4 \ln|x + a_2 y| - \pi C_3 H(x + a_2 y)] \tag{18.54}$$

as can be proved by substitution in (18.53). In the above, one should take into consideration that $\varphi_1 = \operatorname{arctg} a_1 y/x$ is for $y = 0$, $\varphi_1 = 0$ for $x > 0$, and $\varphi_1 = \pi$ for $x < 0$. In (18.54)

$$C_3 = \frac{4a_1 a_2(2 - \alpha_2^2)}{(2 - \alpha_2^2)^4 + 16a_1^2 a_2^2},$$

$$C_4 = \frac{(2 - \alpha_2^2)^3}{(2 - \alpha_2^2)^4 + 16a_1^2 a_2^2}. \tag{18.55}$$

We are now ready to compute all the displacement and stress components according to (18.46), (18.50), and (17.2) to (17.5). After the necessary manipulation

$$u = \frac{P}{\pi G}\left\{C_3 \ln|r_1| + C_4(\pi - \varphi_1) - \frac{2a_1 a_2}{2 - \alpha_2^2}[C_4 \ln|x + a_2 y| - \right.$$
$$\left. - \pi C_3 H(x + a_2 y)]\right\},$$

$$v = \frac{P}{\pi G}\left\{-a_1 C_4 \ln|r_1| + a_1 C_3(\pi - \varphi_1) + \frac{2a_1}{2 - \alpha_2^2}[C_4 \,.\right.$$
$$\left. . \ln|x + a_2 y| - \pi C_3 H(x + a_2 y)]\right\},$$

$$\sigma_x = \frac{P}{\pi}\left\{C_4\left[(2a_1^2 + \alpha_2^2)\frac{a_1 y}{r_1^2} - \frac{4a_1 a_2}{2 - \alpha_2^2}\frac{1}{x + a_2 y}\right] + C_3 \,.\right.$$
$$\left. \cdot\left[(2a_1^2 + \alpha_2^2)\frac{x}{r_1^2} + \frac{4a_1 a_2}{2 - \alpha_2^2}\pi\,\delta(x + a_2 y)\right]\right\},$$

$$\sigma_y = -\frac{P}{\pi}(2 - \alpha_2^2)\left\{C_4\left[\frac{a_1 y}{r_1^2} - \frac{4a_1 a_2}{(2 - \alpha_2^2)^2}\frac{1}{x + a_2 y}\right] + C_3\left[\frac{x}{r_1^2} + \right.\right.$$
$$\left.\left. + \frac{4a_1 a_2}{(2 - \alpha_2^2)^2}\pi\,\delta(x + a_2 y)\right]\right\},$$

$$\tau_{xy} = \frac{2Pa_1}{\pi}\left\{C_3\left[\frac{a_1 y}{r_1^2} - \pi\,\delta(x + a_2 y)\right] + C_4\left(-\frac{x}{r_1^2} + \frac{1}{x + a_2 y}\right)\right\}. \tag{18.56}$$

It should be noted that in checking the correctness of the boundary conditions (18.33) by help of (18.56) one must employ the relations that result from (18.52) rather than those implied by the integrated version of the former, (18.53).

It is clear from (18.56) that the displacement as well as the stress have singularities along the straight line $x = -a_2 y$ (Fig. 18.3).

18.2.3 *Supersonic speed* $c_2 < c_1 < c$

At supersonic speed $c_2 < c_1 < c$, $1 < \alpha_1 < \alpha_2$ and (18.38) may be replaced by

$$n_i^2 = a_i^2 q^2 > 0\,, \quad a_i^2 = \alpha_i^2 - 1 > 0\,, \quad i = 1, 2\,. \tag{18.57}$$

With this notation Eq. (18.37) may be given the form

$$b^2(D^2 + n_1^2)(D^2 + n_2^2)(U, V) = 0 \tag{18.58}$$

and the method outlined in 18.2.1 applied. In view of the characteristic of expression (18.58) the general solution of the system (18.36) will be

$$\begin{aligned} U &= A_1 e^{in_1 y} + A_2 e^{in_2 y}\,, \\ V &= B_1 e^{in_1 y} + B_2 e^{in_2 y}\,. \end{aligned} \tag{18.59}$$

The only wave components taken into consideration from the general solution are those that propagate to the left of the moving load (that is to say, the general solutions with $e^{-in_1 y}$ and $e^{-in_2 y}$ are neglected). Because of the wave character of this case the boundary conditions (18.34) for $x = \pm\infty$ and $y = +\infty$ cannot be satisfied in the absence of damping.

The integration constants $A_{1,2}$, $B_{1,2}$ – dependent on q – are computed from the boundary conditions (18.33) and from the identities after the substitution of (18.59) in Eqs. (18.36). The system of equations thus obtained is

$$\begin{aligned} i(\lambda + 2G)(n_1 B_1 + n_2 B_2) + i\lambda q(A_1 + A_2) &= -P\,, \\ i\,G[n_1 A_1 + n_2 A_2 + q(B_1 + B_2)] &= 0\,, \\ n_1 A_1 &= q B_1\,, \\ q A_2 &= -n_2 B_2\,, \end{aligned}$$

and the solution

$$A_1 = -\mathrm{i}\,\frac{PC_5}{Gq}\,, \qquad A_2 = -\mathrm{i}\,\frac{Pa_2C_6}{Gq}\,,$$

$$B_1 = -\mathrm{i}\,\frac{Pa_1C_5}{Gq}\,, \quad B_2 = \mathrm{i}\,\frac{PC_6}{Gq} \tag{18.60}$$

where

$$C_5 = \frac{2 - \alpha_2^2}{(2 - \alpha_2^2)^2 + 4a_1a_2}\,,$$

$$C_6 = \frac{2a_1}{(2 - \alpha_2^2)^2 + 4a_1a_2}\,. \tag{18.61}$$

What remains is to carry out the inverse transformation of expressions (18.59) in accordance with (18.35) using (18.60) and (18.61) in the operation. In the calculation we shall have need of the integral expressions of the Heaviside and Dirac functions in the complex plane (see [201], Vol. III, Part 2, Sect. 61, and our (1.4))

$$H(x) = \frac{1}{2\pi\mathrm{i}}\int_{-\infty}^{\infty}\frac{\mathrm{e}^{\mathrm{i}xq}}{q}\,\mathrm{d}q\,, \quad \delta(x) = \frac{1}{2\pi}\int_{-\infty}^{\infty}\mathrm{e}^{\mathrm{i}xq}\,\mathrm{d}q\,. \tag{18.62}$$

With these integrals we readily obtain all the displacement and stress components

$$u = \frac{P}{G}\left[C_5\,H(x + a_1y) + a_2C_6\,H(x + a_2y)\right],$$

$$v = \frac{P}{G}\left[a_1C_5\,H(x + a_1y) - C_6\,H(x + a_2y)\right],$$

$$\sigma_x = P[(\alpha_2^2 - 2\alpha_1^2 + 2)\,C_5\,\delta(x + a_1y) + 2a_2C_6\,\delta(x + a_2y)]\,,$$

$$\sigma_y = -P[(2 - \alpha_2^2)\,C_5\,\delta(x + a_1y) + 2a_2C_6\,\delta(x + a_2y)]\,,$$

$$\tau_{xy} = -P[-2a_1C_5\,\delta(x + a_1y) + (2 - \alpha_2^2)\,C_6\,\delta(x + a_2y)]\,. \tag{18.63}$$

As the results (18.63) suggest, both the displacement and the stress have singularities along two planes $x = -a_1y$, $x = -a_2y$ (Fig. 18.3).

18.3 *Motion of a force on elastic half-plane*

In the case where force P moves on the edge of an elastic half-plane or wall (Fig. 18.4) there obtains the so-called plane stress in which

$$\sigma_z = 0\,, \quad \tau_{xz} = 0\,, \quad \tau_{yz} = 0\,.$$

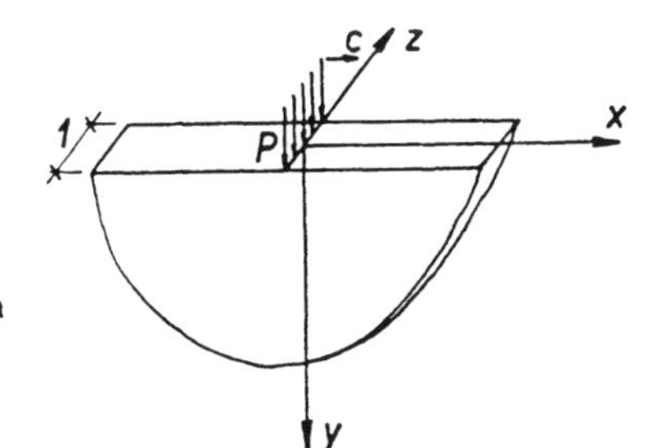

Fig. 18.4. Motion of a force on an elastic half-plane (wall).

Supposing this we may write the equations of motion of a half-plane in the same form as the equations of a half-space (Sect. 18.2). Only Lamé's constant will now be somewhat different than that of the elastic space. It namely holds for plane stress that

$$\lambda' = \frac{(1 - 2\nu)\,\lambda}{1 - \nu} = \frac{\nu E}{1 - \nu^2}\,. \tag{18.64}$$

Therefore the equations applying to the motion of a force on an elastic half-plane will be the same as those in Sect. 18.2, except that constant λ will now be replaced by λ' (18.64) in all the expressions of α_1, a_1, b, etc.; for example

$$c_1 = \left(\frac{\lambda' + 2G}{\varrho}\right)^{1/2}, \quad b^2 = \frac{\lambda' + 2G}{G} = \frac{2}{1 - \nu}\,.$$

18.4 *Application of the theory*

The method evolved in this chapter can be used in calculation of stresses in a half-space, a half-plane or a wall traversed by a body. This is the case of soils and rocks traversed by vehicles, and one comes across it in railway and highway substructures, foundations of airport runways, crane runways, etc., that is to say in foundations of all kinds of ground transport.

In the calculation of stresses in soils, clearly the most important is the information concerning the normal stress in the vertical direction. This stress will obviously reach its maximum underneath the moving load. In the case of plane strain or stress we therefore substitute $x = 0$ in the second of Eqs. (18.43) and get

$$\sigma_y(0, y) = \sigma_{y0}\delta \tag{18.65}$$

where σ_{y0} is the component of the vertical normal stress produced by static immobile load P (cf. [64])

$$\sigma_{y0} = -\frac{2P}{\pi}\frac{1}{y}, \tag{18.66}$$

and δ the dynamic coefficient of stress resulting from the effect of moving force P,

$$\delta = \frac{(2 - \alpha_2^2)^2 - 4a_1^2}{2a_1[(2 - \alpha_2^2)^2 - 4a_1a_2]}. \tag{18.67}$$

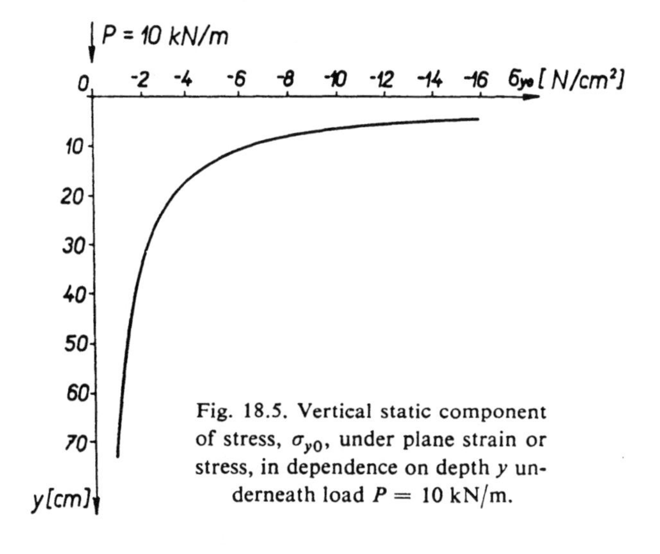

Fig. 18.5. Vertical static component of stress, σ_{y0}, under plane strain or stress, in dependence on depth y underneath load $P = 10$ kN/m.

The distribution of vertical static stress σ_{y0} on a straight line underneath load $P = 10$ kN/m computed from (18.66) is shown in Fig. 18.5. Since the load is concentrated on a small area, the stresses reach very high values in the close neighbourhood of the point of action of the force; elsewhere, of course, they are again quite acceptable to practice.

Formula (18.67) was evaluated for $\nu = 0{\cdot}3$, and the dependence of δ on the speed of force motion, α_2, plotted in Fig. 18.6. The figure refers to the case of plane strain in which the relation between α_1 and α_2 is given by Eq. (17.38), and to the case of plane stress where [cf. (18.64)]

$$\alpha_1^2 = \frac{\alpha_2^2}{2}(1 - \nu)\,. \tag{18.68}$$

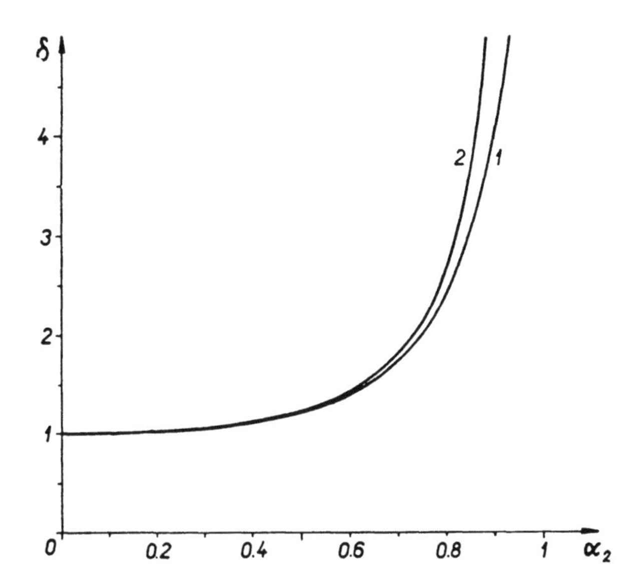

Fig. 18.6. Dynamic coefficient of stress, δ, resulting from the effect of a moving force, in dependence on the force speed, α_2; $\nu = 0{\cdot}3$; 1 — plane strain (a line load on a half-space), 2 — plane stress (a force on a half-plane or wall).

At low velocities, stress $\sigma_y(0, y)$ has nearly a static character, and experiences no growth until speeds approaching the velocity of propagation of transverse waves, $c \to c_2$, are reached.

18.5 *Additional bibliography*

[2, 4, 35, 57, 62, 89, 118, 119, 155, 160, 161, 183, 194, 195, 203, 214, 239, 249, 254].

Part V

SPECIAL PROBLEMS

Part V

19

Load motion at variable speed

So far we have assumed that the load moves on an elastic solid or a structure at constant speed. In some cases, however, the motion is not uniform but a function of time. The problem was first tackled by A. N. Lowan [146], and then treated at length by M. Ya. Ryazanova [192], and A. P. Filippov and S. S. Kokhmanyuk [62], [127] who analyzed the motion of a concentrated mass on mass and massless beams.

In what follows we will derive a solution of the second extreme case, i.e. of the time variable motion of a massless load along a mass beam. We shall also present an approximate solution of the motion of a mass continuous load along a beam.

Our considerations start out from the equations describing the undamped motion of a beam (used already in Chap. 6)

$$EJ \frac{\partial^4 v(x, t)}{\partial x^4} + \mu \frac{\partial^2 v(x, t)}{\partial t^2} = p(x, t) \tag{19.1}$$

where $p(x, t)$ expresses the beam load per unit length at point x and time t. Under arbitrary boundary conditions the solution will be effected by the application of the generalized method of finite integral transformations (6.1), Chap. 6. In the ordinary differential equation thus obtained

$$\ddot{V}(j, t) + \omega_{(j)}^2 V(j, t) = \frac{1}{\mu} p_{(j)}(t) - \frac{EJ}{\mu} z(0, l, t) \tag{19.2}$$

where

$$p_{(j)}(t) = \int_0^l p(x, t)\, v_{(j)}(x)\, \mathrm{d}x \tag{19.3}$$

is the transformation of the load, and $z(0, l, t)$ is function (6.12) dependent on the boundary conditions.

Eq. (19.2) is then solved by the Laplace-Carson transformation; for the initial conditions (6.16), (6.17) and (6.20) this operation gives

$$V^*(j, p) = \frac{1}{p^2 + \omega_{(j)}^2}\left[\frac{1}{\mu} P^*(p) - \frac{EJ}{\mu} Z^*(p) + p^2 G_1(j) + p\, G_2(j)\right] \tag{19.4}$$

where

$$P^*(p) = p \int_0^\infty p_{(j)}(t)\, e^{-pt}\, dt$$

is the Laplace-Carson transformation of function (19.3). Expressions $Z^*(p)$, $G_1(j)$ and $G_2(j)$ are given by Eqs. (6.22) and (6.17).

In view of (19.3), (27.5) and (27.18) the inverse transformations of Eq. (6.4) result in

$$v(x, t) = \sum_{j=1}^{\infty} \left\{\frac{v_{(j)}(x)}{V_j \omega_{(j)}} \int_0^t \left[\int_0^l p(x, \tau)\, v_{(j)}(x)\, dx - EJ\, z(0, l, \tau)\right] . \right.$$

$$. \sin \omega_{(j)}(t - \tau)\, d\tau + \frac{\mu}{V_j} G_1(j)\, v_{(j)}(x) \cos \omega_{(j)} t + \frac{\mu}{V_j \omega_{(j)}} .$$

$$\left. . \, G_2(j)\, v_{(j)}(x) \sin \omega_{(j)} t\right\} . \tag{19.5}$$

Solution (19.5) describes the resultant beam vibration with normal modes $v_{(j)}(x)$ under arbitrary boundary and initial conditions and any load $p(x, t)$.

19.1 *Motion of a concentrated force*

Consider a concentrated force of constant magnitude, P, moving along a beam, with the point of contact x_P describing a function of time

$$x_P = f(t) . \tag{19.6}$$

Then the load has the form

$$p(x, t) = \delta[x - f(t)]\, P . \tag{19.7}$$

If this load is substituted in (19.5), then in view of the property (1.7) of the Dirac function the first component on the right-hand side of the equation will turn out to be

$$v(x, t) = \sum_{j=1}^{\infty} \frac{P v_{(j)}(x)}{V_j \omega_{(j)}} \int_0^t v_{(j)}[f(\tau)] \sin \omega_{(j)}(t - \tau) \, d\tau \,. \qquad (19.8)$$

Eq. (19.8) together with the rest of Eq. (19.5), which describes the effect of the boundary and initial conditions, define the beam deflection at any motion of load P characterized by function $f(t)$. For a general motion of the contact point x_P the integral in (19.8) may be evaluated by numerical integration.

As a *special case* we shall suppose that the motion of the contact point is a quadratic function of time

$$f(t) = x_0 + ct + \frac{at^2}{2}, \qquad (19.9)$$

$$df(t)/dt = c + at, \quad d^2f(t)/dt^2 = a$$

where x_0 – is the point of application of force P at instant $t = 0$,
c – initial speed,
a – constant acceleration of motion.

Function (19.9) expresses a uniformly accelerated $(a > 0)$ or a uniformly decelerated $(a < 0)$ motion.

The integral in Eq. (19.8) will be computed for a simply supported beam of span l, for which it holds that $v_{(j)}(x) = \sin j\pi x/l$, $V_j = \mu l/2$, (1.11). With the aid of the notation

$$\xi_0 = \frac{j\pi x_0}{l}, \qquad \omega = \frac{j\pi c}{l}, \qquad \Omega^2 = \frac{j\pi |a|}{2l}, \qquad (19.10)$$

$$\xi_1 = \xi_0 - \omega_{(j)}t, \quad r_1 = \tfrac{1}{2}(\omega + \omega_{(j)}), \quad b_1 = \frac{r_1}{\Omega},$$

$$\xi_2 = \xi_0 + \omega_{(j)}t, \quad r_2 = \tfrac{1}{2}(\omega - \omega_{(j)}), \quad b_2 = \frac{r_2}{\Omega}, \qquad (19.11)$$

the function in the integrand of (19.8) may be rearranged as follows

$$v_{(j)}[f(\tau)] \sin \omega_{(j)}(t - \tau) = \sin j\pi f(\tau)/l \,.\, \sin \omega_{(j)}(t - \tau) =$$
$$= \tfrac{1}{2}[\cos(\Omega^2\tau^2 + 2r_1\tau + \xi_1) - \cos(\Omega^2\tau^2 + 2r_2\tau + \xi_2)] \,.$$

The integral in (19.8) is then composed of two integrals of the type $\int \cos(\Omega^2\tau^2 + 2r_i\tau + \xi_i)\,d\tau$ whose solution can be found in [100], formula 2.549.4. Through its use the expression may be computed and rearranged to give

$$v(x, t) = \sum_{j=1}^{\infty} \frac{P}{\mu l \omega_{(j)} \Omega} \left(\frac{\pi}{2}\right)^{1/2} \sin\frac{j\pi x}{l} \{\cos(\pm\xi_1 - b_1^2) \cdot$$
$$\cdot [C(\Omega t \pm b_1) - C(\pm b_1)] - \sin(\pm\xi_1 - b_1^2)[S(\Omega t \pm b_1) -$$
$$- S(\pm b_1)] - \cos(\pm\xi_2 - b_2^2)[C(\Omega t \pm b_2) - C(\pm b_2)] +$$
$$+ \sin(\pm\xi_2 - b_2^2)[S(\Omega t \pm b_2) - S(\pm b_2)]\} \tag{19.12}$$

where the upper signs apply to accelerated $(a > 0)$, the lower to decelerated $(a < 0)$ motion. Eq. (19.12) contains the Fresnel integrals defined as

$$S(x) = \left(\frac{2}{\pi}\right)^{1/2} \int_0^x \sin t^2\, dt\,, \quad C(x) = \left(\frac{2}{\pi}\right)^{1/2} \int_0^x \cos t^2\, dt \tag{19.13}$$

which are tabulated, for example, in [113].

We will describe the effect of a uniformly accelerated or a uniformly decelerated motion along the beam by the dimensionless parameter

$$B = \frac{al}{c^2}\,. \tag{19.14}$$

The next paragraphs will be devoted to several special cases of force motion along the beam:

Case 1. — A beam at rest is entered at point $x_0 = 0$ by force P moving with speed c. The motion is uniformly decelerated so that the force stops at the end of the beam, $x = l$ (Fig. 19.1a). The instant t_1 at which the force stops, and the deceleration a are obtained from the conditions

$$f(t_1) = ct_1 + at_1^2/2 = l\,; \quad \dot{f}(t_1) = c + at_1 = 0$$

whence

$$a = -\frac{c^2}{2l}\,, \quad t_1 = \frac{2l}{c}\,, \quad B = -\frac{1}{2}\,.$$

The time variation of the deflection at mid-span of the beam, $v(l/2, t)/v_0$, was computed from Eq. (19.12) for $j = 1$ and an initial dimensionless speed $\alpha = 0{\cdot}5$ [Eq. (1.18)] and is plotted in Fig. 19.2, curve *1*. There v_0 is the static deflection at point $x = l/2$ produced by force P [Eq. (1.21)].

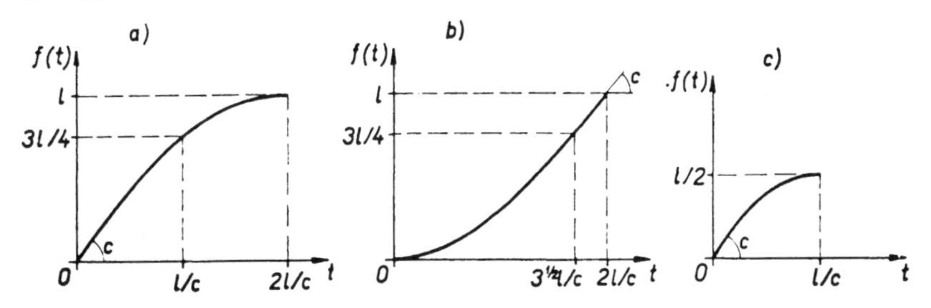

Fig. 19.1. Motion of a force along a beam: a) case 1, b) case 2, c) case 3.

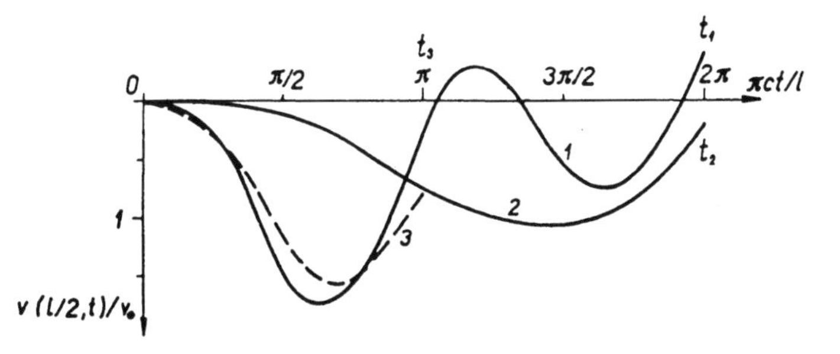

Fig. 19.2. Time variation of the deflection at the centre of a beam traversed by a force at variable speed: 1 — case 1, $B = -1/2$; 2 — case 2, $B = 1/2$; 3 — case 3, $B = -1$.

Case 2. – Force P starts to act on a beam at rest at point $x_0 = 0$. Its motion is uniformly accelerated in a way that it attains speed c at point $x = l$ (Fig. 19.1b). The instant t_2 at which the force arrives to the right-hand end of the beam, and the acceleration a are obtained from the conditions

$$f(t_2) = at_2^2/2 = l\,; \quad \dot{f}(t_2) = at_2 = c$$

whence

$$a = \frac{c^2}{2l}, \quad t_2 = \frac{2l}{c}, \quad B = \frac{1}{2}\,;$$

i.e., Case 2 is just the opposite of Case 1 (as to the sign of the acceleration).

The dependence of the deflection at mid-span of the beam $v(l/2, t)/v_0$ on time $\pi ct/l$ at a dimensionless speed of force P at the end of the beam

$\alpha = 0{\cdot}5$ is illustrated in Fig. 19.2, curve *2*. The values of the diagram were again obtained from (19.12) for $j = 1$.

Case 3. – A beam at rest is entered by force P moving with speed c. The motion is uniformly decelerated so that the force stops at mid-span of the beam, at point $x = l/2$ (Fig. 19.1c). In this case $x_0 = 0$. The instant t_3 at which the force reaches the centre of the beam, and the deceleration a are computed from the conditions

$$f(t_3) = ct_3 + at_3^2/2 = l/2\,; \quad \dot{f}(t_3) = c + at_3 = 0$$

whence

$$a = -\frac{c^2}{l}, \quad t_3 = \frac{l}{c}, \quad B = -1\,.$$

The time variation of the deflection at mid-span of the beam computed from (19.12) for $j = 1$ and initial dimensionless speed $\alpha = 0{\cdot}5$ is in Fig. 19.2, curve *3*.

As Fig. 19.2 reveals, a uniformly decelerated motion of a force along a beam (Cases 1 and 3) results in higher dynamic effects than a uniformly accelerated motion (Case 2). It is of interest, however, that Cases 1 and 3 actually differ very little even though the deceleration in the latter is double that in the former.

19.2 *Arrival of a continuous load*

In this section we shall examine the case of a continuous load p arriving on a beam with a variable speed (Fig. 3.3a)*). Similarly as in (3.22) the load will be

$$p(x, t) = p\{1 - H[x - f(t)]\} \tag{19.15}$$

where $H(x)$ is the Heaviside unit function (3.23).

Substituting (19.15) in the first term on the right-hand side of Eq. (19.5) gives

$$v(x, t) = \sum_{j=1}^{\infty} \frac{p v_{(j)}(x)}{V_j \omega_{(j)}} \int_0^t \sin \omega_{(j)}(t - \tau) \int_0^{f(\tau)} v_{(j)}(x)\, \mathrm{d}x\, \mathrm{d}\tau\,. \tag{19.16}$$

---

*) In this Chapter the letter q used for denoting a continuous load in Chap. 3 is replaced by letter p.

The integrals in Eq. (19.16) will again be evaluated for a simply supported beam and for a contact point moving in accordance with (19.9). Using notation (19.10) and (19.11) we get

$$\int_0^t \sin \omega_{(j)}(t - \tau) \int_0^{f(t)} v_{(j)}(x) \, \mathrm{d}x \, \mathrm{d}\tau = \frac{l}{j\pi} \int_0^t \sin \omega_{(j)}(t - \tau) \, .$$

$$. [1 - \cos j\pi f(\tau)/l] \, \mathrm{d}\tau = \frac{l}{j\pi} \int_0^t \{\sin \omega_{(j)}(t - \tau) - \tfrac{1}{2}[-\sin (\Omega^2\tau^2 +$$

$$+ 2r_1\tau + \xi_1) + \sin (\Omega^2\tau^2 + 2r_2\tau + \xi_2)]\} \, \mathrm{d}\tau \, .$$

In this expression the first integral is elementary, while the other two are obtained through the application of formula 2.549.3 in [100]. After some manipulation we get

$$v(x, t) = \sum_{j=1}^{\infty} \frac{2p \sin j\pi x/l}{j\pi\mu\omega_{(j)}^2} \left\{1 - \cos \omega_{(j)}t - \frac{\omega_{(j)}}{2\Omega} \left(\frac{\pi}{2}\right)^{1/2} \right. .$$

$$. \{\mp \cos (\pm\xi_1 - b_1^2) [\mathrm{S}(\Omega t \pm b_1) - \mathrm{S}(\pm b_1)] \mp \sin (\pm\xi_1 - b_1^2) \, .$$

$$. [\mathrm{C}(\Omega t \pm b_1) - \mathrm{C}(\pm b_1)] \pm \cos (\pm\xi_2 - b_2^2) [\mathrm{S}(\Omega t \pm b_2) -$$

$$\left. - \mathrm{S}(\pm b_2)] \pm \sin (\pm\xi_2 - b_2^2) [\mathrm{C}(\Omega t \pm b_2) - \mathrm{C}(\pm b_2)]\} \right\} \qquad (19.17)$$

where the upper and lower signs apply to accelerated ($a > 0$), and decelerated ($a < 0$) motions, respectively. $\mathrm{S}(x)$ and $\mathrm{C}(x)$ are again Fresnel integrals (19.13).

The arrival on a beam of a continuous load moving with variable speed was analyzed for two cases whose motion is identical with Cases 1 and 2 of Sect. 19.1.:

Case 1. – A continuous load p arrives with speed c on a beam at rest ($x_0 = 0$). Its motion is uniformly decelerated in a way that brings the front of the load to a stop at instant t_1 when it reaches the right-hand end of the beam, $x = l$ (Fig. 19.1a). t_1 and a are computed from the same conditions as in Case 1, Sect. 19.1, with the result

$$a = -\frac{c^2}{2l}, \quad t_1 = \frac{2l}{c}, \quad B = -\frac{1}{2} \, .$$

The variation of the deflection at mid-span of the beam, $v(l/2, t)/v_0$, is plotted in Fig. 19.3, curve *1*. It was computed from Eq. (19.17) for

$j = 1$ and initial dimensionless speed $\alpha = 0{\cdot}5$ [Eq. (1.18)]. v_0 is the static deflection of the centre of the beam span produced by the whole continuous load p [Eq. (3.19)].

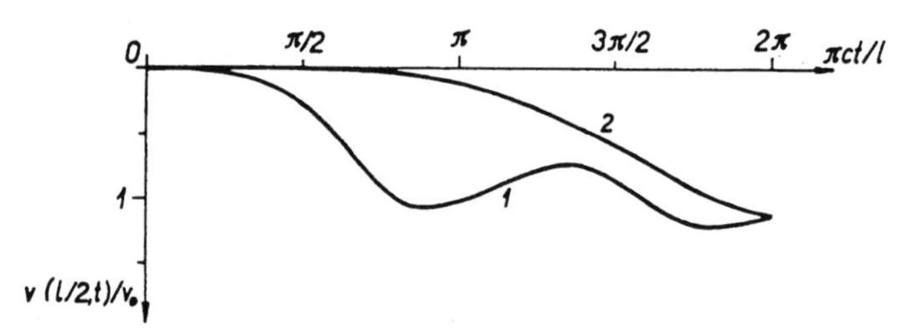

Fig. 19.3. Time variation of the deflection at the centre of a beam for a continuous load arriving at variable speed: 1 — case 1, $B = -1/2$; 2 — case 2, $B = 1/2$.

Case 2. — A continuous load p starts to act on a beam at rest at $x_0 = 0$. Its motion is uniformly accelerated in a way that imparts to the load speed c at instant t_2 when its front reaches the right-hand end of the beam (Fig. 19.1b). t_2 and a are computed from the same conditions as in Case 2, Sect. 19.1, and turn out to be

$$a = \frac{c^2}{2l}, \quad t_2 = \frac{2l}{c}, \quad B = \frac{1}{2}.$$

As to the sign of the acceleration, this case is just the opposite of Case 1.

The deflection $v(l/2, t)/v_0$ at point $x = l/2$ in dependence on time $\pi ct/l$ was again computed from (19.17) for $j = 1$; the dimensionless speed at instant t_2 is $\alpha = 0{\cdot}5$. The deflection is represented in Fig. 19.3, curve 2.

Consulting Fig. 19.3 we see that in these two cases, too, a uniformly decelerated motion results in higher dynamic effects than does a uniformly accelerated motion.

19.3 *The effect of inertial mass of a load moving at variable speed*

If we consider not only the force effects of an external load but its inertial effects as well, then according to d'Alembert's principle the load whose motion follows the law $x_p = f(t)$ is

$$p[x - f(t), t] - \mu_p[x - f(t)] \frac{d^2 v(x_p, t)}{dt^2} \qquad (19.18)$$

where $\mu_p(x)$ is the mass of the moving load $p(x, t)$. The acceleration $\mathrm{d}^2 v(x_p, t)/\mathrm{d}t^2$ of this mass is computed from the total differential of the second order of function $v(x, t)$ with respect to time t, with $x_p = f(t)$:

$$\frac{\mathrm{d}^2 v(x_p, t)}{\mathrm{d}t^2} = \frac{\partial^2 v(x, t)}{\partial t^2} + 2\,\frac{\partial^2 v(x, t)}{\partial x\, \partial t}\,\frac{\mathrm{d}f(t)}{\mathrm{d}t} + \frac{\partial^2 v(x, t)}{\mathrm{d}x^2}\left(\frac{\mathrm{d}f(t)}{\mathrm{d}t}\right)^2 + \\ + \frac{\partial v(x, t)}{\partial x}\,\frac{\mathrm{d}^2 f(t)}{\mathrm{d}t^2}\,. \tag{19.19}$$

For uniformly accelerated or decelerated motions according to (19.9), acceleration (19.19) is in the form

$$\frac{\mathrm{d}^2 v(x_p, t)}{\mathrm{d}t^2} = \frac{\partial^2 v(x, t)}{\partial t^2} + 2(c + at)\,\frac{\partial^2 v(x, t)}{\partial x\, \partial t} + (c + at)^2\,. \\ .\,\frac{\partial^2 (x, t)}{\partial x^2} + a\,\frac{\partial v(x, t)}{\partial x}\,. \tag{19.20}$$

Consider now the case of a continuous load, p, with mass μ_p moving along a beam (the motion at constant speed was solved in Sect. 3.1). The problem is described by the equation

$$EJ\,\frac{\partial^4 v(x, t)}{\partial x^4} + \mu\,\frac{\partial^2 v(x, t)}{\partial t^2} = p - \mu_p\left[\frac{\partial^2 v(x, t)}{\partial t^2} + 2(c + at)\,. \right. \\ \left. .\,\frac{\partial^2 v(x, t)}{\partial x\, \partial t} + (c + at)^2\,\frac{\partial^2 v(x, t)}{\partial x^2} + a\,\frac{\partial v(x, t)}{\partial x}\right]. \tag{19.21}$$

Assume that the load had moved at uniform speed c for an infinitely long time before the change of motion speed has taken place. As a result a steady-state quasi-static deflection was produced in the beam so that in view of (3.18) the initial conditions of the present problem are

$$v(x, t)\big|_{t=0} = v_0 \sum_{j=1,3,5,\ldots}^{\infty} \frac{1}{j^5(1 - \alpha^2 \varkappa/j^2)} \sin\frac{j\pi x}{l}\,, \quad \left.\frac{\partial v(x, t)}{\partial t}\right|_{t=0} = 0\,. \tag{19.22}$$

We shall solve this problem for a simply supported beam with the boundary conditions defined by (1.2).

19.3.1 *Galerkin's method*

Since the problem is very complicated we will solve it by approximate methods. When applying Galerkin's method, described in paragraph 12.1.2, we assume the solution to be in the form of (12.10) and choose $j = n = 1$, $\varphi_1(x) = \sin \pi x/l$

$$v(x, t) = q(t) \sin \frac{\pi x}{l} . \tag{19.23}$$

By (12.9) and (19.21) operator L has the form

$$L = EJv^{\mathrm{IV}} + \mu_p(c + at)^2 v'' + \mu_p a v' + (\mu + \mu_p) \ddot{v} + 2\mu_p(c + at) \dot{v}' .$$

Substituting this expression together with (19.23) in the orthogonality condition (12.11) we get

$$\int_0^l \left[\frac{\pi^4}{l^4} EJ \sin \frac{\pi x}{l} q(t) - \mu_p(c + at)^2 \frac{\pi^2}{l^2} \sin \frac{\pi x}{l} q(t) + \right.$$

$$+ \mu_p a \frac{\pi}{l} \cos \frac{\pi x}{l} q(t) + (\mu + \mu_p) \sin \frac{\pi x}{l} \ddot{q}(t) + 2\mu_p \, .$$

$$\left. . (c + at) \frac{\pi}{l} \cos \frac{\pi x}{l} \dot{q}(t) - p \right] \sin \frac{\pi x}{l} \, \mathrm{d}x = 0 . \tag{19.24}$$

After evaluating the simple integrals

$$\int_0^l \sin^2 \pi x/l \, \mathrm{d}x = l/2 , \qquad \int_0^l \sin \pi x/l \, . \cos \pi x/l \, \mathrm{d}x = 0 ,$$

$$\int_0^l \sin \pi x/l \, \mathrm{d}x = 2l/\pi ,$$

we obtain from (19.24) the ordinary linear differential equation of the second order with variable coefficients for the unknown function $q(t)$

$$\frac{\mathrm{d}^2 q(t)}{\mathrm{d}t^2} + \bar{\omega}_{(1)}^2 \left[1 - \alpha^2 \varkappa \left(1 + \frac{at}{c}\right)^2\right] q(t) = \frac{4p}{\pi \, \mu(1 + \varkappa)} . \tag{19.25}$$

In the above and hereinafter we use notation (1.18) and (3.9) to (3.13), with q again replaced by p. This refers particularly to α, the speed parameter according to (1.18), and to $\varkappa$, the weight parameter of load according to (3.9).

The homogeneous differential equation appertaining to (19.25) is of the type

$$\ddot{q}(t) + (a_1 t^2 + b_1 t + c_1)\, q(t) = 0 \tag{19.26}$$

where

$$a_1 = -\bar{\omega}_{(1)}^2 \alpha^2 \varkappa a^2/c^2\,, \quad b_1 = -2\bar{\omega}_{(1)}^2 \alpha^2 \varkappa a/c\,, \quad c_1 = \bar{\bar{\omega}}_{(1)}^2\,.$$

Eq. (19.26) may be solved exactly by multiple substitution – see [120], Eqs. 2.55, 2.54 and 2.273. Substituting first

$$q(t) = y_2(t)\, e^{st^2} \quad \text{where} \quad s^2 = -a_1/4$$

gives an equation of the type

$$\ddot{y}_2 + a_2 t \dot{y}_2 + (c_2 t + d_2)\, y_2 = 0$$

where

$$a_2 = 4s\,, \quad c_2 = b_1\,, \quad d_2 = c_1 + 2s\,.$$

Substitution in the above of

$$y_2(t) = y_3(t_3)\, e^{-c_2 t/a_2} \quad \text{where} \quad t_3 = (|a_2|)^{1/2}\,(t - 2c_2/a_2^2)$$

results in

$$\ddot{y}_3 \pm t_3 \dot{y}_3 \pm b_3 y_3 = 0 \quad \text{where} \quad b_3 = a_2^{-3}(c_2^2 + a_2^2 d_2)\,.$$

The upper sign applies at $a_2 > 0$, the lower at $a_2 < 0$. To simplify let us merely consider the case of $a_2 > 0$, because for $a_1 < 0$, $s^2 > 0$.

By the substitution

$$y_3(t_3) = t_3^{-1/2} e^{-t^2{}_3/4}\, y_4(t_4)\,, \quad t_4 = t_3^2/2$$

the last differential equation may be changed to Whittaker's differential equation

$$4t_4^2 \ddot{y}_4 = (t_4^2 - 4kt_4 + 4m^2 - 1)\, y_4 \tag{19.27}$$

where

$$k = \frac{b_3}{2} - \frac{1}{4} = \frac{1}{4\alpha[\varkappa(1 + \varkappa)]^{1/2}\, b}\,, \quad m = \frac{1}{4}\,.$$

Here we use the dimensionless parameter of uniformly accelerated or decelerated motions

$$b = \frac{a}{\omega_{(1)}c} = \frac{\alpha B}{\pi}. \tag{19.28}$$

The general solution of the homogeneous equation (19.27) is

$$y_4(t_4) = A_1\, M_{k,m}(t_4) + A_2\, M_{k,-m}(t_4) \tag{19.29}$$

where A_1, A_2 are the integration constants dependent on the initial conditions,

$$M_{k,m}(x) = x^{1/2+m}e^{-1/2.x}\ {}_1F_1(1/2 + m - k,\ 2m + 1,\ x)$$

is the so-called Whittaker function, and

$${}_1F_1(a, b, x) = 1 + \sum_{n=1}^{\infty} \frac{a(a + 1)\ldots(a + n - 1)\, x^n}{b(b + 1)\ldots(b + n - 1)\, n\,!}$$

is Pochhammer's series, also termed the confluent hypergeometric function.

Hence the solution of the homogeneous equation (19.25) is

$$q(t) = t_3^{-1/2}e^{st^2 - c_2t/a_2 - t_3^2/4}\ y_4(t_4) = A_1\varphi_1 + A_2\varphi_2 \tag{19.30}$$

where φ_1 and φ_2 denote the general linearly independent solutions of the homogeneous equation (19.25). The solution of the nonhomogeneous equation (19.25) then is

$$q(t) = \left[A_1 - \frac{4p}{\pi\,\mu(1 + \varkappa)}\int\frac{\varphi_2}{W}\,dt\right]\varphi_1 + \left[A_2 + \frac{4p}{\pi\,\mu(1 + \varkappa)}\int\frac{\varphi_1}{W}\,dt\right]\varphi_2 \tag{19.31}$$

where $W = \varphi_1\varphi_2' - \varphi_1'\varphi_2$.

19.3.2 *The perturbation method*

Though exact, the solution (19.31) of Eq. (19.25) is unsuitable for practical calculations. We shall therefore solve Eq. (19.25) approximately, by the perturbation method described in paragraph 7.3.1.

Note that Eq. (19.25) is easy to solve for $a = 0$. For $j = 1$ and the initial conditions (19.22) it is readily found [see also (3.18)] that

$$q(t) = q_0 = \frac{v_0}{1 - \alpha^2 \varkappa} = \frac{4p}{\pi\mu\, \omega_{(1)}^2 (1 - \alpha^2 \varkappa)} . \tag{19.32}$$

If the motion of the continuous load is accelerated (or decelerated) with a very low acceleration, $a \to 0$, the approximate solution of Eq. (19.25) may be written in the form

$$q(t) = q_0 + a\, q_1(t) + a^2\, q_2(t) + \dots \tag{19.33}$$

and the initial conditions in the form

$$\begin{aligned} q(0) &= q_0 + a\, q_1(0) + a^2\, q_2(0) + \dots = q_0 , \\ \dot{q}(0) &= a\, \dot{q}_1(0) + a^2\, \dot{q}_2(0) + \dots \qquad = 0 . \end{aligned} \tag{19.34}$$

Substituting (19.33) in (19.25) and comparing the coefficients at equal powers of a, result in the following system of differential equations:

$$\begin{aligned} \bar{\bar{\omega}}_{(1)}^2 q_0 &= \frac{4p}{\pi\mu(1 + \varkappa)} , \\ \ddot{q}_1 + \bar{\bar{\omega}}_{(1)}^2 q_1 &= 2\bar{\omega}_{(1)}^2 \alpha^2 \varkappa q_0 t/c , \\ \ddot{q}_2 + \bar{\bar{\omega}}_{(1)}^2 q_2 &= \bar{\omega}_{(1)}^2 \alpha \varkappa q_0 t^2/c^2 + 2\bar{\omega}_{(1)}^2 \alpha^2 \varkappa q_1 t/c , \\ &\dots\dots\dots\dots\dots\dots\dots\dots\dots\dots \end{aligned} \tag{19.35}$$

Confining our considerations to just two terms of series (19.33) we get from the first of Eqs. (19.35) the expression (19.32) and from the second – in view of the initial condition (19.34) –

$$q_1(t) = q_0 \frac{2\alpha^2 \varkappa}{c(1 - \alpha^2 \varkappa)} \left(t - \frac{1}{\bar{\bar{\omega}}_{(1)}} \sin \omega_{(1)} t \right) .$$

Hence the approximate solution of Eq. (19.25) is

$$q(t) = q_0 \left\{ 1 + \frac{2\alpha^2 \varkappa b}{1 - \alpha^2 \varkappa} \left[\omega_{(1)} t - \left(\frac{1 + \varkappa}{1 - \alpha^2 \varkappa} \right)^{1/2} \sin \bar{\bar{\omega}}_{(1)} t \right] \right\} . \tag{19.36}$$

In the absence of damping, it will clearly describe our problem only in a short time period after the change of speed, i.e. in the first period.

Therefore we shall compute the maximum value of (19.36) for $\bar{\bar{\omega}}_{(1)}t_1 = \pi/2$. Then

$$v(l/2, t_1)/v_0 = q(t_1)/q_0 = 1 + \frac{2\alpha^2 \varkappa b}{1 - \alpha^2 \varkappa}\left(\frac{1 + \varkappa}{1 - \alpha^2 \varkappa}\right)^{1/2}\left(\frac{\pi}{2} - 1\right). \quad (19.37)$$

Fig. 19.4 shows expression (19.37) computed for several values of parameters b and $\varkappa$, as a function of speed α. According to it, during the first instants, acceleration causes the deflection to increase — deceleration, to decrease.

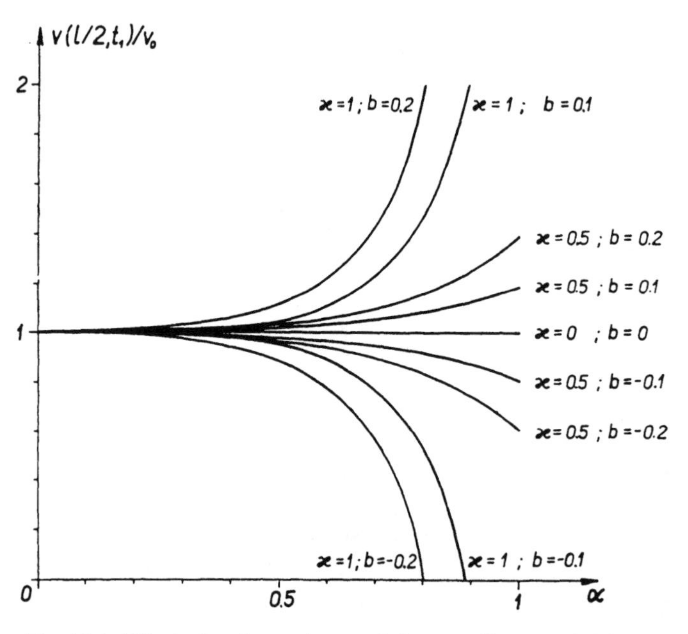

Fig. 19.4. Effect of a change in speed, b, according to (19.28), in motion of a continuous load along a beam.

19.3.3 *Series expansion*

In the short time period following the change of speed of a continuous load motion, the solution of Eq. (19.25) can also be expanded in a power series in t

$$q(t) = q_0(a_0 + a_1 t + a_2 t^2 + a_3 t^3 + \ldots) \quad (19.38)$$

where q_0 is the expression (19.32) and the coefficients a_n, $n = 1, 2, \ldots$ are obtained in part from the initial conditions

$$q(0) = q_0 \, , \quad \dot{q}(0) = 0 \, ,$$

in part from a comparison of the coefficients at equal powers of t following the substitution of (19.38) in (19.25). From the initial conditions $a_0 = 1$, $a_1 = 0$, and from Eq. (19.25)

$$a_2 = 0 \, , \quad a_3 = \frac{\bar{\omega}_{(1)}^2 \alpha^2 \varkappa a}{3c} \, , \quad a_4 = \frac{\bar{\omega}_{(1)}^2 \alpha^2 \varkappa a^2}{12c^2} \, ,$$

$$a_5 = - \frac{\bar{\bar{\omega}}_{(1)}^2 \bar{\omega}_{(1)}^2 \alpha^2 \varkappa a}{60c} \, , \ldots .$$

Hence the solution of (19.25) is

$$q(t) = q_0 \left[1 + \frac{\alpha^2 \varkappa b}{3(1 + \varkappa)} \, \omega_{(1)}^3 t^3 + \frac{\alpha^2 \varkappa b^2}{12(1 + \varkappa)} \, \omega_{(1)}^4 t^4 - \right.$$
$$\left. - \frac{\alpha^2 \varkappa b(1 - \alpha^2 \varkappa)}{60(1 + \varkappa)^2} \, \omega_{(1)}^5 t^5 + \ldots \right] . \qquad (19.39)$$

It is clear to see from Eq. (19.39) that during the first instants following the change of speed of a continuous load motion, acceleration ($b > 0$) causes the beam deflection to increase, deceleration ($b < 0$) to decrease.

This result agrees with the conclusion drawn in the preceding paragraph 19.3.2, and applies to the motion of a continuous load even when its mass is being considered – but only during the very short time period following the change of speed. The case examined in Sect. 19.2 – a continuous load, without consideration given to its mass, entering a beam – is wholly different and that is why one cannot compare the above conclusions with the diagram in Fig. 19.3 which represents the time variation of the whole phenomenon.

19.4 *Application of the theory*

The theory outlined in this chapter serves well in calculations relating to structures traversed by vehicles moving at variable speed. As typical examples of the problem let us name the taking-off and landing of aircrafts on runway, acceleration and braking of automobiles on roadways and highway bridges, braking and acceleration forces in the calculation of rails and railway bridges, braking and acceleration effects of cranes in assessments of the quality of crane runways and factory buildings, etc. So far such forces and effects have been described but approximately, by a portion of the static axle pressures acting horizontally in the direction of the vehicle motion. The chapter has been written in an effort to gain clearer insight into stresses in civil and mechanical engineering structures under actual operating conditions.

19.5 *Additional bibliography*

[62, 127, 129, 146, 151, 192, 194, 198, 239].

20

Beam subjected to an axial force and a moving load

If in addition to the transverse (relative to its axis) load $p(x, t)$ a bar is also subjected to a static, time invariable force N applied at the bar's ends along the axis (Fig. 20.1), the fundamental differential equation takes on the form

$$EJ \frac{\partial^4 v(x, t)}{\partial x^4} - N \frac{\partial^2 v(x, t)}{\partial x^2} + \mu \frac{\partial^2 v(x, t)}{\partial t^2} = p(x, t) . \tag{20.1}$$

We shall take tension for the positive, compression for the negative force N.

The *free vibration* of the bar is established from the homogeneous equation (20.1) assuming that

$$v(x, t) = v_{(j)}(x) \sin \omega_{(j)} t \tag{20.2}$$

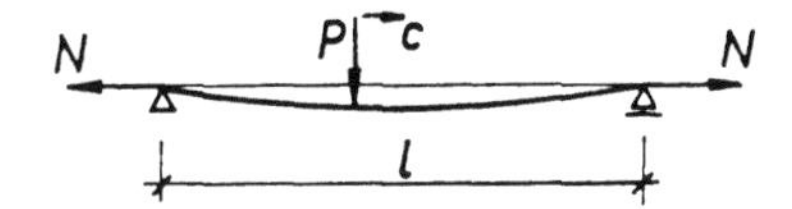

Fig. 20.1. Beam loaded with a static axial force.

where $v_{(j)}(x)$ and $\omega_{(j)}$ are again the normal modes and the natural frequencies of a beam with a static axial force. Substituting (20.2) in the homogeneous equation (20.1) gives the homogeneous equation

$$EJ \frac{d^4 v_{(j)}(x)}{dx^4} - N \frac{d^2 v_{(j)}(x)}{dx^2} - \mu \omega_{(j)}^2 v_{(j)}(x) = 0 . \tag{20.3}$$

The solution of (20.3) will be assumed to be in the form of $e^{\lambda_{(j)} x/l}$, and $\lambda_{(j)}$ computed from the characteristic equation of (20.3)

$$\lambda_{(j)} = \pm i\lambda_1 \; ; \; \pm \lambda_2 \tag{20.4}$$

where

$$\lambda_1 = l\left\{-\frac{N}{2EJ} + \left[\left(\frac{N}{2EJ}\right)^2 + \frac{\mu\omega_{(j)}^2}{EJ}\right]^{1/2}\right\}^{1/2},$$

$$\lambda_2 = l\left\{\frac{N}{2EJ} + \left[\left(\frac{N}{2EJ}\right)^2 + \frac{\mu\omega_{(j)}^2}{EJ}\right]^{1/2}\right\}^{1/2}.$$

With the above notation the normal modes of a bar with a static axial force may be expressed as follows

$$v_{(j)}(x) = \sin\frac{\lambda_1 x}{l} + A_j\cos\frac{\lambda_1 x}{l} + B_j\sinh\frac{\lambda_2 x}{l} + C_j\cosh\frac{\lambda_2 x}{l}. \quad (20.5)$$

The integration constants A_j, B_j and C_j as well as the quantity $\lambda_{(j)}$ are computed from the boundary conditions of the bar, and the natural frequency obtained from the expression

$$\omega_{(j)}^2 = \frac{\lambda_{(j)}^4}{l^4}\frac{EJ}{\mu} - \frac{\lambda_{(j)}^2}{l^2}\frac{N}{\mu}. \quad (20.6)$$

The *forced vibration* is again solved by the generalized method of finite integral transformations according to Sect. 6.1. First, double application of the method of integration by parts will give

$$\int_0^l \frac{\partial^2 v(x,t)}{\partial x^2} v_{(j)}(x)\,\mathrm{d}x = z_N(0,l,t) + \int_0^l v(x,t)\frac{\mathrm{d}^2 v_{(j)}(x)}{\mathrm{d}x^2}\,\mathrm{d}x \quad (20.7)$$

where

$$z_N(0,l,t) = \left[\frac{\partial v(x,t)}{\partial x} v_{(j)}(x) - v(x,t)\frac{\mathrm{d}v_{(j)}(x)}{\mathrm{d}x}\right]_0^l$$

is a function dependent on the boundary conditions and on time, analogous to expression (6.12).

Eq. (20.1) can then be transformed according to (6.1). In view of (6.11), (20.7) and (19.3) this gives

$$EJ\,z(0,l,t) + EJ\int_0^l v(x,t)\frac{\mathrm{d}^4 v_{(j)}(x)}{\mathrm{d}x^4}\,\mathrm{d}x - N\,z_N(0,l,t) - N\,.$$

$$.\int_0^l v(x,t)\frac{\mathrm{d}^2 v_{(j)}(x)}{\mathrm{d}x^2}\,\mathrm{d}x + \mu\ddot{V}(j,t) = p_{(j)}(t)\,. \quad (20.8)$$

Multiplication of Eq. (20.3) by $v(x, t)$ and integration with respect to x from 0 to l will give

$$EJ \int_0^l v(x, t) \frac{\mathrm{d}^4 v_{(j)}(x)}{\mathrm{d}x^4} \mathrm{d}x - N \int_0^l v(x, t) \frac{\mathrm{d}^2 v_{(j)}(x)}{\mathrm{d}x^2} \mathrm{d}x = \mu \omega_{(j)}^2 V(j, t) \quad (20.9)$$

and this when substituted in (20.8) will, after some manipulation, lead to

$$\ddot{V}(j, t) + \omega_{(j)}^2 V(j, t) = \frac{1}{\mu} \left[p_{(j)}(t) - EJ\, z(0, l, t) + N\, z_N(0, l, t) \right] . \quad (20.10)$$

This equation can already be solved by the Laplace-Carson transformation. For the initial conditions (6.16), (6.17) and (6.20) the operation will result in

$$V^*(j, p) = \frac{1}{p^2 + \omega_{(j)}^2} \left[\frac{1}{\mu} P^*(p) - \frac{EJ}{\mu} Z^*(p) + \frac{N}{\mu} Z_N^*(p) + \right.$$
$$\left. + p^2 G_1(j) + p\, G_2(j) \right] \quad (20.11)$$

where

$$Z_N^*(p) = p \int_0^\infty z_N(0, l, t)\, \mathrm{e}^{-pt}\, \mathrm{d}t .$$

Eq. (20.11) resembles (19.4), the meaning of the symbols being the same in both.

Following the inverse transformations, (20.11) gives the deflection in the form

$$v(x, t) = \sum_{j=1}^{\infty} \frac{\mu v_{(j)}(x)}{V_j} \left\{ \frac{1}{\mu \omega_{(j)}} \int_0^t \left[\int_0^l p(x, \tau)\, v_{(j)}(x)\, \mathrm{d}x - EJ\, z(0, l, \tau) + \right. \right.$$
$$\left. + N\, z_N(0, l, \tau) \right] \sin \omega_{(j)}(t - \tau)\, \mathrm{d}\tau + G_1(j) \cos \omega_{(j)} t +$$
$$\left. + \frac{G_{(2)}(j)}{\omega_{(j)}} \sin \omega_{(j)} t \right\} \quad (20.12)$$

where V_j is defined by Eq. (6.6).

20.1 *Beam subjected to a static axial force under a moving load*

To illustrate we shall examine a simply supported beam with a constant axial force subjected to several types of load moving at uniform speed c. On substituting (20.5) in the boundary conditions (1.2) we get $\lambda_1 = j\pi$, $A_j = B_j = C_j = 0$, so that the normal modes and the natural frequencies, respectively, are

$$v_{(j)}(x) = \sin \frac{j\pi x}{l}, \quad j = 1, 2, 3, \ldots$$

$$\omega_{(j)}^2 = \frac{j^4\pi^4}{l^4} \frac{EJ}{\mu} + \frac{j^2\pi^2}{l^2} \frac{N}{\mu}. \tag{20.13}$$

The first term on the right-hand side of (20.13) is the natural frequency of a simple beam without an axial force (1.11). The natural frequency is increased by tension ($N > 0$), reduced by compression ($N < 0$).

For a simply supported beam, functions $z(0, l, t)$ as well as $z_N(0, l, t)$ are zero, because the boundary conditions (1.2) must be satisfied by both $v(x, t)$ and $v_{(j)}(x)$.

In what follows we shall use the notation according to (6.26) and (6.38)

$$\omega = \frac{j\pi c}{l}, \quad \alpha_j = \frac{\omega}{\omega_{(j)}}. \tag{20.14}$$

20.1.1 *Moving concentrated force*

If a beam is traversed by a concentrated force, P, at speed c (Fig. 20.1), the load is

$$p(x, t) = \delta(x - ct)\, P. \tag{20.15}$$

For this load substituted in (20.12) and zero initial conditions, either direct integration or the application of the Laplace-Carson transformations according to (27.32) and (27.35) will give

$$v(x, t) = \sum_{j=1, j\neq n}^{\infty} \frac{2P \sin j\pi x/l}{\mu l \omega_{(j)}^2 (1 - \alpha_j^2)} (\sin \omega t - \alpha_j \sin \omega_{(j)} t) + \\ + \frac{P \sin n\pi x/l}{\mu l \omega_{(n)}^2} (\sin \omega_{(n)} t - \omega_{(n)} t \cos \omega_{(n)} t). \tag{20.16}$$

The first term on the right-hand side of (20.16) applies at $\omega_{(j)} \neq \omega$, $\alpha_j \neq 1$, the second for $j = n$ for which $\omega_{(n)} = \omega$.

It is evident from (20.16) that the deflection is the same as that in Case 1.3.2, Sect. 1.3, Chap. 1. The effect of axial force N does not come into play except in the natural frequency $\omega_{(j)}$ [Eq. (20.13)].

Eq. (20.16) makes it also possible to establish for $\alpha_j = 0$ the *static* beam deflection produced by load P applied at point $s = ct$

$$v(x, s) = \sum_{j=1}^{\infty} \frac{2P \sin j\pi s/l}{\mu l \omega_{(j)}^2} \sin \frac{j\pi x}{l} ; \qquad (20.17)$$

for $s = l/2$, $x = l/2$ [cf. (1.21)]

$$v_0 = v(l/2, l/2) = \sum_{j=1,3,5,\dots}^{\infty} \frac{2P}{\mu l \omega_{(j)}^2} . \qquad (20.18)$$

20.1.2 *Arrival of a continuous load on a beam*

The equation of a continuous load p arriving on a beam as shown in Fig. 3.3a is

$$p(x, t) = p[1 - H(x - ct)] . \qquad (20.19)$$

For this load substituted in (20.12) and zero initial conditions either direct integration or the use of the Laplace-Carson transformations according to formulae (27.73), (27.17) and (27.31), (27.34) will result in

$$v(x, t) = \sum_{j=1, j\neq n}^{\infty} \frac{2p \sin j\pi x/l}{j\pi\mu\omega_{(j)}^2(1 - \alpha_j^2)} [1 - \cos \omega t - \alpha_j^2(1 - \cos \omega_j t)] + $$
$$+ \frac{2p \sin n\pi x/l}{n\pi\mu\omega_{(n)}^2} (1 - \cos \omega_{(n)} t - \tfrac{1}{2}\omega_{(n)} t \sin \omega_{(n)} t) . \qquad (20.20)$$

Like in the preceding case, the first term applies at $\omega_{(j)} \neq \omega$, $\alpha_j \neq 1$, the second for $j = n$ for which $\omega_{(n)} = \omega$.

The result, (20.20), is again the same as that obtained in Sect. 3.2, Chap. 3, and differs from (3.26) merely by the omission of damping and by the natural frequency which now is (20.13).

For $\alpha_j = 0$, Eq. (20.20) gives the *static* beam deflection for the case in which the front of the load is at point $s = ct$

$$v(x, s) = \sum_{j=1}^{\infty} \frac{2p(1 - \cos j\pi s/l)}{j\pi\mu\omega_{(j)}^2} \sin \frac{j\pi x}{l} . \tag{20.21}$$

For $s = l$, $x = l/2$ [cf. (3.19)]

$$v_0 = v(l/2, l) = \sum_{j=1,3,5\ldots}^{\infty} \frac{4p}{j\pi\mu\omega_{(j)}^2} . \tag{20.22}$$

20.1.3 *The effect of mass of a moving continuous load*

The calculation of the effect of mass μ_p of a moving continuous load p starts out from Eqs. (3.2), (3.3) and (3.7)

$$EJ \frac{\partial^4 v(x, t)}{\partial x^4} - N \frac{\partial^2 v(x, t)}{\partial x^2} + \mu \frac{\partial^2 v(x, t)}{\partial t^2} + 2\mu\omega_b \frac{\partial v(x, t)}{\partial t} =$$

$$= p - \mu_p \left[\frac{\partial^2 v(x, t)}{\partial t^2} + 2c \frac{\partial^2 v(x, t)}{\partial x \, \partial t} + c^2 \frac{\partial^2 v(x, t)}{\partial x^2} \right] . \tag{20.23}$$

We will solve Eq. (20.23) by Galerkin's method. Having regard to the boundary conditions (1.2) we shall assume the solution to be in the form

$$v_n(x, t) = \sum_{j=1}^{n} q_j(t) \sin \frac{j\pi x}{l} . \tag{20.24}$$

Substitution of (20.24) in (12.11) will give

$$\int_0^l \left[\sum_{j=1}^{n} \frac{j^4\pi^4}{l^4} EJ \, q_j(t) \sin \frac{j\pi x}{l} + (N - \mu_p c^2) \sum_{j=1}^{n} \frac{j^2\pi^2}{l^2} q_j(t) \, . \right.$$

$$. \sin \frac{j\pi x}{l} + (\mu + \mu_p) \sum_{j=1}^{n} \ddot{q}_j(t) \sin \frac{j\pi x}{l} + 2\mu\omega_b \sum_{j=1}^{n} \dot{q}_j(t) \sin \frac{j\pi x}{l} +$$

$$\left. + 2\mu_p c \sum_{j=1}^{n} \frac{j\pi}{l} \dot{q}_j(t) \cos \frac{j\pi x}{l} - p \right] \sin \frac{k\pi x}{l} \, dx = 0 \, , \quad k = 1, 2, \ldots, n \, . \tag{20.25}$$

In the calculation that follows we shall have need of the integrals:

$$\int_0^l \sin\frac{j\pi x}{l}\sin\frac{k\pi x}{l}\,\mathrm{d}x = \begin{cases} 0 & \text{for} \quad j \neq k \\ l/2 & \text{for} \quad j = k \end{cases}$$

$$\int_0^l \sin\frac{k\pi x}{l}\cos\frac{j\pi x}{l}\,\mathrm{d}x = \begin{cases} 0 & \text{for} \quad j = k \\ 0 & \text{for even } j - k \\ \dfrac{2kl}{\pi(k^2 - j^2)} & \text{for odd } j - k \end{cases}$$

$$\int_0^l \sin\frac{k\pi x}{l}\,\mathrm{d}x = \frac{l}{k\pi}(1 - \cos k\pi)\,. \tag{20.26}$$

With the aid of these integrals, Eq. (20.25) may be rearranged to the system of ordinary differential equations

$$\ddot{q}_k(t) + 2\bar{\omega}_b\,\dot{q}_k(t) + \frac{8\mu_p c}{\bar{\mu}l}\sum_{|j-k|=1,3,5\ldots}^{j=n}\frac{jk}{k^2 - j^2}\,\dot{q}_j(t) + \bar{\omega}_{(k)}^2\,q_k(t) =$$

$$= \frac{2p}{k\pi\bar{\mu}}(1 - \cos k\pi)\,, \quad k = 1, 2, \ldots, n \tag{20.27}$$

where similarly as in (3.9) to (3.13) it is

$$\bar{\mu} = \mu + \mu_p\,, \quad \bar{\omega}_b = \omega_b\mu/\bar{\mu}\,,$$

$$\bar{\omega}_{(j)}^2 = \frac{j^4\pi^4}{l^4}\frac{EJ}{\bar{\mu}} + \frac{j^2\pi^2}{l^2}\frac{N - \mu_p c^2}{\bar{\mu}}\,. \tag{20.28}$$

Eq. (20.27) is easy to solve provided we neglect the effect of the third term on its left-hand side. If we do so, the system degenerates into n separate equations of a wholly identical form. This corresponds to the case of a beam simultaneously traversed by load $p/2$ at speed $+c$ and by load $p/2$ at speed $-c$, for then – as has been shown in Sect. 3.1 – the last-but-one term on the right-hand side of (20.23) vanishes.

In view of (27.20), in such a case the solution of (20.27) for zero initial conditions (1.3) will be

$$q_k(t) = \frac{2p(1 - \cos k\pi)}{k\pi\bar{\mu}\bar{\omega}_{(k)}^2}\left[1 - \mathrm{e}^{-\bar{\omega}_b t}\left(\cos\bar{\omega}'_{(k)}t + \frac{\bar{\omega}_b}{\bar{\omega}'_{(k)}}\sin\bar{\omega}'_{(k)}t\right)\right] \tag{20.29}$$

where

$$\bar{\omega}'^2_{(j)} = \bar{\omega}^2_{(j)} - \bar{\omega}^2_b .$$

After a sufficiently long action of the moving continuous load, the effect of the last two terms in (20.29) will damp out, so that the steady-state quasi-stationary beam deflection [Eq. (20.24)] will be

$$v(x, t)|_{t\to\infty} = \sum_{j=1}^{n} \frac{2p(1 - \cos j\pi)}{j\pi\bar{\mu}\bar{\omega}^2_{(j)}} \sin \frac{j\pi x}{l} . \tag{20.30}$$

Except for the different value of the frequency (20.28), the above is in complete harmony with Eq. (3.18).

On the consideration that $j = k = n = 1$, (20.30) is also an approximate solution of the case of a beam traversed by load p from right to left at speed c, because then the third term on the left-hand side of Eq. (20.27) is zero.

For $n = 2$, (20.27) would turn into the system of two differential equations

$$\ddot{q}_1(t) + 2\bar{\omega}_b\,\dot{q}_1(t) - \frac{16\mu_p c}{3\bar{\mu}l}\,\dot{q}_2(t) + \bar{\omega}^2_{(1)}\,q_1(t) = \frac{4p}{\pi\bar{\mu}},$$

$$\ddot{q}_2(t) + 2\bar{\omega}_b\,\dot{q}_2(t) + \frac{16\mu_p c}{3\bar{\mu}l}\,\dot{q}_1(t) + \bar{\omega}^2_{(2)}\,q_2(t) = 0 . \tag{20.31}$$

It can readily be shown, e.g. by the limit for $p \to 0$ of the Laplace-Carson transformation of system (20.31) that for $t \to \infty$ the asymptotic solution of the system is [see (27.14)]

$$q_1(t)|_{t\to\infty} = \frac{4p}{\pi\bar{\mu}\bar{\omega}^2_{(1)}}, \quad q_2(t)|_{t\to\infty} = 0 \tag{20.32}$$

which is obviously analogous to expressions (3.18) or (20.30) for $j = 1, 2$.

As Eq. (20.30) suggests, at low frequencies $\bar{\omega}_{(j)}$ the beam deflection can attain high values. From the condition $\bar{\omega}^2_{(j)} = 0$ in (20.28) we may even compute the critical speed c_{cr} at which the beam deflections will assume infinite values

$$c^2_{cr} = \left(\frac{j^4\pi^4}{l^4}\frac{EJ}{\mu} + \frac{j^2\pi^2}{l^2}\frac{N}{\mu}\right)\frac{l^2}{j^2\pi^2}\frac{\mu}{\mu_p} = \frac{l^2}{j^2\pi^2\varkappa}\,\omega^2_{(j)} \tag{20.33}$$

where by (3.9) $\varkappa = \mu_p/\mu$, and $\omega^2_{(j)}$ is defined by (20.13).

Our case may also be thought of as that of a beam subjected to the critical Euler's force on the buckling limit

$$S_{cr} = \mu_p c_{cr}^2 - N = \frac{j^2 \pi^2}{l^2} EJ .$$

The action of mass μ_p of moving load p on a beam thus resembles the action of static axial force $N = -\mu_p c^2$.

20.2 *Suspended beams*

Suspended beams*) (Fig. 20.2) composed of stiffening girder *1*, carrying cable *2* and very densely distributed vertical hangers *3*, are a special case of beams with axial forces.

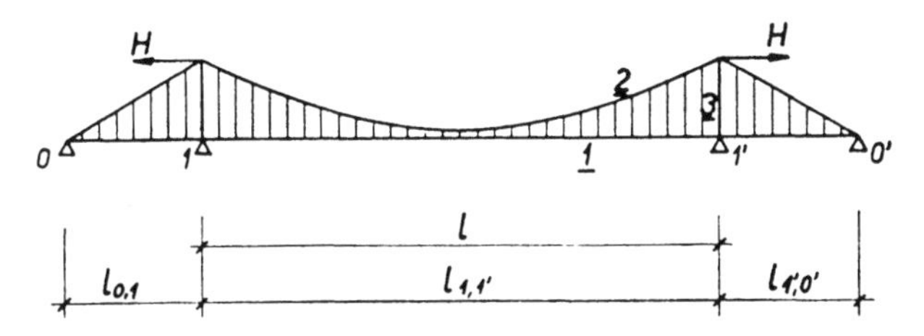

Fig. 20.2. Suspension bridge.

The external forces acting on such systems are:

g_1 – dead load per unit length of beam with mass $\mu_1 = g_1/g$,
$p(x, t)$ – live load per unit length,
g_2 – dead load per unit length of cable with mass $\mu_2 = g_2/g$.

The deflection $v(x, t)$ of a beam with bending rigidity EJ is described by the equation

$$EJ \frac{\partial^4 v(x, t)}{\partial x^4} + \mu_1 \frac{\partial^2 v(x, t)}{\partial t^2} = g_1 + p(x, t) - g_2 . \qquad (20.34)$$

*) Vibrations of suspension highway and railway bridges were dealt with in great detail by K. Klöppel, K. H. Lie: Lotrechte Schwingungen von Hängebrücken. Ingenieur Archiv, 13 (1942), No. 4, pp. 211–266, and further by: A. D. de Pater: Some New Points of View in Calculating Suspension Bridges. Mémoires de l'Association Internationale des Ponts et Charpentes, Zürich, 11 (1951), 41–110, A. G. Pugsley: The Theory of Suspension Bridges. E. Arnold, London, 1957, 7 + 136 pp.

The cable deflection

$$z(x, t) = y(x, t) + v(x, t) \tag{20.35}$$

consists of the original cable sag $y(x, t)$ and of the cable deflection equal – approximately enough – to the beam deflection $v(x, t)$. Hence according to Chap. 14 the differential equation of the cable is

$$-H(t) \frac{\partial^2 z(x, t)}{\partial x^2} + \mu_2 \frac{\partial^2 z(x, t)}{\partial t^2} = g_2 \tag{20.36}$$

where the horizontal tension in the cable

$$H(t) = H_g + H_p(t) \tag{20.37}$$

is composed of the horizontal component H_g of the cable tension resulting from the dead load and of the horizontal component $H_p(t)$ produced by the live load, temperature, etc.

Suspension bridges are usually built so that the stiffening girder is not stressed under the action of dead load. Consequently, the carrying cable takes up the whole dead load of the beam and it may therefore be assumed that

$$-H_g \frac{\partial^2 y(x, t)}{\partial x^2} + \mu_2 \frac{\partial^2 y(x, t)}{\partial t^2} = g_1 \,. \tag{20.38}$$

Even though the equation holds statically, i.e. without the effect of the second term on the left-hand side, we will consider it to approximately apply even dynamically.

If we substitute (20.35) and (20.37) in (20.36) and in doing so also consider Eq. (20.38) we get

$$g_2 = g_1 - H_p y'' - H v'' + \mu_2 \ddot{v} \,.$$

This substituted in (20.34) will give us the vibration equation of a suspension bridge

$$EJ \frac{\partial^4 v(x, t)}{\partial x^4} - H(t) \frac{\partial^2 v(x, t)}{\partial x^2} - H_p(t) \frac{\partial^2 y(x, t)}{\partial x^2} + \\ + \mu \frac{\partial^2 v(x, t)}{\partial t^2} = p(x, t) \tag{20.39}$$

where $\mu = \mu_1 + \mu_2$ is the total mass of the beam and the cable per unit length.

In practice, Eq. (20.39) is simplified still further. First of all, the dynamic component $H_p(t)$ in (20.37) is assumed to be smaller by far than the static component H_g; hence the horizontal component of the cable prestress may approximately be assumed constant, $H(t) \approx H$. Further, the cable deflection curve $y(x, t)$ is assumed to be a parabola in the equilibrium position, and its dynamic component negligible compared with the dynamic beam deflection. Therefore, it is assumed that $\partial^2 y(x, t)/\partial x^2 = y'' \approx$ $\approx$ constant.

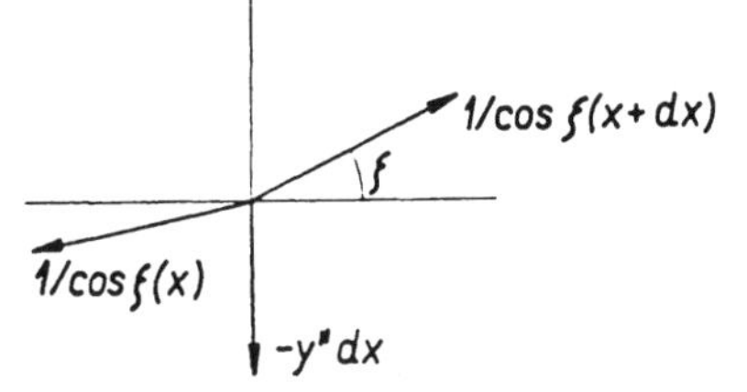

Fig. 20.3. Equilibrium system of forces acting on a cable under unit horizontal force $H_p(t) = 1$ (according to [131]).

The dynamic increment of the horizontal force, $H_p(t)$, is then computed from the principle of virtual work applied to the deformed cable (Fig. 20.3)

$$\int P\delta' \, \mathrm{d}x - \int \frac{NN'}{E_2 F_2} \, \mathrm{d}s = 0 \tag{20.40}$$

where $P = 1$, $\delta' = -y'' v(x, t)$, $N = H_p(t)/\cos \zeta$, $N' = 1/\cos \zeta$, $\mathrm{d}s =$ $= \mathrm{d}x/\cos \zeta$, E_2 is Young's modulus and F_2 the area of the cable, ζ – the slope of the carrying cable at point x, and the integrals include all spans. Hence from (20.40)

$$H_p(t) = -y'' E_2 F_2 \int_0^l v(x, t) \, \mathrm{d}x \Big/ \int \mathrm{d}x/\cos^3 \zeta \, . \tag{20.41}$$

Taking into consideration all the above approximations, the vibration equation of a suspension bridge (20.39) may be written as

$$EJ \frac{\partial^4 v(x, t)}{\partial x^4} - H \frac{\partial^2 v(x, t)}{\partial x^2} + A \int_0^l v(x, t) \, \mathrm{d}x + \mu \frac{\partial^2 v(x, t)}{\partial t^2} = $$

$$= p(x, t) \tag{20.42}$$

where

$$A = \frac{y''^2 E_2 F_2}{\int dx/\cos^3 \zeta}.$$

Eq. (20.42) is usually solved by Galerkin's method (paragraph 12.1.2). Thus, for example, for a *simply supported stiffening girder* the solution is assumed to be of the form (20.24). Similarly as in the derivation of Eqs. (20.25) to (20.27), condition (12.11) applied to Eq. (20.42) will give a system of ordinary differential equations

$$\ddot{q}_k(t) + \left(\frac{k^4\pi^4}{l^4}\frac{EJ}{\mu} + \frac{k^2\pi^2}{l^2}\frac{H}{\mu}\right) q_k(t) + \sum_{j=1}^{n} \frac{2Al}{jk\pi^2\mu} \cdot$$

$$\cdot (1 - \cos j\pi)(1 - \cos k\pi)\, q_j(t) = \frac{2}{\mu l}\int_0^l p(x, t) \sin\frac{k\pi x}{l}\, dx\,,$$

$$k = 1, 2, \ldots, n\,. \tag{20.43}$$

Consider only the first two terms of series (20.24), i.e. $n = 2$. Then the natural frequency of a suspended beam may be obtained from (20.43). For $k = 1$, we get the natural frequency of symmetric vibration

$$\omega_{(1)}^2 = \frac{\pi^4}{l^4}\frac{EJ}{\mu} + \frac{\pi^2}{l^2}\frac{H}{\mu} + \frac{8Al}{\pi^2\mu}, \tag{20.44}$$

and for $k = 2$ the natural frequency of antisymmetric vibration

$$\omega_{(2)}^2 = \frac{16\pi^4}{l^4}\frac{EJ}{\mu} + \frac{4\pi^2}{l^2}\frac{H}{\mu}. \tag{20.45}$$

In the calculation of the effect of a *moving load*, the solution of (20.43) will yield expressions that are exactly like expressions (20.16) to (20.18), (20.20) to (20.22), and (20.29) to (20.32) for $j = 1, 2$. All that is necessary is to substitute there $\omega_{(j)}$ from Eqs. (20.44) and (20.45) and carry out the calculation for $\mu = \mu_1 + \mu_2$. In (20.44) and (20.45) the expression $\bar{\omega}_{(j)}^2 = \omega_{(j)}^2 \mu/\bar{\mu}$ will have the horizontal force $H - \mu_p c^2$ in place of H.

20.3 *Application of the theory*

The theory expounded in Sect. 20.1 can be applied to calculations of prestressed beams. The prestress should, of course, be such as to satisfy the assumptions stated in connection with the derivation of Eq. (20.1). This means that the beam should be compressed by axial force $-N$ at its ends only. To comply with this, the wires of prestressed reinforced concrete bridges would have to be free (i.e. not concreted in) along their whole length, and their stresses invariable during the vehicle traverse.

In actual prestressed reinforced concrete bridges, if grouting has been done properly, reinforcement is bonded to concrete along the whole length of the prestressing wires. The prestress in the wires bears no effect on the potential energy in the beam and therefore causes no change in its natural frequencies. Since the prestress forces are at equilibrium with the concrete compressing forces, the total forces that act on a length element remain unvaried. That is why a beam with grouted prestressing cables is dynamically computed as though it were not subjected to any axial force at all, that is to say, in accordance with Chaps. 1 and 3. Of course, the cross sectional area of the beam must include the whole concrete cross section and the ideal area of reinforcement, and the value of Young's modulus must be that corresponding to the given prestress. The procedure works equally well with pretensioned as with post-tensioned beams.

The theory explained in Sect. 20.2 finds use in dynamic calculations of suspension bridges. Before proceeding any further, let us underscore the fact that in suspension bridges the principal cause of dynamic stresses is the action of wind rather than the action of a moving load. Since suspension bridges are conventionally of the large-span kind, their deadweights far exceed the moving loads and that is why the effect of the latter is so small.

By way of an *example* we shall compute the natural frequencies and the critical speed of the suspension bridge shown in Fig. 20.2. Assume the central span to be a simple beam of length l [131], and further: $l = 750$ m, $E = 21 \times 10^4$ MN/m^2, $J = 13{\cdot}5$ m^4, $\mu = 50$ t/m, $H = 402{\cdot}7$ MN, $E_2 = 16 \times 10^4$ MN/m^2, $F_2 = 1{\cdot}15$ m^2, $y'' = 1/805{\cdot}4$ m^{-1}, $\int dx/\cos^3 \zeta =$ $= 1730$ m.

First we get $A = 0{\cdot}000164$ MN/m^3. From (20·44) the first natural frequency of symmetric vibration $\omega_{(1)} = 1{\cdot}45$ s^{-1}, from (20.45) the first natural frequency of antisymmetric vibration $\omega_{(2)} = 0{\cdot}91\ \mathrm{s}^{-1}$. This frequency is obviously lower than $\omega_{(1)}$.

Since for both (20.30) and (20.32) $q_2(t) = 0$, the critical speed is computed from the symmetric vibration only. By (20.33) $c_{cr}^2 = 1{\cdot}45^2 \times \times 750^2/(\pi^2 \varkappa)$. For $\varkappa = \mu_p/\mu = 1/10$, $c_{cr} = 3950$ km/h, and for $\varkappa = 1$, $c_{cr} = 1245$ km/h. The critical speeds are clearly far higher than those attainable at the present time.

In continuous stiffening girders according to Fig. 20.2, the normal modes may approximately be considered as follows:

symmetric vibration:

end span 0,1

$$v_{(1)}(x) = -\frac{l_{0,1}}{l_{1,1'}} \sin \frac{\pi x}{l_{0,1}},$$

central span 1,1′

$$v_{(1)}(x) = \sin \frac{\pi x}{l_{1,1'}};$$

antisymmetric vibration:

end span 0,1

$$v_{(2)}(x) = -\frac{2l_{0,1}}{l_{1,1'}} \sin \frac{\pi x}{l_{0,1}},$$

central span 1,1′

$$v_{(2)}(x) = \sin \frac{2\pi x}{l_{1,1'}}.$$

Under these approximate assumptions the calculation of the effect of a moving load would be wholly analogous — see also Chap. 11, Sect. 11.2. In [131] continuous suspension bridges are solved by the application of the frequency functions.

20.4 *Additional bibliography*

[17, 108, 131, 227, 250, 252].

21

Longitudinal vibration of bars subjected to a moving load

Longitudinal vibration of bars of constant cross section F is described by the differential equation

$$-EF\frac{\partial^2 u(x,t)}{\partial x^2} + \mu\frac{\partial^2 u(x,t)}{\partial t^2} = p_x(x,t) \tag{21.1}$$

where $u(x, t)$ is the longitudinal (in the direction of axis x – Fig. 21.1) displacement of the bar at point x and time t, and $p_x(x, t)$ is the load on the bar in the direction of axis x.

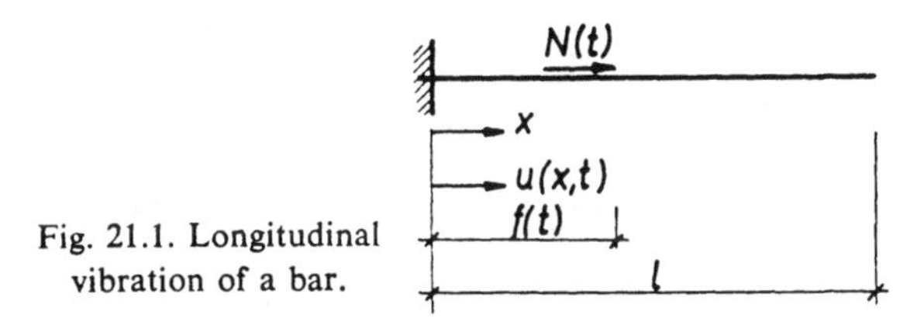

Fig. 21.1. Longitudinal vibration of a bar.

The solution of Eq. (21.1) is analogous to that outlined at the beginning of Chap. 20. The *normal modes* $u_{(j)}(x)$ are obtained by the solution of the ordinary homogeneous equation

$$EF\frac{d^2 u_{(j)}(x)}{dx^2} + \mu\omega_{(j)}^2 u_{(j)}(x) = 0 \tag{21.2}$$

in the form

$$u_{(j)}(x) = \sin\frac{\lambda_j x}{l} + A_j \cos\frac{\lambda_j x}{l} \tag{21.3}$$

where

$$\lambda_j = l\left(\frac{\mu\omega_{(j)}^2}{EF}\right)^{1/2}.$$

Constants λ_j and A_j are computed from the boundary conditions of the bar. Thus, for example, for the cantilever shown in Fig. 21.1 with the boundary conditions

$$u_{(j)}(0) = 0\,, \quad \mathrm{d}u_{(j)}(x)/\mathrm{d}x|_{x=l} = 0\,,$$

they are

$$\lambda_j = (2j - 1)\frac{\pi}{2}\,, \quad A_j = 0\,. \tag{21.4}$$

The natural frequencies then are

$$\omega_{(j)}^2 = \frac{\lambda_j^2}{l^2}\frac{EF}{\mu}\,. \tag{21.5}$$

The *forced vibration* is again analyzed by the generalized method of finite integral transformations (Sect. 6.1 and Chap. 20) and by the Laplace-Carson transformation. Adopting the procedure explained at the beginning of Chap. 20 we get the resultant longitudinal displacement in the form

$$u(x, t) = \sum_{j=1}^{\infty} \frac{\mu\, u_{(j)}(x)}{U_j} \left\{ \frac{1}{\mu\omega_{(j)}} \int_0^t \left[\int_0^l p_x(x, \tau)\, u_{(j)}(x)\, \mathrm{d}x + \right.\right.$$
$$\left. + EF\, z_N(0, l, \tau) \right] \sin \omega_{(j)}(t - \tau)\, \mathrm{d}\tau + G_1(j) \cos \omega_{(j)}t +$$
$$\left. + \frac{G_2(j)}{\omega_{(j)}} \sin \omega_{(j)}t \right\}. \tag{21.6}$$

In Eq. (21.6) denotes

$$U_j = \int_0^l \mu\, u_{(j)}^2(x)\, \mathrm{d}x\,,$$

$$z_N(0, l, t) = \left[\frac{\partial u(x, t)}{\partial x} u_{(j)}(x) - u(x, t) \frac{\mathrm{d}u_{(j)}(x)}{\mathrm{d}x} \right]_0^l,$$

$$G_1(j) = \int_0^l g_1(x)\, u_{(j)}(x)\, \mathrm{d}x\,, \quad G_2(j) = \int_0^l g_2(x)\, u_{(j)}(x)\, \mathrm{d}x\,,$$

$g_1(x)$ and $g_2(x)$ are the initial conditions $u(x, 0) = g_1(x)$ and $\partial u(x, t)/\partial t|_{t=0} = g_2(x)$.

21.1 *Moving load*

The calculation of the effect of a moving longitudinal load is again completely analogous to that outlined in the preceding chapters. The function of time obtained following the integration in Eq. (21.6) is the same as that in Chaps. 1 to 4 and 6, except that $u_{(j)}(x)$ and U_j according to (21.2) to (21.6) now replace $v_{(j)}(x)$ and V_j.

Of importance is the case of longitudinal force $N(t)$ moving at variable speed according to (19.6)

$$p_x(x, t) = \delta[x - f(t)]\, N(t)\,. \tag{21.7}$$

Substitution of (21.7) in (21.6) gives

$$u(x, t) = \sum_{j=1}^{\infty} \frac{u_{(j)}(x)}{U_j\, \omega_{(j)}} \int_0^t N(\tau)\, u_{(j)}[f(\tau)] \sin \omega_{(j)}(t - \tau)\, \mathrm{d}\tau\,. \tag{21.8}$$

To Eq. (21.8) one must, of course, add the terms that express the effect of the boundary and initial conditions, i.e. terms containing $z_N(0, l, t)$, $G_1(j)$ and $G_2(j)$ from (21.6).

21.2 *Bending and longitudinal vibrations of bars*

A longitudinal load on bars is usually accompanied by a transverse load. The displacement of a bar subjected to such loads is described by two independent differential equations

$$EJ\, \frac{\partial^4 v(x, t)}{\partial x^4} + \mu\, \frac{\partial^2 v(x, t)}{\partial t^2} = p(x, t)\,, \tag{21.9}$$

$$-EF\, \frac{\partial^2 u(x, t)}{\partial x^2} + \mu\, \frac{\partial^2 u(x, t)}{\partial t^2} = p_x(x, t) \tag{21.10}$$

where the symbols are the same as in Chaps. 1 and 21.

As an example let us consider the simply supported beam in Fig. 21.2a traversed by horizontal force $N(t)$ moving at distance h from the beam axis. The action of this force is equivalent to the effects of horizontal force $N(t)$ and of bending moment $M(t) = N(t)\, h$ (Fig. 21.2b). Hence the transverse load on the beam is

$$p(x, t) = H_1[x - f(t)]\, M(t)\,, \tag{21.11}$$

and the longitudinal load [Eq. (21.7)]

$$p_x(x, t) = \delta[x - f(t)]\, N(t)$$

assuming a motion at variable speed according to (19.6).

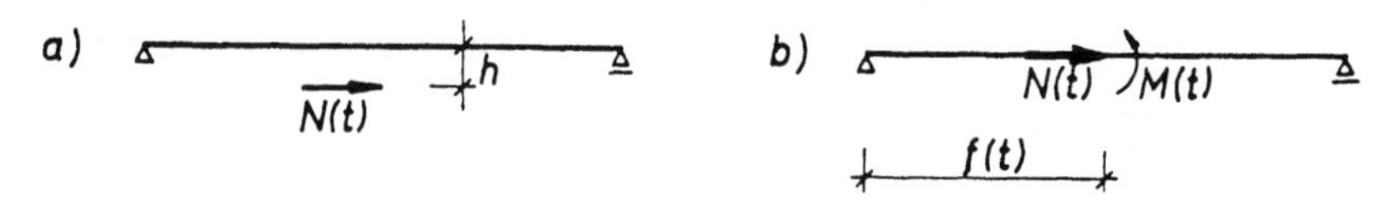

Fig. 21.2. Bending and longitudinal vibrations of a beam.

The longitudinal displacement $u(x, t)$ under load (21.7) was already established by Eq. (21.8). The transverse displacement $v(x, t)$ is computed from (20.12) with (21.11) substituted in. In view of (4.20a) we get (again except for the terms that express the effect of the boundary and initial conditions)

$$v(x, t) = \sum_{j=1}^{\infty} - \frac{v_{(j)}(x)}{V_j \omega_{(j)}} \int_0^t M(\tau) \left. \frac{dv_{(j)}(x)}{dx} \right|_{x = f(\tau)} \sin \omega_{(j)}(t - \tau)\, d\tau\,. \tag{21.12}$$

21.3 *Application of the theory*

The theory presented in this section can be used in calculations of the effects of braking and acceleration forces on structures. Thus, for example, bridges traversed by braking or starting vehicles are subjected to vertical forces as well as to horizontal forces that move along the structure at variable speed. The vertical stresses in a bridge structure may be computed as outlined in Chap. 19. The horizontal forces — acting as a rule outside the gravity axis of the bridge girder — are resolved in a horizontal force and a bending moment (see Sect. 21.2). Stresses in the extreme fibres of the bridge structure are then obtained as the sum of stresses produced by: 1. vertical forces, 2. horizontal longitudinal forces, and 3. bending moments.

21.4 *Additional bibliography*

[61].

22

Thin-walled beams subjected to a moving load

Thin-walled beams are prismatic or cylindrical shells whose dimensions are quantities of different orders of magnitude. Their thickness is small compared to any of the characteristic cross-sectional dimensions, and the cross-sectional dimensions are small against the beam length.

The theory of open-section thin-walled beams was worked out by V. I. Vlasov in [228], that of closed-section thin-walled beams by A. A. Umanskiĭ in [225], [226]. The Vlasov theory of open-section thin-walled beams rests on the following two hypotheses:

1. the beam cross section is non-deformable (rigid) in its plane but free to warp;

2. no shear deformations arise in the middle plane of the beam (i.e. the network of rectangular coordinates on the middle surface continues rectangular even after deformation).

On these assumptions one can deduce the equations of motion of a uniform open-section thin-walled beam with a straight centreline [228], [131]. Having regard to Fig. 22.1 showing the coordinate axes, we write the equations in the form

$$-EF\,\frac{\partial^2 u(x,t)}{\partial x^2} + \mu\,\frac{\partial^2 u(x,t)}{\partial t^2} = p_x(x,t)\,, \tag{22.1}$$

$$EJ_Z\,\frac{\partial^4 v(x,t)}{\partial x^4} - \mu r_Z^2\,\frac{\partial^4 v(x,t)}{\partial x^2\,\partial t^2} + \mu\,\frac{\partial^2 v(x,t)}{\partial t^2} + \mu a_z\,\frac{\partial^2 \Theta(x,t)}{\partial t^2} = p_y(x,t)\,, \tag{22.2}$$

$$EJ_Y\,\frac{\partial^4 w(x,t)}{\partial x^4} - \mu r_Y^2\,\frac{\partial^4 w(x,t)}{\partial x^2\,\partial t^2} + \mu\,\frac{\partial^2 w(x,t)}{\partial t^2} - \mu a_y\,\frac{\partial^2 \Theta(x,t)}{\partial t^2} = p_z(x,t)\,, \tag{22.3}$$

$$EJ_\varphi \frac{\partial^4\Theta(x, t)}{\partial x^4} - GJ_\Theta \frac{\partial^2\Theta(x, t)}{\partial x^2} - \mu r_\varphi^4 \frac{\partial^4\Theta(x, t)}{\partial x^2\, \partial t^2} + \mu r^2 \frac{\partial^2\Theta(x, t)}{\partial t^2} +$$

$$+ \mu a_z \frac{\partial^2 v(x, t)}{\partial t^2} - \mu a_y \frac{\partial^2 w(x, t)}{\partial t^2} = M_A(x, t) . \tag{22.4}$$

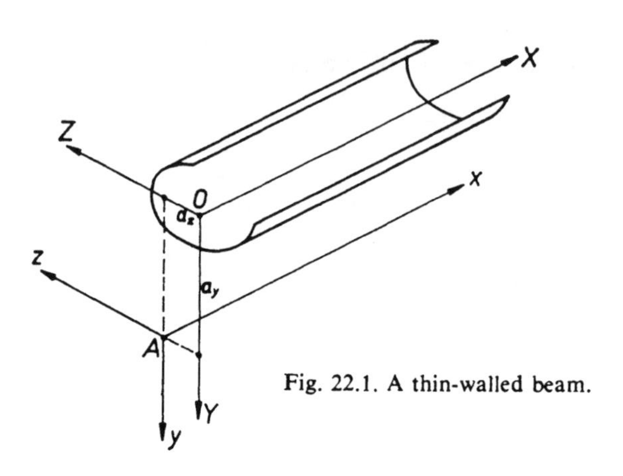

Fig. 22.1. A thin-walled beam.

In the above equations:

$u(x, t)$, $v(x, t)$, $w(x, t)$ – deformations of the axis of bending in the direction of axes x, y, z, respectively, at point x and time t. The axis of bending x passes through the centre of flexure A (Fig. 22.1), not through the centre of gravity of the cross section, O,

$\Theta(x, t)$ – rotations of the beam cross section at point x and time t, about the centre of flexure, A,

$p_x(x, t)$, $p_y(x, t)$, $p_z(x, t)$ – projections of the external load on axes x, y, z, respectively, per unit length of beam,

$M_A(x, t)$ – torsion moment of the external load about the centre of flexure A, per unit length of beam,

E, μ, F – symbols explained in Chaps. 1 and 21,

GJ_Θ – moment of torsional rigidity, G – modulus of elasticity in shear, J_Θ – polar moment of inertia of the cross section about axis x,

J_Y, J_Z – moments of inertia about axes Y, Z, respectively, (Fig. 22.1),

$J_\varphi = \int_F \varphi^2 \, \mathrm{d}F$ – warping constant (sector moment of inertia); φ – sector ordinate equal to double the area of a sector whose pole lies at the centre of flexure, A, and the zero position of the radius vector is given by the condition $\int_F \varphi \, \mathrm{d}F = 0$,

a_y, a_z – coordinates of the centre of flexure A along the principal axes Y, Z (Fig. 22.1),

$r_Y^2 = J_Y/F$, $r_Z^2 = J_Z/F$ – radii of gyration about the principal axes Y, Z,

$r_\varphi^4 = J_\varphi/F$,

$r^2 = a_y^2 + a_z^2 + r_Y^2 + r_Z^2$.

Eq. (22.1) describes the longitudinal vibration of the beam and is completely independent, while Eqs. (22.2) to (22.4) express the composite vibration. Eqs. (22.2) and (22.3) express the beam bending in the principal planes, Eq. (22.4) the beam torsion about the centre of flexure.

Umanskiĭ's theory of closed-section thin-walled beams bases on assumptions similar to those introduced by Vlasov: the cross section does not deform (hence it must be braced) and the warping is the same as in simple twist. Although its equations are formally the same as those of the Vlasov theory of open-section beams, there appears in them the so-called warping coefficient that depends on the polar moment of inertia J_θ and on the directed moment of inertia [63].

Their analyses being similar, we shall consider only thin-walled beams with open section and omit those with closed section.

The problem of thin-walled beams under a moving load has been studied by K. E. Kitaev [125], G. P. Burchak [31] and E. Bielewicz [13] who applied his simplified theory also to bars with curved centrelines [14]. We will now solve several cases of moving load which comes into play in dynamic stresses of bridges.

22.1 *Beam section with vertical axis of symmetry*

Thin-walled beam sections are often symmetrical about the vertical axis. This means that $a_z = 0$ and Eqs. (22.1) to (22.4) simplify as a result. First, we will not consider the longitudinal vibration because Eq. (22.1) with a moving load was already solved in Chap. 21. For $a_z = 0$, Eq. (22.2) becomes completely independent, too, and describes the pure bending vibration in the vertical plane. The composite bending-torsional vibration is defined by Eqs. (22.3) and (22.4).

We will consider the following cases of load moving at constant speed c (Fig. 22.2): a vertical force P acting in the direction of axis y; two harmonic vertical forces, $Q_1 \sin \Omega t$ and $Q_1 \sin(\Omega t + \pi/2)$, acting at distances $\pm b_0/2$ from axis y and shifted 90 degrees in phase. The effect of the

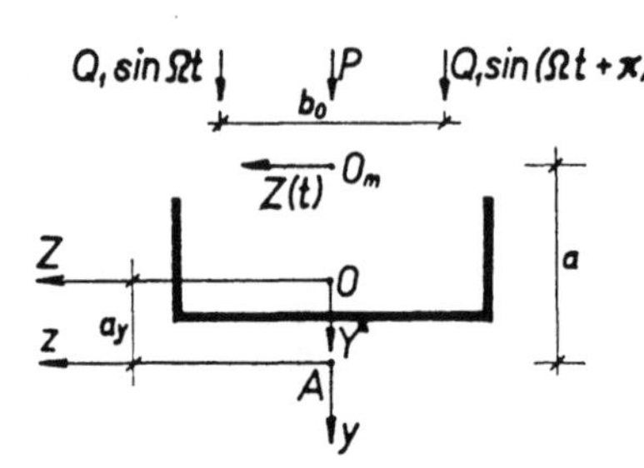

Fig. 22.2. Cross section of a thin-walled beam with vertical axis of symmetry.

latter forces may be replaced by that of vertical force $Q \sin \Omega t$, where $Q = Q_1\, 2^{1/2}$, acting on axis y, and by that of torsional moment $M \sin \Omega t$ where $M = Qb_0/2$. Further, we will consider horizontal force $Z(t) = H \sin Q_1 t$ acting at point O_m in the direction of axis z at distance a from the centre of flexure A (Fig. 22.2). The effect of this force will be replaced by the effect of horizontal force $H \sin \Omega_1 t$ in the direction of axis z acting at the centre of flexure A, and by that of the torsional moment $M_1 \sin \Omega_1 t$ where $M_1 = Ha$.

Accordingly, the loads in Eqs. (22.2) to (22.4) will be

$$p_y(x, t) = \delta(x - ct)\left(P + Q \sin \Omega t\right), \tag{22.5}$$

$$p_z(x, t) = \delta(x - ct)\, H \sin Q_1 t, \tag{22.6}$$

$$M_A(x, t) = \delta(x - ct)\left(M \sin \Omega t + M_1 \sin \Omega_1 t\right). \tag{22.7}$$

The thin-walled beam to be examined has length l, hinges at both ends, no rotation about the longitudinal axis, and zero normal stresses in both end sections; consequently, its boundary conditions are

$$\begin{aligned}
&v(0, t) = 0, \quad v''(0, t) = 0, \quad v(l, t) = 0, \quad v''(l, t) = 0,\\
&w(0, t) = 0, \quad w''(0, t) = 0, \quad w(l, t) = 0, \quad w''(l, t) = 0,\\
&\Theta(0, t) = 0, \quad \Theta''(0, t) = 0, \quad \Theta(l, t) = 0, \quad \Theta''(l, t) = 0,
\end{aligned} \tag{22.8}$$

and the zero initial conditions

$$\begin{aligned} v(x,0) &= 0\,, & \dot{v}(x,0) &= 0\,,\\ w(x,0) &= 0\,, & \dot{w}(x,0) &= 0\,,\\ \Theta(x,0) &= 0\,, & \dot{\Theta}(x,0) &= 0\,. \end{aligned} \tag{22.9}$$

We will now examine the deformations of the thin-walled beam specified above under the action of the various kinds of load defined by Eqs. (22.5) to (22.7).

22.1.1 *Vertical constant force*

In this case Eq. (22.2) takes on the form

$$EJ_Z \frac{\partial^4 v(x,t)}{\partial x^4} - \mu r_Z^2 \frac{\partial^4 v(x,t)}{\partial x^2\,\partial t^2} + \mu \frac{\partial^2 v(x,t)}{\partial t^2} = \delta(x - ct)\,P \tag{22.10}$$

which – except for the omission of damping – differs from (1.1) by only the second term. This term expresses the effect of warping of the cross section. Let us introduce the notation

$$\omega = \frac{j\pi c}{l}\,, \tag{22.11}$$

$$\mu_z = \mu\left(1 + \frac{j^2\pi^2 r_Z^2}{l^2}\right), \tag{22.12}$$

$$\omega_{z(j)}^2 = \frac{j^4\pi^4}{l^4}\,\frac{EJ_Z}{\mu_z}\,, \tag{22.13}$$

and solve Eq. (22.10) by the Fourier finite (sine) and the Laplace-Carson transformations. In view of conditions (22.8) and (22.9) the transformed solution is

$$V^*(j,p) = \frac{P\omega}{\mu_z}\,\frac{p}{(p^2 + \omega_{z(j)}^2)\,(p^2 + \omega^2)} \tag{22.14}$$

and following the inverse transformations (1.9) and (27.32), the required solution turns out to be

$$v(x,t) = \sum_{j=1}^{\infty} \frac{2P \sin j\pi x/l}{\mu_z\, l(\omega_{z(j)}^2 - \omega^2)}\left(\sin \omega t - \frac{\omega}{\omega_{z(j)}} \sin \omega_{z(j)} t\right). \tag{22.15}$$

If for some $j = n$, $\omega_{z(n)} = \omega$, then in view of (27.35) the respective term of series (22.15) becomes

$$v(x, t) = \frac{P \sin n\pi x/l}{\mu_z l \omega_{z(n)}^2} (\sin \omega_{z(n)}t - \omega_{z(n)}t \cos \omega_{z(n)}t) . \qquad (22.16)$$

The form of expressions (22.15) and (22.16) is the same as that of (1.31) and (1.32). This is a clear indication of the fact that section warping affects nothing else but μ_z and $\omega_{z(j)}$ defined by (22.12) and (22.13).

22.1.2 *Vertical harmonic force*

For the vertical resultant of harmonic forces shown in Fig. 22.2, Eq. (22.2) becomes

$$EJ_Z \frac{\partial^4 v(x, t)}{\partial x^4} - \mu r_Z^2 \frac{\partial^4 v(x, t)}{\partial x^2 \partial t^2} + \mu \frac{\partial^2 v(x, t)}{\partial t^2} = \delta(x - ct) \, Q \sin \Omega t . \quad (22.17)$$

Eq. (22.17) with conditions (22.8) and (22.9) is again solved by the method of integral transformations. With the notation

$$r_1 = \Omega + \omega , \quad r_2 = \Omega - \omega \qquad (22.18)$$

and in view of (27.4) and (27.24) the transformed solution is

$$V^*(j, p) = \frac{2Q\Omega\omega}{\mu_z} \frac{p^2}{(p^2 + \omega_{z(j)}^2)(p^2 + r_1^2)(p^2 + r_2^2)} . \qquad (22.19)$$

Inverse transformations (1.9) and (27.55) result in

$$v(x, t) = \sum_{j=1}^{\infty} \frac{Q}{\mu_z l} \frac{\sin j\pi x/l}{(\omega_{z(j)}^2 - r_1^2)(\omega_{z(j)}^2 - r_2^2)} [(\omega_{z(j)}^2 - r_1^2) .$$

$$. \cos r_2 t - (\omega_{z(j)}^2 - r_2^2) \cos r_1 t + 4\Omega\omega \cos \omega_{z(j)}t] . \qquad (22.20)$$

Resonance cases will occur at such $j = n$ for which $\omega_{z(n)} = r_1$ or $\omega_{z(n)} = r_2$ *). In view of (27.61) the respective term in (22.20) has the following

---

*) Since only the first normal mode $j = n = 1$ is of practical importance, and it is usually $\omega \ll \Omega$, the highest effects of a harmonic force arise approximately at $\Omega = \omega_{z(1)}$. This statement is in harmony with Chap. 2. The resonance cases which we shall point out in paragraphs 22.1.3, 22.2.1 and 22.2.2 may be simplified in a similar manner.

form:

for $\omega_{z(n)} = r_2$

$$v(x, t) = \frac{Q}{\mu_z l} \frac{\sin n\pi x/l}{\omega_{z(n)}^2 - r_1^2} \left(\frac{\omega_{z(n)}^2 - r_1^2}{2\omega_{z(n)}^2} \omega_{z(n)} t \sin \omega_{z(n)} t + \cos \omega_{z(n)} t - \cos r_1 t \right), \tag{22.21}$$

for $\omega_{z(n)} = r_1$

$$v(x, t) = \frac{Q}{\mu_z l} \frac{\sin n\pi x/l}{\omega_{z(n)}^2 - r_2^2} \left(- \frac{\omega_{z(n)}^2 - r_2^2}{2\omega_{z(n)}^2} \omega_{z(n)} t \sin \omega_{z(n)} t - \cos \omega_{z(n)} t + \right.$$
$$\left. + \cos r_2 t \right). \tag{22.22}$$

In the absence of damping, Eqs. (22.20) to (22.22) are also a solution of the case treated in Chap. 2. The warping of cross section of thin-walled beams bears effect only on μ_z and $\omega_{z(j)}$.

22.1.3 *Horizontal force and torsion moment*

The calculation of a thin-walled beam subjected to a horizontal force and a torsion moment is carried out simultaneously because under such load the beam performs a composite bending-torsional motion. Under loads (22.6) and (22.7), Eqs. (22.3) and (22.4) have the form

$$EJ_Y \frac{\partial^4 w(x, t)}{\partial x^4} - \mu r_Y^2 \frac{\partial^4 w(x, y)}{\partial x^2 \, \partial t^2} + \mu \frac{\partial^2 w(x, t)}{\partial t^2} - \mu a_y \frac{\partial^2 \Theta(x, t)}{\partial t^2} =$$
$$= \delta(x - ct) \, H \sin \Omega_1 t \,,$$

$$EJ_\varphi \frac{\partial^4 \Theta(x, t)}{\partial x^4} - GJ_\Theta \frac{\partial^2 \Theta(x, t)}{\partial x^2} - \mu r_\varphi^4 \frac{\partial^4 \Theta(x, t)}{\partial x^2 \, \partial t^2} + \mu r^2 \frac{\partial^2 \Theta(x, t)}{\partial t^2} -$$
$$- \mu a_y \frac{\partial^2 w(x, t)}{\partial t^2} = \delta(x - ct) \left(M \sin \Omega t + M_1 \sin \Omega_1 t \right). \tag{22.23}$$

Next to (22.11) and (22.18) we shall introduce the following notation in our solution:

$$\mu_y = \mu\left(1 + \frac{j^2\pi^2 r_Y^2}{l^2}\right), \quad \mu_\varphi = \mu\left(r^2 + \frac{j^2\pi^2 r_\varphi^4}{l^2}\right),$$

$$b_y = \frac{\mu}{\mu_y} a_y, \qquad b_\varphi = \frac{\mu}{\mu_\varphi} a_y,$$

$$\omega_{y(j)}^2 = \frac{j^4\pi^4}{l^4} \frac{EJ_Y}{\mu_y}, \qquad \omega_{\varphi(j)}^2 = \frac{j^4\pi^4}{l^4} \frac{EJ_\varphi}{\mu_\varphi} + \frac{j^2\pi^2}{l^2} \frac{GJ_\theta}{\mu_\varphi},$$

$$\left.\begin{matrix}\omega'^2_{y(j)} \\ \omega'^2_{\varphi(j)}\end{matrix}\right\} = \frac{\omega_{y(j)}^2 + \omega_{\varphi(j)}^2}{2(1 - b_y b_\varphi)} \pm \left[\frac{1}{4}\left(\frac{\omega_{y(j)}^2 + \omega_{\varphi(j)}^2}{1 - b_y b_\varphi}\right)^2 - \frac{\omega_{y(j)}^2\,\omega_{\varphi(j)}^2}{1 - b_y b_\varphi}\right]^{1/2},$$

$$A = \frac{H}{\mu_y} + \frac{M_1 b_y}{\mu_\varphi}, \qquad B = \frac{H\omega_{\varphi(j)}^2}{\mu_y},$$

$$C = \frac{Hb_\varphi}{\mu_y} + \frac{M_1}{\mu_\varphi}, \qquad D = \frac{M_1\omega_{y(j)}^2}{\mu_\varphi},$$

$$r_3 = \Omega_1 + \omega, \qquad r_4 = \Omega_1 - \omega. \tag{22.24}$$

The system of equations (22.23) is again solved by the method of integral transformations. In view of conditions (22.8) and (22.9) the use of (27.24) will give the transformed solution

$$W^*(j, p) = \frac{1}{2(1 - b_y b_\varphi)(p^2 + \omega'^2_{y(j)})(p^2 + \omega'^2_{\varphi(j)})}\left[\frac{Mb_y}{\mu_\varphi} p^4 \left(\frac{1}{p^2 + r_2^2} - \frac{1}{p^2 + r_1^2}\right) + (Ap^4 + Bp^2)\left(\frac{1}{p^2 + r_4^2} - \frac{1}{p^2 + r_3^2}\right)\right],$$

$$\Theta^*(j, p) = \frac{1}{2(1 - b_y b_\varphi)(p^2 + \omega'^2_{y(j)})(p^2 + \omega'^2_{\varphi(j)})}\left[\frac{M}{\mu_\varphi} p^2(p^2 + \omega_{y(j)}^2) \cdot \left(\frac{1}{p^2 + r_2^2} - \frac{1}{p^2 + r_1^2}\right) + (Cp^4 + Dp^2) \cdot \left(\frac{1}{p^2 + r_4^2} - \frac{1}{p^2 + r_3^2}\right)\right]. \tag{22.25}$$

With regard to (1.9), (27.57) and (27.59), the originals of these expressions turn out to be

$$w(x, t) = \sum_{j=1}^{\infty} \frac{\sin j\pi x/l}{l(1 - b_y b_\varphi)(\omega'^2_{y(j)} - \omega'^2_{\varphi(j)})} \left[\frac{M b_y}{\mu_\varphi} \left\{ \frac{1}{(\omega'^2_{\varphi(j)} - r_2^2)(\omega'^2_{y(j)} - r_2^2)} \cdot \right.\right.$$

$$\cdot \left[-r_2^2(\omega'^2_{y(j)} - \omega'^2_{\varphi(j)}) \cos r_2 t + \omega'^2_{\varphi(j)}(\omega'^2_{y(j)} - r_2^2) \cos \omega'_{\varphi(j)} t - \right.$$

$$\left. - \omega'^2_{y(j)}(\omega'^2_{\varphi(j)} - r_2^2) \cos \omega'_{y(j)} t \right] - \frac{1}{(\omega'^2_{\varphi(j)} - r_1^2)(\omega'^2_{y(j)} - r_1^2)} \cdot$$

$$\cdot \left[- r_1^2(\omega'^2_{y(j)} - \omega'^2_{\varphi(j)}) \cos r_1 t + \omega'^2_{\varphi(j)}(\omega'^2_{y(j)} - r_1^2) \cdot \right.$$

$$\left.\left. \cdot \cos \omega'_{\varphi(j)} t - \omega'^2_{y(j)}(\omega'^2_{\varphi(j)} - r_1^2) \cos \omega'_{y(j)} t \right] \right\} +$$

$$+ \frac{1}{(\omega'^2_{\varphi(j)} - r_4^2)(\omega'^2_{y(j)} - r_4^2)} \left[(-A r_4^2 + B)(\omega'^2_{y(j)} - \omega'^2_{\varphi(j)}) \cdot \right.$$

$$\cdot \cos r_4 t + (A\omega'^2_{\varphi(j)} - B)(\omega'^2_{y(j)} - r_4^2) \cos \omega'_{\varphi(j)} t +$$

$$\left. + (-A\omega'^2_{y(j)} + B)(\omega'^2_{\varphi(j)} - r_4^2) \cos \omega'_{y(j)} t \right] -$$

$$- \frac{1}{(\omega'^2_{\varphi(j)} - r_3^2)(\omega'^2_{y(j)} - r_3^2)} \left[(-A r_3^2 + B)(\omega'^2_{y(j)} - \omega'^2_{\varphi(j)}) \cdot \right.$$

$$\cdot \cos r_3 t + (A\omega'^2_{\varphi(j)} - B)(\omega'^2_{y(j)} - r_3^2) \cos \omega'_{\varphi(j)} t +$$

$$\left.\left. + (-A\omega'^2_{y(j)} + B)(\omega'^2_{\varphi(j)} - r_3^2) \cos \omega'_{y(j)} t \right] \right],$$

$$\Theta(x, t) = \sum_{j=1}^{\infty} \frac{\sin j\pi x/l}{l(1 - b_y b_\varphi)(\omega'^2_{y(j)} - \omega'^2_{\varphi(j)})} \left[\frac{M}{\mu_\varphi} \left\{ \frac{1}{(\omega'^2_{\varphi(j)} - r_2^2)(\omega'^2_{y(j)} - r_2^2)} \cdot \right.\right.$$

$$\cdot \left[(\omega^2_{y(j)} - r_2^2)(\omega'^2_{y(j)} - \omega'^2_{\varphi(j)}) \cos r_2 t - (\omega^2_{y(j)} - \omega'^2_{\varphi(j)})(\omega'^2_{y(j)} - r_2^2) \cdot \right.$$

$$\left. \cdot \cos \omega'_{\varphi(j)} t + (\omega^2_{y(j)} - \omega'^2_{y(j)})(\omega'^2_{\varphi(j)} - r_2^2) \cos \omega'_{y(j)} t \right] -$$

$$- \frac{1}{(\omega'^2_{\varphi(j)} - r_1^2)(\omega'^2_{y(j)} - r_1^2)} \left[(\omega^2_{y(j)} - r_1^2)(\omega'^2_{y(j)} - \omega'^2_{\varphi(j)}) \cdot \right.$$

$$\cdot \cos r_1 t - (\omega^2_{y(j)} - \omega'^2_{\varphi(j)})(\omega'^2_{y(j)} - r_1^2) \cos \omega'_{\varphi(j)} t +$$

$$\left.\left. + (\omega^2_{y(j)} - \omega'^2_{y(j)})(\omega'^2_{\varphi(j)} - r_1^2) \cos \omega'_{y(j)} t \right] \right\} +$$

$$+ \frac{1}{\left(\omega'^2_{\varphi(j)} - r_4^2\right)\left(\omega'^2_{y(j)} - r_4^2\right)} \left[\left(-Cr_4^2 + D\right)\left(\omega'^2_{y(j)} - \omega'^2_{\varphi(j)}\right) .\right.$$

$$. \cos r_4 t + \left(C\omega'^2_{\varphi(j)} - D\right)\left(\omega'^2_{y(j)} - r_4^2\right) \cos \omega'_{\varphi(j)} t +$$

$$\left. + \left(-C\omega'^2_{y(j)} + D\right)\left(\omega'^2_{\varphi(j)} - r_4^2\right) \cos \omega'_{y(j)} t\right] -$$

$$- \frac{1}{\left(\omega'^2_{\varphi(j)} - r_3^2\right)\left(\omega'^2_{y(j)} - r_3^2\right)} \left[\left(-Cr_3^2 + D\right)\left(\omega'^2_{y(j)} - \omega'^2_{\varphi(j)}\right) .\right.$$

$$. \cos r_3 t + \left(C\omega'^2_{\varphi(j)} - D\right)\left(\omega'^2_{y(j)} - r_3^2\right) \cos \omega'_{\varphi(j)} t +$$

$$\left. + \left(-C\omega'^2_{y(j)} + D\right)\left(\omega'^2_{\varphi(j)} - r_3^2\right) \cos \omega'_{y(j)} t\right] . \qquad (22.26)$$

Resonant vibration of the beam is apt to occur in the following eight cases:

$$\omega'_{y(j)} = r_1, r_2, r_3, r_4 ,$$

$$\omega'_{\varphi(j)} = r_1, r_2, r_3, r_4 . \qquad (22.27)$$

For any one of these cases the respective term of the series may be computed from formulae (27.63) and (27.64).

22.2 *Beam section with two axes of symmetry*

If a thin-walled beam cross section is symmetric about two axes, the centre of flexure becomes one with the section centroid. Then $a_y = a_z = 0$, and the simultaneous system of equations (22.1) to (22.4) decomposes into four independent equations. The equations describe respectively the longitudinal vibration of the beam, the bending vibration in the vertical and in the horizontal plane, and the torsional vibration about the centre of flexure or the centroid. The motions are independent one of another.

The longitudinal vibration of a beam subjected to a moving load was analyzed in Chap. 21, the bending vibration in the vertical plane in paragraphs 22.1.1 and 22.1.2. The bending vibration in the horizontal plane and the torsional vibration are obtained from (22.26) after the substitution $a_y = a_z = 0$, $b_\varphi = b_y = 0$, $\omega'_{y(j)} = \omega_{y(j)}$, $\omega'_{\varphi(j)} = \omega_{\varphi(j)}$. However, the motions can also be computed directly:

22.2.1 *Horizontal force*

For this load, Eq. (22.6) with (22.3) takes on the form

$$EJ_Y \frac{\partial^4 w(x, t)}{\partial x^4} - \mu r_Y^2 \frac{\partial^4 w(x, t)}{\partial x^2 \partial t^2} + \mu \frac{\partial^2 w(x, t)}{\partial t^2} = \delta(x - ct) H \sin \Omega_1 t . \tag{22.28}$$

With notation (22.24) the transformed solution is

$$W^*(j, p) = \frac{H}{2\mu_y} \frac{p^2}{p^2 + \omega_{y(j)}^2} \left(\frac{1}{p^2 + r_4^2} - \frac{1}{p^2 + r_3^2} \right) \tag{22.29}$$

and by (27.33), its original

$$w(x, t) = \sum_{j=1}^{\infty} \frac{H}{\mu_y l} \sin \frac{j\pi x}{l} \left[\frac{1}{\omega_{y(j)}^2 - r_4^2} (\cos r_4 t - \cos \omega_{y(j)} t) - \right.$$

$$\left. - \frac{1}{\omega_{y(j)}^2 - r_3^2} (\cos r_3 t - \cos \omega_{y(j)} t) \right] . \tag{22.30}$$

In a resonant case, for example at $\omega_{y(n)} = r_4$, the respective term in (22.30) becomes in view of (27.36)

$$w(x, t) = \frac{H}{2\mu_y l \omega_{y(n)}^2} \sin \frac{n\pi x}{l} \omega_{y(n)} t \sin \omega_{y(n)} t . \tag{22.31}$$

Resonance can also occur when $w_{y(n)} = r_3$, i.e. because of the magnitude of r_3, at a higher speed than in the previous case.

22.2.2 *Torsion moment*

For load (22.7), Eq. (22.4) becomes

$$EJ_\varphi \frac{\partial^4 \Theta(x, t)}{\partial x^4} - GJ_\theta \frac{\partial^2 \Theta(x, t)}{\partial x^2} - \iota r_\varphi^4 \frac{\partial^4 \Theta(x, t)}{\partial x^2 \partial t^2} - \mu r^2 \frac{\partial^2 \Theta(x, t)}{\partial t^2} =$$

$$= \delta(x - ct) (M \sin \Omega t + M_1 \sin \Omega_1 t) . \tag{22.32}$$

Using notation (22.18) and (22.24) we get the transform

$$\Theta^*(j, p) = \frac{p^2}{2\mu_\varphi(p^2 + \omega_{\varphi(j)}^2)} \left[M \left(\frac{1}{p^2 + r_2^2} - \frac{1}{p^2 + r_1^2} \right) + \right.$$

$$\left. + M_1 \left(\frac{1}{p^2 + r_4^2} - \frac{1}{p^2 + r_3^2} \right) \right] \tag{22.33}$$

and with a view to (27.33) the original

$$\Theta(x, t) = \sum_{j=1}^{\infty} \frac{\sin j\pi x/l}{\mu_\varphi l} \left\{ M \left[\frac{1}{\omega_{\varphi(j)}^2 - r_2^2} (\cos r_2 t - \cos \omega_{\varphi(j)} t) - \right.\right.$$

$$\left. - \frac{1}{\omega_{\varphi(j)}^2 - r_1^2} (\cos r_1 t - \cos \omega_{\varphi(j)} t) \right] +$$

$$+ M_1 \left[\frac{1}{\omega_{\varphi(j)}^2 - r_4^2} (\cos r_4 t - \cos \omega_{\varphi(j)} t) - \right.$$

$$\left.\left. - \frac{1}{\omega_{\varphi(j)}^2 - r_3^2} (\cos r_3 t - \cos \omega_{\varphi(j)} t) \right] \right\} . \tag{22.34}$$

The beam can start to vibrate in resonance in four cases, namely when

$$\omega_{\varphi(n)} = r_1, r_2, r_3, r_4 .$$

The term of Eq. (22.34) corresponding to each case is computed with the aid of (27.35).

22.3 *Application of the theory*

The fields in which the theory explained in this section can be used to advantage, are the calculations of spatial vibrations, in particular of steel structures of railway bridges.

Large-span bridge structures may be taken for thin-walled beams with a fair degree of approximation. Of the dynamic effects a railway vehicle is apt to produce in them, those considered in the calculation are the motion of force P equal to the vehicle weigth, the motion of two harmonic forces representing the action of counterweights on the driving wheels of a two-cylinder steam locomotive, and the motion of a horizontal harmonic force approximately expressing the so-called lateral impacts of the vehicle. The total stress in the bridge structure is then equal to the sum of stresses produced by these forces.

22.3.1 *The effect of moving mass of a vehicle*

Like in paragraph 1.4.3, the effect of moving mass of a vehicle is obtained on the approximate assumption that the vehicle with mass $m = P/g$ stands immobile at point x_0. For simply supported beams, it is $x_0 = l/2$. In sectional view (Fig. 22.2) the vehicle centroid will be at point O_m, distance a away from the centre of flexure.

By the same calculation as that in paragraph 1.4.3 we shall find that all the equations established earlier in this chapter continue to apply. Only the masses μ_y and μ_φ and the frequencies $\omega_{y(j)}$ and $\omega_{\varphi(j)}$ are now replaced by quantities with bar

$$\bar{\mu}_z = \mu_z + \frac{2m}{l} \sin^2 \frac{j\pi x_0}{l} ,$$

$$\bar{\mu}_y = \mu_y + \frac{2m}{l} \sin^2 \frac{j\pi x_0}{l} , \quad \bar{\mu}_\varphi = \mu_\varphi + \frac{2ma}{l} \sin^2 \frac{j\pi x_0}{l} ,$$

$$\bar{\omega}^2_{z(j)} = \omega^2_{z(j)} \frac{\mu_z}{\bar{\mu}_z} , \quad \bar{\omega}^2_{y(j)} = \omega^2_{y(j)} \frac{\mu_y}{\bar{\mu}_y} , \quad \bar{\omega}^2_{\varphi(j)} = \omega^2_{\varphi(j)} \frac{\mu_\varphi}{\bar{\mu}_\varphi} . \tag{22.35}$$

These values and the notation

$$\bar{b}_y = \frac{\mu}{\bar{\mu}_y} a_y , \quad \bar{b}_\varphi = \frac{\mu}{\bar{\mu}_\varphi} a_y$$

are used in the computation of all the quantities described by (22.24), i.e. of $\bar{\omega}'_{y(j)}$, $\bar{\omega}'_{\varphi(j)}$, $\bar{A}$, $\bar{B}$, $\bar{C}$, $\bar{D}$ as well as of all the deformations in Sects. 22.1 and 22.2.

22.3.2 *The torsional effect of counterweights*

The vertical effects of steam locomotive counterweights were examined in Chap. 2. The amplitude of the resultant Q and frequency Ω are computed from formula (2.14). Since in two-cylinder steam locomotives the counterweights on the right-hand side are turned 90 degrees against those on the left-hand side, there also arises the torsion moment $M \sin \Omega t$ where

$$M = \tfrac{1}{2} Q b_0 \tag{22.36}$$

and b_0 is the track gauge (Fig. 22.3).

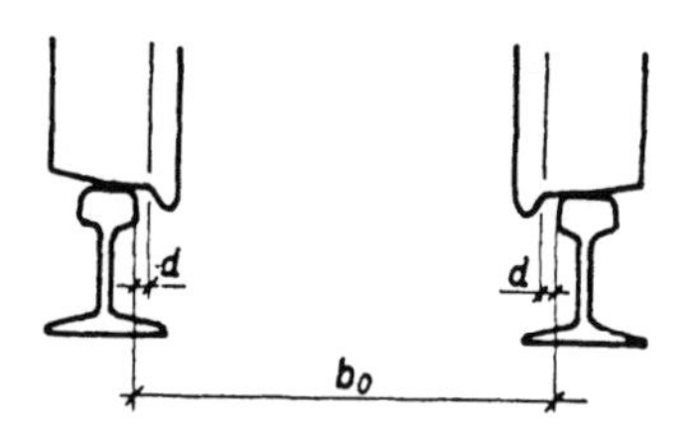

Fig. 22.3. Lateral impacts of railway vehicles.

22.3.3 *Lateral impacts*

Lateral impacts of a railway vehicle give rise to an approximately horizontal force, $Z(t) = H \sin \Omega_1 t$, acting at the vehicle centroid O_m (Fig. 22.2). According to [125] the amplitude H and the frequency Ω_1 of this force, as well as the torsion moment produced by it are computed from the approximate relations

$$H = \frac{md\pi^2 c^2}{D_0^2}, \quad \Omega_1 = \frac{\pi c}{D_0}, \quad M_1 = Ha \tag{22.37}$$

where

$m = P/g$ — mass of the vehicle travelling at speed c,

d — half-width of the gap between the inside edge of rail and the wheel flange (Fig. 22.3),

D_0 — maximum wheel base, i.e. distance between the first and the last wheel of the vehicle,

a — distance between the vehicle centroid O_m and the centre of flexure A (Fig. 22.2).

22.4 *Additional bibliography*

[9, 10, 13, 14, 31, 90, 125, 234, 243, 251, 259, 264, 265, 271].

23

The effect of shear and rotatory inertia

When analyzing beams with height-span ratios larger than about 1/10, or beams made of materials sensitive to shear stresses, one must also give consideration to the effect of shear and rotatory inertia.

In the analysis, the transverse deflection of the beam

$$v(x, t) = v_M(x, t) + v_S(x, t) \tag{23.1}$$

is resolved into component $v_M(x, t)$ arising owing to bending, and component $v_S(x, t)$ caused by shear. The rotation of the cross section produced by the effect of the bending component of deflection is denoted by

$$\psi(x, t) = \frac{\partial v_M(x, t)}{\partial x}. \tag{23.2}$$

The bending moment and the shear force are then computed from the familiar relations

$$M(x, t) = -EJ \frac{\partial^2 v_M(x, t)}{\partial x^2} = -EJ \frac{\partial \psi(x, t)}{\partial x}, \tag{23.3}$$

$$T(x, t) = k^*GF \frac{\partial v_S(x, t)}{\partial x} = k^*GF \left[\frac{\partial v(x, t)}{\partial x} - \psi(x, t)\right]. \tag{23.4}$$

The new symbols introduced in the above in addition to those used in Chap. 1 are:

k^* – a constant dependent on the shape of the section (e.g. $k^* = 2/3$ for a rectangular, $k^* = 3/4$ for a circular cross section),
G – modulus of elasticity in shear,
F – cross sectional area.

Fig. 23.1 shows the forces and moments acting on a length element of a bar subjected to transverse load $p(x, t)$, placed on a Winkler elastic foundation with constant k (see Chap. 13). The bar is made of material with mass ϱ per unit volume ($\mu = \varrho F$). For a bar thus specified the conditions of equilibrium of forces acting in the vertical direction, and of moments are

$$\mu \frac{\partial^2 v(x, t)}{\partial t^2} - \frac{\partial T(x, t)}{\partial x} + kv(x, t) = p(x, t),$$

$$- \frac{\partial M(x, t)}{\partial x} + T(x, t) - J\varrho \frac{\partial^2 \psi(x, t)}{\partial t^2} = 0. \tag{23.5}$$

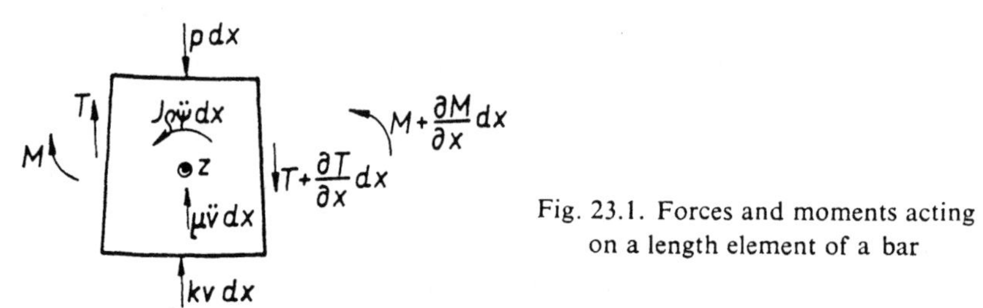

Fig. 23.1. Forces and moments acting on a length element of a bar

On substituting expressions (23.3) and (23.4) in these equations we get a system of partial differential equations for the unknown deflection $v(x, t)$ and for the unknown rotation $\psi(x, t)$

$$\mu \frac{\partial^2 v(x, t)}{\partial t^2} - k^*GF \left[\frac{\partial^2 v(x, t)}{\partial x^2} - \frac{\partial \psi(x, t)}{\partial x} \right] + kv(x, t) = p(x, t),$$

$$EJ \frac{\partial^2 \psi(x, t)}{\partial x^2} + k^*GF \left[\frac{\partial v(x, t)}{\partial x} - \psi(x, t) \right] - J\varrho \frac{\partial^2 \psi(x, t)}{\partial t^2} = 0 \tag{23.6}$$

which together with the boundary and the initial conditions lead to the solution of the beam vibration.

With the aid of the notation

$$c_1^2 = \frac{E}{\varrho}, \quad c_2^2 = \frac{k^*G}{\varrho}, \quad c_1 > c_2, \quad r^2 = \frac{J}{F}, \tag{23.7}$$

representing, respectively, the velocity of propagation of longitudinal, c_1, and of transverse waves, c_2, along the beam, and the radius of gyration, r, system (23.6) may be reduced to the single partial differential equation for the deflection $v(x, t)$*)

$$EJ \frac{\partial^4 v(x, t)}{\partial x^4} - EJ \left(\frac{1}{c_1^2} + \frac{1}{c_2^2} \right) \frac{\partial^4 v(x, t)}{\partial x^2 \, \partial t^2} + \frac{EJ}{c_1^2 c_2^2} \frac{\partial^4 v(x, t)}{\partial t^4} +$$

$$+ \mu \frac{\partial^2 v(x, t)}{\partial t^2} + k \left\{ v(x, t) + r^2 \frac{c_1^2}{c_2^2} \left[\frac{1}{c_1^2} \frac{\partial^2 v(x, t)}{\partial t^2} - \frac{\partial^2 v(x, t)}{\partial x^2} \right] \right\} =$$

$$= p(x, t) + r^2 \frac{c_1^2}{c_2^2} \left[\frac{1}{c_1^2} \frac{\partial^2 p(x, t)}{\partial t^2} - \frac{\partial^2 p(x, t)}{\partial x^2} \right]. \tag{23.8}$$

The beam model described by Eqs. (23.6) or (23.8), in which consideration is given both to the effect of shear and to the effect of rotatory inertia, is called the Timoshenko beam [220]. From there follow three special cases:

1. Shear beam

If the effect of rotatory inertia is neglected and only the effect of shear on the dynamic deflection of beam considered, Eqs. (23.6) reduce to

$$\mu \frac{\partial^2 v(x, t)}{\partial t^2} - k^* GF \left[\frac{\partial^2 v(x, t)}{\partial x^2} - \frac{\partial \psi(x, t)}{\partial x} \right] + k v(x, t) = p(x, t) ,$$

$$EJ \frac{\partial^2 \psi(x, t)}{\partial x^2} + k^* GF \left[\frac{\partial v(x, t)}{\partial x} - \psi(x, t) \right] = 0 . \tag{23.9}$$

This system may again be expressed by the single equation of deflection $v(x, t)$

$$EJ \frac{\partial^4 v(x, t)}{\partial x^4} - \frac{EJ}{c_2^2} \frac{\partial^4 v(x, t)}{\partial x^2 \, \partial t^2} + \mu \frac{\partial^2 v(x, t)}{\partial t^2} + k \left[v(x, t) - r^2 \cdot \right.$$

$$\left. \cdot \frac{c_1^2}{c_2^2} \frac{\partial^2 v(x, t)}{\partial x^2} \right] = p(x, t) - r^2 \frac{c_1^2}{c_2^2} \frac{\partial^2 p(x, t)}{\partial x^2} . \tag{23.10}$$

Expression (23.10) follows from Eq. (23.8) in the case of infinite velocity of propagation of longitudinal waves along the beam ($c_1 \to \infty$).

---

*) Eqs. (23.8) and (23.10) to be deduced presently are not suitable for subsequent computation because the boundary conditions are difficult to express and the right-hand sides of the equations are too complicated. They are given here just to illustrate the effect of the propagation of longitudinal and transverse waves along a beam, which subject will be taken up later on.

2. *Rayleigh beam*

If only the effect of rotatory inertia is considered and the effect of shear neglected, the so-called Rayleigh beam model [185] results. Of course, for such a beam

$$v(x, t) = v_M(x, t)\,, \quad v_S(x, t) = 0\,, \quad \psi(x, t) = \partial v(x, t)/\partial x\,. \quad (23.11)$$

With these values Eqs. (23.3), (23.5) and (23.6) give the single beam deflection equation

$$EJ\,\frac{\partial^4 v(x, t)}{\partial x^4} - \frac{EJ}{c_1^2}\,\frac{\partial^4 v(x, t)}{\partial x^2\,\partial t^2} + \mu\,\frac{\partial^2 v(x, t)}{\partial t^2} + kv(x, t) = p(x, t)\,. \quad (23.12)$$

This equation also follows from (23.8) on the consideration that the transverse waves have an infinitely large velocity of propagation along the beam ($c_2 \to \infty$).

3. *Bernoulli-Euler beam*

If we neglect both the effect of shear and the effect of rotatory inertia we obtain the classical Bernoulli-Euler beam model. Its equation

$$EJ\,\frac{\partial^4 v(x, t)}{\partial x^4} + \mu\,\frac{\partial^2 v(x, t)}{\partial t^2} + kv(x, t) = p(x, t) \quad (23.13)$$

follows either from expressions (23.11), (23.3), (23.5) and (23.6) or from Eq. (23.8) if we assume that the velocities of propagation of longitudinal and transverse waves along the beam are both infinitely large ($c_1 \to \infty$, $c_2 \to \infty$).

We will now analyze the effect of a constant force P, i.e.

$$p(x, t) = \delta(x - ct)\,P\,, \quad (23.14)$$

moving along a beam at constant speed c. The analysis will be carried out for a simply supported beam with span l, without an elastic foundation, as well as for an infinite beam on elastic foundation. In either case we will solve the effect of the moving force on a Timoshenko beam, on a shear beam, and on a Rayleigh beam. The Bernoulli-Euler beam under this kind of load (including the effect of damping) was already analyzed in full in Chaps. 1 and 13.

So far no examination has been made of a Timoshenko beam of finite length subjected to a moving load, whereas the infinite Timoshenko beam on an elastic foundation has been analyzed by S. H. Crandall [39], and J. D. Achenbach and C. T. Sun [3]. Our solution is generalized to include a number of cases.

23.1 *Simply supported beam*

A simply supported beam of finite length l has the boundary conditions

$$v(x, t)\big|_{x=0} = 0\,, \quad v(x, t)\big|_{x=l} = 0\,, \quad \left.\frac{\partial\psi(x, t)}{\partial x}\right|_{x=0} = 0\,,$$

$$\left.\frac{\partial\psi(x, t)}{\partial x}\right|_{x=l} = 0\,, \tag{23.15}$$

and the initial conditions

$$v(x, t)\big|_{t=0} = 0\,, \quad \left.\frac{\partial v(x, t)}{\partial t}\right|_{t=0} = 0\,, \quad \psi(x, t)\big|_{t=0} = 0\,,$$

$$\left.\frac{\partial\psi(x, t)}{\partial t}\right|_{t=0} = 0\,. \tag{23.16}$$

In the solution we will again apply the method of finite Fourier integral transformations (1.9). As to the rotation: in view of (23.15) we shall use the cosine transformation (27.75)

$$\Psi(j, t) = \int_0^l \psi(x, t) \cos\frac{j\pi x}{l}\,\mathrm{d}x\,,$$

$$\psi(x, t) = \frac{2}{l}\sum_{j=1}^{\infty} \Psi(j, t) \cos\frac{j\pi x}{l}\,. \tag{23.17}$$

The Laplace-Carson transformation of function $\Psi(j, t)$ then is

$$\Psi^*(j, p) = p\int_0^{\infty} \Psi(j, t)\,\mathrm{e}^{-pt}\,\mathrm{d}t\,. \tag{23.18}$$

23.1.1 *Timoshenko beam*

Eq. (23.6) with (23.14) for the right-hand side will first be transformed with respect to x; the first of (23.6) will be multiplied by $\sin j\pi x/l$, the second by $\cos j\pi x/l$ and integrated with respect to x from 0 to l. In view of the boundary conditions (23.15) it follows that

$$\int_0^l \psi'(x, t) \sin j\pi x/l \, dx = -\frac{j\pi}{l} \Psi(j, t),$$

$$\int_0^l \psi''(x, t) \cos j\pi x/l \, dx = -\frac{j^2\pi^2}{l^2} \Psi(j, t),$$

$$\int_0^l v'(x, t) \cos j\pi x/l \, dx = \frac{j\pi}{l} V(j, t).$$

Consequently the transformation of Eqs. (23.6) and (23.14) will give

$$\mu \ddot{V}(j, t) - k^*GF\left[-\frac{j^2\pi^2}{l^2} V(j, t) + \frac{j\pi}{l} \Psi(j, t)\right] = P \sin \omega t,$$

$$-\frac{j^2\pi^2}{l^2} EJ\, \Psi(j, t) + k^*GF\left[\frac{j\pi}{l} V(j, t) - \Psi(j, t)\right] - J\varrho\, \ddot{\Psi}(j, t) = 0$$

where again $\omega = j\pi c/l$. In view of (23.16), (27.4) and (27.18) the Laplace-Carson transformation of these equations is

$$\left(p^2 + \frac{j^2\pi^2}{l^2} c_2^2\right) V^*(j, p) - \frac{j\pi}{l} c_2^2\, \Psi^*(j, p) = \frac{P\omega}{\mu} \frac{p}{p^2 + \omega^2},$$

$$-\frac{j\pi}{l} \frac{c_2^2}{r^2} V^*(j, p) + \left(p^2 + \frac{j^2\pi^2}{l^2} c_1^2 + \frac{c_2^2}{r^2}\right) \Psi^*(j, p) = 0.$$

The determinant of this system may be rearranged to the form $(p^2 + \omega_{1(j)}^2)(p^2 + \omega_{2(j)}^2)$ where $\omega_{1(j)}$ and $\omega_{2(j)}$ are the natural circular frequencies of a Timoshenko beam

$$\omega_{1,2(j)}^2 = \omega_{(j)}^2 \frac{l^2}{2j^2\pi^2 r^2}\left\{1 + \frac{c_2^2}{c_1^2} + \frac{c_2^2}{c_1^2} \frac{l^2}{j^2\pi^2 r^2} \mp \right.$$

$$\left. \mp \left[\left(1 + \frac{c_2^2}{c_1^2} + \frac{c_2^2}{c_1^2} \frac{l^2}{j^2\pi^2 r^2}\right)^2 - 4 \frac{c_2^2}{c_1^2}\right]^{1/2}\right\}, \tag{23.19}$$

and $\omega_{(j)}$ is the natural circular frequency of the Bernoulli-Euler beam (1.11).

The transformed solutions then are

$$V^*(j, p) = \frac{P\omega}{\mu} \frac{p(p^2 + \omega_T^2)}{(p^2 + \omega_{1(j)}^2)(p^2 + \omega_{2(j)}^2)(p^2 + \omega^2)},$$

$$\Psi^*(j, p) = \frac{j\pi P\omega c_2^2}{\mu l r^2} \frac{p}{(p^2 + \omega_{1(j)}^2)(p^2 + \omega_{2(j)}^2)(p^2 + \omega^2)}$$

where

$$\omega_T^2 = \frac{j^2\pi^2}{l^2} c_1^2 + \frac{c_2^2}{r^2}.$$

The inverse transformations (1.9), (23.17), (27.58), (27.54), or (27.64) and (27.60) will give the required solutions

$$v(x, t) = \sum_{j=1, j\neq n}^{\infty} \frac{2P}{\mu l} \frac{\sin j\pi x/l}{\omega_{1(j)}\omega_{2(j)}(\omega_{1(j)}^2 - \omega_{2(j)}^2)(\omega_{2(j)}^2 - \omega^2)(\omega_{1(j)}^2 - \omega^2)} \cdot$$
$$\cdot \left[\omega_{1(j)}\omega_{2(j)}(\omega_{1(j)}^2 - \omega_{2(j)}^2)(\omega_T^2 - \omega^2) \sin \omega t - \right.$$
$$- \omega_{1(j)}\, \omega(\omega_{1(j)}^2 - \omega^2)(\omega_T^2 - \omega_{2(j)}^2) \sin \omega_{2(j)} t +$$
$$\left. + \omega_{2(j)}\, \omega(\omega_{2(j)}^2 - \omega^2)(\omega_T^2 - \omega_{1(j)}^2) \sin \omega_{1(j)} t\right] +$$
$$+ \frac{2P}{\mu l} \frac{\sin n\pi x/l}{(\omega_{1(n)}^2 - \omega^2)^2} \left\{\frac{\omega_{1(n)}^2 - \omega^2}{2} \left[\left(1 + \frac{\omega_T^2}{\omega^2}\right) \sin \omega t + \right.\right.$$
$$\left. + \left(1 - \frac{\omega_T^2}{\omega^2}\right) \omega t \cos \omega t\right] + \omega^2 \left(1 - \frac{\omega_T^2}{\omega^2}\right) \sin \omega t -$$
$$\left. - \omega_{1(n)}\omega \left(1 - \frac{\omega_T^2}{\omega_{1(n)}^2}\right) \sin \omega_{1(n)} t\right\},$$

$$\psi(x, t) = \sum_{j=1, j\neq n}^{\infty} \frac{2j\pi P c_2^2}{\mu l^2 r^2} \frac{\cos j\pi x/l}{\omega_{1(j)}\omega_{2(j)}(\omega_{1(j)}^2 - \omega_{2(j)}^2)(\omega_{2(j)}^2 - \omega^2)(\omega_{1(j)}^2 - \omega^2)} \cdot$$
$$\cdot \left[\omega_{1(j)}\omega_{2(j)}(\omega_{1(j)}^2 - \omega_{2(j)}^2) \sin \omega t - \omega_{1(j)}\omega(\omega_{1(j)}^2 - \omega^2) \cdot \right.$$
$$\left. \cdot \sin \omega_{2(j)} t + \omega_{2(j)}\omega(\omega_{2(j)}^2 - \omega^2) \sin \omega_{1(j)} t\right] +$$
$$+ \frac{2n\pi P c_2^2}{\mu l^2 r^2} \frac{\cos n\pi x/l}{(\omega_{1(n)}^2 - \omega^2)^2} \left[\frac{\omega_{1(n)}^2 - \omega^2}{2\omega^2} (\sin \omega t - \omega t \cos \omega t) - \right.$$
$$\left. - \sin \omega t + \frac{\omega}{\omega_{1(n)}} \sin \omega_{1(n)} t\right]. \tag{23.20}$$

The second term on the right-hand sides of Eqs. (23.20) represents the solutions for such $j = n$ for which it just is $\omega_{2(n)} = \omega$. One can get analogous expressions for $\omega_{1(n)} = \omega$ by changing the subscript $1(n)$ to the subscript $2(n)$.

The character of the vibration of the Timoshenko beam subjected to a moving load resembles that of the vibration of the Bernoulli-Euler beam investigated in Chap. 1. In contrast to the latter, the Timoshenko beam has two systems of natural frequencies (23.19) and thus also two speeds at which the deflection grows with time. These speeds are obtained from the conditions that $\omega_{1(n)} = \omega$ and $\omega_{2(n)} = \omega$. Since the beam is of finite length, the process is transient and the deflections fail to attain infinite values in the interval $0 \leqq t \leqq l/c$.

23.1.2 *Shear beam*

We will solve the shear beam by the method indicated in the preceding paragraph. After the Fourier and Laplace-Carson transformations of Eqs. (23.9) and (23.14) we get

$$V^*(j, p) = \frac{P\omega}{\mu} \frac{p}{(p^2 + \omega_{S(j)}^2)(p^2 + \omega^2)},$$

$$\Psi^*(j, p) = \frac{j\pi P\omega}{\mu_S l} \frac{p}{(p^2 + \omega_{S(j)}^2)(p^2 + \omega^2)}, \tag{23.21}$$

where $\omega_{S(j)}$ is the natural circular frequency of the shear beam

$$\omega_{S(j)}^2 = \frac{j^4\pi^4}{l^4} \frac{EJ}{\mu_S},$$

$$\mu_S = \mu\left(1 + \frac{j^2\pi^2 r^2}{l^2} \frac{c_1^2}{c_2^2}\right). \tag{23.22}$$

The inverse transformations of (23.21) in accordance with (1.9), (23.17), (27.32), or (27.35) yield the solutions

$$v(x, t) = \sum_{j=1, j\neq n}^{\infty} \frac{2P}{\mu l} \frac{\sin j\pi x/l}{\omega_{S(j)}^2 - \omega^2}\left(\sin \omega t - \frac{\omega}{\omega_{S(j)}} \sin \omega_{S(j)} t\right) +$$

$$+ \frac{P \sin n\pi x/l}{\mu l \omega_{S(n)}^2}\left(\sin \omega_{S(n)} t - \omega_{S(n)} t \cos \omega_{S(n)} t\right),$$

$$\psi(x, t) = \sum_{j=1, j \neq n}^{\infty} \frac{2Pj\pi}{\mu_S l^2} \frac{\cos j\pi x/l}{\omega_{S(j)}^2 - \omega^2} \left(\sin \omega t - \frac{\omega}{\omega_{S(j)}} \sin \omega_{S(j)} t \right) +$$
$$+ \frac{n\pi P \cos n\pi x/l}{\mu_S l^2 \omega_{S(n)}^2} \left(\sin \omega_{S(n)} t - \omega_{S(n)} t \cos \omega_{S(n)} t \right). \quad (23.23)$$

The vibration of a shear beam is the same as the vibration of the Bernoulli-Euler beam (Chap. 1), the only point of difference being the natural frequency. The particular solutions represented by the second terms of Eqs. (23.23) obtain for such $j = n$ for which $\omega_{S(n)} = \omega$. From the latter condition one can also compute the corresponding speed at which one component of the deflection grows with time.

23.1.3 *Rayleigh beam*

When consideration is given to rotatory inertia, Eq. (23.12) is transformed in accordance with (1.9) and (1.15) to

$$V^*(j, p) = \frac{P\omega}{\mu_R} \frac{p}{(p^2 + \omega_{R(j)}^2)(p^2 + \omega^2)} \quad (23.24)$$

where $\omega_{R(j)}$ is the natural circular frequency of the Rayleigh beam

$$\omega_{R(j)}^2 = \frac{j^4 \pi^4}{l^4} \frac{EJ}{\mu_R},$$
$$\mu_R = \mu \left(1 + \frac{j^2 \pi^2 r^2}{l^2} \right). \quad (23.25)$$

The inverse transformations of (23.24) in accordance with (1.9) and (27.32) or (27.35) give

$$v(x, t) = \sum_{j=1, j \neq n}^{\infty} \frac{2P}{\mu_R l} \frac{\sin j\pi x/l}{\omega_{R(j)}^2 - \omega^2} \left(\sin \omega t - \frac{\omega}{\omega_{R(j)}} \sin \omega_{R(j)} t \right) +$$
$$+ \frac{P \sin n\pi x/l}{\mu_R l \omega_{R(n)}^2} \left(\sin \omega_{R(n)} t - \omega_{R(n)} t \cos \omega_{R(n)} t \right). \quad (23.26)$$

The second term on the right-hand side of (23.26) is again the component of deflection at such $j = n$ at which $\omega_{R(n)} = \omega$. The vibration of the Rayleigh beam is completely analogous to the vibration of the Bernoulli-Euler beam, the respective results differing in the natural frequency only.

23.2 *Infinite beam on elastic foundation*

In the case of an infinite beam, the force P is assumed to move at constant speed c from infinity to infinity with the result that there obtains the so-called quasi-stationary state (see Sect. 13.1) for which we introduce the single independent variable

$$s = \lambda(x - ct) \quad \text{where} \quad \lambda^4 = k/(4EJ) . \tag{23.27}$$

Assuming this we may write the steady-state deflection and rotation in the form

$$v(x, t) = v_0 v(s) , \quad \psi(x, t) = \psi(s) , \quad v_0 = P/(8\lambda^3 EJ) . \tag{23.28}$$

In addition to notations (13.9) and (23.7) we shall introduce the following

$$\begin{aligned} \alpha_1^2 &= \frac{c^2}{c_1^2} = \frac{\varrho c^2}{E} = 4\frac{\alpha_1^2}{\alpha_2^2} m^2\alpha^2 = 4r^2\lambda^2\alpha^2 , \\ \alpha_2^2 &= \frac{c^2}{c_2^2} = \frac{\varrho c^2}{k^*G} = 4m^2\alpha^2 , \quad \alpha_1 < \alpha_2 , \\ m^2 &= \frac{EJ\lambda^2}{k^*GF} = \frac{c_1^2}{c_2^2} r^2\lambda^2 = \frac{\alpha_2^2}{\alpha_1^2} r^2\lambda^2 . \end{aligned} \tag{23.29}$$

By help of (13.2) to (13.9) and (23.27) to (23.29), Eqs. (23.6) and (23.14) may be reduced to the form

$$\begin{aligned} -(1 - \alpha_2^2)\, v''(s) + 4m^2\, v(s) + \frac{1}{v_0\lambda}\, \psi'(s) &= 8m^2\, \delta(s) , \\ \frac{v_0\lambda}{m^2}\, v'(s) + (1 - \alpha_1^2)\, \psi''(s) - \frac{1}{m^2}\, \psi(s) &= 0 \end{aligned} \tag{23.30}$$

where the primes denote the derivatives with respect to s.

In the solution of system (23.30) we will make use of the Fourier integral transformation (13.14) and the other applicable relation

$$\begin{aligned} \Psi(q) &= \int_{-\infty}^{\infty} \psi(s)\, e^{-iqs}\, ds , \\ \psi(s) &= \frac{1}{2\pi}\int_{-\infty}^{\infty} \Psi(q)\, e^{isq}\, dq . \end{aligned} \tag{23.31}$$

In view of (27.79), (27.80) and (27.84), the transformation of system (23.30) will give

$$[(1 - \alpha_2^2)\, q^2 + 4m^2]\, V(q) + \frac{\mathrm{i}q}{v_0\lambda}\, \Psi(q) = 8m^2\,,$$

$$\frac{v_0\lambda}{m^2}\, \mathrm{i}q\, V(q) - \left[(1 - \alpha_1^2)\, q^2 + \frac{1}{m^2}\right] \Psi(q) = 0\,.$$

The solution of this system and the inverse transformations indicated in (13.14) and (23.31) yield the beam deflection and rotation in the integral form

$$v(s) = \frac{4}{\pi} \int_{-\infty}^{\infty} \frac{[1 + m^2(1 - \alpha_1^2)\, q^2]\, \mathrm{e}^{\mathrm{i}sq}}{A^2 q^4 - 4Bq^2 + 4}\, \mathrm{d}q\,,$$

$$\psi(s) = \frac{4v_0\lambda\mathrm{i}}{\pi} \int_{-\infty}^{\infty} \frac{q\mathrm{e}^{\mathrm{i}sq}}{A^2 q^4 - 4Bq^2 + 4}\, \mathrm{d}q \tag{23.32}$$

where

$$A^2 = (1 - \alpha_1^2)(1 - \alpha_2^2)\,,$$

$$B = \alpha^2 - m^2(1 - \alpha_1^2)\,. \tag{23.33}$$

By (23.3) the bending moment is

$$M(x, t) = -EJ\, \frac{\partial^2 v(x, t)}{\partial x^2} = -EJ\lambda\, \frac{\mathrm{d}\psi(s)}{\mathrm{d}s}\,. \tag{23.34}$$

As suggested by the different values of A^2 and B in Table 23.1, (23.32) express in the integral form the deflections and rotations of all models of beams: the shear beam for which $c_1 \to \infty$, $\alpha_1 = 0$, the Rayleigh beam with $c_2 \to \infty$, $\alpha_2 = 0$, $m = 0$, and the Bernoulli-Euler beam for which $c_1 \to \infty$, $c_2 \to \infty$, $\alpha_1 = 0$, $\alpha_2 = 0$, $m = 0$ [cf. Eq. (13.17)]. From these conditions also follow the values of A^2 and B of the various types of beams – see Table 23.1.

Integrals (23.32) depend on the poles of the function of the complex variable in the integrand. The poles – obtained as the roots of the biquadratic equation

$$A^2 q^4 - 4Bq^2 + 4 = 0$$

Table 23.1 Characteristics of various beam models

Characteristics \ Beam model	Bernoulli-Euler $\alpha_1 = \alpha_2 = m = 0$	Rayleigh $\alpha_2 = m = 0$	Shear beam $\alpha_1 = 0$	Timoshenko
A^2	1	$1 - \alpha_1^2$	$1 - \alpha_2^2$	$(1 - \alpha_1^2)(1 - \alpha_2^2)$
B	α^2	α^2	$\alpha^2 - m^2$	$\alpha^2 - m^2(1 - \alpha_1^2)$
α_0	1	$\left\{2r^2\lambda^2\left[-1 + \left(1 + \frac{1}{4r^4\lambda^4}\right)^{\frac{1}{2}}\right]\right\}^{\frac{1}{2}}$	$(1 - m^2)^{\frac{1}{2}}$	$\left\{\frac{m^2 D}{C^2}\left[1 \pm \left(1 + \frac{1 - m^4}{m^4}\frac{C^2}{D^2}\right)^{\frac{1}{2}}\right]\right\}^{\frac{1}{2}}$
$\alpha_1'\ (\alpha_1 = 1)$	∞	$\frac{1}{2r\lambda}$	∞	$\frac{\alpha_2}{\alpha_1}\frac{1}{2m}$
$\alpha_2'\ (\alpha_2 = 1)$	∞	∞	$\frac{1}{2m}$	$\frac{1}{2m}$
α_3	0	0	m	$m/(1 + 4\alpha_1^2 m^4/\alpha_2^2)^{\frac{1}{2}}$

are in the form

$$q^2 = \frac{2B}{A^2}\left[1 \pm \left(1 - \frac{A^2}{B^2}\right)^{1/2}\right]. \tag{23.35}$$

Generally, they are the complex quantities $q = a + ib$.

Consequently, the position of the poles depends on the magnitude of A^2 and B. Since it is possible that $A^2 \gtreqless 0$, $B \gtreqless 0$, $B^2 \geqq 0$, we must compute the characteristic dimensionless speeds α (13.9) at which those quantities reach special values:

$A^2 = 0$ can occur at $\alpha_1 = 1$ as well as at $\alpha_2 = 1$, i.e. when the speed of the moving force equals the velocity of propagation of longitudinal or transverse waves. Substitution of these values in (23.29) gives the appertaining values of α (marked with prime and the respective subscript)

$$\alpha_1' = \frac{\alpha_2}{\alpha_1}\frac{1}{2m}, \quad \alpha_2' = \frac{1}{2m}, \quad \alpha_1' > \alpha_2'. \tag{23.36}$$

$B = 0$ occurs for $\alpha = \alpha_3$ where

$$\alpha_3 = \frac{m}{(1 + 4\alpha_1^2 m^4/\alpha_2^2)^{1/2}} \,. \tag{23.37}$$

$A^2 = B^2$ occurs for $\alpha = \alpha_0$. This case is of importance because it is accompanied by the appearance of multiple poles.

A survey of the values of α_0, α_1', α_2' and α_3 of the various kinds of beam is in Table 23.1. An analysis of curves A^2 and B^2 at the points of intersection of which lies α_0 must, of course, be made separately for each type of beam.

23.2.1 *Timoshenko beam*

Two of the several types of curves A^2 and B^2 of a Timoshenko beam are shown in Fig. 23.2. After the substitution of (23.29) the condition that $A^2 = B^2$ may be given the form of the biquadratic equation in α

$$C^2\alpha^4 - 2m^2 D\alpha^2 + m^4 - 1 = 0 \tag{23.38}$$

where

$$C = 1 - 4m^4\alpha_1^2/\alpha_2^2 \,,$$

$$D = \frac{2\alpha_1^2}{\alpha_2^2}(2m^4 - 1) - 1 \,.$$

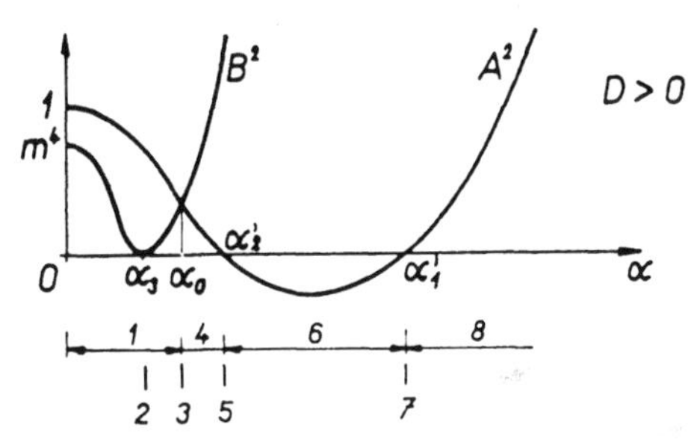

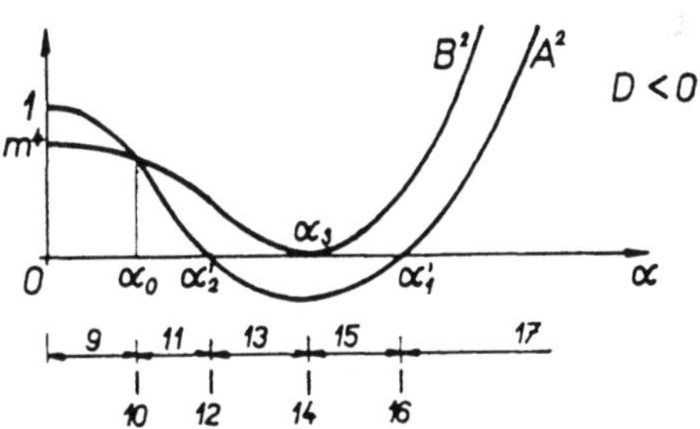

Fig. 23.2. Curves A^2 and B^2 of a Timoshenko beam for $m < 1$ and $D > 0$ or $D < 0$.

The roots of Eq. (23.38) are

$$\alpha_0^2 = \frac{m^2 D}{C^2}\left[1 \pm \left(1 + \frac{1 - m^4}{m^4}\frac{C^2}{D^2}\right)^{1/2}\right] \tag{23.39}$$

and generally they are complex numbers $\alpha_0 = x + iy$. Since it may be $m \gtreqqless 0$, $C^2 \geqq 0$, $D \gtreqqless 0$, the magnitude of the real and the imaginary parts depends above all on the value of m and D. The positions of roots α_0 relative to α_1', α_2' and α_3 for various values of m and D are shown schematically in Fig. 23.3.

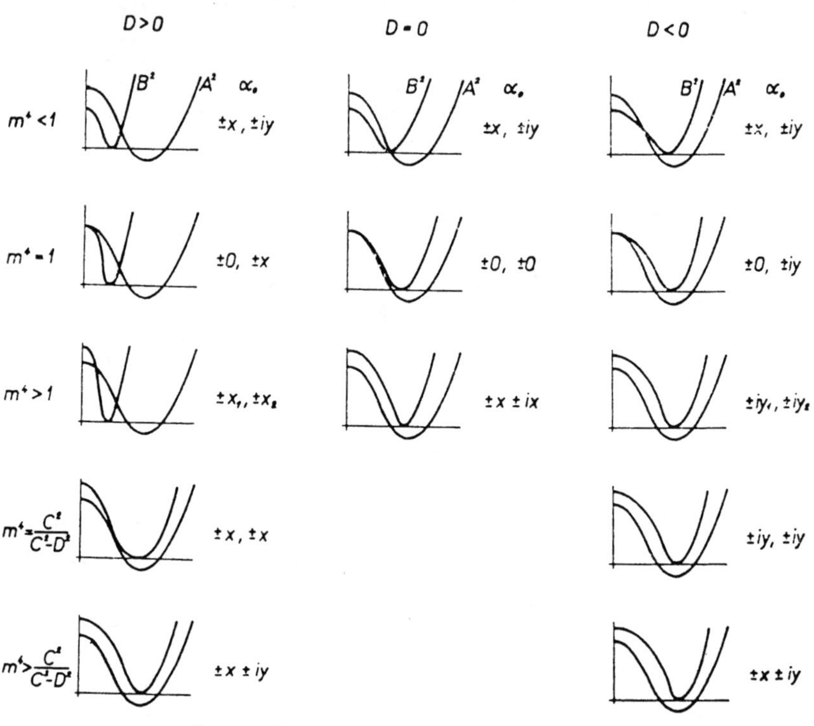

Fig. 23.3. Curves A^2 and B^2 and their points of intersection $\alpha_0 = x + iy$ for various values of m and D.

We are now ready to undertake the analysis of the poles defined by (23.35). The magnitude of the real and the imaginary parts of poles $q = a + ib$ will clearly depend on whether $A^2 > B^2$, $A^2 = B^2$, $B^2 > A^2$ or possibly $A^2 = 0$ or $B = 0$. The analysis shown in Table 23.2 has been made for the cases of $D > 0$ and $D < 0$ and $m^4 < 1$, i.e. for the two limiting

Table 23.2 Poles and the range of validity of the various cases, Timoshenko beam

$A^2 \gtreqless B^2$	$\frac{2B}{A^2}$	A^2	B	Poles $q = a + ib$	$m^4 < 1$			
					$D > 0$		$D < 0$	
					Range of α	Case No.	Range of α	Case No.
$A^2 > B^2$		>0		$\pm a \pm ib$	$0 \leqq \alpha < \alpha_0$	1	$0 \leqq \alpha < \alpha_0$	9
$A^2 = B^2$	>0	>0	>0	$\pm a, \pm a$	$\alpha = \alpha_0$	3	—	—
	<0	>0	<0	$\pm ib, \pm ib$	—	—	$\alpha = \alpha_0$	10
$B^2 > A^2$	>0	>0	>0	$\pm a_1, \pm a_2$	$\alpha_0 < \alpha < \alpha_2'$	4		
					$\alpha_1' < \alpha$	8	$\alpha_1' < \alpha$	17
	>0	<0	<0	$\pm a, \pm ib$	—	—	$\alpha_2' < \alpha < \alpha_3$	13
	<0	<0	>0	$\pm a, \pm ib$	$\alpha_2' < \alpha < \alpha_1'$	6	$\alpha_3 < \alpha < \alpha_1'$	15
	<0	>0	<0	$\pm ib_1, \pm ib_2$	—	—	$\alpha_0 < \alpha < \alpha_2'$	11
$A^2 = 0$		0	>0	$\pm a$	$\alpha = \alpha_2'$	5	$\alpha = \alpha_1'$	16
					$\alpha = \alpha_1'$	7		
		0	<0	$\pm ib$	—	—	$\alpha = \alpha_2'$	12
$B^2 = 0$		>0	0	$\pm a \pm ia$	$\alpha = \alpha_3$	2	—	—
		<0	0	$\pm a, \pm ia$	—	—	$\alpha = \alpha_3$	14
$A^2 = B^2 = 0$		0	0		$\alpha = \alpha_2' = \alpha_3$	18	$\alpha = \alpha_2' = \alpha_3$	18

cases in the first row of Fig. 23.3. In the remaining cases the calculation would be completely analogous.

If we know the types of poles $q \equiv A_i = a + ib$, $i = 1, 2, 3, 4$, we get their real and imaginary parts from a comparison of the coefficients at like powers of q in the expression

$$
\begin{aligned}
A^2q^4 - 4Bq^2 + 4 &= A^2(q - A_1)(q - A_2)(q - A_3)(q - A_4) = \\
&= A^2\{q^4 - (A_1 + A_2 + A_3 + A_4)\, q^3 + \\
&+ [A_1A_2 + (A_1 + A_2)(A_3 + A_4) + A_3A_4]q^2 - \\
&- [A_1A_2(A_3 + A_4) + A_3A_4(A_1 + A_2)]\, q + A_1A_2A_3A_4\}\,. \quad (23.40)
\end{aligned}
$$

Integrals (23.32) are evaluated with the aid of the residue theorem (cf. Sect. 13.3). In the paragraphs that follow we shall present the results of calculation of a, b, deflection, rotation and bending moments of various cases arranged in the order of growing dimensionless speed α (13.9), made first for $D > 0$, then for $D < 0$, with m always less than unity. The ordinal numbers of the various cases correspond to Table 23.2 and Fig. 23.2.

a) $m < 1$, $D > 0$

Case 1 $(0 \leqq \alpha < \alpha_0)$

The poles are complex conjugate, $\pm a \pm ib$. From (23.40)

$$a^2 = \frac{1}{A}\left(1 + \frac{B}{A}\right), \quad b^2 = \frac{1}{A}\left(1 - \frac{B}{A}\right).$$

Following the integration in the complex plane we get

$$v(s) = \frac{e^{-b|s|}}{Aab}\left\{a\left[1 + \frac{2m^2(1-\alpha_1^2)}{A}\right]\cos as + b\left[1 - \frac{2m^2(1-\alpha_1^2)}{A}\right]. \right.$$
$$\left. . \sin a|s|\right\},$$

$$\psi(s) = -\frac{2v_0\lambda}{A^2ab}\, e^{-b|s|} \sin as\,,$$

$$M(s) = \frac{M_0}{A^2ab}\, e^{-b|s|}\,(a\cos as - b\sin a|s|) \tag{23.41}$$

where

$$M_0 = \frac{P}{4\lambda}\,.$$

Case 2 $(\alpha = \alpha_3)$

Since $\alpha_3 < \alpha_0$, α_3 lies in the range of Case 1. And in fact, the poles are complex conjugate as in Case 1; their real part, however, is equal to the imaginary part $\pm a \pm ia$

$$a^2 = b^2 = \frac{1}{A}\,.$$

With these values substituted in, the solution becomes the same as in Case 1, i.e. (23.41).

Case 3 $(\alpha = \alpha_0)$

The poles are double and lie on the real axis $(\pm a)$. There exists no solution to this case in the absence of damping. We may imagine that the beam will lose stability as though acted upon by the axial critical Euler force.

Case 4 $(\alpha_0 < \alpha < \alpha_2')$

Poles $\pm a_1$, $\pm a_2$ lie on the real axis. As revealed by an analysis of the effect of damping in the foundation [3], poles $\pm a_1$ have a small positive imaginary part while poles $\pm a_2$ have a small negative imaginary part. That is the reason why in the absence of damping, poles $\pm a_1$ are included in the upper, poles $\pm a_2$ in the lower half-plane in the limit.

Thus we get

$$a_1^2 = \frac{2B}{A^2}\left[1 + \left(1 - \frac{A^2}{B^2}\right)^{1/2}\right], \quad a_2^2 = \frac{2B}{A^2}\left[1 - \left(1 - \frac{A^2}{B^2}\right)^{1/2}\right],$$

$$v(s) = -\frac{4[1 + a_1^2 m^2(1 - \alpha_1^2)]}{a_1(A^2 a_1^2 - 2B)} \sin a_1 s \quad \text{for} \quad s > 0\,,$$

$$v(s) = \frac{4[1 + a_2^2 m^2(1 - \alpha_1^2)]}{a_2(A^2 a_2^2 - 2B)} \sin a_2 s \quad \text{for} \quad s < 0\,,$$

$$\psi(s) = -\frac{4v_0\lambda}{A^2 a_1^2 - 2B} \cos a_1 s \quad \text{for} \quad s > 0\,,$$

$$\psi(s) = \frac{4v_0\lambda}{A^2 a_2^2 - 2B} \cos a_2 s \quad \text{for} \quad s < 0\,,$$

$$M(s) = -\frac{2M_0 a_1}{A^2 a_1^2 - 2B} \sin a_1 s \quad \text{for} \quad s > 0\,,$$

$$M(s) = \frac{2M_0 a_2}{A^2 a_2^2 - 2B} \sin a_2 s \quad \text{for} \quad s < 0\,. \qquad (23.42)$$

Case 5 $(\alpha = \alpha_2')$

When the velocity of propagation of transverse waves along the beam is attained, poles $\pm a_1$ of the preceding case move to infinity; there therefore remain only poles $\pm a_2$, which we shall denote $\pm a$, and count to the lower half-plane.

From (23.40)

$$a^2 = 1/B\,.$$

Proceeding in the same way as in the integration of case 4 we get

for $s > 0$

$$v(s) = 0\,, \quad \psi(s) = 0\,, \quad M(s) = 0\,,$$

and for $s < 0$

$$\begin{aligned} v(s) &= -2a[1 + a^2m^2(1 - \alpha_1^2)]\sin as\,, \\ \psi(s) &= -2v_0\lambda a^2\cos as\,, \\ M(s) &= -M_0a^3\sin as\,. \end{aligned} \tag{23.43}$$

Case 6 $(\alpha_2' < \alpha < \alpha_1')$

Like in the preceding case, the two poles, $\pm a$, on the real axis are included in the lower half-plane (as the limit case with damping considered), next to them there exist two poles $\pm ib$ on the imaginary axis.

Expression (23.40) results in

$$a^2 = -\frac{2B}{A^2}\left[-1 + \left(1 - \frac{A^2}{B^2}\right)^{1/2}\right],$$

$$b^2 = -\frac{2B}{A^2}\left[1 + \left(1 - \frac{A^2}{B^2}\right)^{1/2}\right],$$

and the integration in the complex plane gives

$$v(s) = -\frac{2[1 - b^2m^2(1 - \alpha_1^2)]}{b(A^2b^2 + 2B)}\,e^{-bs} \qquad \text{for} \quad s > 0\,,$$

$$v(s) = -\frac{2[1 - b^2m^2(1 - \alpha_1^2)]}{b(A^2b^2 + 2B)}\,e^{bs} + \frac{4[1 + a^2m^2(1 - \alpha_1^2)]}{a(A^2a^2 - 2B)}\sin as \qquad \text{for} \quad s < 0\,,$$

$$\psi(s) = \frac{2v_0\lambda}{A^2b^2 + 2B}\,e^{-bs} \qquad \text{for} \quad s > 0\,,$$

$$\psi(s) = -\frac{2v_0\lambda}{A^2b^2 + 2B}\,e^{bs} + \frac{4v_0\lambda}{A^2a^2 - 2B}\cos as \qquad \text{for} \quad s < 0\,,$$

$$M(s) = \frac{M_0 b}{A^2 b^2 + 2B} \mathrm{e}^{-bs} \qquad \text{for} \quad s > 0,$$

$$M(s) = M_0 \left(\frac{b}{A^2 b^2 + 2B} \mathrm{e}^{bs} + \frac{2a}{A^2 a^2 - 2B} \sin as \right) \qquad \text{for} \quad s < 0. \tag{23.44}$$

Case 7 $(\alpha = \alpha_1')$

The case of the attainment of the velocity of propagation of longitudinal waves along the beam is exactly the same as Case 5. Poles $\pm a$ are computed from $a^2 = 1/B$, and Eq. (23.43) applies also.

Case 8 $(\alpha_1' < \alpha)$

At very high speeds, poles $\pm a_1$, $\pm a_2$ lie on the real axis. In the presence of damping all poles have a small negative imaginary part, and that is why in the absence of damping they are in the limit included in the lower half-plane. The values of a_1, a_2 are the same as in Case 4 but the beam deformations are as follows:

for $s > 0$:

$$v(s) = 0, \quad \psi(s) = 0, \quad M(s) = 0,$$

for $s < 0$:

$$v(s) = \frac{4[1 + a_1^2 m^2 (1 - \alpha_1^2)]}{a_1 (A^2 a_1^2 - 2B)} \sin a_1 s + \frac{4[1 + a_2^2 m^2 (1 - \alpha_1^2)]}{a_2 (A^2 a_2^2 - 2B)} \sin a_2 s,$$

$$\psi(s) = 4 v_0 \lambda \left(\frac{1}{A^2 a_1^2 - 2B} \cos a_1 s + \frac{1}{A^2 a_2^2 - 2B} \cos a_2 s \right),$$

$$M(s) = 2 M_0 \left(\frac{a_1}{A^2 a_1^2 - 2B} \sin a_1 s + \frac{a_2}{A^2 a_2^2 - 2B} \sin a_2 s \right). \tag{23.45}$$

b) $m < 1$, $D < 0$.

Case 9 $(0 \leqq \alpha < \alpha_0)$

This case is the same as Case 1 (see Table 23.2) and therefore Eq. (23.41) applies also.

Case 10 $(\alpha = \alpha_0)$

The double poles $\pm ib$ lie on the imaginary axis. Eq. (23.40) gives

$$b^2 = -2/B \quad \text{or} \quad b^4 = 4/A^2$$

and the integration in the complex plane

$$\begin{aligned} v(s) &= \tfrac{1}{2}b\,{}^{-b|s|}\{1 + b^2m^2(1 - \alpha_1^2) + b|s|\,[1 - b^2m^2(1 - \alpha_1^2)]\}\,, \\ \psi(s) &= -\tfrac{1}{2}v_0\lambda b^2 \mathrm{e}^{-b|s|}bs\,, \\ M(s) &= \tfrac{1}{4}M_0b^3\mathrm{e}^{-b|s|}(1 - b|s|)\,. \end{aligned} \tag{23.46}$$

Case 11 $(\alpha_0 < \alpha < \alpha_2')$

All four poles lie on the imaginary axis, $\pm ib_1$, $\pm ib_2$. From (23.40)

$$b_1^2 = -\frac{2B}{A^2}\left[1 + \left(1 - \frac{A^2}{B^2}\right)^{1/2}\right],$$

$$b_2^2 = -\frac{2B}{A^2}\left[1 - \left(1 - \frac{A^2}{B^2}\right)^{1/2}\right].$$

In this case the beam deformations are

$$\begin{aligned} v(s) &= -2\left[\frac{1 - b_1^2m^2(1 - \alpha_1^2)}{b_1(A^2b_1^2 + 2B)}\,\mathrm{e}^{-b_1|s|} + \frac{1 - b_2^2m^2(1 - \alpha_1^2)}{b_2(A^2b_2^2 + 2B)}\,\mathrm{e}^{-b_2|s|}\right], \\ \psi(s) &= \pm 2\,v_0\lambda\left(\frac{1}{A^2b_1^2 + 2B}\,\mathrm{e}^{-b_1|s|} + \frac{1}{A^2b_2^2 + 2B}\,\mathrm{e}^{-b_2|s|}\right) \quad \text{for} \quad s \gtrless 0\,, \\ M(s) &= M_0\left(\frac{b_1}{A^2b_1^2 + 2B}\,\mathrm{e}^{-b_1|s|} + \frac{b_2}{A^2b_2^2 + 2B}\,\mathrm{e}^{-b_2|s|}\right). \end{aligned} \tag{23.47}$$

Case 12 $(\alpha = \alpha_2')$

When the velocity of propagation of transverse waves along the beam is reached, poles $\pm ib_1$ move to infinity and only poles $\pm ib_2$ remain; here we denote them by $\pm ib$ where

$$b^2 = -1/B\,.$$

After integration,

$$\begin{aligned} v(s) &= b[1 - b^2m^2(1 - \alpha_1^2)]\,\mathrm{e}^{-b|s|}\,, \\ \psi(s) &= \mp v_0\lambda b^2\mathrm{e}^{-b|s|} \quad \text{for} \quad s \gtrless 0\,, \\ M(s) &= -\tfrac{1}{2}M_0b^3\mathrm{e}^{-b|s|}\,. \end{aligned} \tag{23.48}$$

Case 13 $(\alpha_2' < \alpha < \alpha_3)$

The poles now are on the imaginary axis $\pm ib$ and on the real axis $\pm a$. The latter are included in the lower half-plane. Expression (23.40) gives

$$a^2 = \frac{2B}{A^2}\left[1 + \left(1 - \frac{A^2}{B^2}\right)^{1/2}\right], \quad b^2 = \frac{2B}{A^2}\left[-1 + \left(1 - \frac{A^2}{B^2}\right)^{1/2}\right],$$

and the integration

$$v(s) = -\frac{2[1 - b^2m^2(1 - \alpha_1^2)]}{b(A^2b^2 + 2B)}\, e^{-bs} \qquad \text{for} \quad s > 0\,,$$

$$v(s) = -\frac{2[1 - b^2m^2(1 - \alpha_1^2)]}{b(A^2b^2 + 2B)}\, e^{bs} + \frac{4[1 + a^2m^2(1 - \alpha_1^2)]}{a(A^2a^2 - 2B)} \sin as \qquad \text{for} \quad s < 0\,,$$

$$\psi(s) = \frac{2v_0\lambda}{A^2b^2 + 2B}\, e^{-bs} \qquad \text{for} \quad s > 0\,,$$

$$\psi(s) = -\frac{2v_0\lambda}{A^2b^2 + 2B}\, e^{bs} + \frac{4v_0\lambda}{A^2a^2 - 2B} \cos as \qquad \text{for} \quad s < 0\,,$$

$$M(s) = \frac{M_0 b}{A^2b^2 + 2B}\, e^{-bs} \qquad \text{for} \quad s > 0\,,$$

$$M(s) = M_0\left(\frac{b}{A^2b^2 + 2B}\, e^{bs} + \frac{2a}{A^2a^2 - 2B} \sin as\right) \qquad \text{for} \quad s < 0\,. \tag{23.49}$$

Case 14 $(\alpha = \alpha_3)$

Poles $\pm a$, $\pm ia$ now have equal real and imaginary parts; this means that in the preceding Case 13 we set $a = b$ with

$$a^4 = -4/A^2\,.$$

With this notation the beam deformations are the same as (23.49).

Case 15 $(\alpha_3 < \alpha < \alpha_1')$

Poles $\pm a$, $\pm ib$ have the same form as those of Case 13; in view of the magnitude of A^2 and B, their notation, however, is somewhat different

$$a^2 = -\frac{2B}{A^2}\left[-1+\left(1-\frac{A^2}{B^2}\right)^{1/2}\right], \quad b^2 = -\frac{2B}{A^2}\left[1+\left(1-\frac{A^2}{B^2}\right)^{1/2}\right].$$

The applicable equation is (23.49) with these values of a and b substituted in.

Case 16 $(\alpha = \alpha_1')$

According to Table 23.2, this case with poles $\pm a$ is exactly the same as Case 7 and hence also as Case 5. The applicable equations therefore are (23.43).

Case 17 $(\alpha_1' < \alpha)$

At very high speeds the situation is as that in Case 8; the applicable equations therefore are (23.45) – see also Table 23.2.

Case 18 $(\alpha = \alpha_2' = \alpha_3)$

Irrespective of the magnitude of m and D, another interesting case occurs whenever the dimensionless speed becomes equal to the velocity of propagation of transverse waves along the beam, which is simultaneously equal to α_3. Such a situation arises just when curves A^2 and B^2 intersect on axis α, i.e. when $A^2 = B^2 = 0$.

Then the integration in Eqs. (23.32) can be carried out with the aid of generalized functions of higher orders and their integral representations in the complex plane (according to [201], Vol. III/2. Sect. 61, p. 226):

$$\delta(x) = \frac{1}{2\pi}\int_{-\infty}^{\infty} e^{ixz}\,dz\,,$$

$$H_1(x) = \frac{d\delta(x)}{dx} = -\frac{1}{2\pi i}\int_{-\infty}^{\infty} z e^{ixz}\,dz\,,$$

$$H_2(x) = \frac{dH_1(x)}{dx} = \frac{d^2\delta(x)}{dx^2} = -\frac{1}{2\pi}\int_{-\infty}^{\infty} z^2 e^{ixz}\,dz\,. \qquad (23.50)$$

Using formulae (23.50) we get from (23.32) and (23.34)

$$v(s) = 2[\delta(s) - m^2(1-\alpha_1^2)\,H_2(s)]\,,$$
$$\psi(s) = 2v_0\lambda\,H_1(s)\,,$$
$$M(s) = -M_0\,H_2(s)\,. \qquad (23.51)$$

Table 23.3 Poles and the range of validity of the various cases, shear beam

$A^2 \gtreqless B^2$	$\frac{2B}{A^2}$	A^2	B	Poles $q = a + ib$	$m^4 < 1$			
					$m^2 < \frac{1}{2}$		$m^2 > \frac{1}{2}$	
					Range of α	Case No.	Range of α	Case No.
$A^2 > B^2$		>0		$\pm a \pm ib$	$0 \leqq \alpha < \alpha_0$	19=1	$0 \leqq \alpha < \alpha_0$	25=1=9
$A^2 = B^2$	>0 <0	>0 >0	>0 <0	$\pm a, \pm a$ $\pm ib, \pm ib$	$\alpha = \alpha_0$ —	21=3 —	— $\alpha = \alpha_0$	— 26=10
$B^2 > A^2$	>0 >0 <0 <0	>0 <0 >0 <0	>0 <0 <0 >0	$\pm a_1, \pm a_2$ $\pm a, \pm ib$ $\pm ib_1, \pm ib_2$ $\pm a, \pm ib$	$\alpha_0 < \alpha < \alpha_2'$ — — $\alpha_2' < \alpha$	22=4 — — 24=6	— $\alpha_2' < \alpha < \alpha_3$ $\alpha_0 < \alpha < \alpha_2'$ $\alpha_3 < \alpha$	— 29=13 27=10 31=15
$A^2 = 0$		0 0	>0 <0	$\pm a$ $\pm ib$	$\alpha = \alpha_2'$ —	23=5 —	— $\alpha = \alpha_2'$	— 28=12
$B = 0$		>0 <0	0 0	$\pm a \pm ia$ $\pm a, \pm ia$	$\alpha = \alpha_3$ —	20=2 —	— $\alpha = \alpha_3$	— 30=14
$A^2 = B^2 = 0$		0	0		$\alpha = \alpha_0 = \alpha_2' = \alpha_3$ $m^2 = \frac{1}{2}$	31=18	$\alpha = \alpha_0 = \alpha_2' = \alpha_3$	31=18

23.2.2 *Shear beam*

The values of A^2, B, α_0, α_2' and α_3 computed for the shear beam are set out in Table 23.1. The poles of expressions (23.32) depend on the course of curves A^2 and B^2. Two of the possible types of these curves are shown in Fig. 23.4 ($m^2 < 1/2$ and $1/2 < m^2 < 1$). The type of poles and the range of validity of the various cases can be determined from an analysis of curves A^2 and B^2 and from the magnitude of A^2 and B. The pertinent data are summarized in Table 23.3 (see also Fig. 23.4).

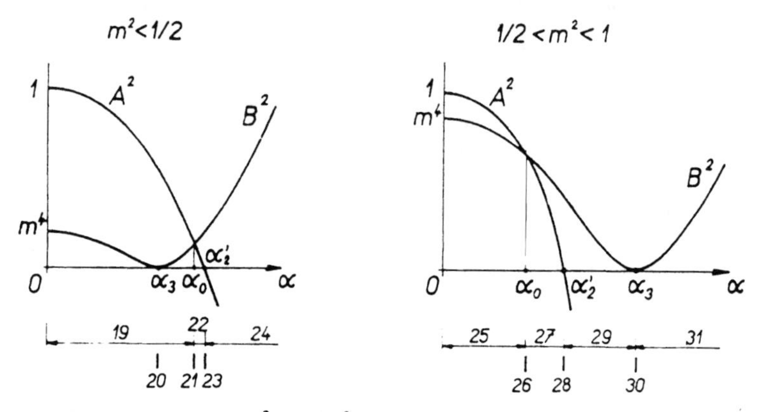

Fig. 23.4. Curves A^2 and B^2 of a beam with shear considered, for $m^2 < 1/2$ and $1/2 < m^2 < 1$.

The deformations of the shear beam in the various cases considered are the same as those of the corresponding cases of the Timoshenko beam (cf. Table 23.3). As indicated in Table 23.1, the values to be used in the respective formulae are $\alpha_1 = 0$, $A^2 = 1 - \alpha_2^2$, $B = \alpha^2 - m^2$.

23.2.3 *Rayleigh beam*

The values of A^2, B, α_0, α_1' of the Rayleigh beam are listed in Table 23.1, the curves A^2 and B^2 drawn in Fig. 23.5. The type of poles and the range of validity of the various cases are again established on the basis of an analysis of the curves and from the magnitude of A^2 and B. The data thus obtained are shown in Table 23.4 and Fig. 23.5.

The deflections and bending moments of the Rayleigh beam are again the same as those of the Timoshenko beam in the corresponding cases (see Table 23.4). The values to be substituted in the appertaining formulae are $m = 0$, $A^2 = 1 - \alpha_1^2$, $B = \alpha^2$ (Table 23.1).

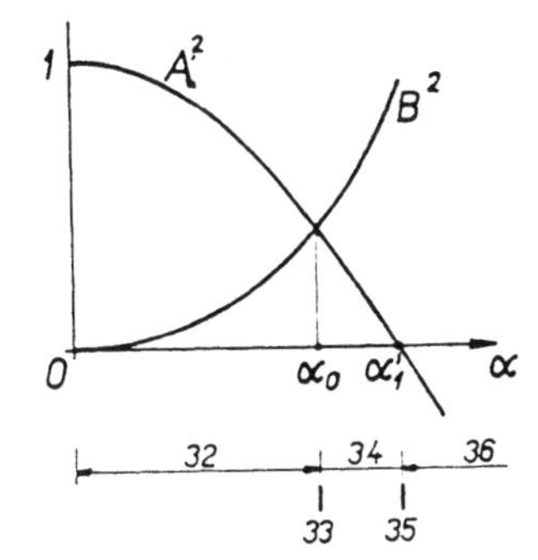

Fig. 23.5 Curves A^2 and B^2 for a Rayleigh beam.

Table 23.4 Poles and the range of validity of the various cases, Rayleigh beam

$A^2 \gtrless B^2$	$\frac{2B}{A^2}$	A^2	B	Poles $q = a + ib$	Range of α	Case No.
$A^2 > B^2$		>0		$\pm a \pm ib$	$0 \leqq \alpha < \alpha_0$	32=1
$A^2 = B^2$	>0	>0	>0	$\pm a, \pm a$	$\alpha = \alpha_0$	33=3
$B^2 > A^2$	>0 <0	>0 <0	>0 >0	$\pm a_1, \pm a_2$ $\pm a, \pm ib$	$\alpha_0 < \alpha < \alpha_1'$ $\alpha_1' < \alpha$	34=4 36=6
$A^2 = 0$		0	>0	$\pm a$	$\alpha = \alpha_1'$	35=7=5

23.2.4 *Bernoulli-Euler beam*

The Bernoulli-Euler beam, for which $A^2 = 1$, $B = \alpha^2$, and $\alpha_0 = 1$, would be analyzed in an analogous manner (Table 23.1). If the pertinent formulae are evaluated for $m = 0$, $\alpha_1 = 0$, $\alpha_2 = 0$, we get Case 1 for $0 \leqq \alpha < \alpha_0$, Case 3 for $\alpha = \alpha_0$, and Case 4 for $\alpha_0 < \alpha$. This is in complete agreement with the results of Chap. 13.

23.3 *Application of the theory*

The theory discussed in this section can be used in calculations relating to deep beams, materials sensitive to shear stresses, etc., especially when very high speeds of the moving force are involved.

So far as beams of finite length are concerned, there exist for the Timoshenko beam two speeds and for the simpler beam models but one speed at which the deflections grow with time. Since in this instance the processes are transient, the amplitudes will never reach dangerous values within a finite time interval.

As to infinite beams on elastic foundations under steady-state vibrations produced by a moving force: For all types of beams there exists a speed, α_0, at which the beams lose stability. When the speed equals the velocity of propagation of longitudinal waves (α_1) or of transverse waves (α_2) along the beam, the beam is not stressed in front of the moving load. This is so even in the case of a Timoshenko beam at speeds higher than α_1'. However, beams of the simpler types are stressed in front of the moving load no matter how high the speed is.

It is of interest to note that with materials sensitive to shear (large m) the beams do not lose stability even at speed α_0 — cf. Case 10. This is due to the fact that shear stresses cause a break in the deflection line underneath the load. In the transition between this case (large m) and the case of beam losing stability (at small m) lies the special Case 18.

23.4 *Additional bibliography*

[3, 39, 66, 112, 175, 212, 265].

24

Finite beams subjected to a force moving at high speed

We have seen in the preceding chapters that finite beams possess solution in Fourier series. The series are highly effective at low speeds of load motion. At high speeds of load motion it is already necessary to take a fairly large number of terms of the series. The deflections of finite beams are always finite because the moving force acts on them in a limited time interval only.

Steady-state vibration of infinite beams, on the other hand, are solved for the force moving from infinity to infinity. The solution is very simple and not expressed in series. In the absence of damping there exists here a certain critical speed at which the beam deflections grow beyond all limits.

An attempt to reconcile the two basic methods of solution has been made by C. R. Steele [205] whose method is based on Poisson's summation formula [38], [165]:

Consider function $f(s, x, t)$ of continuous independent variable s and parameters x and t (which are not essential for what follows. They are mentioned here for the sole reason that the beam deflections are functions of x and t). In each finite interval, function $f(s, x, t)$ possesses bounded variation and for $s \to +\infty$, $s \to -\infty$, satisfies one of the following conditions:

a) it is monotonic and absolutely integrable,

b) it is integrable and has absolutely integrable derivative with respect to s.

The Fourier integral transformation of this function will be formed somewhat differently than the conventional (13.14)

$$F(q, x, t) = \int_{-\infty}^{\infty} f(s, x, t)\, e^{-2\pi i q s}\, ds\,. \tag{24.1}$$

Poisson's summation formula then is

$$\sum_{j=-\infty}^{\infty} f(j, x, t) = \sum_{n=-\infty}^{\infty} F(n, x, t) . \tag{24.2}$$

In the above the continuous variables s and q are changed to integer discrete variables j and n in functions $f(s, x, t)$ and $F(q, x, t)$, respectively. The series on the left-hand side of (24.2) is transformed in another series whose terms are the Fourier integrals.

24.1 *Finite beams*

Vibration of finite beams is usually expressed by series (6.2) which represents expansion of beam deflections in normal modes. This series can easily be expanded in the Fourier series in complex form

$$v(x, t) = \sum_{j=1}^{\infty} \frac{\mu}{V_j} V(j, t)\, v_{(j)}(x) = \sum_{j=-\infty}^{\infty} h(j, t)\, \mathrm{e}^{\mathrm{i}j\pi x/l} \tag{24.3}$$

where

$$h(j, t) = \frac{1}{2l} \int_{-l}^{l} v(x, t)\, \mathrm{e}^{-\mathrm{i}j\pi x/l}\, \mathrm{d}x .$$

The right-hand side of Eq. (24.3) has already the form analogous to that of the left-hand side of Eq. (24.2). Assuming that the above-stated requirements on function $h(j, t)\, \mathrm{e}^{\mathrm{i}j\pi x/l}$ are satisfied, one can apply Poisson's summation formula (24.2) with (24.1) substituted in. The result is the infinite series of improper integrals

$$v(x, t) = \sum_{n=-\infty}^{\infty} \int_{-\infty}^{\infty} h(s, t)\, \mathrm{e}^{\mathrm{i}\pi s(x/l - 2n)}\, \mathrm{d}s . \tag{24.4}$$

One finds that the term of series (24.4) for $n = 0$ is the solution of the given case for a semi-infinite beam, i.e. the case in which the boundary conditions are satisfied at the left-hand end and the beam is infinitely long. The other terms of the series for $n \neq 0$ then express the effect of the way the beam is supported on the right-hand side. If the beams are long and the speeds of force motion high, series (24.4) converges very fast.

24.2 *Simply supported beam*

The deflection of a simply supported beam subjected to force P moving at speed c [see (1.31) or (6.27)] is

$$v(x, t) = \frac{2}{l} \sum_{j=1}^{\infty} V(j, t) \sin \frac{j\pi x}{l} \tag{24.5}$$

where

$$V(j, t) = \begin{cases} \dfrac{P}{\mu} \dfrac{1}{\omega_{(j)}^2 - \omega^2} \left(\sin \omega t - \dfrac{\omega}{\omega_{(j)}} \sin \omega_{(j)} t \right) & \text{for } 0 \leqq t \leqq l/c = T \\[2ex] \dfrac{P}{\mu} \dfrac{\omega}{\omega_{(j)}(\omega_{(j)}^2 - \omega^2)} \left[\cos j\pi \sin \omega_{(j)}(t - T) - \sin \omega_{(j)} t \right] & \text{for } \quad t \geqq T , \end{cases}$$

$$\omega = \frac{j\pi c}{l}, \quad \omega_{(j)}^2 = \frac{j^4 \pi^4}{l^4} a^2 , \quad a^2 = \frac{EJ}{\mu} .$$

The sine series (24.5) is put in the complex form

$$v(x, t) = \frac{2}{l} \sum_{j=-\infty}^{\infty} c(j, t) \, \mathrm{e}^{\mathrm{i} j \pi x / l} \tag{24.6}$$

where the Fourier coefficients

$$c(j, t) = \tfrac{1}{2}[a(j, t) - \mathrm{i} V(j, t)] , \quad c(-j, t) = \tfrac{1}{2}[a(j, t) + \mathrm{i} V(j, t)]$$

are computed from the coefficients of the appertaining sine series $V(j, t)$ and the cosine series $a(j, t)$, the latter coefficients being of course zero $[a(j, t) = 0]$ in this case. Substituting the coefficients in (24.6) yields

$$v(x, t) = -\frac{\mathrm{i}}{l} \sum_{j=-\infty}^{\infty} V(j, t) \, \mathrm{e}^{\mathrm{i} j \pi x / l} . \tag{24.7}$$

Applying Poisson's formula (24.2) with the simultaneous substitution of (24.1) in the right-hand side of Eq. (24.7) yields

$$v(x, t) = -\frac{\mathrm{i}}{l} \sum_{n=-\infty}^{\infty} \int_{-\infty}^{\infty} V(s, t) \, \mathrm{e}^{\mathrm{i} \pi s (x/l - 2n)} \, \mathrm{d}s . \tag{24.8}$$

Since in formulae (24.5) j always appears in connection with $j\pi/l$, we will substitute in (24.8) a new variable, $q = s\pi/l$ [q here is obviously different from that in (24.1)]. Consequently the resultant deflection is

$$v(x, t) = \frac{1}{\pi i} \sum_{n=-\infty}^{\infty} \int_{-\infty}^{\infty} V(q, t) \, e^{iq(x-2nl)} \, dq \tag{24.9}$$

where

$$V(q, t) = \frac{P}{\mu a} \frac{1}{q^3(c^2 - a^2 q^2)} (c \sin atq^2 - aq \sin ctq)$$

follows from formula (24.5) for $V(j, t)$ at $0 \leqq t \leqq T$ in which the discrete variable $j\pi/l$ has been changed to continuous variable q.

We shall now prove that for $n = 0$, formula (24.9) will give the solution of a semi-infinite beam subjected to a moving force. Let us solve this case directly from the equation

$$EJ \frac{\partial^4 v(x, t)}{\partial x^4} + \mu \frac{\partial^2 v(x, t)}{\partial t^2} = \delta(x - ct) \, P \, . \tag{24.10}$$

The boundary conditions now are

$$v(x, t)|_{x=0} = 0 \, , \quad v''(x, t)|_{x=0} = 0 \, , \quad v(x, t)|_{x\to\infty} = 0 \, ,$$
$$v'(x, t)|_{x\to\infty} = 0 \, , \quad v''(x, t)|_{x\to\infty} = 0 \, , \quad v'''(x, t)|_{x\to\infty} = 0 \, ,$$

and the initial conditions are (1.3).

In the solution we shall use the Fourier sine transformations (27.85) to (27.87)

$$V(q, t) = \int_0^\infty v(x, t) \sin qx \, dx \, , \quad v(x, t) = \frac{2}{\pi} \int_0^\infty V(q, t) \sin xq \, dq \, . \tag{24.11}$$

With (24.11) and the boundary conditions, Eq. (24.10) will give

$$EJq^4 \, V(q, t) + \mu \, \ddot{V}(q, t) = P \sin qct \, .$$

In view of the initial conditions and (27.4), (27.18) the Laplace-Carson transformation will result in

$$V^*(q, p) = \frac{Pqc}{\mu} \frac{p}{(p^2 + q^2 c^2)(p^2 + q^4 a^2)}$$

and the inverse transformation (27.32) and (24.11) in

$$v(x, t) = \frac{2}{\pi} \int_0^\infty V(q, t) \sin xq \, dq \, . \tag{24.12}$$

Since according to (24.9) function $V(q, t)$ is an odd one, integral (24.12) can also be expressed by integral (24.9) for $n = 0$. It namely holds for odd functions $f(-x) = -f(x)$ that

$$\int_{-\infty}^{\infty} f(z) \, e^{iaz} \, dz = 2i \int_0^\infty f(x) \sin ax \, dx \, .$$

Even though the integrals in (24.9) are quite complicated, C. R. Steele succeeded in evaluating them. For details of the solution the reader is referred to Steele's study [205].

24.3 *Application of the theory*

The theory outlined in the foregoing is useful in problems involving large-span, slender beams subjected to forces moving at high speeds, i.e. for $cl/(\pi a) \gg 1$. As proved by C. R. Steele [205], the convergence of series (24.9) is very fast in such cases. We may also think of the problem as of one in which the load moves at a speed so high that all the beam experiences is an impulse of a sort. This impulse causes the maximum stresses in the beam to come into being only after the departure of the force, i.e. at $t > T$.

24.4 *Additional bibliography*

[119, 205].

25

Non-elastic properties of materials

The real properties of materials used in civil and mechanical engineering for structures intended to support moving loads, are very complicated. They depend not only on the macrophysical characteristics of the structure and its material but on the physical and chemical microstructure of matter as well. A detailed treatment of this topical subject is outside the scope of our book. All we plan to do here is to offer an analysis of beams whose materials possess some very simple non-elastic properties, i.e. examine several cases of viscoelastic and plastic properties of beams.

25.1 *Viscoelastic beam subjected to a moving load*

So far we have been dealing with truly elastic materials, that is to say with materials whose linear stress-strain relation and material constants E, G, λ, ν (see Chap. 17) are independent of time, magnitude of deformation, location, temperature, etc. Some structural materials (e.g. concrete, plastics) have, however, distinctly rheologic properties that cause some of the constants to be dependent on time. Here we will ignore their temperature-induced changes and consider constants E, G, λ to be functions of time alone.

According to W. Nowacki [170], the stress-strain relation of a viscoelastic body can in general be expressed by the linear differential operator relations

$$P_1(D)\, s_{ij} = P_2(D)\, e_{ij}\,, \quad i = 1, 2, 3\,, \quad j = 1, 2, 3\,,$$
$$P_3(D)\, s = P_4(D)\, \Theta \tag{25.1}$$

where

$$s_{ij} = \sigma_{ij} - \tfrac{1}{3} s \delta_{ij},$$
$$e_{ij} = \varepsilon_{ij} - \tfrac{1}{3} \Theta \delta_{ij},$$

$s \quad = \sigma_{11} + \sigma_{22} + \sigma_{33}$,

$\Theta \quad = \varepsilon_{11} + \varepsilon_{22} + \varepsilon_{33}$,

σ_{ij} – components of normal $(i = j)$ or tangential $(i \neq j)$ stress,

ε_{ij} – corresponding components of strain,

$\delta_{ij} = \begin{cases} 1 \text{ for } i = j \\ 0 \text{ for } i \neq j \end{cases}$ Kronecker delta symbol,

$P_i(D)$ – linear differential operators (in a general case they may be linear as well as nonlinear differential and integral operators),

D – symbol of the derivative with respect to time.

For several types of very simple viscoelastic solids, operators $P_i(D)$ are summarized in Table 25.1.

Table 25.1 Properties of viscoelastic solids

Solid	$P_1(D)$	$P_2(D)$	$P_3(D)$	$P_4(D)$
Hooke	1	$2G$	1	$3K$
Kelvin	1	$2G(1 + \eta D)$	D	$3KD$
Maxwell	$\eta_1 + D$	$2GD$	D	$3KD$
Standard linear	$1 + \eta_1 D$	$2G(1 + \eta D)$	D	$3KD$

For a *perfectly elastic Hooke solid* it is then according to Table 25.1 and (25.1)

$$s_{ij} = 2Ge_{ij}, \quad s = 3K\Theta$$

where

$$K = \frac{2G}{3} \frac{1 + \nu}{1 - 2\nu}. \tag{25.2}$$

The model of the Hooke solid – a spring with no damping – and its strain-time relation for suddenly applied constant stress σ_0 are shown in Fig. 25.1a.

The model of the *Kelvin* (*Voigt*) *solid* consists of a spring and a viscous damper connected in parallel. Its material constants are E (or G) and η. The model and its strain-time dependence (creep curve) under suddenly applied constant stress σ_0 are shown in Fig. 25.1b.

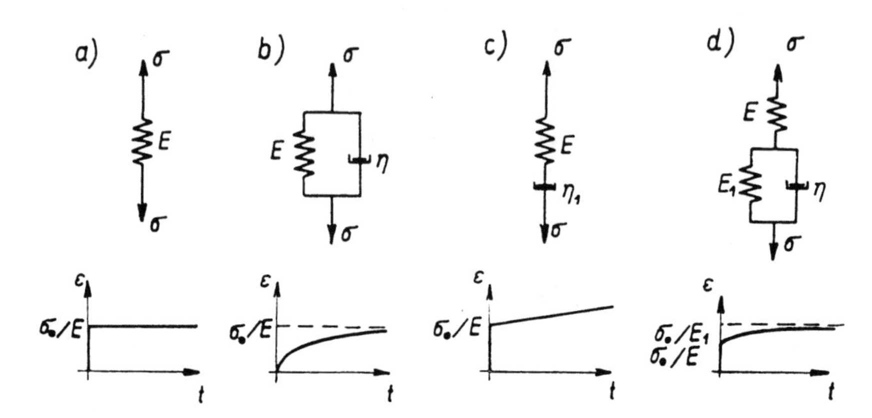

Fig. 25.1. Models of viscoelastic solids, and the dependence of strain ε on time t (creep curves) for suddenly applied constant stress σ_0. a) Hooke solid, b) Kelvin solid, c) Maxwell solid, d) standard linear solid.

The *Maxwell solid* (Fig. 25.1c) is represented by a spring and a viscous damper connected in series; its material constants are E (or G) and η_1.

The *standard linear model* of a viscoelastic solid is obtained by connecting a Kelvin unit with a spring in series (Fig. 25.1d). Its material constants are E, E_1, η (or in the Table 25.1 G, η, η_1).

To simplify calculations one frequently introduces the incompressible viscoelastic solid for which

$$P_3(D) = 0\,, \quad \nu = 1/2\,, \quad K \to \infty, \quad \lambda \to \infty\,, \quad E = 3G\,. \tag{25.3}$$

The stress-strain relations of all the above solids are computed from Eqs. (25.1) with the substitution of the respective operators from Table 25.1.

More complicated viscoelastic solids with a number of springs and viscous dampers connected in series or in parallel can also be obtained by combining the simple models described.

Having introduced the assumptions stated, W. Nowacki [170] deduced directly the relation between Young's modulus $E(t)$ and operators

$P_i(D)$. For beams with $\sigma_{11} = E(t)\,\varepsilon_{11}$, $\sigma_{22} = \sigma_{33} = \sigma_{12} = \sigma_{23} = \sigma_{31} = = 0$, he expressed this relation in the Laplace-Carson transformation

$$E^*(p) = \frac{1}{E}\,\frac{3P_2^*(p)\,P_4^*(p)}{2\,P_1^*(p)\,P_4^*(p) + P_2^*(p)\,P_3^*(p)} \tag{25.4}$$

where $P_i^*(p)$ are the Laplace-Carson transforms of operators $P_i(D)$ and E is the initial Young's modulus.

Using notation (25.4) we can write the expression of the Laplace-Carson and Fourier (sine) finite integral transformations of beam deflection. For a simply supported viscoelastic beam traversed by force P moving at constant speed c, the expression is analogous to Eq. (1.17) in the absence of damping ($\omega_b = 0$):

$$V^*(j,p) = \frac{P\omega}{\mu}\,\frac{p}{(p^2 + \omega^2)\,[p^2 + \omega_{(j)}^2 E^*(p)]} \tag{25.5}$$

where $\omega = j\pi c/l$, $\omega_{(j)}^2$ is according to (1.11) the natural circular frequency of a perfectly elastic beam, and $E^*(p)$ is defined by Eq. (25.4). Eq. (25.5) applies only during the force travel over the beam, i.e. $0 \leqq t \leqq l/c$.

We will now solve the beam deflection (25.5) for the basic models of viscoelastic bodies:

25.1.1 *Perfectly elastic Hooke solid*

For a perfectly elastic Hooke solid, Eq. (25.4), the data of Table 25.1 and the formulae at the beginning of Chap. 17 will give

$$E^*(p) = 1\,;$$

accordingly, after inverse transformations (25.5) the beam deflection is the same as in Eq. (1.31) for an undamped beam.

25.1.2 *Kelvin solid*

a) Incompressible solid

Eqs. (25.3), (25.4) and Table 25.1 give

$$E^*(p) = 1 + \eta p\,.$$

With the substitution of the above expression in (25.5) the transform becomes the same as in Eq. (1.17) except that now

$$\omega_b = \tfrac{1}{2}\eta\omega_{(j)}^2 .$$

With this notation the beam deflection is the same as in Eq. (1.24) which gives consideration to viscous damping.

b) Compressible solid

For the Kelvin solid, Eq. (25.4) and Table 25.1 will result in

$$E^*(p) = \frac{1}{E} \frac{9\,GK(1 + \eta p)}{3K + G(1 + \eta p)} .$$

With this expression transformation (25.5) may be written as follows:

$$V^*(j, p) = \frac{P\omega}{\mu} \frac{p(p + d)}{(p^2 + \omega^2)(p^3 + a_2 p^2 + a_1 p + a_0)} \tag{25.6}$$

where

$$d = \frac{1}{\eta}\left(1 + \frac{3K}{G}\right),$$

$$a_0 = \frac{9K\omega_{(j)}^2}{E\eta}, \quad a_1 = \frac{9K\omega_{(j)}^2}{E}, \quad a_2 = \frac{G + 3K}{G\eta} .$$

The third-degree polynomial $p^3 + a_2 p^2 + a_1 p + a_0$ can be resolved in root factors. Since all coefficients a_i are positive we can conclude that according to Descartes' rule of sign [186, Sect. 31.2] not a single root of the cubic equation $p^3 + a_2 p^2 + a_1 p + a_0 = 0$ will be positive. Then there exist four possible solutions of that equation:

1. One positive root $(-c)$*) and two complex conjugate roots $(-a + ib, -a - ib)$. The resolution in root factors then is

$$(p + c)\left[(p + a)^2 + b^2\right] . \tag{25.7}$$

2. All the roots are real $(-a, -b, -c)$. The resolution is

$$(p + a)(p + b)(p + c) . \tag{25.8}$$

---

*) Root $-c$ is not identical with the symbol of speed c contained only in the expression $\omega = j\pi c/l$.

3. A zero triple root with the resolution

$$p^3 . \tag{25.9}$$

4. All the roots are real but two of them are equal $(-a, -a, -c)$. The resolution is

$$(p + c)(p + a)^2 . \tag{25.10}$$

The roots a, b, c of the cubic equation will be obtained by the familiar methods of solution [186, Sect. 1.20].

The resolution of the polynomial will also control the solution of the inverse Laplace-Carson transformation of Eq. (25.6) – see (27.49) to (27.52).

The only beam deflection we shall write here will be that with the resolution according to (25.7). By (1.9) and (27.49) it is

$$\begin{aligned} v(x, t) = \sum_{j=1}^{\infty} \frac{2P}{\mu l} \sin \frac{j\pi x}{l} \Bigg\{ & \frac{1}{A_1^2 + B_1^2} \left[(dA_1 + \omega B_1) \cos \omega t - \right. \\ & \left. - (\omega A_1 - dB_1) \sin \omega t\right] + \frac{\omega}{b} \frac{\mathrm{e}^{-at}}{A_2^2 + B_2^2} \cdot \\ & \cdot \left[((d - a) A_2 + bB_2) \cos bt - (bA_2 - (d - a) B_2) \sin bt\right] + \\ & + \frac{\omega(d - c)\, \mathrm{e}^{-ct}}{(\omega^2 + c^2)\left[(a - c)^2 + b^2\right]} \Bigg\} \end{aligned} \tag{25.11}$$

where

$$\begin{aligned} A_1 &= \omega(\omega^2 - 2ac - a^2 - b^2) , \\ A_2 &= b(b^2 + 2ac - \omega^2 - 3a^2) , \\ B_1 &= c(a^2 + b^2) - \omega^2(c + 2a) , \\ B_2 &= a(3b^2 - a^2 - \omega^2) + c(a^2 - b^2 + \omega^2) . \end{aligned}$$

The beam deflections for the remaining cases of the resolution, (25.8) to (25.10), would be obtained analogously from Eq. (1.9) and from (27.50) to (27.52).

Deflection (25.11) differs from the deflection of a perfectly elastic beam with viscous damping (1.4) by the term containing e^{-ct}, which expresses the viscoelastic growth of the transient phenomenon that comes into being as force P crosses the beam.

25.1.3 *Maxwell solid*

a) Incompressible solid

(25.3), (25.4) and Table 25.1 result in

$$E^*(p) = \frac{p}{p + \eta_1}$$

which substituted in Eq. (25.2) gives the transformed solution

$$V^*(j, p) = \frac{P\omega}{\mu} \frac{p + \eta_1}{(p^2 + \omega^2)\left[(p + a)^2 + b^2\right]} \tag{25.12}$$

where $a = \eta_1/2$, $b^2 = \omega_{(j)}^2 - \eta_1^2/4$.

In view of (1.9), (27.40) and (27.41) the inverse transformations of (25.12) yield the deflection

$$\begin{aligned} v(x, t) = \sum_{j=1}^{\infty} \frac{2P}{\mu l} \frac{\sin j\pi x/l}{(\omega_{(j)}^2 - \omega^2)^2 + 4a^2\omega^2} \Bigg\{ (\omega_{(j)}^2 - \omega^2 - \eta_1^2) \sin \omega t + \\ + \frac{\eta_1}{\omega} \omega_{(j)}^2 (1 - \cos \omega t) + \frac{\omega}{b} \left[-b^2 + a^2 + \omega^2 + \frac{\eta_1^2}{2\omega_{(j)}^2} \cdot \right. \\ \left. . (3b^2 - a^2 - \omega^2) \right] e^{-at} \sin bt - \eta_1 \omega \left[1 - \frac{1}{\omega_{(j)}^2} \cdot \right. \\ \left. . (3a^2 - b^2 + \omega^2) \right] (1 - e^{-at} \cos bt) \Bigg\} . \end{aligned} \tag{25.13}$$

b) Compressible solid

By (25.4) and Table 25.1, for the Maxwell model

$$E^*(p) = \frac{1}{E} \frac{9Gp}{(3 + G/K)\, p + 3\eta_1}$$

so that the transformed solution (25.5) takes on the form

$$V^*(j, p) = \frac{P\omega}{\mu} \frac{p + d}{(p^2 + \omega^2)\left[(p + a)^2 + b^2\right]} \tag{25.14}$$

where

$$d = \frac{3\eta_1}{3 + G/K},$$

$$a = d/2, \quad b^2 = \frac{9G\omega_{(j)}^2}{E(3 + G/K)} - \frac{d^2}{4}.$$

For a small value of d, i.e. for $b^2 > 0$, the inverse transformations of (25.14) according to (1.9), (27.40) and (27.41) give the deflection

$$v(x, t) = \sum_{j=1}^{\infty} \frac{2P}{\mu l} \frac{\sin j\pi x/l}{(a^2 + b^2 - \omega^2)^2 + 4a^2\omega^2} \Big\{(a^2 + b^2 - \omega^2 - 2ad) .$$

$$. \sin \omega t - 2a\,\omega\,(\cos \omega t - e^{-at} \cos bt) + \frac{d}{\omega}\,(a^2 + b^2 - c^2) .$$

$$. (1 - \cos \omega t) - \frac{\omega}{b}\Big[b^2 - a^2 - \omega^2 - \frac{ad}{a^2 + b^2} .$$

$$. (3b^2 - a^2 - \omega^2)\Big] e^{-at} \sin bt + \frac{d\omega}{a^2 + b^2} .$$

$$. (3a^2 - b^2 + \omega^2)(1 - e^{-at} \cos bt)\Big\} . \tag{25.15}$$

Except for different constants, the deflections of incompressible and compressible Maxwell beams are the same. It is interesting to note that the viscoelastic action of such beams is analogous to the case of continuous load $q = Pd/c$ arriving on a beam with viscous damping. The use of (27.73) together with the action of moving force P will namely result in a transformation that is formally the same as (25.14).

25.1.4 *Standard linear solid*

a) Incompressible solid

Eqs. (25.3), (25.4) and Table 25.1 give

$$E^*(p) = \frac{1 + \eta p}{1 + \eta_1 p}$$

with which the transformed solution (25.5) will turn out in the same form as (25.6). Coefficients d and a_i now are, of course, different, namely

$$d = 1/\eta_1 ,$$

$$a_0 = \omega_{(j)}^2/\eta_1 , \quad a_1 = \eta\omega_{(j)}^2/\eta_1 , \quad a_2 = 1/\eta_1 .$$

For these coefficients the deflection is defined by Eq. (25.11); for the other kinds of resolution in root factors, it is computed from (27.50) to (27.52).

b) Compressible solid

For the standard linear solid, (25.4) and Table 25.1 give

$$E^*(p) = \frac{1}{E} \frac{9GK(1 + \eta p)}{(1 + \eta_1 p) 3K + (1 + \eta p) G} .$$

Like in the preceding case, with this expression the transformed solution will be in the form of Eq. (25.6); its coefficients, of course, will be different, namely

$$d = \frac{G + 3K}{G\eta + 3K\eta_1} ,$$

$$a_0 = \frac{9GK\omega_{(i)}^2}{E(G\eta + 3K\eta_1)} , \quad a_1 = \frac{9GK\omega_{(j)}^2\eta}{E(G\eta + 3K\eta_1)} , \quad a_2 = \frac{G + 3K}{G\eta + 3K\eta_1} .$$

For these coefficients the deflection is defined by Eq. (25.11); for the other kinds of resolution in root factors [(25.8) to (25.10)], it is computed with the aid of relations (27.50) to (27.52).

Like in the Maxwell model, in this case, too, we can compare the viscoelastic action of the beam material to the combined action of a moving force and a continuous load, the latter arriving on the beam at the same time as the load does (Fig. 3.3). As soon as the load departs from the beam, the vibration dies away. In the viscoelastic case the after-effects are slower than in perfectly elastic materials. As we already know from the solution of beam deflection, viscoelasticity of material turns the beam vibration into a damped one. In this sense viscoelasticity and viscous damping are nearly equivalent. The only other solid for which this is precisely so, is the Kelvin incompressible solid.

25.2 *Rigid-plastic beam subjected to a moving load*

The stress-strain diagrams of some materials feature a clearly defined plastic portion which, too, must be taken into account in structural design according to the limit states. A typical diagram of this sort is in Fig. 25.2 which shows strains in a test rod under growing tensile stresses. If the stress pattern is of an oscillatory character, the stress-strain relation is more complicated still.

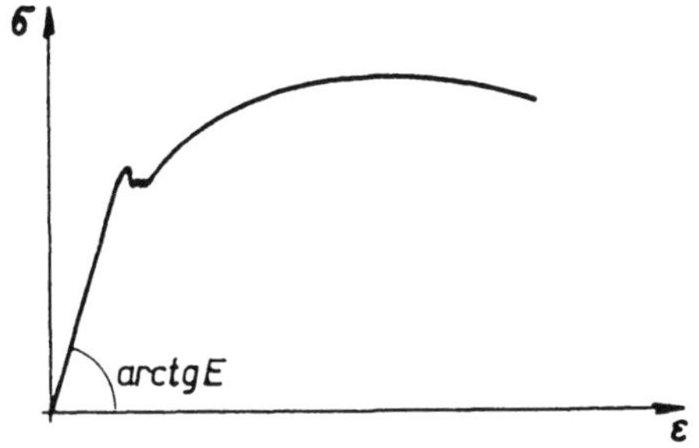

Fig. 25.2. Stress-strain diagram of mild steel (relation between stresses and strains in material under pure tension).

That is why in calculations that are carried out up to and including plastic strains of the structure, the stress-strain diagrams are usually simplified to a considerable extent. For beams, the dependence of bending moment M on curvature K is idealized in the way depicted in Fig. 25.3 which, in effect, is a highly simplified version of the stress-strain diagram of Fig. 25.2 for flexural members. Symbol M_p in Fig. 25.3 denotes the bending moment on the limit of plastic deformation of beam cross section.

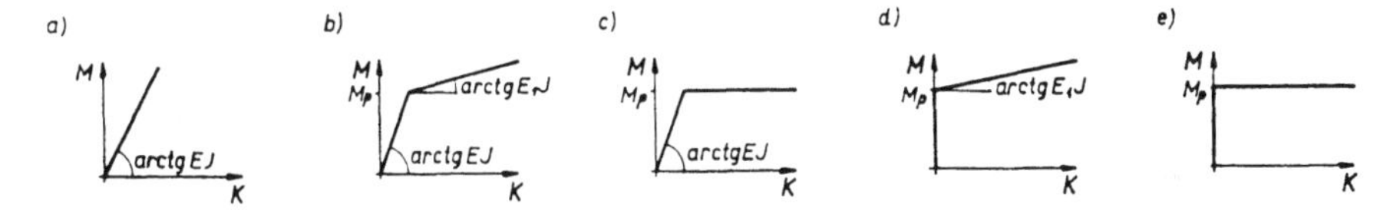

Fig. 25.3. Relations between bending moment M and curvature K, used in the theory of beam bending: a) ideal elastic beam, b) elasto-plastic beam with linear strain hardening, c) ideal elasto-plastic beam, d) rigid-plastic beam with linear strain hardening, e) ideal rigid-plastic beam.

If consideration is given to plastic reserves in the material, the problem of a load moving on a structure becomes very involved. Let us prove this by considering the action of a concentrated load during its travel on a beam: at the beginning, soon after the arrival of the load, the beam behaves like a perfectly elastic one. This goes on until the instant one of the beam cross sections reaches the limit moment M_p. As the load travels on, this cross section in plastic state enlarges into a plastic region which, too, becomes dependent on load motion. The rest of the beam remains elastic. When the load departs, the beam continues to be permanently deformed; this state depends on the actual relation between stresses and strains under the given stress variation in the various cross sections.

To somewhat simplify, the dependence of bending moment on curvature is idealized to such a degree (Fig. 25.3) that in the limit case one arrives at a perfectly rigid-plastic beam according to Fig. 25.3e (cf. J. Henrych [104], I. L. Dikovich [47]). The problem of loads travelling on beams with plastic reserves in the material taken into account, has been dealt with in just three studies: E. W. Parkes [177] solved the motion of a mass on a massless rigid-plastic beam while P. S. Symonds and B. G. Neal [211] extended his solution to a mass beam. So far the most complete numerical solution of the case has been offered by T. G. Toridis and R. K. Wen [221] who considered an elasto-plastic material with linear strain hardening according to Fig. 25.3b. In what follows we will tackle only the case of a mass moving on a mass rigid-plastic beam, the plastic hinge being assumed to be stationary. This is analogous to the case solved in [211]; however, our calculation will encompass all phases of the motion as well as the motion of a force on a mass beam and of a mass on a massless beam.

25.2.1 *Perfectly rigid beam with plastic hinge*

Consider a perfectly rigid simply supported beam appreciably weakened in cross section x_p (Fig. 25.4). According to Fig. 25.4d), the weakened cross section can transmit the constant limit bending moment M_p (Fig. 25.3e) as well as the shear force $T(t)$.

I. L. Dikovich [47] writes the equation of equilibrium of an ideally rigid-plastic beam as follows:

$$\mu x_p(t) \frac{\partial^2 v(x, t)}{\partial t^2} + \frac{\mu\, x_p(t)\, \omega(t)}{2} \left(- \frac{\mathrm{d}x_{p-}(t)}{\mathrm{d}t} + \frac{\mathrm{d}x_{p+}(t)}{\mathrm{d}t} \right) = \sum P \,, \quad (25.16)$$

$$\frac{\mu\, x_p^3(t)}{12} \frac{\mathrm{d}\omega(t)}{\mathrm{d}t} = \sum M \,, \quad (25.17)$$

$$\mu \frac{\partial^2 v(x, t)}{\partial t^2} = p(x, t)\,, \tag{25.18}$$

$$\frac{\mathrm{d}v(x_{p-}(t), t)}{\mathrm{d}t} = \frac{\mathrm{d}v(x_{p+}(t), t)}{\mathrm{d}t}\,. \tag{25.19}$$

Eqs. (25.16) and (25.17) apply to the rigid portion of variable length $x_p(t)$ and express the condition of equilibrium of vertical forces P, including the reactions, and of moments M about the centre of the rigid segments, respectively. Eq. (25.18) describes the behaviour of the plastic portion of the beam under load $p(x, t)$; it resembles Eq. (14.1) defining the deflection of a string under $N \to 0$. The condition that applies on the boundary between the rigid and the plastic portions is Eq. (25.19).

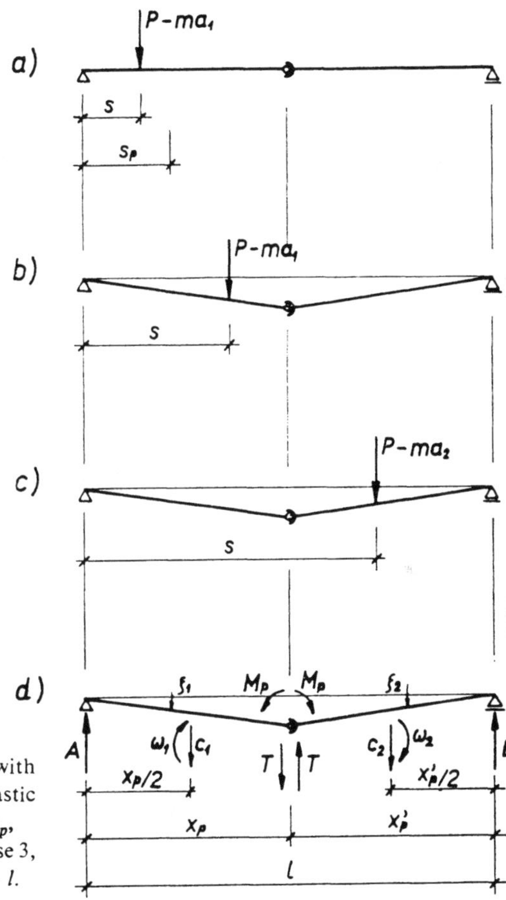

Fig. 25.4. Motion of load P with mass m on an ideal rigid-plastic beam: a) phase 1, $0 < s < s_p$, b) phase 2, $s_p < s < x_p$, c) phase 3, $x_p < s < l$, d) phase 4, $s > l$.

Next to the conventional or familiar symbols, Eqs. (25.16) to (25.19) contain the following:

$\omega(t)$ – angular velocity of rotation of the rigid portion of the beam,

$$x_{p-} = x_p - \varepsilon\,, \quad x_{p+} = x_p + \varepsilon\,, \quad \varepsilon \to 0\,.$$

Eqs. (25.16) to (25.19) are simplified by assuming $x_p(t)$ to be constant. In Eq. (25.16) this causes the expression in parentheses to become zero, and Eq. (25.18) to drop out entirely because the plastic portion is thus reduced to a plastic cross section in position x_p.

It is further assumed that the beam deflections at the left-hand end, $v_1(x, t)$, and at the right-hand end, $v_2(x, t)$, are proportional to beam rotations in the supports, $\zeta_1(t)$ and $\zeta_2(t)$ (Fig. 25.4d):

$$\begin{aligned} v_1(x, t) &= x\,\zeta_1(t) && \text{for} \quad 0 \leqq x \leqq x_p\,, \\ v_2(x, t) &= (l - x)\,\zeta_2(t) && \text{for} \quad x_p \leqq x \leqq l\,. \end{aligned} \tag{25.20}$$

On the above assumptions, condition (25.19) takes on the form

$$x_p \frac{\mathrm{d}\zeta_1(t)}{\mathrm{d}t} = x_p' \frac{\mathrm{d}\zeta_2(t)}{\mathrm{d}t}\,. \tag{25.21}$$

Let us further denote the velocity of the centroid of, respectively, the left- and right-hand side by $c_1(t) = \mathrm{d}v(x_p/2, t)/\mathrm{d}t$ and $c_2(t) = \mathrm{d}v(x_p'/2, t)/\mathrm{d}t$, and the appertaining angular velocities of rotation of these two rigid portions (the positive directions of ζ_1, ζ_2, ω_1, ω_2, c_1, c_2 are marked out in Fig. 25.4d) by

$$\omega_1(t) = \frac{\mathrm{d}\zeta_1(t)}{\mathrm{d}t}\,, \quad \omega_2(t) = -\frac{\mathrm{d}\zeta_2(t)}{\mathrm{d}t} = -\frac{x_p}{x_p'}\frac{\mathrm{d}\zeta_1(t)}{\mathrm{d}t}\,. \tag{25.22}$$

The condition that the beam velocities

$$\frac{\mathrm{d}v_1(x, t)}{\mathrm{d}t} = c_1(t) + \omega_1(t)\,(x - x_p/2)\,,$$

$$\frac{\mathrm{d}v_2(x, t)}{\mathrm{d}t} = c_2(t) + \omega_2(t)\,(x - x_p - x_p'/2)$$

should be zero in the supports, i.e. $\mathrm{d}v_1(0, t)/\mathrm{d}t = 0$, $\mathrm{d}v_2(l, t)/\mathrm{d}t = 0$, results in

$$c_1(t) = \omega_1(t)\,x_p/2\,, \quad c_2(t) = -\omega_2(t)\,x_p'/2\,.$$

Using (25.22) and (25.21) then gives

$$c_1(t) = \frac{x_p}{2}\,\frac{\mathrm{d}\zeta_1(t)}{\mathrm{d}t}\,, \quad c_2(t) = \frac{x_p}{2}\,\frac{\mathrm{d}\zeta_1(t)}{\mathrm{d}t}\,. \tag{25.23}$$

What the assumptions introduced above mean in technical practice is that elastic strains of the beam are neglected against the plastic ones and that at a specified location x_p the beam has a weakened cross section capable of transmitting only the given limit bending moment M_p and the hitherto unknown shear force $T(t)$.

We are now ready to divide the problem of load P with mass $m = P/g$ moving on a beam at constant speed c into several time phases*):

Phase 1 $(0 \leqq s \leqq s_p)$

Phase 1 continues from the instant the load arrives on the left-hand support of the beam until instant $t_p = s_p/c$ when the load at point s_p just produces the limit bending moment M_p in position x_p (Fig. 25.4a). Coordinate s_p is computed from the known x_p and M_p using the second of Eqs. (5.9)

$$s_p = \frac{M_p l}{P x'_p} \tag{25.24}$$

where $x'_p = l - x_p$.

In phase 1 the beam continues undeformed, has no deflection nor velocity (Fig. 25.4a).

Phase 2 $(s_p < s < x_p)$

In the second phase, plastic deformations already set in the location x_p, and the beam is acted upon by forces and moments indicated in Figs. 25.4b and d. The beam may be thought of as divided in two portions in position x_p; the effect of one portion of the beam on the other is represented by bending moment M_p and shear force $T(t)$ (Fig. 25.4d).

The other forces acting here are the external forces of support reactions $A(t)$ and $B(t)$. The vertical acceleration a_1 of moving mass m is

$$a_1 = \frac{\mathrm{d}^2(s\zeta_1)}{\mathrm{d}t^2} = s\,\ddot{\xi}_1(t) + 2c\,\dot{\xi}_1(t) \tag{25.25}$$

*) The time origin $t = 0$ of each phase lies always at the beginning of that particular phase; the length coordinates s of the point at which the load is just acting are, on the other hand, measured from the left support.

where $s = s_p + ct$ because according to (25.20) the deflection underneath the load is $v(s, t) = s\,\zeta_1(t)$.

We will now set up the equations of equilibrium (25.16) and (25.17) for the two portions of the beam:

$$\begin{aligned}
\mu x_p\,\dot{c}_1(t) &= -A(t) + P - ma_1 + T(t)\,,\\
\tfrac{1}{12}\mu x_p^3\,\dot{\omega}_1(t) &= [A(t) + T(t)]\,\frac{x_p}{2} + (P - ma_1)\left(s - \frac{x_p}{2}\right) - M_p\,,\\
\mu x_p'\,\dot{c}_2(t) &= -T(t) - B(t)\,,\\
\tfrac{1}{12}\mu x_p'^3\,\dot{\omega}_2(t) &= M_p + [T(t) - B(t)]\,\frac{x_p'}{2}\,.
\end{aligned} \tag{25.26}$$

Elimination of $A(t)$, $B(t)$ and $T(t)$ from Eqs. (25.26) and substitution of (25.22) and (25.23) leads to the linear ordinary differential equation of the second order in function $\zeta_1(t)$ or of the first order in function $\omega_1(t)$

$$\frac{d\omega_1(t)}{dt} + \frac{2(t + t_p)\,\omega_1(t)}{t^2 + 2t_pt + b^2} = \frac{g}{c}\,\frac{t - a}{t^2 + 2t_pt + b^2} \tag{25.27}$$

where

$$t_p = s_p/c\,, \quad a = \frac{M_pl}{Pcx_p'} - t_p\,, \quad b^2 = \frac{x_p^2}{3\varkappa c^2} + t_p^2\,, \quad \varkappa = \frac{P}{G} = \frac{m}{\mu l}\,.$$

For the initial condition $\omega_1(0) = 0$, the solution of Eq. (25.27) is easy enough — see [120, Sect. A 4.3]*). Following integration we get

$$\omega_1(t) = \frac{gt}{2c}\,\frac{t - 2a}{t^2 + 2t_pt + b^2}\,. \tag{25.28}$$

Rotation $\zeta_1(t)$ is obtained by integration from Eq. (25.22) for the initial condition $\zeta_1(t) = 0$, see [120, Sect. A 4.1]**)

$$\zeta_1(t) = \frac{g}{2c}\left\{t - (a + t_p)\ln\frac{t^2 + 2t_pt + b^2}{b^2} + \frac{2t_p(t_p + a) - b^2}{(b^2 - t_p^2)^{1/2}}\,\cdot\right.$$
$$\left.\cdot\left[\operatorname{arctg}\frac{t + t_p}{(b^2 - t_p^2)^{1/2}} - \operatorname{arctg}\frac{t_p}{(b^2 - t_p^2)^{1/2}}\right]\right\}. \tag{25.29}$$

---

*) The solution of the equation $y'(x) + f(x)\,y(x) = g(x)$, which passes through point (ξ, η) is $y = e^{-F}[\eta + \int_\xi^x g(x)\,e^F\,dx]$ where $F = \int_\xi^x f(x)\,dx$.

**) The solution of the equation $y'(x) = f(x)$, which passes through point (ξ, η) is $y = \eta + \int_\xi^x f(x)\,dx$.

The angular velocity $\omega_2(t)$, rotation $\zeta_2(t)$ and the deflection of the right-hand part of the beam, $v_2(x, t)$, are then readily obtained from Eqs. (25.20) to (25.22).

Phase 2 ends at the instant

$$t_3 = x_p/c - t_p \tag{25.30}$$

when the load reaches x_p. We shall denote the rotations and angular velocities at that instant by

$$\omega_{23} = \omega_2(t_3), \quad \zeta_{23} = \zeta_2(t_3)\,. \tag{25.31}$$

Phase 3 $(x_p \leqq s \leqq l)$

In the third phase the load is in the right-hand part of the beam (Fig. 25.4c). The forces and moments acting on the beam are shown in Fig. 25.4c,d. It is tacitly assumed that although $l - s_p \leqq s \leqq l$, the bending moment M_p continues to act in the weakened cross section. The vertical acceleration a_2 of the moving mass m now is

$$a_2 = \frac{\mathrm{d}^2 v(s, t)}{\mathrm{d}t^2} = \frac{\mathrm{d}^2[(l - s)\,\zeta_2(t)]}{\mathrm{d}t^2} = (l - s)\,\ddot{\xi}_2(t) - 2c\,\dot{\xi}_2(t) \tag{25.32}$$

where

$$s = x_p + ct\,, \quad l - s = x_p' - ct\,.$$

The conditions of equilibrium (25.16) and (25.17) for the two portions of the beam now are

$$\begin{aligned}
\mu x_p\,\dot{c}_1(t) &= -A(t) + T(t)\,,\\
\tfrac{1}{12}\mu x_p^3\,\dot{\omega}_1(t) &= [A(t) + T(t)]\,\frac{x_p}{2} - M_p\,,\\
\mu x_p'\,\dot{c}_2(t) &= -T(t) + P - ma_2 - B(t)\,,\\
\tfrac{1}{12}\mu x_p'^3\,\dot{\omega}_2(t) &= [T(t) - B(t)]\,\frac{x_p'}{2} + (P - ma_2)\left(ct - \frac{x_p'}{2}\right) + M_p\,.
\end{aligned} \tag{25.33}$$

Following the elimination of the shear force and of the reactions, and the substitution of (25.22) and (25.23) we again get the differential equation of the second order in $\zeta_2(t)$ or of the first order in $\omega_2(t)$:

$$\frac{\mathrm{d}\omega_2(t)}{\mathrm{d}t} + \frac{2(t - t_3')\,\omega_2(t)}{t^2 - 2t_3' t + b'^2} = \frac{g}{c}\,\frac{t + a'}{t^2 - 2t_3' t + b'^2} \tag{25.34}$$

where $t'_3 = x'_p/c$,

$$a' = \frac{M_p l}{Pcx_p} - t'_3\,, \quad b'^2 = \frac{x'^2_p}{3\varkappa c^2} + t'^2_3\,.$$

For the initial conditions (25.31) the solution of Eq. (25.34) is

$$\omega_2(t) = \frac{b'^2}{t^2 - 2t'_3 t + b'^2}\left[\omega_{23} + \frac{gt}{2cb'^2}(t + 2a')\right] \tag{25.35}$$

and similarly so

$$\zeta_2(t) = \zeta_{23} - \frac{g}{2c}\left\{t + (a' + t'_3)\ln\frac{t^2 - 2t'_3 t + b'^2}{b'^2} + \frac{1}{(b'^2 - t'^2_3)^{1/2}}\,\cdot\right.$$

$$\cdot\left[2t'_3(t'_3 + a') - b'^2 + \frac{2c\omega_{23}b'^2}{g}\right]\left[\operatorname{arctg}\frac{t - t'_3}{(b'^2 - t'^2_3)^{1/2}} +\right.$$

$$\left.\left. + \operatorname{arctg}\frac{t'_3}{(b'^2 - t'^2_3)^{1/2}}\right]\right\}. \tag{25.36}$$

The other particulars about the beam motion are again obtained from Eqs. (25.20) to (25.22). Phase 3 ends at the instant $t_4 = x'_p/c = t'_3$ when the load reaches the right-hand end of the beam. The end values of angular velocity and rotation of the first portion of the beam are

$$\omega_{14} = \omega_1(t_4)\,, \quad \zeta_{14} = \zeta_1(t_4)\,. \tag{25.37}$$

If it were not for the inertia forces, the beam motion would cease for $s = l - s_p$ in the third phase. As it is, another phase will take place.

Phase 4 $(s \geqq l)$

Phase 4 takes place as soon as the load leaves the beam. The forces and moments acting on the beam are shown in Fig. 25.4d. The equations of equilibrium (25.16) and (25.17) for the two portions of the beam now are

$$\begin{aligned}
\mu x_p\,\dot{c}_1(t) &= -A(t) + T(t)\,,\\
\tfrac{1}{12}\mu x_p^3\dot{\omega}_1(t) &= [A(t) + T(t)]\frac{x_p}{2} - M_p\,,\\
\mu x'_p\,\dot{c}_2(t) &= -T(t) - B(t)\,,\\
\tfrac{1}{12}\mu x'^3_p\,\dot{\omega}_2(t) &= [T(t) - B(t)]\frac{x'_p}{2} + M_p\,.
\end{aligned} \tag{25.38}$$

Using the same procedure as before we get from this set of equations the simple differential equation

$$\frac{d\omega_1(t)}{dt} = -\frac{g}{c}\frac{a}{b^2} \tag{25.39}$$

where

$$a = \frac{M_p l}{Pcx'_p}, \quad b^2 = \frac{x_p^2}{3\varkappa c^2}.$$

For the initial conditions (25.37) its solution gives

$$\omega_1(t) = \omega_{14} - \frac{g}{c}\frac{a}{b^2}t,$$

$$\zeta_1(t) = \zeta_{14} + \omega_{14}t - \frac{gat^2}{2cb^2} \tag{25.40}$$

while the other information on the beam motion is obtained from (25.20) to (25.22).

25.2.2 *Moving force*

The special case of force P moving on a mass beam follows from the preceding paragraph 25.2.1 for load mass m negligibly small against the beam mass μl. Accordingly:

Phase 1 is the same as in paragraph 25.2.1.

Phase 2

Neglecting mass m we may rearrange Eq. (25.26) to the form

$$\frac{d\omega_1(t)}{dt} = \frac{g}{c}\frac{t-a}{b^2}$$

where

$$b^2 = \frac{x_p^2}{3\varkappa c^2},$$

and a is the same as in (25.27). For zero initial conditions the solution of this equation is

$$\omega_1(t) = \frac{gt}{2cb^2}(t-2a), \quad \zeta_1(t) = \frac{gt^2}{6cb^2}(t-3a). \tag{25.41}$$

Phase 3

Proceeding as above we get from (25.33)

$$\frac{d\omega_2(t)}{dt} = \frac{g}{c}\frac{t + a'}{b'^2}$$

where $b'^2 = x_p'^2/(3\varkappa c^2)$, and a' is the same as in (25.34).

For the initial conditions (25.31), the solution of this equation gives

$$\omega_2(t) = \omega_{23} + \frac{gt}{2cb'^2}(t + 2a')\,,$$

$$\zeta_2(t) = \zeta_{23} - \omega_{23}t - \frac{gt^2}{6cb'^2}(t + 3a')\,. \tag{25.42}$$

Phase 4 is the same as in paragraph 25.2.1.

Example: Consider a simple beam with a weakened cross section capable of transmitting the limit bending moment $M_p = Pl/8$ in position $x_p = l/2$.

Phase 1 — undeformed beam — continues to the instant force P is in position $s_p = l/4$; this is computed from (25.24).

Phase 2 — Eqs. (25.20) to (25.22) and (25.41) together with the notation $\tau = ct/l$ give

$$a = 0\,,\quad b^2 = \frac{l^2}{12\varkappa c^2}\,,$$

$$\omega_1(t) = \frac{6\varkappa g}{c}\tau^2\,,\quad \zeta_1(t) = \frac{2\varkappa lg}{c^2}\tau^3\,,$$

$$v_1(l/2, t) = \frac{\varkappa l^2 g}{c^2}\tau^3\,, \tag{25.43}$$

$$t_3 = \frac{l}{4c}\,,\quad \omega_{13} = \frac{3}{8}\frac{\varkappa g}{c} = -\omega_{23}\,.$$

$$\zeta_{13} = \frac{1}{32}\frac{\varkappa lg}{c^2} = \zeta_{23}\,.$$

Phase 3 – From (25.42)

$$t'_3 = \frac{l}{2c}, \quad a' = -\frac{l}{4c}, \quad b'^2 = \frac{l^2}{12\varkappa c^2},$$

$$\omega_2(t) = -\frac{3}{8}\frac{\varkappa g}{c} + \frac{6\varkappa g\tau}{c}\left(\tau - \frac{1}{2}\right),$$

$$\zeta_2(t) = \frac{\varkappa l g}{c^2}\left[\frac{1}{32} + \frac{3}{8}\tau - 2\tau^2\left(\tau - \frac{3}{4}\right)\right],$$

$$v_2(l/2, t) = \frac{\varkappa l^2 g}{c^2}\left[\frac{1}{64} + \frac{3}{16}\tau - \tau^2\left(\tau - \frac{3}{4}\right)\right], \tag{25.44}$$

$$t_4 = \frac{l}{2c}, \quad \omega_{24} = -\frac{3}{8}\frac{\varkappa g}{c} = -\omega_{14}, \quad \zeta_{24} = \frac{11}{32}\frac{\varkappa l g}{c^2} = \zeta_{14}.$$

Phase 4 – Here (25.40) gives

$$a = \frac{l}{4c}, \quad b^2 = \frac{l^2}{12\varkappa c^2},$$

$$\zeta_1(t) = \frac{\varkappa l g}{c^2}\left(\frac{11}{32} + \frac{3}{8}\tau - \frac{3}{2}\tau^2\right),$$

$$v_1(l/2, t) = \frac{\varkappa l^2 g}{c^2}\left(\frac{11}{64} + \frac{3}{16}\tau - \frac{3}{4}\tau^2\right). \tag{25.45}$$

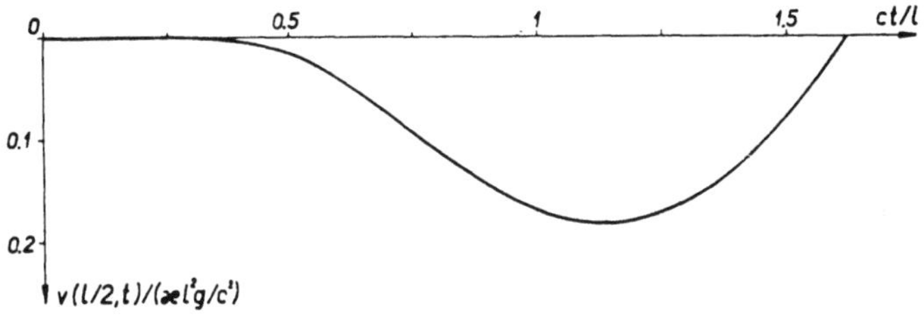

Fig. 25.5. Time variation of deflection $v(l/2, t)$ at the mid-span of an ideal rigid-plastic beam traversed by force P at speed c. The plastic hinge is in position $x_p = l/2$; $s_p = l/4$, $M_p = Pl/8$.

The deflection at the mid-span of the beam, $v(l/2, t)$, according to Eqs. (25.43) to (25.45), is shown in Fig. 25.5. Until instant $t_p = l/(4c)$ the beam is undeformed. Thereafter it starts to deform in position $x_p = l/2$.

The maximum deflection is not reached until after the force departs from the beam. The deflection is directly proportional to the magnitude of force P and to span l, and inversely proportional to beam mass μ and to the square of speed, c^2, because $\varkappa l^2 g/c^2 = Pl/(\mu c^2)$.

25.2.3 *Massless beam*

If the beam mass is negligible against the load mass, $\varkappa \to 0$. For this limit all the equations of paragraph 25.2.1 continue to apply.

In *phase 2*, $b^2 = t_p^2$, and for this value $\omega_1(t)$ is the same as (25.28). Rotation $\zeta_1(t)$ now is

$$\zeta_1(t) = \frac{g}{2c}\left[t - 2(a + t_p)\ln\frac{t + t_p}{t_p} + \frac{2t(a + t_p) + t_p^2}{t + t_p}\right]. \quad (25.46)$$

In *phase 3*, $b'^2 = t_3'^2$, and with this notation $\omega_2(t)$ is the same as (25.35); the rotation is

$$\zeta_2(t) = \zeta_{23} - \frac{g}{2c}\left\{t + 2(a' + t_3')\ln\frac{t - t_3'}{t_3'} + \frac{1}{t - t_3'}\cdot\right.$$
$$\left.\cdot\left[t_3'^2 - 2(a' + t_3')\,t - \frac{2c\omega_{23}t_3't}{g}\right]\right\}. \quad (25.47)$$

The motion of a massless beam ceases at the end of phase 3.

25.3 *Application of the theory*

The theory presented in this chapter can be applied to structures whose materials possess distinct non-elastic properties. In our discussion we have noted but two aspects of the question — namely the viscoelastic and the plastic properties of materials.

Under a moving load whose effect on a structure is relatively of very short duration, viscoelasticity, a property just on the borderline between pure elastic and inelastic behaviour of material, exerts no pronounced effect during the load traverse. It manifests itself solely by an increase of damping which particularly affects the free vibration of the structure after the vehicle departs from it.

Utilization of plastic reserves in materials, on the other hand, has remarkable effects also on structures subjected to moving loads. We have found in paragraph 25.2.2 that even comparatively large loads can cross a beam provided they move at a high enough speed. The restrictive factor in this process is the beam rotation in the supports. This finding could be made use of in extreme cases, as for example, when a bridge structure damaged at some place must be crossed by a heavy vehicle. Theoretically, this can be done whenever the vehicle speed is sufficiently high. But to the best of our knowledge, so far such a feat has been accomplished only in the fantasy of Jules Verne who described it in his novel "Round the World in 80 Days". A theoretical explanation of that daring and dangerous crossing of a damaged bridge that did not collapse until after the fast-running locomotive had left it, was presented by P. S. Symonds and B. G. Neal [211].

Up to this time questions concerning the effects of moving loads on structures stressed beyond the elastic limit have not been answered in full. That is why there is still a scarcity of data on which one could base designs according to the limit states. Theoretical and experimental studies of the problem have become the order of the day because they would undoubtedly provide scientific information on how to proceed when designing structures under moving loads for some of the limit states in which advantage could be taken of the plastic reserves in the materials.

25.4 *Additional bibliography*

[43, 47, 101, 123, 140, 160, 170, 175, 177, 211, 221, 237, 248, 254, 255, 256, 267, 270].

26

Moving random loads

So far in this book we have assumed that the load was specified in advance, i.e. that it had the form, for example, of a known function of time or of a quantity governed by a known functional dependence (so-called deterministic processes). Sometimes, however, such an assumption is far from reality; under certain circumstances actual loads, no less than the strength properties of structural materials, obey laws that are random to a greater or lesser degree (so-called stochastic processes).

In moving loads and structures serving in transport, it is particularly the effect of track irregularities, the effect of foundation supporting the structures, the effect of vehicle engines, etc., that sometimes have a very appreciable random component. In other cases there exists a large number of deterministic causes each of which separately brings about comparatively small dynamic stresses in the structure, but whose simultaneous action frequently results in noteworthy effects. The random simultaneous action of a large number of causes follows the laws of probability and that is why the whole process is considered stochastic.

Thus, for example, random stresses in railway and highway bridges are caused by random track irregularities, by composition and speed of trains or traffic flow, random motions of vehicles, effects of vehicle engines, etc. In rails these causes of dynamic stresses are augmented by random variable properties of the substructure. Since the random causes add up to quite a number, the moving load is comprehensively assumed to be of a random character. Of the many analogous instances let us just quote that of stresses in space ships produced by the effect of random turbulent flow.

Random, predominantly stationary vibrations of mechanical systems have been studied in detail by V. V. Bolotin [22] and S. H. Crandall [40], [41], whereas problems involving the effects of moving random loads have not been solved so far. Only V. V. Bolotin ([22], Sect. 54) and J. D. Robson ([188], Sect. 5.6) gave a rough outline of a procedure for solving the effect of a moving continous load on a beam.

In this chapter we will solve the following three very important problems: the motion of a random variable force and of an infinite random strip load on a simple beam, and an infinite beam on random elastic foundation subjected to a moving force randomly variable in time.

26.1 *General theory*

In general, the static and dynamic deflections of structures are described by the linear differential equation

$$L[v(x, t)] = p(x, t) \tag{26.1}$$

where $v(x, t)$ denotes the deflection, $p(x, t)$ the load. The random variation of $p(x, t)$ is assumed to be not only with respect to the time coordinate t but also with respect to the position coordinate x; moreover, load $p(x, t)$ is regarded as a nonstationary Gaussian random process of the non-Markov type.

L represents a linear differential operator of the type

$$L = L_0 + \mu \frac{\partial^2}{\partial t^2} + 2\mu\omega_b \frac{\partial}{\partial t} \tag{26.2}$$

where L_0 is a self-adjoint linear operator in space coordinate x, μ – mass per unit length, and ω_b – circular frequency of viscous damping – similarly as in Chap. 1.

Elastic systems described by Eqs. (26.1) and (26.2) are conveniently solved by the normal-mode analysis

$$v(x, t) = \sum_{j=1}^{\infty} v_{(j)}(x)\, q_{(j)}(t)\,, \tag{26.3}$$

$$p(x, t) = \sum_{j=1}^{\infty} \mu\, v_{(j)}(x)\, Q_{(j)}(t) \tag{26.4}$$

where $v_{(j)}(x)$ are the normal modes of vibration obtained with regard to the boundary conditions from the equations

$$L_0[v_{(j)}(x)] = \mu\omega_{(j)}^2\, v_{(j)}(x)\,. \tag{26.5}$$

$\omega_{(j)}$ is the natural circular frequency of the system,

$$Q_{(j)}(t) = \frac{1}{V_{(j)}} \int_0^l p(x, t)\, v_{(j)}(x)\, \mathrm{d}x \tag{26.6}$$

is the generalized force,

$$V_{(j)} = \int_0^l \mu\, v_{(j)}^2(x)\, \mathrm{d}x\,, \qquad \int_0^l \mu\, v_{(j)}(x)\, v_{(k)}(x)\, \mathrm{d}x = 0 \quad \text{for} \quad j \neq k\,, \tag{26.7}$$

and $q_{(j)}(t)$ is the generalized deflection obtained with regard to the initial conditions from the equation

$$\ddot{q}_{(j)}(t) + 2\omega_b\, \dot{q}_{(j)}(t) + \omega_{(j)}^2\, q_{(j)}(t) = Q_{(j)}(t)\,. \tag{26.8}$$

For zero initial conditions the solution of Eq. (26.8) is

$$q_{(j)}(t) = \int_0^t h_{(j)}(t - \tau)\, Q_{(j)}(\tau)\, \mathrm{d}\tau = \int_{-\infty}^{\infty} h_{(j)}(\tau)\, Q_{(j)}(t - \tau)\, \mathrm{d}\tau \tag{26.9}$$

where $h_{(j)}(t)$ denotes the impulse function

$$h_{(j)}(t) = \begin{cases} \dfrac{1}{\omega'_{(j)}}\, \mathrm{e}^{-\omega_b t} \sin \omega'_{(j)} t & \text{for} \quad t \geqq 0 \\ 0 & \text{for} \quad t < 0\,, \end{cases} \tag{26.10}$$

and $\omega'^2_{(j)} = \omega^2_{(j)} - \omega^2_b$. Since $Q_{(j)}(t - \tau) = 0$ for $\tau > t$ and $h_{(j)}(\tau) = 0$ for $\tau < 0$, respectively, the limits of integration in (26.9) may be extended to ∞ and $-\infty$, respectively.

Functions $h_{(j)}(t)$ and $v_{(j)}(x)$ are deterministic, functions $q_{(j)}(t)$, $Q_{(j)}(t)$, $v(x, t)$ and $p(x, t)$ are random.

26.1.1 *Correlation analysis*

In probability analyses one must know the statistic characteristics of the input

$$p(x, t) = E[p(x, t)] + \mathring{p}(x, t)\,, \tag{26.11}$$

$$K_{pp}(x_1, x_2, t_1, t_2) = E[\mathring{p}(x_1, t_1)\, \mathring{p}(x_2, t_2)] \tag{26.12}$$

where E denotes the mean value linear operator, $\mathring{p}(x, t)$ – the centred value of the load, and $K_{pp}(x_1, x_2, t_1, t_2)$ – the covariance of the non-stationary function $p(x, t)$.

As the definition of the covariance (26.12) implies, the covariance of the generalized deflection may be established from (26.9)

$$K_{q(j)q(k)}(t_1, t_2) = \int_{-\infty}^{\infty} \int_{-\infty}^{\infty} h_{(j)}(\tau_1)\, h_{(k)}(\tau_2)\, K_{Q(j)Q(k)}(t_1 - \tau_1, t_2 - \tau_2)\, .\, d\tau_1\, d\tau_2\,, \tag{26.13}$$

the covariance of the deflection from (26.3)

$$K_{vv}(x_1, x_2, t_1, t_2) = \sum_{j=1}^{\infty} \sum_{k=1}^{\infty} v_{(j)}(x_1)\, v_{(k)}(x_2)\, K_{q(j)q(k)}(t_1, t_2)\,, \tag{26.14}$$

and the covariance of the load from (26.4)

$$K_{pp}(x_1, x_2, t_1, t_2) = \sum_{j=1}^{\infty} \sum_{k=1}^{\infty} \mu^2\, v_{(j)}(x_1)\, v_{(k)}(x_2)\, K_{Q(j)Q(k)}(t_1, t_2)\,. \tag{26.15}$$

In Eqs. (26.13) and (26.15) the covariance of the generalized force is computed from (26.6)

$$K_{Q(j)Q(k)}(t_1, t_2) = \frac{1}{V_{(j)} V_{(k)}} \int_0^l \int_0^l v_{(j)}(x_1)\, v_{(k)}(x_2)\, K_{pp}(x_1, x_2, t_1, t_2)\, .\, dx_1\, dx_2\,. \tag{26.16}$$

26.1.2 *Spectral density analysis*

The spectral density of a nonstationary function is defined in [41]; for the generalized deflection the Wiener-Khinchine relations between the spectral density and the covariance are as follows:

$$S_{q(j)q(k)}(\omega_1, \omega_2) = \frac{1}{4\pi^2} \int_{-\infty}^{\infty} \int_{-\infty}^{\infty} K_{q(j)q(k)}(t_1, t_2)\, e^{-i(\omega_2 t_2 - \omega_1 t_1)}\, dt_1\, dt_2\,, \tag{26.17}$$

$$K_{q(j)q(k)}(t_1, t_2) = \int_{-\infty}^{\infty} \int_{-\infty}^{\infty} S_{q(j)q(k)}(\omega_1, \omega_2)\, e^{i(\omega_2 t_2 - \omega_1 t_1)}\, d\omega_1\, d\omega_2\,. \tag{26.18}$$

For the spectral density analysis it is also convenient to introduce the transfer function

$$H_{(j)}(\omega) = \int_{-\infty}^{\infty} h_{(j)}(t)\, e^{-i\omega t}\, dt = \frac{1}{\omega_{(j)}^2 - \omega^2 + 2i\omega_b\omega} \tag{26.19}$$

as the Fourier integral transformation of $h_{(j)}(t)$ given by (26.10).

The spectral density of the generalized deflection may then be obtained as a function of the spectral density of the generalized force

$$S_{q_{(j)}q_{(k)}}(\omega_1, \omega_2) = \bar{H}_{(j)}(\omega_1)\, H_{(k)}(\omega_2)\, S_{Q_{(j)}Q_{(k)}}(\omega_1, \omega_2) \tag{26.20}$$

where $\bar{H}_{(j)}(\omega)$ is the complex conjugate function of $H_{(j)}(\omega)$.

In the above we have used the spectral density $S_{Q_{(j)}Q_{(k)}}(\omega_1, \omega_2)$ of the generalized force defined similarly as in (26.17). In view of (26.16) this can be rewritten to

$$S_{Q_{(j)}Q_{(k)}}(\omega_1, \omega_2) = \frac{1}{4\pi^2}\int_{-\infty}^{\infty}\int_{-\infty}^{\infty} K_{Q_{(j)}Q_{(k)}}(t_1, t_2)\, e^{-i(\omega_2 t_2 - \omega_1 t_1)}\, dt_1\, dt_2 =$$

$$= \frac{1}{V_{(j)}V_{(k)}}\int_0^l\int_0^l v_{(j)}(x_1)\, v_{(k)}(x_2)\, S_{pp}(x_1, x_2, \omega_1, \omega_2)\,.$$

$$.\, dx_1\, dx_2\,, \tag{26.21}$$

$$K_{Q_{(j)}Q_{(k)}}(t_1, t_2) = \int_{-\infty}^{\infty}\int_{-\infty}^{\infty} S_{Q_{(j)}Q_{(k)}}(\omega_1, \omega_2)\, e^{i(\omega_2 t_2 - \omega_1 t_1)}\, d\omega_1\, d\omega_2\,. \tag{26.22}$$

Then in view of (26.14) the spectral density of the deflection is

$$S_{vv}(x_1, x_2, \omega_1, \omega_2) = \sum_{j=1}^{\infty}\sum_{k=1}^{\infty} v_{(j)}(x_1)\, v_{(k)}(x_2)\, S_{q_{(j)}q_{(k)}}(\omega_1, \omega_2)\,; \tag{26.23}$$

from there the covariance of the deflection can be calculated similarly as in (26.18) and (26.14).

26.2 *Moving random force*

To illustrate we will solve a simple beam of span l traversed by massless force $P(t) = P + \mathring{P}(t)$ with constant mean value $E[P(t)] = P$, moving at constant velocity c (Fig. 26.1). The analogous deterministic case was examined in Chap. 1. Since the solution obtained there represents the mean value of the present one, $E[v(x, t)]$, all we shall do is to analyze the stochastic case.

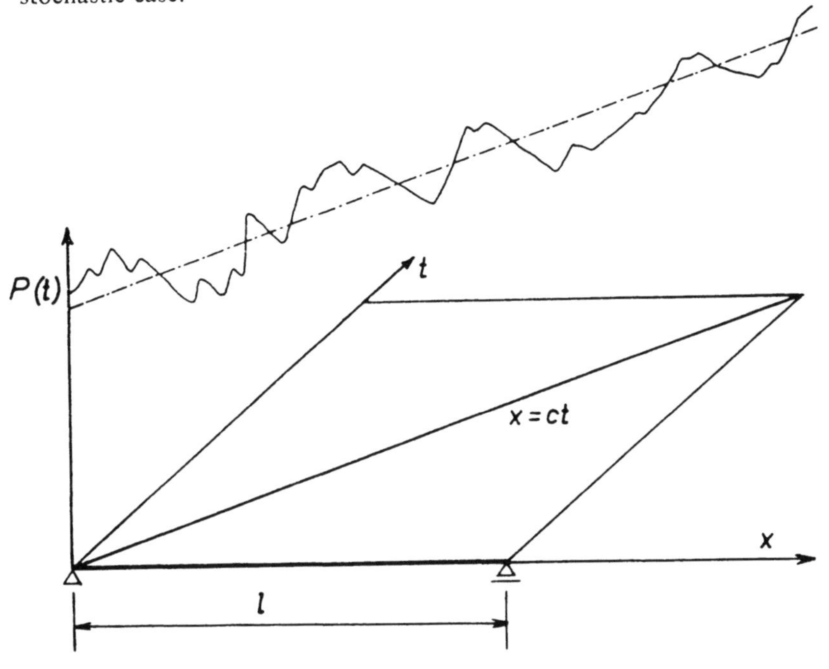

Fig. 26.1. Motion of a random variable force along a beam.

The load per unit length and its mean and centred values are

$$\begin{aligned} p(x, t) &= \delta(x - ct)\, P(t), \\ E[p(x, t)] &= \delta(x - ct)\, P, \\ \mathring{p}(x, t) &= \delta(x - ct)\, \mathring{P}(t) \end{aligned} \tag{26.24}$$

where $\delta(x)$ again represents the Dirac delta function. The covariance of the load can be calculated from (26.12)

$$K_{pp}(x_1, x_2, t_1, t_2) = \delta(x_1 - ct_1)\, \delta(x_2 - ct_2)\, K_{PP}(t_1, t_2) \tag{26.25}$$

where $K_{PP}(t_1, t_2)$ is the known covariance of load $P(t)$. Substituting (26.25) in (26.16) and recalling the well known properties of the Dirac function we get the covariance of the generalized force

$$K_{Q_{(j)}Q_{(k)}}(t_1, t_2) = \frac{1}{V_{(j)} V_{(k)}} v_{(j)}(ct_1)\, v_{(k)}(ct_2)\, K_{PP}(t_1, t_2) \qquad (26.26)$$

and with its aid, from (26.13) and (26.14), the covariance of the deflection.

Let us assume, for *example*, that the covariance of force $P(t)$ has the form

$$K_{PP}(t_1, t_2) = K_{PP}(t_2 - t_1) = 2\pi S_P\, \delta(t_2 - t_1) \qquad (26.27)$$

where S_P is the constant spectral density (white noise spectrum). Then Eq. (26.13) gives

$$\begin{aligned} K_{q_{(j)}q_{(k)}}(t_1, t_2) &= \frac{1}{V_{(j)} V_{(k)}} \int_{-\infty}^{\infty} \int_{-\infty}^{\infty} h_{(j)}(\tau_1)\, h_{(k)}(\tau_2)\, v_{(j)}[c(t_1 - \tau_1)]\, . \\ &\quad . \, v_{(k)}[c(t_2 - \tau_2)]\, K_{PP}(t_1 - \tau_1, t_2 - \tau_2)\, d\tau_1\, d\tau_2 = \\ &= \frac{2\pi S_P}{V_{(j)} V_{(k)}} \int_{-\infty}^{\infty} h_{(j)}(\tau_1)\, h_{(k)}(\tau_1 + t_2 - t_1)\, . \\ &\quad . \, v_{(j)}[c(t_1 - \tau_1)] v_{(k)}[c(t_1 - \tau_1)]\, d\tau_1\, . \end{aligned} \qquad (26.28)$$

If for the sake of simplification we neglect the cross-correlations of the generalized deflection, i.e. $K_{q_{(j)}q_{(k)}}(t_1, t_2) = 0$ for $j \neq k$ (see [22]), the variance of the deflection can be obtained from (26.14)

$$\begin{aligned} \sigma_v^2(x, t) &= K_{vv}(x, x, t, t) = \sum_{j=1}^{\infty} v_{(j)}^2(x)\, K_{q_{(j)}q_{(j)}}(t, t) = \\ &= \sum_{j=1}^{\infty} \frac{2\pi S_P}{V_{(j)}^2} v_{(j)}^2(x) \int_{-\infty}^{\infty} h_{(j)}^2(\tau_1)\, v_{(j)}^2[c(t - \tau_1)]\, d\tau_1\, . \end{aligned} \qquad (26.29)$$

For a simple beam of span l and bending stiffness EJ (see Chaps. 1 and 6) it holds true:

$$v_{(j)}(x) = \sin \frac{j\pi x}{l}, \quad V_{(j)} = \frac{\mu l}{2}, \quad \omega_{(j)}^2 = \frac{j^4 \pi^4}{l^4} \frac{EJ}{\mu}\, . \qquad (26.30)$$

Substituting (26.30) and (26.10) in (26.29) yields (note that the limits of integration may be changed as $h_{(j)}(\tau_1) = 0$ for $\tau_1 < 0$ and $v_{(j)}[c(t - \tau_1)] = 0$ for $\tau_1 > t$):

$$\sigma_v^2(x, t) = \sum_{j=1}^{\infty} \frac{8\pi S_P}{\mu^2 l^2 \omega_{(j)}'^2} \sin^2 \frac{j\pi x}{l} \int_0^t \left[e^{-\omega_b \tau_1} \sin \omega_{(j)}' \tau_1 \sin \frac{j\pi c}{l} (t - \tau_1) \right]^2 .$$

$$. \, d\tau_1 = \sum_{j=1}^{\infty} \frac{8\pi S_P}{\mu^2 l^2 \omega_{(j)}'^2} \sin^2 \frac{j\pi x}{l} \frac{1}{16} \left\{ \frac{\omega_{(j)}' + j\pi c/l}{(\omega_{(j)}' + j\pi c/l)^2 + \omega_b^2} . \right.$$

$$. \left[\sin 2j\pi ct/l + e^{-2\omega_b t} \sin 2\omega_{(j)}' t + \frac{\omega_b}{\omega_{(j)}' + j\pi c/l} (\cos 2j\pi ct/l - \right.$$

$$\left. - e^{-2\omega_b t} \cos 2\omega_{(j)}' t) \right] + \frac{\omega_{(j)}' - j\pi c/l}{(\omega_{(j)}' - j\pi c/l)^2 + \omega_b^2} \left[- \sin 2j\pi ct/l + \right.$$

$$+ e^{-2\omega_b t} \sin 2\omega_{(j)}' t + \frac{\omega_b}{\omega_{(j)}' - j\pi c/l} (\cos 2j\pi ct/l -$$

$$\left. - e^{-2\omega_b t} \cos 2\omega_{(j)}' t) \right] - \frac{2\omega_b}{j^2 \pi^2 c^2/l^2 + \omega_b^2} \left(\frac{j\pi c/l}{\omega_b} \sin 2j\pi ct/l + \right.$$

$$\left. + \cos 2j\pi ct/l - e^{-2\omega_b t} \right) - \frac{2\omega_b}{\omega_{(j)}^2} \left[1 - e^{-2\omega_b t} \left(\cos 2\omega_{(j)}' t - \right. \right.$$

$$\left. \left. \left. - \frac{\omega_{(j)}'}{\omega_b} \sin 2\omega_{(j)}' t \right) \right] + \frac{2}{\omega_b} (1 - e^{-2\omega_b t}) \right\} . \tag{26.31}$$

Since the variance of the deflection is a function of time, the resulting beam vibration turns out to be a nonstationary process even though the motion considered was that of a stationary random force.

For subcritical speed ($c < c_{cr}$) the largest static and dynamic effects of a moving force will appear at about the instant the force crosses the beam centre. Therefore the coefficient of variation defined as

$$C_v(x, t) = \sigma_v(x, t)/E[v(x, t)]$$

should be computed for $x = l/2$ and $t = T/2 = l/(2c)$. It then represents the relative dynamic increment of the deflection produced by the moving random force, and takes the form [from (26.31) approximately for $j = 1$]

$$C_v(l/2, T/2) = C_P C_{vP} . \tag{26.32}$$

In the above, $C_P = (\pi S_P \omega_{(1)})^{1/2}/P$ is the analogous coefficient of variation of force $P(t)$; C_{vP} is drawn in Fig. 26.2 as a function of parameters

α and β (α — the speed parameter, β — the damping parameter — cf. Chap. 1):

$$\alpha = c/c_{cr}\,; \quad c_{cr} = (\pi/l)\,(EJ/\mu)^{1/2}\,; \quad \beta = \omega_b/\omega_{(1)}\,. \tag{26.33}$$

The random effects of the load considered decrease with increasing speed and damping.

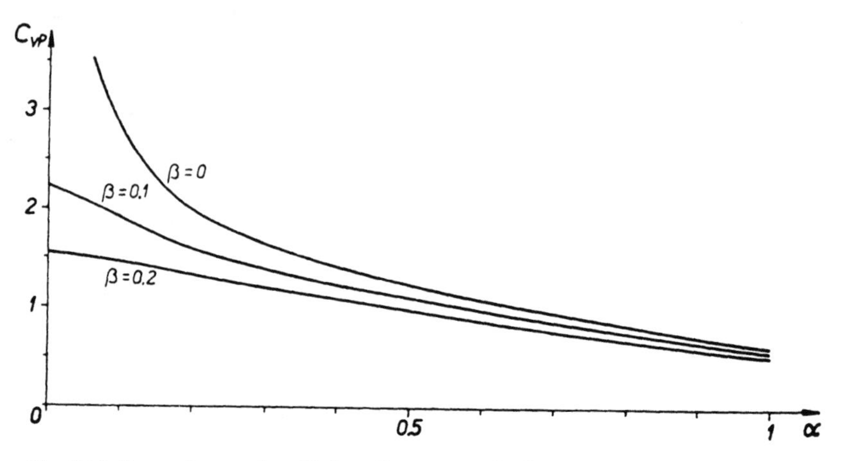

Fig. 26.2. Dependence of coefficient C_{vP} on speed α for various values of damping β.

The spectral density analysis outlined in paragraph 26.1.2 would yield the same results; however, in this case the load [Eq. (26.24)] must be considered a function of time alone.

26.3 *Moving random continuous load*

The next important case is that of an infinite random strip of load moving on a simple beam at constant speed c (Fig. 26.3). The analogous deterministic case was analyzed in Chap. 3; there we considered the motion of continuous load p (per unit length) as well as the effects of its inertia mass $\mu_p = p/g$. In the examination that follows the load mass is considered constant and hence deterministic.

The load is assumed to be of the form

$$p(x, t) = q(\xi)\, r(t)\,. \tag{26.34}$$

The first component, $q(\xi)$, is a random variable in the moving coordinate system $\xi = x - ct$, the second, $r(t)$, a random function of time. The mean values of the functions are assumed to be constant,

$$E[p(x, t)] = p, \quad E[q(\xi)] = p, \quad E[r(t)] = 1; \tag{26.35}$$

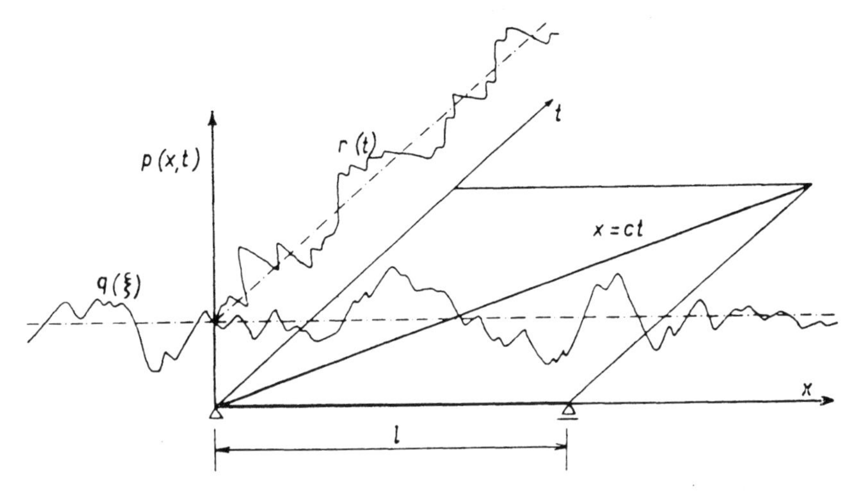

Fig. 26.3. An infinite random strip load moving on a beam.

hence the load (26.34) may be written as

$$p(x, t) = p + \mathring{p}(x, t) = [p + \mathring{q}(\xi)]\,[1 + \mathring{r}(t)] \tag{26.36}$$

where

$$\mathring{p}(x, t) = \mathring{q}(\xi) + p\mathring{r}(t) + \mathring{q}(\xi)\,\mathring{r}(t).$$

In view of (26.12) the covariance of the load is

$$\begin{aligned} K_{pp}(x_1, x_2, t_1, t_2) = {} & K_{qq}(\xi_1, \xi_2) + p^2 K_{rr}(t_1, t_2) + pK_{qr}(\xi_1, t_2) + \\ & + pK_{qr}(\xi_2, t_1) + K_{qqr}(\xi_1, \xi_2, t_1) + \\ & + K_{qqr}(\xi_1, \xi_2, t_2) + pK_{qrr}(\xi_1, t_1, t_2) + \\ & + pK_{qrr}(\xi_2, t_1, t_2) + K_{qqrr}(\xi_1, \xi_2, t_1, t_2) \end{aligned} \tag{26.37}$$

where

$$\xi_i = x_i - ct_i, \quad i = 1, 2.$$

If the random components of the load defined by (26.36) are very small

$$p(x, t) = [p + \varepsilon\mathring{q}(\xi)]\,[1 + \varepsilon\mathring{r}(t)],$$

where $\varepsilon \ll 1$ is a small parameter, we find from an analysis of the covariance (26.37) that the cross-correlation functions of the third (K_{qqr}, K_{qrr}) and of the fourth (K_{qqrr}) order are very small quantities, of the order of ε^3 and ε^4. For that reason they are completely neglected. To simplify, assume that functions $q(\xi)$ and $r(t)$ have no cross-correlation ($K_{qr} = 0$). Then Eq. (26.37) will reduce to

$$K_{pp}(x_1, x_2, t_1, t_2) = K_{qq}(\xi_1, \xi_2) + p^2 K_{rr}(t_1, t_2) \tag{26.38}$$

where $K_{qq}(\xi_1, \xi_2)$ is the covariance of the load function in the moving coordinate system ξ, and $K_{rr}(t_1, t_2)$ the covariance in the time coordinate.

As an *example*, assume the covariance of those functions to have the form

$$K_{qq}(\xi_1, \xi_2) = 2\pi S_q\, \delta(\xi_2 - \xi_1)\,, \quad K_{rr}(t_1, t_2) = 2\pi S_r\, \delta(t_2 - t_1) \tag{26.39}$$

where S_q and S_r are the constant spectral densities (white noise spectra). The covariance of the generalized deflection is obtained from (26.13) with (26.38) and (26.39) substituted in:

$$\begin{aligned} K_{q(j)q(k)}(t_1, t_2) = {} & \frac{1}{V_{(j)}V_{(k)}} \int_{-\infty}^{\infty} \int_{-\infty}^{\infty} \int_0^l \int_0^l h_{(j)}(\tau_1)\, h_{(k)}(\tau_2)\, v_{(j)}(x_1)\, v_{(k)}(x_2)\, . \\ & . \left\{2\pi S_q\, \delta[x_2 - x_1 - c(t_2 - \tau_2 - t_1 + \tau_1)] + \right. \\ & \left. + 2\pi S_r p^2\, \delta(t_2 - t_1 - \tau_2 + \tau_1)\right\} \mathrm{d}x_1\, \mathrm{d}x_2\, \mathrm{d}\tau_1\, \mathrm{d}\tau_2\, . \end{aligned} \tag{26.40}$$

Because of (26.10) and the fact that the motion of the load is of infinitely long duration, the limits of integration with respect to time τ are taken to be zero and infinity.

Neglecting the cross-correlation $K_{q(j)q(k)}(t_1, t_2) = 0$ for $j \neq k$ (see [22]), the variance may be obtained from (26.40) and (26.41)

$$\begin{aligned} \sigma_v^2(x, t) = K_{vv}(x, x, t, t) = {} & \sum_{j=1}^{\infty} v_{(j)}^2(x)\, K_{q(j)q(j)}(t, t) = \\ = {} & \sum_{j=1}^{\infty} \frac{4\pi S_q c^2}{\mu^2(1 + \varkappa)^2\, l^2} \frac{1}{\overline{\overline{\omega}}{}_{(j)}^2 \overline{\overline{\omega}}{}'_{(j)} \overline{\omega}_b D} \sin^2 \frac{j\pi x}{l} \left[\!\left[\frac{\overline{\overline{\omega}}{}_{(j)}^2 \overline{\overline{\omega}}{}'_{(j)} \overline{\omega}_b l}{c} + \right.\right. \\ & + \frac{j^2\pi^2c^2/l^2}{D} \left\{\overline{\overline{\omega}}{}'_{(j)}[D + 4\overline{\omega}_b^2(\overline{\omega}_b^2 - 3\overline{\overline{\omega}}{}'^2_{(j)} + j^2\pi^2c^2/l^2)]\right. . \end{aligned}$$

$$. \left(1 - e^{-\bar{\omega}_b l/c} \cos \bar{\bar{\omega}}'_{(j)} l/c \,.\, \cos j\pi\right) -$$
$$- \bar{\omega}_b \left[D + 4\bar{\bar{\omega}}'^2_{(j)} \left(\bar{\bar{\omega}}'^2_{(j)} - 3\bar{\omega}_b^2 - j^2\pi^2 c^2/l^2\right)\right] e^{-\bar{\omega}_b l/c} \sin \bar{\bar{\omega}}'_{(j)} l/c \,.$$
$$. \cos j\pi \Big\} \Bigg] + \sum_{j=1}^{\infty} \frac{4S_r\, p^2(1 - \cos j\pi)}{\mu^2 j^2 \pi \bar{\bar{\omega}}^2_{(j)} \bar{\omega}_b (1 + \varkappa)^2} \sin^2 \frac{j\pi x}{l} \qquad (26.41)$$

where

$$\bar{\bar{\omega}}^2_{(j)} = \omega^2_{(j)}\left(1 - \alpha^2 \varkappa/j^2\right)/(1 + \varkappa)\,, \quad \bar{\bar{\omega}}'^2_{(j)} = \bar{\bar{\omega}}^2_{(j)} - \bar{\omega}_b^2\,,$$
$$\bar{\omega}_b = \omega_b/(1 + \varkappa)\,, \quad D = \left(\bar{\bar{\omega}}^2_{(j)} - j^2\pi^2c^2/l^2\right)^2 + 4j^4 \bar{\omega}_b^2 \pi^2 c^2/l^2 \qquad (26.42)$$

similarly as in (3.9) to (3.13). Since (26.41) does not depend on time, the beam vibration is a random process stationary in time.

The coefficient of variation can be computed by the procedure outlined in Sect. 26.2.

26.4 *Infinite beam on random elastic foundation traversed by a random force*

The last example we are going to solve is the case of an infinite beam resting on an elastic foundation, the foundation being a random function of the length coordinate x

$$k(x) = k + \varepsilon\, \mathring{k}(x)\,, \quad E[k(x)] = k\,. \qquad (26.43)$$

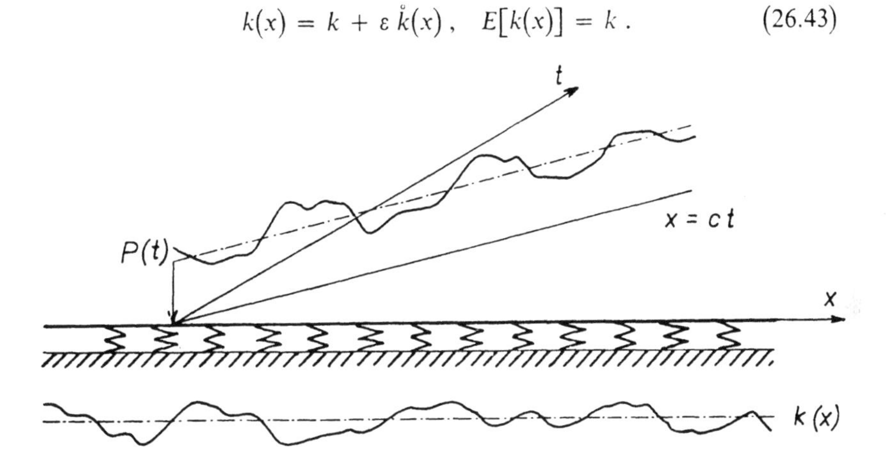

Fig. 26.4. Infinite beam on random elastic foundation traversed by a moving random force.

In writing the above we assume that the mean value $E[k(x)] = k$ is constant, function $\mathring{k}(x)$ is a random stationary ergodic function with zero mean value, and ε is the small parameter (see Sect. 26.3).

The concentrated force that moves along the beam from infinity to infinity at constant speed c, is a random function of time, likewise with a constant mean value (Fig. 26.4)

$$P(t) = P + \varepsilon \mathring{P}(t) , \quad E[P(t)] = P . \tag{26.44}$$

The random function of time $\mathring{P}(t)$ has again zero mean value.

Analogous to Chap. 13, in the absence of damping the case is solved by the differential equation

$$EJ \frac{\partial^4 v(x, t)}{\partial x^4} + \mu \frac{\partial^2 v(x, t)}{\partial t^2} + k(x) v(x, t) = \delta(x - ct) P(t) . \tag{26.45}$$

Eq. (26.45) with (26.43) and (26.44) lends itself to solution whenever parameter ε is zero. This, of course, brings us to the deterministic concept which we have examined in detail in Chap. 13. What the introduction of small parameter ε means is that the nonuniformity of the foundation stiffness and of the force is statistically small, and therefore the probability of large deviations from the mean values is sufficiently low.

As this consideration implies, the best suited for the solution of our problem will be the perturbation method. We will assume that the solution of Eq. (26.45) will have the form of the series

$$v(x, t) = v_0(x, t) + \varepsilon v_1(x, t) + \varepsilon^2 v_2(x, t) + \dots \tag{26.46}$$

On substituting relations (26.43), (26.44) and (26.46) in Eq. (26.45) and comparing the terms with like powers of parameter ε, we get the sequence of equations

$$EJ \frac{\partial^4 v_0(x, t)}{\partial x^4} + \mu \frac{\partial^2 v_0(x, t)}{\partial t^2} + k v_0(x, t) = \delta(x - ct) P , \tag{26.47}$$

$$\begin{aligned} EJ \frac{\partial^4 v_1(x, t)}{\partial x^4} + \mu \frac{\partial^2 v_1(x, t)}{\partial t^2} + k v_1(x, t) &= \delta(x - ct) \mathring{P}(t) - \mathring{k}(x) v_0(x, t) , \\ EJ \frac{\partial^4 v_2(x, t)}{\partial x^4} + \mu \frac{\partial^2 v_2(x, t)}{\partial t^2} + k v_2(x, t) &= - \mathring{k}(x) v_1(x, t) , \end{aligned} \tag{26.48}$$

. . . .

The first equation is deterministic and its solution was obtained in Chap. 13

$$v_0(x, t) = v_0\, v(s) \tag{26.49}$$

where v_0 is defined by Eq. (13.5), s is the dimensionless moving coordinate (13.2), and $v(s)$ is described by (13.54).

The solution of the second, (26.48) and of the subsequent equations is assumed to have the form (ξ and τ are new length and time variables, respectively)

$$v_j(x, t) = \int_{-\infty}^{\infty}\int_{-\infty}^{\infty} p(\xi, \tau)\, h(x - \xi, t - \tau)\, \mathrm{d}\xi\, \mathrm{d}\tau\,, \quad j = 1, 2, \ldots \tag{26.50}$$

where $p(x, t)$ are the right-hand sides of Eqs. (26.48), that is, the load per unit length, and $h(x, t)$ the influence impulse function, i.e. the response of the system to a time-spatial impulse in position $x = 0$ and at time $t = 0$

$$L[h(x, t)] = \delta(x)\, \delta(t)\,. \tag{26.51}$$

Here the operator L according to (26.1) schematically represents the left-hand sides of Eqs. (26.48).

We will consider here only the approximate solution of Eq. (26.51)

$$h(x, t) \approx \frac{v_0}{P}\, v(s)\, \delta(t) \tag{26.52}$$

which describes the steady-state component of vibration and which can be obtained by the method expounded in [68]*).

We are now ready to undertake the calculation of functions $v_j(x, t)$, $j = 1, 2, \ldots$ In (26.48) the load is

$$p(x, t) = \delta(x - ct)\, \mathring{P}(t) - \mathring{k}(x)\, v_0(x, t)\,. \tag{26.53}$$

Substitution of (26.52) and (26.53) in (26.50), integration and a bit of handling will result in

$$v_1(x, t) = v_1(s, t) = \frac{v_0}{P}\, v(s)\, \mathring{P}(t) - \frac{v_0^2}{\lambda P}\int_{-\infty}^{\infty} \mathring{k}(\xi)\, v(\xi)\, v(s - \xi)\, \mathrm{d}\xi\,. \tag{26.54}$$

---

*) Paper [68] also contains a method of calculating the error committed when using the method.

In the calculation of Eq. (26.54) use was made of the ergodic properties of function $\mathring{k}(x) = \mathring{k}(s)$ and of notation (13.3); ξ is a new auxiliary dimensionless variable. An analogous procedure would be used for obtaining $v_2(x, t)$, etc.

This preliminary work done, we can approach the analysis of the statistical characteristic of the response. From (26.46) the mean value of the beam deflection is

$$E[v(x, t)] = E[v_0(x, t)] + \varepsilon\, E[v_1(x, t)] + \dots \tag{26.55}$$

and from (26.49) $E[v_0(x, t)] = v_0\, v(s)$. The mean values of the higher perturbations are zero, $E[v_j(x, t)] = 0$ for $j = 1, 2, \dots$ as implied by (26.54) and by the right-hand sides of Eqs. (26.48) which all contain but functions $\mathring{P}(t)$ and $\mathring{k}(x)$ with zero mean values. Therefore the mean value of the beam deflection is equal to the deterministic solution of the given case

$$E[v(x, t)] = v_0\, v(s)\,. \tag{26.56}$$

The computation of higher statistical moments will be restricted to the first perturbation only. The centred value of the deflection then is

$$\mathring{v}(x, t) = v(x, t) - E[v(x, t)] \approx \varepsilon\, v_1(s, t)\,. \tag{26.57}$$

With regard to definition (26.12) the covariance of the deflection is obtained from (26.57) and (26.54) as

$$\begin{aligned} K_{vv}(x_1, x_2, t_1, t_2) &= K_{vv}(s_1, s_2, t_1, t_2) = E[\varepsilon\, v_1(s_1, t_1)\, \varepsilon\, v_1(s_2, t_2)] = \\ &= \varepsilon^2 \left(\frac{v_0}{\lambda P}\right)^2 \Bigg\{\lambda^2\, v(s_1)\, v(s_2)\, K_{PP}(t_1, t_2) - v_0 \lambda \Bigg[\int_{-\infty}^{\infty} v(\xi_1)\, v(s_2)\, v(s_1 - \xi_1)\, . \\ &\quad . \, K_{kP}(\xi_1, t_2)\, \mathrm{d}\xi_1 + \int_{-\infty}^{\infty} v(s_1)\, v(\xi_2)\, v(s_2 - \xi_2)\, K_{kP}(\xi_2, t_1)\, \mathrm{d}\xi_2\Bigg] + \\ &\quad + v_0^2 \int_{-\infty}^{\infty}\int_{-\infty}^{\infty} v(\xi_1)\, v(\xi_2)\, v(s_1 - \xi_1)\, v(s_2 - \xi_2)\, K_{kk}(\xi_1, \xi_2)\, \mathrm{d}\xi_1\, \mathrm{d}\xi_2 \Bigg\}\,. \end{aligned} \tag{26.58}$$

In the above, notation $K_{PP}(t_1, t_2)$, $K_{kk}(s_1, s_2)$ and $K_{kP}(s, t)$ refers respectively to the covariance of the force, to the covariance of the elastic foundation, and to the cross-correlation function of both quantities.

By way of an *example*, consider the following form of the covariances

$$\begin{aligned} K_{kk}(s_1, s_2) &= 2\pi S_k\, \delta(s_1 - s_2)\,, \\ K_{kP}(s_1, t_2) &= 2\pi S_{kP}\, \delta(s_1 - \lambda c t_2)\,, \\ K_{kP}(s_2, t_1) &= 2\pi S_{kP}\, \delta(s_2 - \lambda c t_1) \end{aligned} \tag{26.59}$$

which represent the white noise spectra S_k and S_{kP}.

For the above the covariance of the beam response may be given the form

$$\begin{aligned} K_{vv}(s_1, s_2, t_1, t_2) = \varepsilon^2 \left(\frac{v_0}{\lambda P}\right)^2 \Big\{ & \lambda^2\, v(s_1)\, v(s_2)\, K_{PP}(t_1, t_2) - 2\pi S_{kP} v_0 \lambda\,. \\ & . \left[v(\lambda c t_2)\, v(s_2)\, v(s_1 - \lambda c t_2) + v(s_1)\, v(\lambda c t_1)\, v(s_2 - \lambda c t_1)\right] + \\ & + 2\pi S_k v_0^2 \int_{-\infty}^{\infty} v^2(\xi_1)\, v(s_1 - \xi_1)\, v(s_2 - \xi_1)\, \mathrm{d}\xi_1 \Big\}\,. \end{aligned} \tag{26.60}$$

The variance of the deflection is obtained from the covariance for $s_1 = s_2 = s$, $t_1 = t_2 = t$

$$\begin{aligned} \sigma_v^2(s, t) = K_{vv}(s, s, t, t) = \varepsilon^2 \left(\frac{v_0}{\lambda P}\right)^2 \Big[& \lambda^2\, v^2(s)\, \sigma_P^2(t) - 4\pi S_{kP} v_0 \lambda\,. \\ & . \, v(s)\, v(\lambda c t)\, v(s - \lambda c t) + 2\pi S_k v_0^2 \int_{-\infty}^{\infty} v^2(\xi_1)\, v^2(s - \xi_1)\, \mathrm{d}\xi_1 \Big] \end{aligned} \tag{26.61}$$

where $\sigma_P^2(t) = K_{PP}(t, t)$ is the variance of force $P(t)$.

The coefficient of variation

$$C_v(s, t) = \frac{\sigma_v(s, t)}{E[v(s, t)]}$$

can readily be computed for point $s = 0$, the point of action of the moving force, viz.

$$\begin{aligned} C_v(0, t) = \frac{\sigma_v(0, t)}{v_0\, v(0)} = \varepsilon \Big[& C_P^2(t) - 4\pi S_{kP} \frac{v_0}{\lambda P^2}\, v^2(\lambda c t) + \\ & + 2\pi S_k \frac{v_0^2}{\lambda^2 P^2} \int_{-\infty}^{\infty} v^2(\xi_1)\, v^2(-\xi_1)\, \mathrm{d}\xi_1 \Big]^{1/2} \end{aligned} \tag{26.62}$$

where $C_P(t) = \sigma_P(t)/P$ is the coefficient of variation of the force.

For $\alpha < 1$, the integral in (26.62) evaluated from (13.54) is

$$\int_{-\infty}^{\infty} v^2(\xi_1)\, v^2(-\xi_1)\, \mathrm{d}\xi_1 = \frac{41 - 50\alpha^2 + 12\alpha^4}{8(5 - 3\alpha^2)(1 - \alpha^2)^{5/2}}, \qquad (26.63)$$

i.e. dependent on the dimensionless speed parameter α (13.9).

26.5 *Application of the theory*

The probabilistic method of structural design — a highly advanced approach, particularly to the problems of moving load — is still in the early stages of development. Though our discussion has been confined to the computation of the correlation functions, spectral density, variance and coefficient of variations, work is in progress on the experimental aspects of the problems, which — it is hoped — will enable the project engineer to estimate the service life of structures from data on probable damage to structures (fatigue, brittle fracture, etc.). Our exposition presented in the foregoing sections, is an attempt to lay theoretical foundations to this new line of research in the field of structures traversed by moving loads.

The example solved in Sect. 26.2 is typical of problems involving short-span bridges and longitudinal beams (Fig. 26.1), both structures that are usually subjected to loads produced by one vehicle axle only, see [88]. The random variable axle pressure results from the random variability of the initial conditions at the instant the axle arrives on the bridge, from random motions of the vehicle, from random roadway unevenesses and various irregularities of the road bed, etc.

The case solved in Sect. 26.3 is illustrative of the other extreme, large-span bridges (Fig. 26.3, see [88]). Such bridges are usually loaded with a number of axles, and that is why their stresses unquestionably depend on the traffic flow or on the train whose length is assumed to be larger by far than the bridge span. In idealizing this case we arrive at an infinite random continuous strip of load that represents the random arrangement of the various vehicles or wagons with different axle pressures on the bridge, the random changes in additional loads on the axles, the random character of irregularities, etc. In Eq. (26.35), function $q(x - ct)$ expresses the random variation of static axle pressures along the length of the traffic flow or train, while function $r(t)$ represents the random variation of the axle loads in time.

The theory explained in Sect. 26.4 is best illustrated by a computation of dynamic stresses in rails, see also [87]. A rail is conventionally idealized by an infinite beam on an elastic foundation. However, the properties of the foundation are not homogeneous along the length of the rail but vary more or less at random from one place to another. The stochastic inhomogeneity of the foundation affects the motion of vehicles whose dynamic effects on the rails are therefore random, too.

Another problem of importance for engineering practice is the inverse problem which the author attempted to solve in [86]. It is namely frequently possible to measure directly the statistical characteristics of deflection, stress, etc. This makes for a new problem because so far we have assumed the "input" to the system to be known, while the sought-for quantities are the characteristics of the "output".

26.6 *Additional bibliography*

[9, 22, 40, 41, 50, 51, 86, 87, 88, 126, 143, 188, 238, 241, 244, 245, 246, 247, 253, 260, 263, 266, 268, 269].

Part VI

APPENDIX AND BIBLIOGRAPHY

Part VI

27

Tables of integral transformations

Symbols used in the tables correspond to the conventional notation of mathematical literature, e.g. [32, 48, 49, 53, 171, 204]. Thus a, b, c, d, f, α, ω denote the positive real constants, $n = 0, 1, 2, 3, \ldots$, $j = 1, 2, 3, \ldots$, etc. To the functions (originals) $f(t)$, $f(x)$ or $f(r)$ belong the integral transformations (transforms) $F(p)$, $F(j)$ or $F(q)$.

Table 27.1 Laplace - Carson integral transformation

Transform	Original	Equation No.
$F(p) = p \int_0^\infty f(t)\, e^{-pt}\, dt$	$f(t) = \frac{1}{2\pi i} \int_{a-i\infty}^{a+i\infty} \frac{F(p)}{p} e^{tp}\, dp$	(27.1)
a	a	(27.2)
$\sum_{j=1}^{n} a_j F_j(p)$	$\sum_{j=1}^{n} a_j f_j(t)$	(27.3)
$p^n F(p) - p^n f(0_+) - p^{n-1} \frac{df(0_+)}{dt} - \ldots - p \frac{d^{n-1} f(0_+)}{dt^{n-1}}$	$\frac{d^n f(t)}{dt^n}$	(27.4)
$\frac{1}{p} F_1(p) F_2(p)$	$\int_0^t f_1(\tau) f_2(t - \tau)\, d\tau$	(27.5)
e^{-ap}	$H(t - a)$	(27.6)
pe^{-ap}	$\delta(t - a)$	(27.7)
$p^2 e^{-ap}$	$H_1(t - a)$	(27.8)

$p^3 e^{-ap}$	$H_2(t-a)$	(27.9)
$e^{-ap} F(p)$	$f(t-a)\,H(t-a)$	(27.10)
$\frac{p}{2i}\left[\frac{F(p-ia)}{p-ia} - \frac{F(p+ia)}{p+ia}\right]$	$f(t)\sin at$	(27.11)
$\frac{p}{2}\left[\frac{F(p-ia)}{p-ia} + \frac{F(p+ia)}{p+ia}\right]$	$f(t)\cos at$	(27.12)
$\lim_{p\to\infty} F(p)$	$\lim_{t\to 0_+} f(t)$	(27.13)
$\lim_{p\to 0_+} F(p)$	$\lim_{t\to\infty} f(t)$	(27.14)
$\frac{1}{p^n}$	$\frac{t^n}{n!}$, $n = 0, 1, 2, 3, \ldots$	(27.15)
$\frac{p}{p+a}$	e^{-at}	(27.16)
$\frac{a^2}{p^2+a^2}$	$1 - \cos at$	(27.17)
$\frac{p}{p^2+a^2}$	$\frac{1}{a}\sin at$	(27.18)

Table 27.1 — *continued*

$F(p)$	$f(t)$	Equation No.
$\frac{p^2}{p^2 + a^2}$	$\cos at$	(27.19)
$\frac{1}{(p + a)^2 + b^2}$	$\frac{1}{a^2 + b^2}\left[1 - e^{-at}\left(\frac{a}{b}\sin bt + \cos bt\right)\right]$	(27.20)
$\frac{p}{(p + a)^2 + b^2}$	$\frac{1}{b}e^{-at}\sin bt$	(27.21)
$\frac{p(p + 2a)}{(p + a)^2 + b^2}$	$e^{-at}\left(\cos bt + \frac{a}{b}\sin bt\right)$	(27.22)
$\frac{1}{(p + a)^2 - b^2}$	$\frac{1}{b^2 - a^2}\left[e^{-at}\left(\frac{a}{b}\sinh bt + \cosh bt\right) - 1\right]$	(27.23)
$\frac{p^2}{2}\left(\frac{1}{p^2 + r_2^2} - \frac{1}{p^2 + r_1^2}\right)$	$\sin at \sin bt$, where $r_1 = a + b$, $r_2 = a - b$	(27.24)
$\frac{p}{2}\left(\frac{r_2}{p^2 + r_2^2} + \frac{r_1}{p^2 + r_1^2}\right)$	$\sin at \cos bt$, where $r_1 = a + b$, $r_2 = a - b$	(27.25)
$\frac{1}{p(p + a)}$	$\frac{1}{a^2}(e^{-at} + at - 1)$	(27.26)

$\dfrac{1}{(p+a)^2}$	$\dfrac{1}{a^2}\left[1-\mathrm{e}^{-at}(1+at)\right]$	(27.27)
$\dfrac{p}{\left[(p+a)^2+b^2\right]\left(1-\mathrm{e}^{-pT}\right)}$	$\dfrac{\mathrm{e}^{-at}}{b(1-2\mathrm{e}^{aT}\cos bT+\mathrm{e}^{2aT})}\left[(1-\mathrm{e}^{aT}\cos bT)\sin bt-\right.$ $\left.-\mathrm{e}^{aT}\sin bT\cos bt\right]+$ $+\dfrac{2}{T}\sum_{n=0}^{\infty}\dfrac{1}{\left(a^2+b^2-\dfrac{4n^2\pi^2}{T^2}\right)^2+\dfrac{16n^2\pi^2a^2}{T^2}}\,.$ $.\left[\left(a^2+b^2-\dfrac{4n^2\pi^2}{T^2}\right)\cos\dfrac{2n\pi t}{T}+\dfrac{4n\pi a}{T}\sin\dfrac{2n\pi t}{T}\right]$	(27.28)
$\dfrac{p^2}{\left[(p+a)^2+b^2\right]\left(1-\mathrm{e}^{-pT}\right)}$	$\dfrac{\mathrm{e}^{-at}}{b(1-2\mathrm{e}^{aT}\cos bT+\mathrm{e}^{2aT})}\left\{\left[a\mathrm{e}^{aT}\sin bT+b(1-\mathrm{e}^{aT}\cos bT)\right].\right.$ $\left..\cos bt+\left[b\mathrm{e}^{aT}\sin bT-a(1-\mathrm{e}^{aT}\cos bT)\right]\sin bt\right\}-$ $-\sum_{n=1}^{\infty}\dfrac{4n\pi}{T^2}\dfrac{1}{\left(a^2+b^2-\dfrac{4n^2\pi^2}{T^2}\right)^2+\dfrac{16n^2\pi^2a^2}{T^2}}\,.$ $.\left[\left(a^2+b^2-\dfrac{4n^2\pi^2}{T^2}\right)\sin\dfrac{2n\pi t}{T}-\dfrac{4n\pi a}{T}\cos\dfrac{2n\pi t}{T}\right]$	(27.29)

Table 27.1 — *continued*

$F(p)$	$f(t)$	Equation No.
$\frac{1}{p(p^2 + a^2)}$	$\frac{1}{a^3}(at - \sin at)$	(27.30)
$\frac{1}{(p^2 + a^2)(p^2 + b^2)}$	$\frac{1}{a^2b^2}\left[1 + \frac{1}{a^2 - b^2}(b^2 \cos at - a^2 \cos bt)\right]$	(27.31)
$\frac{p}{(p^2 + a^2)(p^2 + b^2)}$	$\frac{1}{ab(a^2 - b^2)}(a \sin bt - b \sin at)$	(27.32)
$\frac{p^2}{(p^2 + a^2)(p^2 + b^2)}$	$\frac{1}{a^2 - b^2}(\cos bt - \cos at)$	(27.33)
$\frac{1}{(p^2 + a^2)^2}$	$\frac{1}{a^4}(1 - \cos at - \frac{1}{2}at \sin at)$	(27.34)
$\frac{p}{(p^2 + a^2)^2}$	$\frac{1}{2a^3}(\sin at - at \cos at)$	(27.35)
$\frac{p^2}{(p^2 + a^2)^2}$	$\frac{1}{2a^2} at \sin at$	(27.36)
$\frac{p(p^2 - a^2)}{(p^2 + a^2)^2}$	$t \cos at$	(27.37)

$\dfrac{p}{(p^2 + a^2)(p + b)^2}$	$\dfrac{1}{(a^2 + b^2)^2} \left\{ \dfrac{b^2 - a^2}{a} \sin at - 2b \cos at + e^{-bt}[(a^2 + b^2) t + 2b] \right\}$	(27.38)
$\dfrac{p}{(p^2 + a^2)(p + b)(p + c)}$	$\dfrac{1}{a(a^2 + b^2)(a^2 + c^2)} \left\{ (bc - a^2) \sin at - a(b + c) \cos at + \dfrac{a}{b - c} [(a^2 + b^2) e^{-ct} - (a^2 + c^2) e^{-bt}] \right\}$	(27.39)
$\dfrac{1}{(p^2 + c^2)[(p + a)^2 + b^2]}$	$\dfrac{1}{(a^2 + b^2 - c^2)^2 + 4a^2c^2} \left[\dfrac{a^2 + b^2 - c^2}{c^2} (1 - \cos ct) + \dfrac{3a^2 - b^2 + c^2}{a^2 + b^2} (1 - e^{-at} \cos bt) - \dfrac{2a}{c} \sin ct + \dfrac{a(3b^2 - a^2 - c^2)}{b(a^2 + b^2)} e^{-at} \sin bt \right]$	(27.40)
$\dfrac{p}{(p^2 + c^2)[(p + a)^2 + b^2]}$	$\dfrac{1}{(a^2 + b^2 - c^2)^2 + 4a^2c^2} \left[\dfrac{a^2 + b^2 - c^2}{c} \sin ct - \dfrac{b^2 - a^2 - c^2}{b} e^{-at} \sin bt - 2a(\cos ct - e^{-at} \cos bt) \right]$	(27.41)

Table 27.1 — *continued*

$F(p)$	$f(t)$	Equation No.
$\dfrac{p^2}{(p^2 + c^2)\,[(p + a)^2 + b^2]}$	$\dfrac{1}{(a^2 + b^2 - c^2)^2 + 4a^2c^2}\Big[(a^2 + b^2 - c^2)(\cos ct - e^{-at}\cos bt) + 2ac\sin ct - \dfrac{a}{b}(a^2 + b^2 + c^2)\,e^{-at}\sin bt\Big]$	(27.42)
$\dfrac{p^3}{(p^2 + c^2)\,[(p + a)^2 + b^2]}$	$\dfrac{1}{(a^2 + b^2 - c^2)^2 + 4a^2c^2}\Big\{-c(a^2 + b^2 - c^2)\sin ct + \dfrac{1}{b}[(a^2 + b^2)^2 + c^2(a^2 - b^2)]\,e^{-at}\sin bt + 2ac^2(\cos ct - e^{-at}\cos bt)\Big\}$	(27.43)
$\dfrac{p^4}{(p^2 + c^2)\,[(p + a)^2 + b^2]}$	$\dfrac{1}{(a^2 + b^2 - c^2)^2 + 4a^2c^2}\Big\{-c^2(a^2 + b^2 - c^2)\cos ct + [(a^2 + b^2)^2 + c^2(3a^2 - b^2)]\,e^{-at}\cos bt - 2ac^3\sin ct - \dfrac{a}{b}[(a^2 + b^2)^2 + c^2(a^2 - 3b^2)]\,e^{-at}\sin bt\Big\}$	(27.44)
$\dfrac{p}{[(p + d)^2 + f^2]\,[(p + a)^2 + b^2]}$	$\dfrac{1}{[(d - a)^2 + f^2 - b^2]^2 + 4(d - a)^2\,b^2}\Big[\dfrac{(d - a)^2 + f^2 - b^2}{b}\,.$	

	$. \mathrm{e}^{-at} \sin bt + \dfrac{(d-a)^2 - f^2 + b^2}{f} \mathrm{e}^{-dt} \sin ft - 2(d-a) . (\mathrm{e}^{-at} \cos bt - \mathrm{e}^{-dt} \cos ft)\Big]$	(27.45)
$\dfrac{p^2}{[(p+d)^2 + f^2][(p+a)^2 + b^2]}$	$\dfrac{1}{[(d-a)^2 + f^2 - b^2]^2 + 4(d-a)^2 b^2} \Big\{(a^2 + b^2 - d^2 - f^2) . (\mathrm{e}^{-dt} \cos ft - \mathrm{e}^{-at} \cos bt) + \Big[2af - \dfrac{d}{f}((d-a)^2 + f^2 + b^2)\Big] \mathrm{e}^{-dt} \sin ft - \Big[\dfrac{a}{b}((d-a)^2 + f^2 + b^2) - 2bd\Big] . \mathrm{e}^{-at} \sin bt\Big\}$	(27.46)
$\dfrac{p^3}{[(p+d)^2 + f^2][(p+a)^2 + b^2]}$	$\dfrac{1}{[(d-a)^2 + f^2 - b^2]^2 + 4(d-a)^2 b^2} \Big\{\Big[b(b^2 - d^2 - f^2) - 2ab(d-a) + \dfrac{a^2}{b}((d-a)^2 + f^2)\Big] \mathrm{e}^{-at} \sin bt + \Big[-f(a^2 + b^2 - f^2) + 2df(d-a) + \dfrac{d^2}{f}((d-a)^2 + b^2)\Big] \mathrm{e}^{-dt} \sin ft + 2[d(a^2 + b^2) - a(d^2 + f^2)] . (\mathrm{e}^{-at} \cos bt - \mathrm{e}^{-dt} \cos ft)\Big\}$	(27.47)

Table 27.1 — *continued*

$F(p)$	$f(t)$	Equation No.
$\dfrac{p^4}{[(p+d)^2+f^2]\,[(p+a)^2+b^2]}$	$\dfrac{1}{[(d-a)^2+f^2-b^2]^2+4(d-a)^2\,b^2}\Big[\!\Big[\{-f^2(a^2+b^2-f^2)+d^2[(d-a)^2+b^2]+2d[d(a^2+b^2)-a(d^2+f^2)+f^2(d-a)]\}\,\mathrm{e}^{-dt}\cos ft+\{b^2(b^2-d^2-f^2)+a^2[(d-a)^2+f^2]-2a[d(a^2+b^2)-a(d^2+f^2)+b^2(d-a)]\}\,\mathrm{e}^{-at}\cos bt+\Big\{df(a^2+b^2-f^2)-\dfrac{d^3}{f}[(d-a)^2+b^2]+2f[d(a^2+b^2)-a(d^2+f^2)-d^2(d-a)]\Big\}\,\mathrm{e}^{-dt}\sin ft+\Big\{-ab(b^2-d^2-f^2)-\dfrac{a^3}{b}[(d-a)^2+f^2]-2b[d(a^2+b^2)-a(d^2+f^2)-a^2(d-a)]\Big\}\,\mathrm{e}^{-at}\sin bt\Big]\!\Big]$	(27.48)

$\dfrac{p(p+d)}{(p^2+\omega^2)(p+c)[(p+a)^2+b^2]}$ where $A_1 = \omega(\omega^2 - 2ac - a^2 - b^2)$ $A_2 = b(b^2 + 2ac - \omega^2 - 3a^2)$ $B_1 = c(a^2 + b^2) - \omega^2(c + 2a)$ $B_2 = a(3b^2 - a^2 - \omega^2) +$ $+ c(a^2 - b^2 + \omega^2)$	$\dfrac{1}{\omega(A_1^2 + B_1^2)}[(dA_1 + \omega B_1)\cos\omega t - (\omega A_1 - dB_1)\sin\omega t] +$ $+ \dfrac{e^{-at}}{b(A_2^2 + B_2^2)}\{[(d-a)A_2 + bB_2]\cos bt - [bA_2 -$ $- (d-a)B_2]\sin bt\} + \dfrac{(d-c)e^{-ct}}{(\omega^2 + c^2)[(a-c)^2 + b^2]}$	(27.49)
$\dfrac{p(p+d)}{(p^2+\omega^2)(p+a)(p+b)(p+c)}$ where $A = \omega(\omega^2 - ab - ac - bc)$ $B = abc - \omega^2(a + b + c)$	$\dfrac{1}{\omega(A^2 + B^2)}[(dA + \omega B)\cos\omega t - (\omega A - dB)\sin\omega t] +$ $+ \dfrac{(d-a)e^{-at}}{(a^2+\omega^2)(b-a)(c-a)} +$ $+ \dfrac{(d-b)e^{-bt}}{(b^2+\omega^2)(a-b)(c-b)} +$ $+ \dfrac{(d-c)e^{-ct}}{(c^2+\omega^2)(a-c)(b-c)}$	(27.50)

Table 27.1 — *continued*

$F(p)$	$f(t)$	Equation No.
$\dfrac{p(p+d)}{(p^2+\omega^2)\,p^3}$	$\dfrac{1}{\omega^4}\left[-d(1-\cos\omega t)-\omega\sin\omega t+\omega^2 t\left(1+\dfrac{dt}{2}\right)\right]$	(27.51)
$\dfrac{p(p+d)}{(p^2+\omega^2)(p+c)(p+a)^2}$ where $A=\omega(\omega^2-2ac-a^2)$, $B=c(a^2-\omega^2)-2a\omega^2$	$\dfrac{1}{\omega(A^2+B^2)}[(dA+\omega B)\cos\omega t-(\omega A-dB)\sin\omega t]+$ $+\dfrac{(d-c)\,e^{-ct}}{(c^2+\omega^2)(a-c)^2}+\dfrac{e^{-at}}{(a^2+\omega^2)^2(c-a)^2}\cdot$ $\cdot\{(a^2+\omega^2)(c-a)[1+t(d-a)]-$ $-(d-a)[-2a(c-a)+a^2+\omega^2]\}$	(27.52)
$\dfrac{1}{(p^2+a^2)(p^2+b^2)(p^2+c^2)}$	$\dfrac{1}{abc(a^2-b^2)(b^2-c^2)(a^2-c^2)}\left[\dfrac{ab}{c}(a^2-b^2)\cdot\right.$ $\cdot(1-\cos ct)-\dfrac{ac}{b}(a^2-c^2)(1-\cos bt)+$ $\left.+\dfrac{bc}{a}(b^2-c^2)(1-\cos at)\right]$	(27.53)

$\dfrac{p}{(p^2 + a^2)(p^2 + b^2)(p^2 + c^2)}$	$\dfrac{1}{abc(a^2 - b^2)(b^2 - c^2)(a^2 - c^2)} [ab(a^2 - b^2) \sin ct - ac(a^2 - c^2) \sin bt + bc(b^2 - c^2) \sin at]$	(27.54)
$\dfrac{p^2}{(p^2 + a^2)(p^2 + b^2)(p^2 + c^2)}$	$\dfrac{1}{(a^2 - b^2)(b^2 - c^2)(a^2 - c^2)} [(a^2 - b^2) \cos ct - (a^2 - c^2) \cos bt + (b^2 - c^2) \cos at]$	(27.55)
$\dfrac{p^3}{(p^2 + a^2)(p^2 + b^2)(p^2 + c^2)}$	$\dfrac{1}{(a^2 - b^2)(b^2 - c^2)(a^2 - c^2)} [-c(a^2 - b^2) \sin ct + b(a^2 - c^2) \sin bt - a(b^2 - c^2) \sin at]$	(27.56)
$\dfrac{p^4}{(p^2 + a^2)(p^2 + b^2)(p^2 + c^2)}$	$\dfrac{1}{(a^2 - b^2)(b^2 - c^2)(a^2 - c^2)} [-c^2(a^2 - b^2) \cos ct + b^2(a^2 - c^2) \cos bt - a^2(b^2 - c^2) \cos at]$	(27.57)
$\dfrac{p(p^2 + \alpha^2)}{(p^2 + a^2)(p^2 + b^2)(p^2 + c^2)}$	$\dfrac{1}{abc(a^2 - b^2)(b^2 - c^2)(a^2 - c^2)} [ab(a^2 - b^2)(\alpha^2 - c^2) \cdot \sin ct - ac(a^2 - c^2)(\alpha^2 - b^2) \sin bt + bc(b^2 - c^2) \cdot (\alpha^2 - a^2) \sin at]$	(27.58)

Table 27.1 — *continued*

$F(p)$	$f(t)$	Equation No.
$\dfrac{p^2(p^2+\alpha^2)}{(p^2+a^2)(p^2+b^2)(p^2+c^2)}$	$\dfrac{1}{(a^2-b^2)(b^2-c^2)(a^2-c^2)}\left[(\alpha^2-c^2)(a^2-b^2)\cos ct - -(\alpha^2-b^2)(a^2-c^2)\cos bt + (\alpha^2-a^2)(b^2-c^2)\cos at\right]$	(27.59)
$\dfrac{p}{(p^2+a^2)(p^2+b^2)^2}$	$\dfrac{1}{(a^2-b^2)^2}\left[\dfrac{a^2-b^2}{2b^3}(\sin bt - bt\cos bt) - \dfrac{1}{b}\sin bt + + \dfrac{1}{a}\sin at\right]$	(27.60)
$\dfrac{p^2}{(p^2+a^2)(p^2+b^2)^2}$	$\dfrac{1}{(a^2-b^2)^2}\left(\dfrac{a^2-b^2}{2b^2}bt\sin bt - \cos bt + \cos at\right)$	(27.61)
$\dfrac{p^3}{(p^2+a^2)(p^2+b^2)^2}$	$\dfrac{1}{(a^2-b^2)^2}\left[\dfrac{a^2-b^2}{2b}(\sin bt + bt\cos bt) + b\sin bt - -a\sin at\right]$	(27.62)
$\dfrac{p^4}{(p^2+a^2)(p^2+b^2)^2}$	$\dfrac{1}{(a^2-b^2)^2}\left[-\dfrac{a^2-b^2}{2}bt\sin bt + a^2(\cos bt - \cos at)\right]$	(27.63)

$\dfrac{p(p^2+\alpha^2)}{(p^2+a^2)(p^2+b^2)^2}$	$\dfrac{1}{(a^2-b^2)^2}\left\{\dfrac{a^2-b^2}{2b}\left[\left(1+\dfrac{\alpha^2}{b^2}\right)\sin bt+\left(1-\dfrac{\alpha^2}{b^2}\right)\cdot bt\cos bt\right]+b\left(1-\dfrac{\alpha^2}{b^2}\right)\sin bt-a\left(1-\dfrac{\alpha^2}{a^2}\right)\sin at\right\}$	(27.64)
$\dfrac{p(p^2-b^2)}{(p^2+a^2)(p^2+b^2)^2}$	$\dfrac{1}{(a^2-b^2)^2}\left(\dfrac{a^2-b^2}{b}bt\cos bt+2b\sin bt-\dfrac{a^2+b^2}{a}\sin at\right)$	(27.65)
$\dfrac{p^2}{(p^2+c^2)^2[(p+a)^2+b^2]}$ where $A=(a^2+b^2-c^2)^2+4a^2c^2$	$\dfrac{1}{A}\left\{\dfrac{a^2+b^2-c^2}{2c^2}ct\sin ct-\dfrac{a}{c}ct\cos ct+\dfrac{1}{A}[(a^2+b^2-c^2)(3a^2-b^2+c^2)+4a^2c^2](\cos ct-e^{-at}\cos bt)-\dfrac{a}{cA}[(a^2+b^2-c^2)^2+4c^2(b^2-c^2)]\sin ct-\dfrac{a}{bA}[(a^2-b^2+c^2)^2-4b^2(b^2-c^2)]e^{-at}\sin bt\right\}$	(27.66)

Table 27.2 Fourier sine finite integral transformation

Original	Transform	Equation No.
$f(x) = \frac{2}{l} \sum_{j=1}^{\infty} F(j) \sin \frac{j\pi x}{l}$	$F(j) = \int_0^l f(x) \sin \frac{j\pi x}{l} \, dx$	(27.67)
$\frac{d^2 f(x)}{dx^2}$ for $f(0) = f(l) = 0$	$-\frac{j^2\pi^2}{l^2} F(j)$	(27.68)
$\frac{d^4 f(x)}{dx^4}$ for $f(0) = f(l) = f''(0) = f''(l) = 0$	$\frac{j^4\pi^4}{l^4} F(j)$	(27.69)
$a\, f(x)$	$a\, F(j)$	(27.70)
a	$\frac{al}{j\pi}(1 - \cos j\pi) = \begin{cases} \frac{2al}{j\pi} & \text{for } j = 1, 3, 5, \ldots \\ 0 & \text{for } j = 2, 4, 6, \ldots \end{cases}$	(27.71)
$H(x - a)$	$\frac{l}{j\pi}\left(\cos \frac{j\pi a}{l} - \cos j\pi\right)$	(27.72)
$b[1 - H(x - a)]$	$\frac{bl}{j\pi}\left(1 - \cos \frac{j\pi a}{l}\right)$	(27.73)
$\delta(x - a)$	$\sin \frac{j\pi a}{l}$	(27.74)

Table 27.3 Fourier cosine finite integral transformation

Original	Transform	Equation No.
$f(x) = \frac{1}{l} F(0) + \frac{2}{l} \sum_{j=1}^{\infty} F(j) \cos \frac{j\pi x}{l}$	$F(j) = \int_0^l f(x) \cos \frac{j\pi x}{l} \, dx$	(27.75)
$\frac{d^2 f(x)}{dx^2}$ for $f'(0) = f'(l) = 0$	$-\frac{j^2 \pi^2}{l^2} F(j)$	(27.76)
$\frac{d^4 f(x)}{dx^4}$ for $f'(0) = f'(l) = f'''(0) = f'''(l) = 0$	$\frac{j^4 \pi^4}{l^4} F(j)$	(27.77)

Table 27.4 Fourier complex integral transformation

Original	Transform	Equation No.
$f(x) = \frac{1}{2\pi}\int_{-\infty}^{\infty} F(q)\, e^{ixq}\, dq$	$F(q) = \int_{-\infty}^{\infty} f(x)\, e^{-iqx}\, dx$	(27.78)
$\frac{df(x)}{dx}$ for $f(\pm\infty) = 0$	$iq\, F(q)$	(27.79)
$\frac{d^2f(x)}{dx^2}$ for $f(\pm\infty) = f'(\pm\infty) = 0$	$-q^2\, F(q)$	(27.80)
$\frac{d^3f(x)}{dx^3}$ for $f(\pm\infty) = f'(\pm\infty) = f''(\pm\infty) = 0$	$-iq^3\, F(q)$	(27.81)
$\frac{d^4f(x)}{dx^4}$ for $f(\pm\infty) = f'(\pm\infty) = f''(\pm\infty) = f'''(\pm\infty) = 0$	$q^4\, F(q)$	(27.82)
$a\, f(x)$	$a\, F(q)$	(27.83)
$\delta(x)$	1	(27.84)

Table 27.5 Fourier sine integral transformation

Original	Transform	Equation No.
$f(x) = \frac{2}{\pi}\int_0^\infty F(q)\sin xq\,\mathrm{d}q$	$F(q) = \int_0^\infty f(x)\sin qx\,\mathrm{d}x$	(27.85)
$\frac{\mathrm{d}^2 f(x)}{\mathrm{d}x^2}$ for $f(0) = f(\infty) = f'(\infty) = 0$	$-q^2 F(q)$	(27.86)
$\frac{\mathrm{d}^4 f(x)}{\mathrm{d}x^4}$ for $f(0) = f''(0) = 0$, $f(\infty) = f'(\infty) = f''(\infty) = f'''(\infty) = 0$	$q^4 F(q)$	(27.87)

Table 27.6 Fourier cosine integral transformation

Original	Transform	Equation No.
$f(x) = \frac{2}{\pi}\int_0^\infty F(q)\cos xq\,\mathrm{d}q$	$F(q) = \int_0^\infty f(x)\cos qx\,\mathrm{d}x$	(27.88)
$\frac{\mathrm{d}^2 f(x)}{\mathrm{d}x^2}$ for $f'(0) = f(\infty) = f'(\infty) = 0$	$-q^2 F(q)$	(27.89)
$\frac{\mathrm{d}^4 f(x)}{\mathrm{d}x^4}$ for $f'(0) = f'''(0) = 0$, $f(\infty) = f'(\infty) = f''(\infty) = f'''(\infty) = 0$	$q^4 F(q)$	(27.90)

Table 27.7 Hankel integral transformation

Original	Transform	Equation No.
$f(r) = \int_0^\infty F(q)\, q\, \mathrm{J}_0(rq)\, \mathrm{d}q$	$F(q) = \int_0^\infty f(r)\, r\, \mathrm{J}_0(qr)\, \mathrm{d}r$	(27.91)
$\left(\frac{\partial^2}{\partial r^2} + \frac{1}{r}\frac{\partial}{\partial r}\right) f(r)$ for $r = 0$ and $r \to \infty$ it must hold $r\, f(r) = r\, \frac{\mathrm{d}f(r)}{\mathrm{d}r} = 0$	$-q^2\, F(q)$	(27.92)
$\frac{\delta(r)}{r}$	$\mathrm{J}_0(0) = 1$	(27.93)
$-\frac{1}{\lambda^2}\,\mathrm{kei}\,(\lambda r)$ for $r > 0$, $\lvert\arg \lambda\rvert < \pi/4$	$\frac{1}{q^4 + \lambda^4}$	(27.94)

BIBLIOGRAPHY

[1] The AASHO Road Test. Report 4, Bridge Research. Highway Research Board, Special Report 61 D. National Academy of Sciences, National Research Council, Washington, 1962, Publ. 953.

[2] ABUBAKAR I.: Buried Moving Sources in an Elastic Halfspace. Pure and Appl. Geophysics, **57** (1964), No. 1, 117—124.

[3] ACHENBACH J. D., SUN C. T.: Moving Load on a Flexibly Supported Timoshenko Beam. Intern. J. Solids Structures, **1** (1965), No. 4, 353—370.

[4] ACHENBACH J. D., KESHAWA S. P., HERRMANN G.: Moving Load on a Plate Resting on an Elastic Half Space. Journal Appl. Mech., **34** (1967), No. 4, 910—914.

[5] AYRE R. S., FORD G., JACOBSEN L. S.: Transverse Vibration of a Two-Span Beam under the Action of a Moving Constant Force. Journ. Appl. Mech., **17** (1950), No. 1, 1—12.

[6] AYRE R. S., JACOBSEN L. S.: Transverse Vibration of a Two-Span Beam under the Action of a Moving Alternating Force. Journ. Appl. Mech., **17** (1950), No. 3, 283—290.

[7] AYRE R. S., JACOBSEN L. S., HSU C. S.: Transverse Vibration of One- and Two-Span Beams under the Action of a Moving Mass Load. Proc. First U. S. National Congress of Applied Mechanics, 81—90.

[8] BABAKOV I. M.: Theory of Vibration (Teoriya kolebaniĭ). Second edition, Nauka, Moscow 1965.

[9] BARSHTEĬN M. F., ZUBKOV A. N.: Statistical Analysis of Lateral Forces Arising in the Motion of a Bridge Crane (in Russian). Mekhanika i raschet sooruzheniĭ, **8** (1966), No. 2, 17—23.

[10] BERNSHTEĬN S. A.: Theory of Transverse Vibration of Steel Bridges (in Russian). Sbornik instituta inzhenernȳkh issledovaniĭ, Transpechat', Moscow 1930, 5—46.

[11] BERNSHTEĬN S. A., KEROPYAN K. K.: Determination of the Frequencies of Vibration of Frame Systems by the Method of Spectral Function (Opredelenie chastot kolebaniĭ sterzhnevȳkh sistem metodom spektral'noĭ funktsii). Gosstroĭizdat, Moscow, 1960.

[12] BETZHOLD CH.: Erhöhung der Beanspruchung des Eisenbahnoberbaues durch Wechselwirkung zwischen Fahrzeug und Oberbau. Glasers Annalen, **81** (1957), No. 3, 76—82, No. 4, 108—115, No. 5, 137—145.

[13] BIELEWICZ E.: Spatial Vibration of One-Span Bridges (in Polish). Rozprawy inżynierskie, **113** (1959), 558—602.

[14] BIELEWICZ E., MIKOŁAJCZAK H.: A Case of the Effect of a Moving Load on a Curved Bar (in Polish). Archiwum inżynierii lądowej, 7 (1961), No. 4, 523—534.

[15] BIGGS J. M., SUER H. S.: Vibration Measurements on Simple-Span Bridges. Highway Res. Board Bull., 1956, No. 124, 1—15.

[16] BIGGS J. M., SUER H. S., LOUW J. M.: Vibration on Simple-Span Highway Bridges. Trans. Amer. Soc. Civ. Engrs., **124** (1959), No. 2979, 291—318.

[17] BLEICH J., ROSENKRANS R., VINCENT G. S., COLLOUGH C. B.: The Mathematical Theory of Vibration in Suspension Bridges. U.S. Gov. Print. Office, Washington 1950.

[18] BOGACZ R., KALISKI S.: Stability of Motion of Nonlinear Oscillators Moving along a Beam on an Elastic Foundation. Proc. Vibrat. Probl., Warsaw, **5** (1964), No. 4, 279—296.

[19] BOLOTIN V. V.: On the Effect of a Moving Load on Bridges (in Russian). Trudȳ Moskovskogo instituta inzhenerov zheleznodorozhnogo transporta, Vol. 74, 1950, 269—296.

[20] BOLOTIN V. V.: On Dynamic Calculations of Railway Bridges with Consideration Given to the Mass of the Moving Load (in Russian). Trudȳ Moskovskogo instituta inzhenerov zheleznodorozhnogo transporta, Vol. 76, 1952, 87—107.

[21] BOLOTIN V. V.: Problem of Bridge Vibration under the Action of a Moving Load (in Russian). Izvestiya AN SSSR, Mekhanika i mashinostroenie, 1961, No. 4, 109—115.

[22] BOLOTIN V. V.: Statistical Methods in Structural Mechanics (Statisticheskie metodȳ v stroitel'noĭ mekhanike). Second edition, Stroĭizdat, Moscow 1965.

[23] BONDAR' N. G.: Dynamic Calculations of Beams Subjected to a Moving Load (in Russian). Issledovaniya po teorii sooruzheniĭ, Vol. 6. Stroĭizdat, Moscow 1954, 11—23.

[24] BONDAR' N. G., KAZEĬ I. I., LESOKHIN B. F., KOZ'MIN YU. G.: Dynamics of Railway Bridges (Dinamika zheleznodorozhnȳkh mostov). Transport, Moscow 1965.

[25] BONDAR' N. G., DENISHENKO YU. N.: Application of the Method of Variable Time Scale to the Solution of Problems Relating to the Dynamic Effect of a Moving Load on Structures (in Russian). Issledovaniya po teorii sooruzheniĭ, Vol. 14, Stroĭizdat, Moscow 1965, 73—91.

[26] BOOLE G.: A Treatise on Differential Equations, Mc. Millan and Co., Cambridge 1859.

[27] Bridge Subcommittee Reports. Government of India Central Publication Branch, Calcutta, Tech. Paper, 1926, No. 247.

[28] BRÜCKMANN B.: Brückenmesswesen, Brückenschwingungen und Brückenbelastbarkeit. Der Eisenbahnbau, **3** (1950), No. 10, 230—235, No. 11, 250—254.

[29] Brückmann B.: Brückenschwingungen unter Verkehrslasten. 5. Kongress Internat. Vereinigung f. Brückenbau u. Hochbau, 1955, 181—198.

[30] Burak Ya. I., Podstrigach Ya. S.: Moving Rotational Force Effects in an Unbounded Elastic Space (in Russian). Sbornik Voprosȳ mekhaniki real'nogo tverdogo tela, Vol. 2, Kiev 1964, 114—124.

[31] Burchak G. P.: Spatial Vibration of Girder Bridges under the Action of Moving Loads (in Russian). Trudȳ Moskovskogo instituta inzhenerov zheleznodorozhnogo transporta, Vol. 92/11, Transzheldorizdat, Moscow 1957, 74—104.

[32] Campbell G. A., Foster, R. M.: Fourier Integrals for Practical Applications. New York 1948.

[33] Cassé M.: La détermination des effects dynamiques dans les ponts. Rev. gén. Chem. de fer, **77** (1958), No. 7/8, 406—408.

[34] Chilver A. H.: A Note on the Mise-Kunii Theory of Bridge Vibration. Quart. Journ. Mech. Appl. Math., **9** (1956), No. 2, 207—211.

[35] Cole J., Huth J.: Stresses Produced in a Half Plane by Moving Loads. Journ. Appl. Mech., **25** (1958), No. 4, 433—436.

[36] Collatz L.: Natürliche Schrittweite bei numerischer Integration von Differentialgleichungssystemen. Zeit. angew. Math. Mech., **22** (1942), No. 4, 216—225.

[37] Collatz L.: Numerische Behandlung von Differentialgleichungen. 2. Aufgabe. Springer-Verlag, Berlin—Göttingen—Heidelberg 1955.

[38] Courant R., Hilbert D.: Methoden der mathematischen Physik. Band I. Springer-Verlag, Berlin 1924.

[39] Crandall S. H.: The Timoshenko Beam on an Elastic Foundation. Proc. Third Midwest. Conf. Solid Mech. (1957, Ann Arbor, Mich.) Ann Arbor Univ. Mich. Press 1957, 146—159.

[40] Crandall S. H. (editor): Random Vibration. Vol. I, II, The M.I.T. Press, Cambridge, Mass., 1958, 1963.

[41] Crandall S. H., Mark W. D.: Random Vibration in Mechanical Systems. Academic Press, New York, London 1963.

[42] Czitary E.: Über die Schwingungen des Zugseiles von Seilschwebebahnen. Österr. Ingenieur-Archiv, **15** (1961), 34—53.

[43] Chernov M. L.: Effect of a Moving Load on Deformation of Steel Girders beyond the Elasticity Limit (in Russian). Prikladnaya mekhanika, **10** (1964), No. 1, 46—54.

[44] Chikov E. A.: Behaviour of Rectangular Plates under the Action of Vertical Forces Moving in Various Directions (in Russian). Izvest. vȳssh. uchebnȳkh zavedeniĭ, seriya Stroitel'stvo i arkhitektura, **8** (1965), No. 1, 50—53.

[45] Dähn J.: Schwingungen der Trägermitte des beiderseitig frei aufliegenden Trägers unter rollender schwingungfähig Last. Wiss. Zeit. Humboldt Univ. Berlin, Math.—naturwiss. Reihe, **13** (1964), No. 5, 869—880.

[46] Delpuech P.: Flexion dynamique et oscillations des ponts. Ann. Ponts Chauss., **121** (1951), No. 1, 1—41, No. 2, 225—247, No. 3, 321—344.

[47] Dikovich I. L.: Dynamics of Elasto-Plastic Beams (Dinamika uprugo-plasticheskikh balok). Sudpromgiz, Leningrad 1962.

[48] Ditkin V. A., Kuznetsov P. I.: Handbook of Operator Calculus (Spravochnik po operatsionnomu ischisleniyu). Gostekhizdat, Moscow, Leningrad 1951.

[49] Ditkin V. A., Prudnikov A. P.: Integral Transformations and Operator Calculus (Integral'nȳe preobrazovaniya i operatsionnoe ischislenie). Fizmatgiz, Moscow 1961.

[50] Ditlevsen O.: Statistical Description of Traffic Loads on Structures. Acta Polytechnica Scandinavica, Civ. Engr. and Build. Construction Series, 1964, No. 28.

[51] Ditlevsen O.: Correlation Functions for Traffic Load Processes. Acta Polytechnica Scandinavica, Civ. Eng. and Build. Construction Series, 1967, No. 44.

[52] Dmitriev A. S.: Vertical Vibration of a Simple Beam Subjected to a Moving Force and Systems of Forces with Consideration Given to Track Irregularities (in Russian). Sbornik Trudov Leningradskogo instituta inzhenerov zheleznodorozhnogo transporta, Vol. 251, Transport, Moscow, Leningrad 1966, 169—179.

[53] Doetsch G.: Handbuch der Laplace-Transformation. Band I—IV. Birkhäuser Verlag, Basel 1950—1956.

[54] Dörr J.: Der unendliche, federnd gebettete Balken unter dem Einfluss einer gleichförmig bewegten Last. Ingenieur-Archiv, **14** (1943), No. 3, 167—192; **16** (1948), Nos. 5, 6, 287—298.

[55] Eason G., Fulton J., Sneddon I. N.: The Generation of Waves in an Infinite Elastic Solid by Variable Body Forces. Phil. Trans. Roy. Soc. London, Ser. A **248** (1955/1956), No. A 955, 575—607.

[56] Edgerton R. C., Beecroft G. W.: Dynamic Stresses in Continuous Plate Girder Bridges. Trans. Amer. Soc. Civ. Engrs., 1958, No. 123, 266—292.

[57] Elasticity and Plasticity. Encyclopedia of Physics, Vol. VI (editor S. Flügge). Springer-Verlag, Berlin—Göttingen—Heidelberg 1958.

[58] Ergebnisse der experimentellen Brückenuntersuchungen in der USSR. Forschungsarbeiten des Wissenschaftlich-technischen Komitees des Volkskomissariats für Verkehrswesen, Band 89, Transpetschat, Moskau 1928.

[59] Esmeijer W. L.: On the Dynamic Behaviour of an Elastically Supported Beam of Infinite Length Loaded by a Concentrated Force. Applied Scientific Research, A **1** (1948), No. 2, 151—168.

[60] Ferrari P.: Contributo allo studio in regime dinamico delle pavimentazioni di strade e aeroporti. Strade, **42** (1962), Nos. 8—9, 392—395.

[61] Filippov A. P.: Vibrations of Mechanical Systems (Kolebaniya mekhanicheskikh sistem). Naukova dumka, Kiev 1965.

[62] FILIPPOV A. P., KOKHMANYUK S. S.: The Dynamic Effects of Moving Loads on Bars (Dinamicheskoe vozdeĭstvie podvizhnȳkh nagruzok na sterzhni). Naukova dumka, Kiev 1967.

[63] FILONENKO-BORODICH M. M., IZYUMOV S. M., OLISOV B. A., KUDRYAVTSEV I. N., MAL'GINOV L. I.: Course in Strength of Materials (Kurs soprotivleniya materialov). Vol. 1, 2. 4-th ed., Gostekhizdat, Moscow 1955, 1956.

[64] FILONENKO-BORODICH M. M.: Theory of Elasticity (Teoriya uprugosti). Fizmatgiz, Moscow 1959.

[65] FLEMING J. F., ROMUALDI J. P.: Dynamic Response of Highway Bridges. J. Struct. Div., Proc. Amer. Soc. Civ. Engrs. **87** (1961), No. 7, 31—61; Discussion **88** (1962), No. 3, 323—324; **89** (1963), No. 2, 221.

[66] FLORENCE A. L.: Travelling Force on a Timoshenko Beam. Journ. Appl. Mech., **32** (1965), No. 2, 351—358; Discussion **33** (1966), No. 1, 233—234.

[67] FORESTIER M. R.: Contribution à l'étude des coefficients dynamiques des poutres de roulement des ponts roulants. Contribution à l'étude du réglage rationnel des freins de ponts roulants. Ann. Inst. techn. bâtim. et trav. public, **16** (1963), No. 182, 259—286.

[68] FRÝBA L.: Infinite Beam on an Elastic Foundation Subjected to a Moving Load (in Czech.). Aplikace matematiky, **2** (1957), No. 2, 105—132.

[69] FRÝBA L.: Schwingungen des unendlichen, federnd gebetteten Balkens unter der Wirkung eines unrunden Rades. Zeit. angew. Math. Mech., **40** (1960), No. 4, 170—184.

[70] FRÝBA L.: Dynamic Characteristics of Steel Railway Bridges (in Czech). Inženýrské stavby, **8** (1960), No. 12, 457—466.

[71] FRÝBA L.: Approximate Calculation of Dynamic Stresses in Crane Runways (in Czech). Inženýrské stavby, **9** (1961), No. 8, 305—311.

[72] FRÝBA L.: Dynamics of Superstructure. Infinite Beam on an Elastic Foundation Loaded By a Flat Wheel (in Czech). Paper III in: KOLOUŠEK V. et al.: Dynamics of Engineering Structures, Part III, Selected Topics. SNTL, Prague 1961, 99—160.

[73] FRÝBA L.: Some Data Obtained in Dynamic Tests of Metal Railway Bridges (in Russian). Trudȳ Moskovskogo instituta inzhenerov zheleznodorozhnogo transporta, Vol. 155, Transzheldorizdat, Moscow 1962, 30—40.

[74] FRÝBA L.: Les efforts dynamiques des pont-rails métalliques. Bulletin mensuel de l'Assoc. Intern. Congrès Chem. fer. **40** (1963), No. 5, 367—403. Also in the English (No. 1, 1—37), German (No. 3, 192—228) and Russian (No. 3, 57—76) edition of this journal.

[75] FRÝBA L.: Railway Bridge with Prestressed Bolts and its Loading Tests (in Czech). Inženýrské stavby, **11** (1963), No. 3, 101—105.

[76] FRÝBA L.: Experimental Investigation of Dynamic Effects of New Types of Locomotives on Large-Span Metal Bridge Structures (in Russian). Acta technica ČSAV, **9** (1964), No. 1, 67—95.

[77] FRÝBA L.: Dynamic Calculations of Bridge Structures (in Russian). Stroitel'naya mekhanika i raschet sooruzheniĭ, **6** (1964), No. 2, 31—35.

[78] Frýba L.: Vibration of a Beam with an Elastic Layer, Resulting from the Motion of a System with Two Degrees of Freedom (in Czech). Rozpravy ČSAV, technical science series, NČSAV, Prague, **74** (1964), No. 9.

[79] Frýba L.: Static and Dynamic Investigations of a Six-Span Continuous Truss Girder (in Czech). Stavebnícky časopis SAV, **13** (1965), No. 10, 581—600.

[80] Frýba L.: Vibration of a Beam under the Action of a Moving Mass System. Acta Technica Academiae Scientiarum Hangaricae, **55** (1966), No. 1—2, 213—240.

[81] Frýba L.: Dynamik des Balkens mit federnder Schicht veränderlicher Steifigkeit unter der Wirkung eines bewegten Systems. Applied Mechanics. Proceedings of the 11th Inter. Congress of Applied Mechanics. Munich (Germany) 1964. Edited by H. Görtler. Springer-Verlag, Berlin—Heidelberg—New York 1966, 245—251.

[82] Frýba L.: The Cross-Beam Effect of Steel Railway Bridges (in Czech). Inženýrské stavby, **14** (1966), No. 3, 115—118.

[83] Frýba L.: Superstructure (Sect. 14.3, 581—608). Cable runways (Sect. 14.4, 608—617). Conveyor bridges (Sect. 14.5, 617—618). Cable cranes (Sect. 14.6, 618—622). Nettings and netting bridges (Sect. 14.7, 622—634). Protective walls and galleries (Sect. 14.8, 634—646) (in Slovak). In: Koloušek V. et al.: Civil Engineering Structures Subjected to Dynamic Effects. SVTL, Bratislava 1967.

[84] Frýba L.: Dynamischer Einfluss der unrunden Räder auf die Eisenbahnbrücken. Monatschrift der Intern. Eisenbahn-Kongress-Vereinigung, **44** (1967), No. 5, 353—389. Also in the English (No. 7, 477—512), French (No. 6, 375—411) and Russian (**45**, (1968), No. 10, 42—57) edition of the journal.

[85] Frýba L.: Impacts of Two-Axle System Traversing a Beam. Intern. Journ. Solids Structures, **4** (1968), No. 11, 1107—1123.

[86] Frýba L : The Inverse Problem in Stochastic Processes. Zeit. angew. Math. Mech., **48** (1968), No. 8 u. Sonderheft, T10—T12.

[87] Frýba L.: Vibration of Solids under Random Load Nonstationary in Space and Time. 12th Intern. Congress of Applied Mechanics, Stanford 1968.

[88] Frýba L.: Non-Stationary Vibration of Bridges under Random Moving Load. 8th Congress of the International Association for Bridge and Structural Engineering. New York 1968. Final Report published by the Secretariat of IABSE in Zürich, pp. 1223—1236.

[89] Galin L. A.: Contact Problems of the Theory of Elasticity (Kontaktnȳe zadachi teorii uprugosti). Gostekhizdat, Moscow 1953.

[90] Gerasimov I. S.: Effects of a Moving Load on Conical Shells (in Russian). Izvestiya vȳsshikh uchebn. zavedeniĭ, seria Matematika, 1962, No. 4, 33—37.

[91] Gesund H., Young D.: Dynamic Response of Beams to Moving Loads. Mém. Assoc. intern. ponts et charpentes, **21** (1961), 95—110.

[92] Głomb J.: Some Problems of Highway Bridges (in Polish). Politechnika Śląska, Gliwice 1962.

[93] GŁOMB J.: Effect of Roadway Irregularities on Dynamic Stresses in Highway Bridges (in Polish). Inżynieria i Budownictwo, **19** (1962), No. 8, 325—328.

[94] GŁOMB J.: On Dynamic Work of Highway Bridges (in Polish). Archiwum inżynierii lądowej, **10** (1964), No. 1, 19—33.

[95] GŁOMB J.: On the Possibility of Using Dynamic Calculations in Bridge Design (in Polish). Inżynieria i Budownictwo, **22** (1965), No. 4, 127—130.

[96] GOLDSMITH W.: Impact. The Theory and Physical Behaviour of Colliding Solids. Edward Arnold, London 1960.

[97] GOL'DENBLAT I. I.: Some New Problems of Structural Dynamics (in Russian). Izvestiya AN SSSR, Otd. tekhn. nauk, 1950, No. 6, 819—833.

[98] GOLOSKOKOV E. G., FILIPPOV A. P.: Steady-State Vibration of Beams on Elastic Foundation under the Action of Load Moving at Constant Speed (in Russian). Trudȳ laboratorii gidravlicheskikh mashin AN USSR, 1962, No. 10, 19—26.

[99] GOSH R. K., LAL R., VIJAYARAGHAVAN S. R.: An Approximate Analysis for Kinetic Response of Triangular Cantilever Overhang. Austral. Road Res., **3** (1967), No. 2, 55—81.

[100] GRADSHTEĬN I. S., RȲZHIK I. M.: Tables of Integrals, Sums, Series and Products (Tablitsȳ integralov, summ, ryadov i proizvedeniĭ). 4th ed. Fizmatgiz, Moscow, 1963.

[101] GROSS O., PRAGER W.: Minimum-Weight Design for Moving Loads. Proc. 4th U.S. Nat. Congr. Appl. Mech., Berkeley, Cal., 1962, Vol. 2, Pergamon Press, Oxford—London—New York—Paris 1962, 1047—1051.

[102] HART R. C. DE: Response of a Rigid Frame to a Distributed Transient Load. J. Struct. Div., Proc. Amer. Soc. Civ. Engrs., 1956, ST 5, 1056, 1—23.

[103] HART R. C. DE: Dynamic Effect of a Moving Load on a Rigid Frame. J. Engng. Mech. Div., Proc. Amer. Soc. Civ. Engrs., **84** (1958), No. 4, 1794, 1—25.

[104] HENRYCH J.: Deformations of Beams Subjected to Intensive Dynamic Loads (in Czech.). Stavebnícky časopis SAV, **13** (1965), No. 8, 487—501.

[105] HIEKE M.: Die Saite unter bewegter Last. Wiss. Zeit. Martin Luther-Univ., Halle—Wittenberg, Math.-naturwiss. Reihe, **12** (1963), No. 1, 71—72.

[106] HILLERBORG A.: A Study of Dynamic Influences of Moving Loads on Girders. 3rd Congress, Intern. Assoc. for Bridge and Struct. Engng., Preliminary Publ. 1948, 661—667.

[107] HILLERBORG A.: Dynamic Influences of Smoothly Running Loads on Simply Supported Girders. Kungl. Tekn. Högskolan, Stockholm 1951.

[108] HIRAI A., ITO M.: Impact on Long-Span Suspension Bridges. Proc. Symp. High-Rise and Long-Span Struct., 1963, Japan. Soc. Promot. Sci., Tokyo 1964, 107—118.

[109] HORÁK V.: Die Eigenschwingungen eines elastisch gelagerten, durch eine statische Axialkraft belasteten Stabes. Acta Technica ČSAV, **1** (1956), No. 3, 187—208.

[110] Il'yasevich S. A.: To the Problem of Vibration of Steel Bridges (K voprosu o kolebaniyakh stal'nȳkh mostov). Izdanie Voenno-inzhenernoĭ akademii Krasnoĭ armii im. V. V. Kuĭbȳsheva, Moscow 1940.

[111] Inglis C. E.: A Mathematical Treatise on Vibration in Railway Bridges. The University Press, Cambridge, 1934.

[112] Jahanshahi A., Monzel F. J.: Effects of Rotatory Inertia and Transverse Shear on the Response of Elastic Plates to Moving Forces. Ingenieur-Archiv, **34** (1965), No. 6, 401—410.

[113] Jahnke - Emde - Lösch: Tafeln höherer Funktionen. 7. Auflage. B. G. Teubner Verlagsgesellschaft, Stuttgart 1966.

[114] Jakowluk A., Ziemba S.: Present-Day Directions of Research in Bridge Dynamics (in Polish). Przeglad kolejowy, **12** (1960), No. 12, 441—446.

[115] Jeffcott H. H.: On the Vibration of Beams under the Action of Moving Loads. Philosoph. Magazine, ser. **7**, **8** (1929), No. 48, 66—97.

[116] Kączkowski Z.: Vibration of a Beam under a Moving Load. Proc. Vibr. Probl., Warszaw, **4** (1963), 357—373.

[117] Kączkowski Z.: Instationäre Schwingungen eines Brückenbalkens unter der Wirkung der verschiebbaren Belastungen. Wiss. Zeitsch. Hochschule Archit. u. Bauwesen, Weimar, **12** (1965), No. 5—6, 428—433.

[118] Kaliski S.: Self-Excited Vibration of a System of Oscillators Moving on the Surface of an Elastic Semi-Space. Proc. Vibrat. Probl., Warsaw, **5** (1964), No. 1, 3—18.

[119] Kaliski S.: Travelling Surface Waves in Finite Systems. Proc. Vibrat. Probl., Warsaw, **7** (1966), No. 4, 387—402.

[120] Kamke E.: Differentialgleichungen, Lösungsmethoden und Löşungen. I. Gewöhnliche Differentialgleichungen. 5. Auflage. Akademische Verlagsgesellschaft Geest und Portig K.-G., Leipzig 1956.

[121] Kazeĭ I. I.: Dynamic Calculations of Railway Bridge Structures (Dinamicheskiĭ raschet proletnȳkh stroeniĭ zheleznodorozhnȳkh mostov). Transzheldorizdat, Moscow 1960.

[122] Kenney J. T.: Steady-State Vibrations of Beams on Elastic Foundation for Moving Load. Journal Appl. Mech., **21** (1954), No. 4, 359—364; Discussion **22** (1955), No. 3, 436.

[123] Kiselev V. A.: On the Calculation of Statically Determinate Systems under Moving Loads, with Consideration Given to the Plastic Properties of Materials (in Russian). Stroitel'naya mekhanika i raschet sooruzheniĭ, 1961, No. 4, 4—9.

[124] Kiselev V. A.: Dynamic Influence Surfaces of Displacements and Internal Forces of Orthotropic Plates Resting on an Elastic Foundation with Two Coefficients of Embedding (in Russian). Issledovaniya po teorii sooruzheniĭ, Vol. 12, 1963, 43—64.

[125] Kitaev K. E.: Bending-Torsional Vibration of Railway Bridge Structures (in Russian). Trudȳ Moskovskogo instituta inzhenerov zheleznodorozhnogo transporta, Vol. 76. Transzheldorizdat Moscow, 1952, 123—134.

[126] KNOWLES J. K.: On the Dynamic Response of a Beam to a Randomly Moving Load. Journal Appl. Mech., **35** (1968), No. 1, 1–6.

[127] KOKHMANYUK S. S., FILIPPOV A. P.: Dynamic Effects on a Beam of a Load Moving at Variable Speed (in Russian). Stroitel'n. mekhanika i raschet sooruzhenii, **9** (1967), No. 2, 36–39.

[128] KOLESNIK I. A.: On the Dynamic Effect of Moving Loads on a Combined System Composed of an Elastic Arch with a Stiffening Girder (in Russian). Prikl. mekhanika, **10** (1964), No. 4, 360–367.

[129] KOLESNIK I. A., YAROVOĬ V. E.: Vibration of Combined Systems under the Action of a Load Moving at Variable Speed (in Russian). Prikl. mekhanika, **2** (1966), No. 8, 71–74.

[130] KOLOUŠEK V.: Dynamics of Civil Engineering Structures. Part I — General Problems, second edition; Part II — Continuous Beams and Frame Systems, second edition; Part III — Selected Topics (in Czech.) SNTL, Prague 1967, 1956, 1961.

[131] KOLOUŠEK V. et al.: Civil Engineering Structures Subjected to Dynamic Loads (in Slovak). SVTL, Bratislava 1967.

[132] KONASHENKO S. I.: On the Critical Speed of Motion of a Sprung Mass on a Beam (in Russian). Izvestiya vȳsshikh uchebn. zaveden., seriya Stroitel'stvo i arkhitektura, 1963, No. 2, 19–25.

[133] KORENEV B. G.: Some Problems of the Theory of Elasticity and Heat Conduction Solved by the Bessel Functions (Nekotorȳe zadachi teorii uprugosti i teploprovodnosti reshaemȳe v Besselevȳkh funktsiyakh). Fizmatgiz, Moscow 1960.

[134] KORENEV B. G.: On the Motion of a Load on a Plate Resting on Elastic Foundation (in Russian). Stroitel'naya mekhanika i raschet sooruzhenii, **7** (1965), No. 6, 28–32.

[135] KOWALCZYK R., RÖSLI A.: Dynamische Versuche an der Glatt-Brücke in Oerlikon. Schweiz. Arch. angew. Wiss. u. Techn., **29** (1963), No. 2, 38–67.

[136] KOŽEŠNÍK J.: Machine Dynamics (in Czech). SNTL, Prague 1958.

[137] KOZ'MIN YU. G., NEVZOROV I. N.: Dynamic Effects of Trains Drawn by Electric Locomotives on Metal Structures of Railway Bridges (in Russian). Sbornik trudov Nauchno-issledovatel'skogo instituta mostov Leningradskogo in-ta inzh. zh.-d. transp. Vol. 7. Transzheldorizdat, Moscow 1962, 102–128.

[138] National Scientific Conference on Bridge Dynamics (in Polish), Gliwice 1965. Zeszyty naukowe Politechniki Śląskiej, No. 201, seria Budownictwo, No. 20, 1967, 1–224.

[139] KRȲLOV A. N.: Mathematical Collection of Papers of the Academy of Sciences, Vol. 61. (Matematicheskii sbornik Akademii Nauk). Peterburg 1905. KRILOFF A. N.: Über die erzwungenen Schwingungen von gleichförmigen elastischen Stäben. Mathematische Annalen, **61** (1905), 211.

[140] LAZAN B. J.: Damping of Materials and Members in Structural Mechanics. Pergamon Press, London 1968.

[141] LESOKHIN B. F., MEL'NIKOV YU. L., POL'EVKO B. P., KHROMETS YU. N.: Metal Bridges (in Russian). Trudȳ Vses. nauchn.-issled. ins-ta transportnogo stroitel'stva, Vol. 29, Transzheldorizdat, Moscow 1959.

[142] LICARI J. S., WILSON E. N.: Dynamic Response of a Beam Subjected to a Moving Forcing System. Proc. 4th U.S. Nat. Congr. Appl. Mech., Berkeley, Calif. 1962, Vol. I, Pergamon Press, Oxford—London—NewYork—Paris 1962, 419—425.

[143] LIN Y. K.: Probabilistic Theory of Structural Dynamics. McGraw-Hill, New York 1967.

[144] LINGER D. A., HULSBOS C. L.: Forced Vibration of Continuous Highway Bridges. Highway Res. Board Bull. 1962, No. 339, 1—22.

[145] LOONEY C. T. G.: High-Speed Computer Applied to Bridges Impact. J. Struct. Div., Proc. Amer. Soc. Civ. Engrs. **84** (1958), ST5, 1759, 1—41.

[146] LOWAN A. N.: On Transverse Oscillations of Beams under the Action of Moving Variable Loads. Philosophical Magazine, ser. 7., **19** (1935), No. 127, 708—715.

[147] LUR'E A. I.: Three-Dimensional Problems of the Theory of Elasticity (Prostranstvennȳe zadachi teorii uprugosti). Gostekhizdat, Moscow, 1955.

[148] L'VOVSKIĬ V. M.: On Uniform Motion of a Concentrated Load on an Infinite Beam Resting on an Elastic, Mass, Two-Layered Foundation (in Russian). Izvestiya vȳsshikh ucheb. zaveden., seriya Lesn. zh. 1964, No. 2, 81—88.

[149] MAĬZEL' YU. M.: Vibration of Frames with Displacing Nodes under the Effect of a Moving Load (in Russian). Sbornik nauchn. trudov Dnepropetrov. metallurg. in-ta, Vol. 34, 1958, 161—175.

[150] MARQUARD E.: Zur Berechnung von Brückenschwingungen unter rollenden Lasten. Ingenieur-Archiv, **23** (1955), No. 1, 19—35.

[151] MASCIA L.: Sul moto vibratorio di un ponte percorso da un carico mobile a velocità variable. Atti Accad. ligure sci. e lettere, **16** (1959/1960), 326—334.

[152] MATHEWS P. M.: Vibrations of a Beam on Elastic Foundation. I, II. Zeitschr. angew. Math. Mech., **38** (1958), No. 3/4, 105—115; **39** (1959), No. 1/2, 13—19.

[153] MIKHLIN S. G.: Integral Equations and Their Application to Some Problems of Mechanics, Mathematical Physics and Engineering (Integral'nȳe uravneniya i ikh prilozheniya k nekotorȳm problemam mekhaniki, matematicheskoĭ fiziki i tekhniki). 2-nd ed. Gostekhizdat, Moscow, Leningrad 1949.

[154] MIKUSIŃSKI J., SIKORSKI R.: The Elementary Theory of Distributions (I). Rozprawy matematyczne, **12** (1957), 1—54.

[155] MILES J. W.: Response of a Layered Half-Space to a Moving Load. Journal Appl. Mech., **33** (1966), No. 3, 680—681; Discussion **34** (1967), No. 1, 247—248.

[156] MISE K., KUNII S.: A Theory for the Forced Vibrations of a Railway Bridge under the Action of Moving Loads. Quart. Journ. Mech. Applied Math., **9** (1956), No. 2, 195—206.

[157] Mitchell G. R.: Dynamic Stresses in Cast Iron Girder Bridges. Nat. Bldg. Stud. Research Paper No. 19. H. M. Stationery Office, London 1954.

[158] Morgaevskiĭ A. B.: On Vibration of Arches under the Effect of Moving Loads (in Russian). Stroitel'naya mekhanika i raschet sooruzheniĭ, 1960, No. 2, 1—6.

[159] Morgaevskiĭ A. B.: On the Influence of Springs on the Magnitude of the Dynamic Effect of Moving Loads (in Russian). Sbornik Issledovaniya po teorii sooruzheniĭ, Vol. 14, Stroĭizdat, Moscow 1965, 67—72.

[160] Morland L. W.: Dynamic Stress Analysis for Viscoelastic Half-Plane Subjected to Moving Surface Tractions. Proc. London Math. Soc. **13** (1963), No. 51, 471—492.

[161] Morley L. S. D.: Stresses Produced in an Infinite Elastic Plate by the Application of Loads Travelling with Uniform Velocity along the Bounding Surface. Aeronaut. Res. Council Repts. and Mem., 1962, No. 3266.

[162] Muchnikov V. M.: Some Methods of Computing Vibration of Elastic Systems Subjected to Moving Loads (Nekotorȳe metodȳ rascheta uprugikh sistem na kolebaniya pri podvizhnoĭ nagruzke), Gosstroĭizdat, Moscow 1953.

[163] Muravskiĭ G. B.: Nonstationary Vibration of Beams Resting on an Elastic Foundation and Subjected to the Action of a Moving Load (in Russian). Izvestiya AN SSSR, OTN, Mekhanika i mashinostroenie, 1962, No. 1, 91—97.

[164] Muravskiĭ G. B.: Nonstationary Vibration of Infinite Plates Resting on Elastic Foundations and Subjected to the Action of a Moving Load (in Russian). Trudȳ Moskovskogo instituta inzhenerov zheleznodorozhnogo transporta, Vol. 193, 1964, 166—171.

[165] Naas J., Schmid H. L.: Mathematisches Wörterbuch. Band I, II. 3. Auflage. Akademie-Verlag GMBH, Berlin, B. G. Teubner Verlagsgesellschaft, Leipzig 1967.

[166] Naleskiewicz J.: On the Dynamics of Bridge Girders (in Polish). Archiwum mechaniki stosowanej, **5** (1953), No. 4. 517—544.

[167] Naruoka M., Hirai I.: Dynamic Influence of Rectangular Plates under the Force of Moving Loads. Proc. 5th Japan. Nat. Congr. Appl. Mech. 1955, 411—414.

[168] Nieto-Ramirez J. A., Veletsos A. S.: Response of Three-Span Continuous Highway Bridges to Moving Vehicles. Univ. Ill. Engng. Experim. Stat. Bull., 1966, No. 489.

[169] Noreĭko S. S.: The Effect of Roadway on Dynamic Deflections of Main Girders of Railway Bridges (in Russian). Sbornik trudov Leningradskogo instituta inzhenerov zheleznodorozhnogo transporta, Vol. 164, Transzheldorizdat, Moscow 1959, 69—83.

[170] Nowacki W.: Dynamics of Structures (Dynamika budowli). Arkady Budownictwo—Sztuka—Architektura, Warsaw 1961.

[171] Oberhettinger F.: Tabellen zur Fourier Transformation. Springer-Verlag, Berlin—Heidelberg—New York 1957.

[172] Ödman S. T. A.: Differential Equation for Calculation of Vibrations Produced in Load-Bearing Structures by Moving Load. 3rd Congr. Intern. Assoc. Bridge Structural Engng., Liège, 1948, Preliminary Publ. 669—680.

[173] Ohchi Y.: Der Schwingbeiwert bei Eisenbahnbrücken. Monatschr. der internat. Eisenbahn-Kongress-Vereinigung, **42** (1965), No. 6, 413—419.

[174] Pailloux H.: Charges roulantes. Bull. scien. math. **80** (1956), No. 3—4, 46—61.

[175] Pan H. H.: Vibration of a Viscoelastic Timoshenko Beam. J. Engng. Mech. Div., Proc. Amer. Soc. Civ. Engrs. **92** (1966), No. 2, 213—234.

[176] Panovko Ya. G., Gubanova I. I.: Stability and Vibration of Elastic Systems (Ustoïchivost' i kolebaniya uprugikh sistem). 2-nd ed. Nauka, Moscow 1967.

[177] Parkes E. W.: How to Cross an Unsafe Bridge. A Diversion in Dynamic Plasticity. Engineering, **186** (1958), No. 4835, 606—608.

[178] Pater A. D., De: Calcul de barres de longueur infinie sur appuis élastiques. Bull. mensuel Assoc. intern. Congrès Chem. fer, **24** (1947), No. 11, 949—969.

[179] Payton R. G.: The Response of a Thin Elastic Plate to a Moving Pressure Point. Zeitsch. angew. Math. Physik, **18** (1967), No. 1, 1—12.

[180] Piszczek K.: The Possibility of Dynamic Stability Loss under Moving Concentrated Loads. Archiwum mechaniki stosowanej, **10** (1958), No. 2, 195—210.

[181] Pitloun R.: Entwurf und Nachrechnung schwingungsempfindlicher Strassenbrückensysteme. Die Strasse, **8** (1968), No. 7, 327—334.

[182] Polsoni G.: Il metodo delle rotazioni concentrate, come carichi nodali nell' analogia di Mohr, per le linea d'influenza, come linee elastiche. Ingegneria ferroviaria, **16** (1961), No. 9, 793—800.

[183] Radok J. R. M.: On the Solution of Problems of Dynamic Plane Elasticity. Quart. Appl. Math. **14** (1956), No. 3, 289—298.

[184] Raske T. F., Schlack A. L. jr.: Dynamic Response of Plates due to Moving Loads. J. Acoust. Soc. America, **42** (1967), No. 3, 625—635.

[185] Rayleigh J. W. S.: Theory of Sound. McMillan and Co., Ltd., London 1894.

[186] Rektorys et al.: Review of Applied Mathematics (Přehled užité matematiky). SNTL, Prague 1963.

[187] Report of the Bridge Stress Committee. Depart. of Scientific and Industrial Research, H.M. Stationery Office, London 1928.

[188] Robson J. D.: An Introduction to Random Vibration. Edinburgh Univ. Press, Edinburgh, Elsevier Publishing Co., Amsterdam, London, New York 1964.

[189] Romualdi J. P., D'Appollonia E., Stelson T. E.: Dynamic Response of Floating Bridges to Transient Loads. Proc. 3rd. U.S. Nat. Congr. Appl. Mech., 1958, 227—232.

[190] RÖSLI A.: Über das dynamische Verhalten von vorgespannten Brücken. 6. Kongress, Intern. Vereinigung für Brückenbau und Hochbau, Stockholm 1960, Vorbericht, 693—706.

[191] RUBLE E. J.: Impact in Railroad Bridges. J. Struct. Div., Proc. Amer. Soc. Civ. Engrs., **81** (1955), No. 736, 1—36.

[192] RYAZANOVA M. YA.: Vibration of Beams Produced by the Action of Load Moving on Them (in Russian). Dopovidi AN URSR, **20** (1958), No. 2, 157—161.

[193] RYAZANOVA M. YA., FILINENKO G. G.: Vibration of Beams on Elastic Foundations under Moving Loads, with Consideration Given to Energy Dissipation (in Russian). Prikl. mekhanika, **1** (1965), No. 8, 128—130.

[194] RYAZANOVA M. YA., FILINENKO G. G.: On Simultaneous Vibrations of an Elastic Half-Space and a Layer under the Effect of Moving Load (in Russian). Dopovidi AN USSR, **28** (1966), No. 3, 318—322.

[195] SACKMAN J. L.: Uniformly Progressing Surface Pressure on a Viscoelastic Half Plane. Proc. 4th U.S. Nat. Congr. Appl. Mech., Berkeley, Calif. 1962, Vol. 2. Pergamon Press, Oxford—London—New York—Paris 1962, 1067 to 1074.

[196] SALLER H.: Einfluss bewegter Last auf Eisenbahnoberbau und Brücken. Kreidels Verlag, Berlin 1921.

[197] SCHALLENKAMP A.: Schwingungen von Trägern bei bewegten Lasten. Ingenieur-Archiv, **8** (1937), 182—198.

[198] SCHLACK A. L.: Resonance of Beams due to Cyclic Moving Loads. Journ. Engng. Mech. Div., Proc. Amer. Soc. Civ. Engrs., **92** (1966), No. 6, 175—184.

[199] SCHULZE H.: Das dynamische Zusammenwirken von Lokomotive und Brücke. Deutsche Eisenbahntechnik, **7** (1959), No. 5, 229—236.

[200] SEKHNIASHVILI E. A., BYUS I. E., SARKISOV YU. S.: Work of Girder Bridge Structures under Dynamic Loads (in Russian). Beton i zhelezobeton, 1961, No. 11, 491—494.

[201] SMIRNOV V. I.: Course in Higher Mathematics (Kurs vȳssheĭ matematiki). Vol. I, 15-th ed., 1954; Vol. II, 14-th ed., 1956, Vol. III/1, 7th ed., 1956, Vol. III/2, 6th ed., 1956; Vol. IV, 3rd ed., 1957; Vol. V, 1st ed., 1960. Gostekhizdat, Vol. V Fizmatgiz, Moscow.

[202] SMITH C. E.: Motions of a Stretched String Carrying a Moving Mass Particle. Journ. Appl. Mech., **31** (1964), No. 1, 29—37.

[203] SNEDDON I. N.: Stress Produced by a Pulse of Pressure Moving Along the Surface of a Semi-Infinite Solid. Rendiconti Circolo Matematico di Palermo, **2** (1952), January—April, 57—62.

[204] SNEDDON I. N.: Fourier Transforms. McGraw-Hill, New York, 1951.

[205] STEELE C. R.: The Finite Beam with a Moving Load. Journ. Appl. Mech., **34** (1967), No. 1, 111—118.

[206] STEUDING H.: Die Schwingungen von Trägern bei bewegten Lasten. I, II. Ingenieur-Archiv, **5** (1934), No. 4, 275—305, **6** (1935), No. 4, 265—270.

[207] STOKES G. G.: Discussion of a Differential Equation Relating to the Breaking of Railway Bridges. Trans. Cambridge Philosoph. Soc., **8** (1849), Part 5, 707–735. Reprinted in: Mathematical and Physical Papers, **2** (1883), 178 to 220.

[208] STÜSSI F.: Trägerschwingungen unter bewegter Last. Abhandlungen, Intern. Vereinigung f. Brückenbau u. Hochbau, Band **13**, 1953, 339–355.

[209] Summary of Tests on Steel Girder Spans. Amer. Railway Engng. Assoc. – Bulletin, **61** (1959), No. 551, 51–78.

[210] SWOPE R. D., AMES W. F.: Vibrations of a Moving Threadline. J. Franklin Inst. **275** (1963), No. 1, 36–55.

[211] SYMONDS P. S., NEAL B. G.: Travelling Loads on Rigid-Plastic Beams. J. Engng. Mech. Div., Proc. Amer. Soc. Civ. Engrs., **86** (1960), EM 1, 2337, 79–89.

[212] TANG S.: A Solution to the Timoshenko Beam under a Moving Force. AIAA Journal, **4** (1966), No. 4, 711–713.

[213] TANG S.: Response of a Finite Tube to Moving Pressure. J. Engng. Mech. Div., Proc. Amer. Soc. Civ. Engrs., **93** (1967), No. 3, 239–256.

[214] TERENT'EV V. N., FILIPPOV A. P.: Steady-State Vibration of Infinite Beams Resting on an Elastic Half-Space (in Russian). Prikl. mekhanika, **1** (1965), No. 9, 107–114.

[215] TIMOSHENKO S. P.: Forced Vibration of Prismatic Bars (in Russian). Izvestiya Kievskogo politekhnicheskogo instituta, 1908. Also in German: Erzwungene Schwingungen prismatischer Stäbe. Zeitsch. f. Mathematik u. Physik, **59** (1911), No. 2, 163–203.

[216] TIMOSHENKO S. P.: On the Forced Vibration of Bridges. Philosoph. Magazine, Ser. 6, **43** (1922), 1018.

[217] TIMOSHENKO S. P.: Statical and Dynamical Stresses in Rails. Proc. Intern. Congr. Appl. Mech., Zürich, 1926, 407–418.

[218] TIMOSHENKO S. P.: Strength of Materials. Part I, II., 2nd ed., D. Van Nostrand Co., New York 1940, 1941.

[219] TIMOSHENKO S. P.: History of the Strength of Materials. D. Van Nostrand Co., New York 1953.

[220] TIMOSHENKO S. P., YOUNG D. H.: Vibration Problems in Engineering. 3rd ed., D. Van Nostrand Co., New York 1955.

[221] TORIDIS T. G., WEN R. K.: Inelastic Response of Beams to Moving Loads. J. Engng. Mech. Div., Proc. Amer. Soc. Civ. Engrs., **92** (1966), EM 6, 43–62; Discussion **93** (1967), EM 4, 182–186.

[222] Papers of the Dnepropetrovsk Institute of Railway Engineers (Trudȳ Dnepropetrovskogo instituta inzhenerov zheleznodorozhnogo transporta). Vol. 25 (1956), 28 (1958), 31 (1961), 32 (1961), 38 (1962), 45 (1963), 49 (1965), 73 (1968). Transport, Moscow.

[223] Tseng C.: Moving Mass on an Infinite Beam. Proc. 4th U.S. Nat. Congr. Appl. Mech., Berkeley, Cal., 1962. Vol. 1, Pergamon Press, Oxford—London—New York—Paris 1962, 411—418.

[224] Tung T. P., Goodman L. E., Chen T. Y., Newmark N. M.: Highway-Bridge Impact Problems. Highway Research Board Bull., 1956, No. 124, 111—134.

[225] Umanskiĭ A. A.: Torsion and Bending of Thin-Walled Aircraft Structures (Kruchenie i izgib tonkostennȳkh aviakonstruktsiĭ). Oborongiz, Moscow 1939.

[226] Urban I. V.: Theory of Calculation of Thin-Walled Frameworks (Teoriya rascheta sterzhnevȳkh tonkostennȳkh konstruktsiĭ). Transzheldorizdat, Moscow 1955.

[227] Vellozzi J.: Vibration of Suspension Bridges under Moving Loads. J. Struct. Div., Proc. Amer. Soc. Civ. Engrs. **93** (1967), No. 4, 123—138; Discussion **94** (1968), No. 3, 848—850.

[228] Vlasov V. Z.: Thin-Walled Elastic Beams (Tenkostennȳe uprugie sterzhni). 2-nd ed. Fizmatgiz, Moscow 1959.

[229] Walker W. H., Veletsos A. S.: Response of Simple-Span Highway Bridges to Moving Vehicles. Univ. Ill., Engng. Exper. Stat. Bull., 1966, No. 486.

[230] Wen R. K.: Dynamic Response of Beams Traversed by Two-Axle Loads. J. Engng. Mech. Div., Proc. Amer. Soc. Civ. Engrs., **86** (1960), EM 5, 2624, 91—111; Discussion **87** (1961), EM 3, 61, EM 5, 85.

[231] Wen R. K., Veletsos A. S.: Dynamic Behaviour of Simple-Span Highway Bridges. Highway Research Board Bull., 1962, No. 315, 1—26.

[232] Wen R. K., Toridis T.: Dynamic Behaviour of Cantilever Bridges. J. Engng. Mech. Div., Proc. Amer. Soc. Civ. Engrs., **88** (1962), EM 4, 27—43.

[233] Willis R. et al.: Preliminary Essay to the Appendix B.: Experiments for Determining the Effects Produced by Causing Weights to Travel over Bars with Different Velocities. In: Grey G. et al.: Report of the Commissioners Appointed to Inquire into the Application of Iron to Railway Structures. W. Clowes and Sons, London 1849. Reprinted in: Barlow P.: Treatise on the Strength of Timber, Cast Iron and Malleable Iron. London 1851.

[234] Wilson E. N., Tsirk A.: Dynamic Behaviour of Rectangular Plates and Cylindrical Shells. New York University, Dept. of Civil Engng., Report No. S-67-7, 1967.

[235] Zagustin E. A.: Response of Beams to Travelling Loads. Dissertation, Stanford University 1964.

[236] Zimmermann H.: Die Schwingungen eines Trägers mit bewegter Last. Centralblatt der Bauverwaltung, **16** (1896), No. 23, 249—251; No. 23 A, 257—260; No. 24, 264—266; No. 26, 288.

[237] Zuk W.: Bridge Vibrations as Influenced by Elastomeric Bearings. Highway Research Board Bull., 1962, No. 315, 27—34.

SUPPLEMENTARY BIBLIOGRAPHY

[238] BARCHENKOV A. G., KOTUKOV A. N., SAFRONOV V. S.: Application of the Correlation Theory for Dynamic Calculation of Bridges (in Russian). Stroitel'naya mekhanika i raschet sooruzheniĭ, **12** (1970), No. 4 (70), 43—48.

[239] BEITIN K. I.: Response of an Elastic Half Space to a Decelerating Surface Point Load. Trans. ASME, E **36** (1969), No. 4, 819—826.

[240] BIGGERS S. B., WILSON J. F.: Dynamic Interactions of High Speed Tracked Air Cushion Vehicles with their Guideways — a Parametric Study. AIAA Paper, 1971, No. 386, 12 p.

[241] BUCHETICH I. I.: Random Vibration of Railway Cars Moving on Elastic Track (in Russian). Trudȳ Vsesoyuznogo nauchno-issledovatel'skogo instituta vagonostroeniya, 1970, Volume 12, 94—106.

[242] BYERS W. G.: Impact from Railway Loading on Steel Girder Spans. J. Struct. Div., Proc. ASCE, **96** (1970), No. 6, 1083—1103, errata **97** (1971), ST 4, p. 1365.

[243] CHRISTIANO P. P., CULVER C. G.: Horizontally Curved Bridges Subject to Moving Load. J. Struct. Div., Proc. ASCE, **95** (1969), No. 8, 1615—1643. Discussion: **96** (1970), No. 4, 861—863; No. 5, 995—997; No. 11, 2524—2527.

[244] DITLEVSEN O.: Extremes and First Passage Times with Applications in Civil Engineering. Some Approximative Results in the Theory of Stochastic Processes. Thesis in the Fulfilment of the Requirements for the Degree of Doctor Technices. Technical University of Denmark, Copenhagen 1971.

[245] DMITRIEV A. S.: Effect of Initial Irregularities of Track on Dynamics of a Beam Under Moving Load (in Russian). Stroitel'naya mekhanika i rashet sooruzheniĭ, 1970, No. 5 (71), 65—66.

[246] EFIMOV P. P.: Spectral Analysis of Forced Vibration of Short-Span Bridges Under Moving Automobiles (in Russian). Izvestiya vȳsshikh uchebnȳkh zavedeniĭ, Stroitel'stvo i arkhitektura, 1970, No. 1, 148—152.

[247] FRÝBA L.: Statistical Distribution of Axle-Loads and Stresses in Steel Railway Bridges. Symposium on Concepts of Safety of Structures and Methods of Design. Final Report, AIPC, London 1969, pp. 127—128.

[248] FRÝBA L.: Non-Elastic Behaviour of Bridges Under Moving Loads. Stavebnický časopis SAV, **19** (1971), No. 1, 3—22.

[249] Gakenheimer D. C., Miklowitz J.: Transient Excitation of an Elastic Half Space by a Point Load Travelling on the Surface. Trans. ASME, E **36** (1969), No. 3, 505—515.

[250] Indeĭkin A. V.: On the Dynamic Stability of Bars of Railway Girder Bridges (in Russian). Izvestiya vȳsshikh uchebnȳkh zavedeniĭ, Stroitel'stvo i arkhitektura, 1969, No. 7, 9—15.

[251] Isada N. M.: Dynamics of Cambered Rapid-Transit Bridge. J. Struct. Div., Proc. ASCE, **95** (1969), No. 17, 1553—1568.

[252] Ito M., Kitagawa M.: Response of Multi-Span Suspension Bridges to a Moving Concentrated Load. Annu. Rept. Eng. Res. Inst. Fac. Eng. Univ. Tokyo, **29** (1970), 11—18.

[253] Knowles J. K.: A Note on the Response of a Beam to a Randomly Moving Force. Journ. Appl. Mech., Trans. ASME, E **37** (1970), No. 4, 1192—1194.

[254] Léger P. H.: Massif viscoélastique indéfini soumis à une charge roulante. Ann. Ponts et Chaussés, **138** (1968), No. 1, 13—24.

[255] Lewis K. H., Harr M. E.: Analysis of Concrete Slabs on Ground Subjected to Warping and Moving Loads. Highway Res. Rec., 1969, No. 291, 194—211.

[256] L'vovskiĭ V. M.: Vibration of a Plate Resting on a Non-linear Elastic Mass Foundation Under Moving Load (in Russian). Izvestiya vȳsshikh uchebnȳkh zavedeniĭ, Stroitel'stvo i arkhitektura, 1970, No. 9, 37—45.

[257] Matsuura A.: A Theoretical Analysis of Dynamic Response of Railway Bridge to the Passage of Rolling Stock. Quarterly Report of Railway Technical Research Institute, Tokyo, **11** (1970), No. 1, 18—21.

[258] Ore D 23 "Determination of Dynamic Forces in Bridges". Reports Nos. 1—17, Utrecht, 1955—1970.

[259] Pang T. C., Sidney S.: Response of Horizontally Curved Bridge to Moving Load. J. Struct. Div., Proc. ASCE, **94** (1968), No. 9, 2135—2151.

[260] Pirner M., Švejda J.: The Response of the Beam to Moving Random Loads. Sborník prací VŠD a VÚD, Vol. 28, 1970, 115—129. Nadas, Praha 1970.

[261] Pitloun R.: Beispiele dynamischer Untersuchungen von Strassenbrücken. Die Strasse, **11** (1971), No. 9, 414—423.

[262] Reipert Z.: Vibration of Frames Under Moving Load. Arch. inż. ląd., **16** (1970), No. 3, 419—447.

[263] Schulze H.: Dynamische Kraft zwischen Rad und Schiene unter Berücksichtigung zufälliger Gleisunebenheiten. Dissertation, Technische Hochschule Otto von Guericke, Magdeburg, 1969.

[264] Stassen H. G.: Random Lateral Motions of Railway Vehicles. Thesis in Fulfillment of the Requirements for the Degree of Doctor of Technical Sciences. Technische Hogeschool Delft, 1967.

[265] Steele C. R.: Beams and Shells with Moving Loads. Int. J. Solids Structures, **7** (1971), No. 9, 1171—1198.

[266] SVETLITSKIĬ V. A.: Vibration of a Beam Under Moving Random Forces the Modulus of Which is Limited (in Russian). Izvestiya vȳsshikh uchebnȳkh zavedeniĭ, Mashinostroenie, 1970, No. 6, 25–29.

[267] TANG S. C., YEN D. H. Y.: A Note on the Non-Linear Response of an Elastic Beam on a Foundation to a Moving Load. Int. J. Solids and Struct., **6** (1970), No. 11, 1451–1461.

[268] TUNG C. C.: Random Response of Highway Bridges to Vehicle Loads. J. Engng. Mech. Div., Proc. ASCE, **93** (1967), No. 5, 79–94, errata **94** (1968), No. 2, 710.

[269] TUNG C. C.: Response of Highway Bridges to Renewal Traffic Loads. Journ. Engng. Mech. Div., Proc. ASCE, **95** (1969), EM 1, 41–57, errata **96** (1970), EM 1, 88.

[270] UVAROV B. V.: Beam of Finite Length on Visco-Elastic Mass Foundation Under Moving Load. (in Russian). Izvestiya vȳsshikh uchebnȳkh zavedeniĭ, Lesn. zh., 1971, No. 1, 31–37.

[271] VASHI K. M., SCHELLING D. R., HEINS C. P. jr.: Impact Factors for Curved Highway Bridges. Highway Res. Rec., 1971, No. 302, 49–61.

[272] VELETSOS A. S., HUANG T.: Analysis of Dynamic Response of Highway Bridges. Journ. Engng. Mech. Div., Proc. ASCE, **96** (1971), EM 5, 593 to 620.

Author index

Subject index